METAL IONS IN GENE REGULATION

Cover figure: Structure of the pig heart *mitochondrial aconitase* (*see* Fig. 7.5, p.185): The protein secondary structure is represented by yellow coils for helices and red ribbons for strands within the β sheets. The N terminus is at the lower right and the C terminus at the upper right. The [Fe-4S] cluster and bound isocitrate (stick models) are in the center. The yellow (S) and red (Fe) spheres of the cluster and black (C) and red (O) spheres of isocitrate are shown. Four polypeptide domains are best seen in the figure by viewing their bundles of red β strands. Large blue spheres marked with arrows at the left represent the location of Gly 108 (bottom) and Leu 590 (top) which, based on sequence alignment, represent the approximate locations of the phosphorylation sites of IRP-1 at Ser138 and Ser711. For more information, *see* Fig. 7.5 on p.185, which includes a schematic view of the IRE RNA molecule positioned next to aconitase. The cover figure was supplied by C. D. Stout (Scripps Institute). The reader is referred to Chapter 7 for futher information.

Chapman & Hall Microbiology Series

Physiology / Ecology / Molecular Biology / Biotechnology

SERIES EDITORS

C.A. Reddy, Editor-in-Chief
Department of Microbiology
Michigan State University
East Lansing, MI 48824-1101

A.M. Chakrabarty
Department of Microbiology and Immunology
University of Illinois Medical Center
835 S. Wolcott Avenue
Chicago, IL 60612

Arnold L. Demain
Department of Biology, Rm 68-223
Massachusetts Institute of Technology
Cambridge, MA 02139

James M. Tiedje
Center for Microbial Ecology
Department of Crop and Soil Sciences
Michigan State University
East Lansing, MI 48824

Other Publications in the Chapman & Hall Microbiology Series

Methanogenesis; James G. Ferry, ed.
Acetogenesis; Harold L. Drake, ed.
Gastrointestinal Microbiology, Volume 1; Roderick I. Mackie and Bryan A. White, eds.
Gastrointestinal Microbiology, Volume 2; Roderick I. Mackie, Bryan A. White, and
   Richard E. Isaacson, eds.
Bacteria in Oligotrophic Environments; Richard Y. Morita
Mathematical Modeling in Microbial Ecology; Arthur L. Koch, Joseph A. Robinson,
   and George A. Milliken, eds.

Forthcoming in the Chapman & Hall Microbiology Series

Oxygen Regulation of Gene Regulation in Bacteria; Rob Gunsalus, ed.

METAL IONS IN GENE REGULATION

Edited by

Simon Silver

Department of Microbiology and Immunology,
University of Illinois College of Medicine, Chicago, Illinois

William Walden

Department of Microbiology and Immunology,
University of Illinois College of Medicine, Chicago, Illinois

CHAPMAN & HALL

 **International Thomson Publishing**
Thomson Science
New York • Albany • Bonn • Boston • Cincinnati • Detroit • London • Madrid • Melbourne
Mexico City • Pacific Grove • Paris • San Francisco • Singapore • Tokyo • Toronto • Washington

JOIN US ON THE INTERNET
WWW: http://www.thomson.com
EMAIL: findit@kiosk.thomson.com

thomson.com is the on-line portal for the products, services and resources available from International Thomson Publishing (ITP). This Internet kiosk gives users immediate access to more than 34 ITP publishers and over 20,000 products. Through *thomson.com* Internet users can search catalogs, examine subject-specific resource centers, and subscribe to electronic discussion lists. You can purchase ITP products from your local bookseller, or directly through *thomson.com*.

Visit Chapman & Hall's Internet Resource Center for information on our new publications, links to useful sites on the World Wide Web and the opportunity to join our e-mail mailing list. Point your browser to: **http://www.chaphall.com** or **http://www.thomson.com/chaphall/lifesce.html** for Life Sciences

A service of I(T)P

Cover design: Trudi Gershenov

For more information, contact:

Chapman & Hall
115 Fifth Avenue
New York, NY 10003

Thomas Nelson Australia
102 Dodds Street
South Melbourne, 3205
Victoria, Australia

International Thomson Editores
Campos Eliseos 385, Piso 7
Col. Polanco
11560 Mexico D. F.
Mexico

International Thomson Publishing Asia
221 Henderson Road #05-10
Henderson Building
Singapore 0315

Chapman & Hall
2-6 Boundary Row
London SE1 8HN
England

Chapman & Hall GmbH
Postfach 100 263
D-69442 Weinheim
Germany

International Thomson Publishing - Japan
Hirakawacho-cho Kyowa Building, 3F
1-2-1 Hirakawacho-cho
Chiyoda-ku, 102 Tokyo
Japan

1 2 3 4 5 6 7 8 9 10 XXX 01 00 99 98
Library of Congress Cataloging-in-Publication Data

    Metal ions in gene regulation / Simon Silver and William Walden, editors.
      p. cm.
    Includes bibliographical references and index.
    ISBN 0-412-05331-4 (alk. paper).
    1. Metal ions--Metabolism. 2. Genetic regulation.
    3. Metallothionein--Physiological effect. I. Silver, Simon
    II. Walden, William
    QP532.M473 1997
    572'52--dc21
                           96-4469
                            CIP

British Library Cataloguing in Pubication Data available

To order this or any other Chapman & Hall book, please contact **International Thomson Publishing, 7625 Empire Drive, Florence, KY 41042**. Phone: (606) 525-6600 or 1-800-842-3636. Fax: (606) 525-7778. e-mail: order@chaphall.com.

For a complete listing of Chapman & Hall's titles, send your request to **Chapman & Hall, Dept. BC, 115 Fifth Avenue, New York, NY 10003**.

Contents

Contributors

Professor Helmut Beinert
Institute for Enzyme Research
University of Wisconsin
1710 University Avenue
Madison, WI 53705-4098

Dr. Amanda J. Bird
Department of Biochemistry and
    Genetics
University of Newcastle Medical
    School
Newcastle-upon-Tyne, NE2 4HH
    U.K.

Professor Volkmar Braun
Lehrstuhl Mikrobiologie II
Universität Tübingen
Auf der Morgenstelle 28
D-72076 Tübingen, Germany

Professor Jean-François Briat
Laboratoire de Biochimie et
    Physiologie Végétale
ENSA/INRA
Place Vialla
F34060 Montpellier Cedex 1, France

Professor Nigel L. Brown
School of Biological Sciences
University of Birmingham
POB 363
Birmingham B15 2TT U.K.

Dr. James Camakaris
Department of Genetics
University of Melbourne
Parkville Victoria 3052, Australia

Professor David Eide
Nutritional Sciences Program
University of Missouri-Columbia
Columbia, MO 65211

Professor Richard S. Eisenstein
Department of Nutritional Sciences
University of Wisconsin
1415 Linden Drive
Madison, WI 53706-1571

Dr. Rohan Farrell
Department of Genetics
University of Melbourne
Parkville Victoria 3052, Australia

Professor Anthony P. Fordham-
Skelton
Department of Biochemistry and
Genetics
University of Newcastle Medical
School
Newcastle-upon-Tyne, NE2 4HH
U.K.

Dr. Lisa S. Goessling
Department of Biology
Washington University
St. Louis, MO 63130-4899

Professor Peter B. Goldsbrough
Department of Horticulture
Purdue University
West Lafayette, IN 47907

Dr. Klaus Hantke
Lehrstuhl Mikrobiologie II
Auf der Morgenstelle 28
D-7400 Tübingen, Germany

Professor John D. Helmann
Section of Microbiology
Cornell University
Wing Hall, W123
Ithaca, NY 14853-8101

Professor M. Claire Kennedy
Biochemistry Department
Medical College of Wisconsin
8701 Watertown Plank Road
Milwaukee, WI 53226-0509

Dr. Simon A.B. Knight
University of Pennsylvania Medical
Center
Division of Hematology/Oncology
Stellar-Chance Laboratories
422 Curie Boulevard
Philadelphia, PA 19104

Dr. Keith A. Koch
Department of Biological Chemistry
University of Michigan Medical
School
Medical Sciences I, Room 4311
Ann Arbor, MI 48109-0606

Dr. Simon Labbé
Centre de recherche en cancérologie
Université Laval
l'Hôtel-Dieu de Québec
Québec City, Québec G1R 2J6
Canada

Dr. Stéphane Lobréaux
Laboratoire de Biochimie et
Physiologie Végétale
ENSA-M/INRA
Place Pierre Viaca
F34060 Montpellier Cedex 2, France

Professor David P. Mascotti
Department of Chemistry
Richard Stockton College
NAMS Office
Jim Leeds Road
Pomona, NJ 08240

Dr. Julian F.B. Mercer
Murdoch Institute
Royal Children's Hospital
Parkville, Victoria, Australia 3052

Professor Sabeeha Merchant
Department of Chemistry and
Biochemistry
University of California
5086B Young Hall
405 Hilgard Avenue
Los Angeles, CA 90024-1569

Professor Dietrich Nies
Institute for Microbiology
Martin-Luther University of Halle
Weinbergweg 16a
D-06099 Halle, Germany

Professor Nigel J. Robinson
Department of Biochemistry and
  Genetics
University of Newcastle Medical
  School
Newcastle-upon-Tyne, NE2 4HH
  U.K.

Dr. Diane Rup
Department of Biology
Washington University
St. Louis, MO 63130

Professor Carl Séguin
Centre de recherche en cancérologie
Université Laval
l'Hôtel-Dieu de Québec
Québec City, Québec G1R 2J6
  Canada

Dr. Andrew K. Sewell
Hematology-Oncology Division
University of Utah
4C-314 Medical Center
Salt Lake City, UT 84132-0001

Professor Simon Silver
Department of Microbiology and
  Immunology
University of Illinois at Chicago
M/C 790, Room 703
835 South Wolcott Avenue
Chicago, IL 60612-7344

Dr. Carl Simard
Département de physiologie
Faculté de médecine
Université Laval
Québec, G1K 7P4, Canada

Professor Robert E. Thach
Department of Biology
Washington University
St. Louis, MO 63130-4899

Professor Elizabeth C. Theil
Department of Biochemistry
North Carolina State University
339 Polk Hall/Box 7622
Raleigh, NC 27695-7622

Professor Dennis J. Thiele
Department of Biological Chemistry
University of Michigan Medical
  School
Medical Sciences I, Room 4311
Ann Arbor, MI 48109-0606

Dr. Joanne L. Thorvaldsen
Hematology-Oncology Division
University of Utah
4C-314 Medical Center
Salt Lake City, UT 84132-0001

Dr. Jennifer S. Turner
Department of Biochemistry and
  Genetics
University of Newcastle Medical
  School
Newcastle-upon-Tyne, NE2 4HH
  U.K.

Professor William Walden
Department of Microbiology and
  Immunology
University of Illinois at Chicago
M/C 790, Room 703
835 South Wolcott Avenue
Chicago, IL 60612-7344

Professor Dennis R. Winge
Hematology-Oncology Division
University of Utah
4C-314 Medical Center
Salt Lake City, UT 84132-0001

Dr. Wei Yu
Hematology-Oncology Division
University of Utah
4C-314 Medical Center
Salt Lake City, UT 84132-0001

1

Overview of Cellular Inorganic Metabolism and the Need for Gene Regulation

Simon Silver and William Walden

"For want of a nail, a shoe is lost; for want of a shoe, the horse is lost; for want of the horse, the rider is lost, for want of the rider the battle is lost; and for want of the battle the Kingdom is lost." (George Herbert, 1593–1633. *Barlett's Famous Quotes*). Clearly iron is important for maintenance of our way of life.

Not only is iron required for all living organisms (a very few bacteria being an exception), but other metal ions such as are described, chapter by chapter in this monograph, are also essential for all living organisms. To acquire and use metal ions, a wide range of physiological and biochemical processes are needed, organism by organism, and cation by cation. The processes are (a) uptake, (b) storage, (c) redox or organometallic chemistry, (d) incorporation into proteins, and (e) energy dependent efflux from the cell. This has become such a large area of biological understanding that one series of monographs, *Metal Ions in Biological Systems* (edited by H. Siegel, the First volume appeared in 1973), is now into well over 30 volumes. There is also now a journal called *BioMetals*. Frequently, more focused monographs, for example *Metal Ion Homeostasis: Molecular Biology and Chemistry* (Hamer and Winge 1989) have approached topics similar to those here. When we planned this effort, entitled *Metal Ions in Gene Regulation*, it was clear that gene regulation alone was as much as could be handled. Inevitably, individual authors (e.g., Robinson and Goldsbrough) have moved from gene regulation to the physiological processes for uptake, storage, and use of metal cations.

Because it reflects the way we think, animal systems are considered along with those of plants and bacteria. For gene regulation of metal cation metabolism, the specific mechanisms (and the proteins and nucleic acids involved) are frequently different for the same cation in plants, animals, and bacteria. However, the overall logic seems quite uniform.

The first part of this monograph concerns bacterial systems. Iron comes first. Bacteria, fungi, and to a lesser extent plants (but not animal cells) contain and regulate the genes to synthesize a remarkable series of highly specific, high-affinity iron chelators, called siderophores, and highly specific transport systems for uptake of iron-complexed siderophores. The major regulatory protein for these systems in bacteria is the transcriptional repressor Fur (for ferric uptake regulation), which is described by Hantke and Braun in Chapter 2. Molecular geneticists prefer their genes (and proteins) to be in families for related functions, but Fur repressors from different bacteria constitute a family alone today, without cousins that regulate other cations.

In Chapter 3 John Helmann discusses the range of gene regulators of cation metabolism in Gram-positive bacteria, from a vantage point of effects on and control with RNA polymerase. Helmann considers Fur and mercury regulatory proteins (MerR) proteins of Gram-positive bacteria, as well as the functions of regulatory genes for iron-related bacterial toxins, manganese, cadmium, and copper. This makes sense as the other chapters here emphasize detailed studies with Gram-negatives, and especially *E. coli,* to the virtual exclusion of consideration of the systems in Gram-positive cells. In Chapter 4 Dietrich Nies and Nigel Brown introduce us to the first metal-regulating examples of bacterial "two-component" regulatory systems, consisting of a membrane-embedded sensor protein (which measures external cation and then autophosphorylates itself at an invariant histidine residue) plus a response-regulator protein that is transphosphorylated (at an invariant aspartyl residue) from the sensor protein. The response-regulator then associates with the operator DNA to control transcription. The new copper, cadmium, cobalt, and zinc two-component sensor-response-regulator join a very large class of more than 100 structurally and functionally related gene (and protein) pairs (Hoch and Silhavy 1995), which were initially found in bacteria, but have been recently identified in yeast and plant systems as well (e.g., Posas et al. 1996). The potassium sensor/response-regulator pair KdpD and KdpE have not been considered, but relatively little is known about their function (Walderhaug et al. 1992).

Phosphate is so central for life processes, and the study of phosphate metabolism has been so useful in bringing about understanding of other systems that we have declared phosphorous an honorary metal for the purposes of this monograph. In Chapter 5 Barry Wanner discusses the Pho "stimulon" of more than 400 *E. coli* genes that are regulated in response to phosphate starvation (about half induced and half repressed; VanBogelen et al. 1996). The regulators of phosphate transport and metabolism include additional genes beyond the two-component class (although PhoB and PhoR are indeed members of the sensor/response-regulator class) and demonstrate regulation at a higher level of complexity.

Part II covers animal metal metabolism. The most extensively investigated cases of metal ion regulated gene expression in animals are those of proteins of iron storage and transport, and Cd^{2+}- and Zn^{2+}-regulated metallothionein expression. The contrast between these phenomena is also one of the best illustrations of how organisms evolve very different mechanisms to deal with what, at the surface, appears to be a common problem, namely metal ion homeostasis. Iron regulation and homeostasis are central to animal life and to the continuing battle between animal cells and invading bacterial pathogens. Excess iron, on the other hand, can be toxic, as is excess zinc (Cd^{2+} by contrast is only toxic). Despite the potential toxicity and the benefits of both zinc and iron, animal cells have evolved very different mechanisms of dealing with these cations. Both uptake and storage of iron are regulated by iron at the level of mRNA metabolism; i.e., at translation or mRNA stability. Excess zinc (and cadmium) induce the expression of the metal binding protein metallothionein, which presumably serves to sequester these metal ions in a nontoxic form. Metal ion regulation of metallothionein occurs at the level of transcription, however.

Three chapters deal with aspects of iron regulation of genes in animals. In Chapter 6 Elizabeth Thiel introduces regulation of ferritin and transferrin receptor synthesis at the level of translation or mRNA stability, respectively. The messenger RNAs for ferritin and transferrin receptor are synthesized, but whether they are translated or degraded is determined by interaction between a family of mRNA binding proteins called IRP (iron regulatory protein) and its cognate mRNA binding site (called IRE for iron responsive element). In binding to IRE located at the 5′ end of mRNAs, IRPs repress translation, whereas when bound at the 3′ end of transferrin receptor mRNA they stabilize the message and thus increase the synthesis of protein. The result is reciprocal but coordinated regulation of iron uptake and storage in animal cells. It is a useful lesson that not all genetic regulation takes place at the level of transcription. In Chapter 7 Eisenstein et al. describe the amazing discovery that IRPs are homologous to a central enzyme of the tricarboxylic acid cycle, aconitase. One family member, IRP-1, is the cytosolic isoform of aconitase and thus functions in two very different roles in metabolism. As an enzyme containing a 4Fe-4S cluster, which predominates in iron replete conditions, cytosolic aconitase functions enzymatically. Under low-iron conditions, the iron-sulfur cluster is lost, and the same polypeptide functions as an mRNA binding protein, regulating translation and mRNA stability. Chapter 8 by Mascotti et al. emphasizes the role of alternative pathways for regulating IRP/IRE interaction and hence gene regulation.

Animal metallothionein, discovered some 30 years ago, remains somewhat a protein seeking a function, though it is considered to be either a storage protein for intracellular Zn^{2+} (and sometimes Cu^{2+}, especially in yeast) needed for nutrition or as a detoxification system (again by sequestration) of toxic cations such as Cd^{2+}. Labbé et al. discuss the regulation of the synthesis of metallothionein in mouse cells, though not the ultimate functions of the protein, in Chapter 9. Metallothionein biosynthesis is regulated through *cis* elements called MREs (metal response elements), but these are DNA sequences that bind transcriptional regulatory proteins, quite different from the situation with iron. Multiple DNA elements (six upstream from one metallothionein gene) and multiple proteins are involved in this regulation.

Copper metabolism and transport in animals has been advanced by the recent discovery of genes for copper efflux pumps defective in patients with either of the two human hereditary disorders of copper metabolism, Menkes' disease and Wilson's disease. This explosion of understanding is described by Mercer and Camakaris in Chapter 10. In this chapter still another type of genetic regulatory mechanism for metal cation metabolism is described with the Menkes ATPase pump. At high available copper, cultured mammalian cells move the Menkes ATPase to the cell surface where it might function in efflux. When transferred to copper-starvation conditions, the Menkes ATPase moves back to the cellular interior. This is an example of regulation by posttranslational "protein trafficking", and we may expect to see more of these types of control mechanisms in future work.

Part III of this monograph is concerned with lower eukaryotes (mostly yeasts) and higher plants. For convenience, the metallothionein regulation of the prokaryotic cyanobacteria is included here with an overview by Robinson et al. (Chapter 14). The cyanobacterial metallothionein could have been discussed with other prokaryotic cell systems (but it is unrelated in function) or together with the animal cell metallothionein (although it appears to be independent in evolutionary origin and perhaps in function from these proteins). The fit seems better here, however, after reports of the thoroughly studied and understood systems for regulation of metal cation stress (and metallothioneins) in yeast by Winge et al. in Chapter 11 and Knight et al. in Chapter 12. Metallothionein functions as a copper buffer in bakers' yeast *Saccharomyces cerevisiae*. Regulation of yeast metallothionein synthesis involves both copper regulation of a transcriptional factor and gene amplification (multiple copies of the genes being produced on prolonged exposure to high copper both in yeast and in cyanobacteria). Otherwise the components of gene regulation in cyanobacteria and yeast seem quite different. Different yeast such as the fission yeast *Schizosaccharomyces pombe* do not make metallothionein proteins but instead produce polyglutathione complexes that are called phytochelatins, since they are found in higher plants as well as in yeast. Phytochelatins are synthesized via unique pathways that are metal ion regulated. Winge et al. start a discussion of this, which is then picked up in more detail by Goldsbrough et al. in Chapter 15. Flowering plants can also synthesize metallothioneins; that is, small, high cysteine content metal binding proteins. It seems from sequence homologies, motifs, and functions as if the same solution (sequestration by high cysteine content, small proteins) has evolved four times separately in cyanobacteria, lower eukaryotes, animals, and plants. Knight et al. detail the best understood of these systems in terms of gene regulation, those of yeast cells, including a third key group, *Candida.* In Chapter 13 David Eide expands the consideration of regulation of cation transport in yeast, from copper to now iron and zinc. Some yeast take up iron using siderophores (similar to the systems of bacteria, and indeed some bacteria can utilize yeast siderophores). However, *Saccharomyces cerevisiae,* as described by Eide, lack siderophores but use cell-surface transporters for iron and for copper that involve reduction from Fe^{3+} or Cu^{2+} to Fe^{2+} or Cu^+ prior to transport. The reductases and transport proteins need to be carefully regulated at the gene level to maintain metal homeostasis. Eide's work has lead to the first discovery of highly specific zinc transporters, initially in *S. cerevisiae,* but quickly then found in higher plants as well. We anticipate that such transporters of zinc (and their regulation) will quickly be found in animal cells and prokaryotes as well, again justifying the scholars of one set of organisms paying careful attention to advances in very different life forms. In Chapter 16 Briat and Lobréaux discuss plant ferritins, which function as iron storage proteins, but are regulated at the level of transcription, totally unlike that of animal cell ferritin. And in the final chapter, 17, Sabeeha Merchant details the elegant regulation of copper metabolism associated with photosynthesis in algae and compares this to copper regulation in cyanobacteria.

Summary

Other than the simple statements that metals are important, and every process is regulated in all biological systems, what else is there to say? The 16 expert and technical chapters that follow include many of those that are best understood in terms of molecular genetics and function. Others obviously are missing. Of particular note is regulation of mercury detoxification through the mercury regulatory protein MerR. MerR is an activator of RNA transcription, which is more unusual than the negative repressor role of transcriptional regulators such as Fur. Although MerR started as a family by itself, "cousins" involved in the regulation of other genes [including SoxR, the regulator of defense mechanisms against oxidative stress (Ding and Demple 1996) and a few regulators involved in other biological processes] share with MerR sequence homologies indicative of structural similarities and functionally related processes. MerR is the best understood regulator of the genes for resistance to a toxic heavy metal cation (in this case Hg^{2+}), expanding the range of genes for regulation of metal cations from those needed for nutrition to most toxic ions that are abundant in environments to which plants, bacteria, and animals are exposed. We encourage those readers who are interested in mercury resistance and gene regulation to see recent reviews on this subject (e.g., Hobman and Brown 1997).

As is illustrated in this monograph, there are regulatory genes for a surprisingly wide range of the cations formed by elements of the periodic table. Not all could be handled here and note for example the absence of consideration of regulation of Mg^{2+}, Ca^{2+}, and Na^+. These gaps reflect lack of space and gaps in understanding. Frequently (for example, in the case of bacterial magnesium and animal calcium) it is an absence of understanding of genetic regulation that has determined what is covered and what is missing. The large topic of zinc as a structural component of transcriptional regulators of processes unrelated to zinc metabolism (e.g., zinc finger proteins; Berg and Shi 1996) has not been covered, as the zinc here appears to be a structural component of a regulatory protein (i.e., transcription factors) rather than genetic regulation of zinc transport and metabolism, or regulation induced directly by zinc. The authors of the separate chapters interact frequently at scientific meetings and elsewhere, which is a measure of their belief that they have something to learn from one another. We hope that the common threads will be apparent to readers as well. This is the first advanced monograph on the topic of regulatory genes for biological metal cations. We expect that it will facilitate further efforts in these directions (and on additional metal cations that have not been covered here). It seems likely that a monograph in ten years' time on topics in this area would need to be much more focused, not for intellectual but for practical reasons.

There is a global question that might be addressed in this introduction as well. That is the distinction (almost unique to metal cations) between nutrient and toxic, and how these borders break down in real environments. Are the uptake, storage, metabolism, and homeostatic control of cellular levels of cations regulated at the gene level similarly or differently in response to their differing biological roles? When one thinks of the major cations of living organisms, they are the "macronutrients," potassium and magnesium, plus the unique solubility problems and importance of iron. In addition there are numerous "micronutrients," such as cobalt, copper, manganese, nickel, and zinc. The situations in multicellular plants and animals differ from those of single-cell microbes (both prokaryotic and eukaryotes). Multicellular organisms get cations from foodstuffs or by absorption directly from the environment.

Therefore specific distribution systems for spreading the cation throughout the organisms (themselves regulated) are essential for each. The internal cells and organs of the body might be thought to live in a rather metal-buffered constant environment. Nevertheless, the newly found genes for intracellular copper transport (see the chapter by Mercer and Camakaris) and the sophisticated machinery for intracellular iron homeostasis (see Chapters 6, 7, and 8) indicate that regulation of cations is essential, cell by cell, in animals and that "surface" regulation is not sufficient. Plants need specific systems for uptake in the roots and these are starting to be understood. As far as we know, specific gene products for transport, storage, and homeostasis of needed cations elsewhere within the plant have not been studied. Single-cell eukaryotes transport and store cations in the cytoplasm and in organelles, importantly the vacuoles and mitochondria, both of which are well known to have roles in cation metabolism, although this has not been covered in any of our chapters. We expect to see more progress in these directions in the near future.

For toxic cations, it seems logical to develop physiology and biochemistry to exclude them, but in fact more often mechanisms (using binding components such as phytochelatins and metallothioneins) are used for short-term storage, or in bacteria a wide range of efflux pumps have been described for energy-dependent elimination of toxic cations after they have entered the cell (Silver and Phung 1996). Each of these systems is regulated at the transcriptional level by specific activator or repressor proteins (such as those described by Helmann and Nies and Brown). There is a difference in the need for regulation for cations such as Hg^{2+} and Cd^{2+} that are always or only toxic in contrast with Fe^{3+}, Cu^{2+}, and Mn^{2+}, which are required at trace levels but are toxic at excess levels. For the pure toxics, the logic seems to be to regulate the genes so that transcription only occurs when the level of environmental toxic cation exceeds a threshold level, and therefore the cell requires protective resistance. For the trace nutrients that are required at very low levels but toxic in excess, careful transcriptional control of genes for uptake transporters, intracellular "buffer" chelates, and efflux transporters need to be coordinated for cellular homeostasis over a wide range of exposures. Nevertheless, at this stage of our understanding there are no ways of recognizing the nature of the cell roles (nutrients, trace nutrient, or toxin) from the pattern of gene regulation.

Another global question is similarity or differences between equivalent regulatory systems in plants, animals, and bacteria. The emphasis in this monograph is on how much we have to learn from one another. However, it is equally clear that different molecular genetic regulatory systems have evolved in different cell types, sometimes achieving the same overall homeostasis by completely dissimilar mechanisms.

References

Berg, J. M., and Y. Shi. 1996. The galvanization of biology: A growing appreciation for the roles of zinc. *Science* 271:1081–1085.

Ding, H., and B. Demple. 1996. Glutathione-mediated destabilization in vitro of [2Fe-2S] centers in the SoxR regulatory protein. *Proc. Natl. Acad. Sci. U.S.A.* 92:9449–9453.

Hamer, D. H., and D. R. Winge. 1989. Metal ion homeostasis. In *Molecular Biology and Chemistry*. 490 pp. Alan R. Liss, New York.

Hobman, J. L., and N. L. Brown. 1997. Bacterial mercury-resistance genes. *Met. Ions Biol. Syst.* 34:527–568.

Hoch, J. A., and T. J. Silhavy. 1995. *Two-Component Signal Transduction.* pp. 488. ASM Press, Washington, D.C.

O'Halloran, T. V. 1993. Transition metals in control of gene expression. *Science* 261:715–725.

Posas, F., S. M. Wurgler-Murphy, T. Maeda, E. A. Witten, T. C. Thai, and H. Saito. 1996. Yeast HOG1 MAP kinase cascade is regulated by a multistep phosphorelay mechanism in the SLN1-YPD1-SSK1 "Two Component" osmoregulator. *Cell* 86:865–875.

Sigel, H., and A. Sigel. 1973. *Metal Ions in Biological Systems.* Marcel Dekker, New York.

Silver, S., and L. T. Phung. 1996. Bacterial heavy metal resistance: New surprises. *Ann. Rev. Microbiol.* 50:753–789.

Van Bogelen, R. A., E. R. Olson, B. L. Wanner, and F. C. Neidhardt. 1996. Global analysis of proteins synthesized during phosphorus restriction in *Escherichia coli. J. Bacteriol.* 178:4344–4366.

Walderhaug, M. O., J. W. Polarek, P. Voelkner, J. M. Daniel, J. E. Hesse, K. Altendorf, and W. Epstein. 1992. KdpD and KdpE, proteins that control expression of the kdpABC operon, are members of the two-component sensor-effector class of regulators. *J. Bacteriol.* 174:2152–2159.

I

BACTERIAL METAL METABOLISM

2

Control of Bacterial Iron Transport by Regulatory Proteins

Klaus Hantke and Volkmar Braun

1. Introduction

It has been postulated that pyrite (FeS_2) metabolism was the basis for the early evolution of life (Wächtershäuser 1990). As a consequence, iron is found today in many enzymes, mainly redox enzymes, as a cofactor. Iron is required by nearly all organisms, although it has two unfavorable properties: (1) iron(II) catalyzes, in the presence of O_2, the life-threatening generation of radicals (Halliwell and Gutteridge 1984), and (2) at neutral pH, the oxidation product iron(III) is practically insoluble and precipitates as iron hydroxide (Braun and Hantke 1991). Although iron is a very abundant metal, its insolubility poses a problem for the iron supply of living organisms. Many microbes have developed the strategy of secreting specific iron chelators, called siderophores, which have a high affinity for iron(III). Specific transport systems enable the cells to take up the iron-siderophore complexes to satisfy their iron demands (Braun and Hantke 1991).

The characterization of these transport systems began in 1973 with the attempt to characterize phage and colicin receptor proteins in the outer membrane of *Escherichia coli* (Braun and Wolff 1973). Because *tonB* mutants are resistant to phage T1, the putative T1 receptor was expected to be defective or absent from these cells. Unexpectedly, the *tonB* mutants synthesized additional proteins in the range from 74 to 83 kDa. Later, it was found that these same proteins are also strongly expressed under low-iron growth conditions (Braun et al. 1976). *tonB* mutants are unable to take up iron(III) siderophores and, as a consequence, an iron demand inside the cell induces the formation of iron-regulated sidero-phores, which are secreted, and receptors, which are targeted to the outer membrane. The same type of iron-regulated outer membrane proteins are also detected in other Gram-negative bacteria (Braun and Hantke 1991).

The first regulatory mutation was identified in *Salmonella typhimurium* in a gene that was called *fur* (ferric iron uptake regulation; Ernst et al. 1978). The *fur* mutant produces enterochelin (enterobactin), the siderophore of many enterobacteria, under iron-rich conditions. The mutant is constitutive for the uptake of enterochelin and ferrichrome (a fungal siderophore). A similar regulatory mutation was found in *E. coli* K-12. Genetic evidence indicated that Fur acted as a repressor of genes for iron uptake (Hantke 1981). The mutant was used to map, clone, and sequence the *fur* gene (Hantke 1984; Schäffer et al. 1985). The derived amino acid sequence of the Fur protein shows that it is a histidine-rich protein with 148 amino acid residues and a molecular mass of about 17 kDa (Schäffer et al. 1985).

2. The Fur Protein

2.1 Fur-Binding Studies and the Identification of a Consensus Sequence

The cloned gene and the high histidine content of Fur allowed the isolation and purification of the Fur protein by zinc-chelate affinity chromatography (Wee et al. 1988; Braun et al. 1990). deLorenzo et al. (1988) showed by footprinting experiments that Fur binds to two regions of the promoter of the aerobactin biosynthesis operon. Binding of Fur to DNA is dependent on the presence of Fe^{2+}, indicating that Fe^{2+} is the corepressor *in vivo*. Other divalent metal ions also mediate Fur binding to DNA in the order $Co^{2+} \geq Mn^{2+} \geq Cd^{2+} \geq Cu^{2+}$. In most experiments, Mn^{2+} was used because Fe^{2+} is rapidly oxidized to Fe^{3+}. Also *in vivo*, Mn^{2+} is able to replace Fe^{2+} in the Fur protein, and the complex can repress or induce iron-regulated genes (Hantke 1987).

Similar sequences are found upstream of other iron-regulated genes in *E. coli* and also in front of iron-regulated genes from other bacteria. A consensus sequence, the Fur box, was confirmed in several *in vitro* studies with the *E. coli* promoters of *cir* (catecholate receptor protein; Griggs and Konisky 1989), *fur* (de Lorenzo et al. 1987), *fepB-entC* (divergent promoter for enterochelin biosynthesis and enterochelin-uptake genes; Brickman et al. 1990), *sodA* (superoxide dismutase; Tardat and Touati 1993), *fepA-fes* (divergent promoter for enterochelin receptor and enterochelin esterase genes; Hunt et al. 1994), and the *hly* operon (hemolysin; Frechon and Le Cam 1994) and with a promoter of a *Pseudomonas* iron-assimilation gene (O'Sullivan et al. 1994). Remarkably, even binding of *E. coli* Fur to a Fur box of the promoter of *fbpA*, a gene for a ferric-iron binding protein from the distantly related *Neisseria gonorrhoeae* was observed (Desai et al. 1996).

In some cases, an additional Fur box is found in the neighborhood of the main Fur box at the promoter region (Martinez et al. 1994). Such secondary Fur binding sites may play a role in the formation of a repression loop. In other negatively regulated systems (*lac, gal, deo*), this type of loop formation stabilizes repression (Ptashne 1986). Such a cooperative Fur binding could enhance the efficiency of transcriptional repression in the presence of iron.

LacZ has been used as a reporter gene for the study of iron-regulated genes since the Mud1 phage became available (Casadaban and Cohen 1979). After random mutagenesis with MudI, mutations in iron-regulated genes are detected with 2,2′-bipyridine, which complexes the iron in a form that is unavailable to *E. coli*. Cells containing *lacZ* fusions to iron-regulated genes form white colonies on iron-rich MacConkey agar plates and red colonies in the diffusion zone of 2,2′-bipyridine. The reporter gene technique helped to identify the gene for the coprogen receptor (*fhuE*) and the gene for the catecholate receptor (*fiu;* Hantke 1983). The *fur* gene was cloned by complementing an *E. coli fur* mutant with an iron-regulated *lacZ* operon fusion from *E. coli* (Hantke 1984), *Legionella pneumophila* (Hickey and Cianciotto 1994), *Pseudomonas aeruginosa* (Prince et al. 1993), *Yersinia pestis* (Staggs and Perry 1992), and *Vibrio anguillarum* (Tomalsky et al. 1994).

In a comparison of various iron-regulated promoters, it was found that the promoter of the *fhuF* gene is exceptionally sensitive to slight changes in the iron concentration of the medium. Under conditions where *fhuF-lacZ* is derepressed and the cells produce red colonies (MacConkey agar supplemented with 5 µM iron), *lacZ* fusions to *cir, fiu,* or *fhuA* genes are still repressed and the cells produce white colonies. The high sensitivity of *fhuF-lacZ* to small changes in the Fur-Fe concentration was used as a powerful enrichment technique to clone Fur binding sites (Stojiljkovic et al. 1994). A high-copy-number plasmid containing a binding site for Fur was shown to titrate the Fe^{2+}-Fur complex in the cell, resulting in a derepression of the *fhuF-lacZ* fusion and formation of red colonies on iron-rich MacConkey plates, whereas cells with plasmids containing no Fur box produce white colonies. The titration effect is dependent on the copy number of the plasmid used and on the sequence of the Fur box used. A deviation from the strongly conserved adenine in position 5 of the Fur box is tolerated, whereas the additional exchange of the T in position 16 in the other half of the palindromic sequence is not tolerated. Not only DNA with Fur boxes are selected by this approach; genes for iron-storage proteins are also selected because these proteins also lower the concentration of active Fe^{2+}-Fur repressor complex (Stojiljkovic et al. 1994).

2.2 fur *Genes from Other Gram-negative Bacteria*

In nearly all Gram-negative bacteria studied, derepression of outer membrane proteins in the range of 70–100 kDa is observed under low-iron growth conditions

(Braun and Hantke 1991). The structural genes of some of these proteins were cloned and sequenced (Figure 2.1). In many cases, a Fur box is found in the promoter region, and Fur homologs were cloned from several organisms. Remarkably, most of these Fur proteins are also active in *E. coli,* where they suppress the expression of an iron-regulated reporter gene.

Fur proteins were cloned and sequenced mainly from pathogenic organisms because many virulence determinants are iron regulated. Iron-uptake and/or iron-regulated toxins are important for the pathogenicity of bacteria (for a review, see Litwin and Calderwood 1993b). The 18 currently known sequences were compared and divided into two groups (Figure 2.1), one containing the highly similar sequences from Gram-negative bacteria (Figure 2.2), the other containing the more distantly related sequences from *Campylobacter* and *Synechococcus* and the Fur-like sequences from Gram-positive organisms.

The most obvious difference in the first group is that the C-terminal 14 amino acids are missing in the *Pseudomonas* Fur proteins, including a well-conserved pair of cysteine residues (position 133/138 in the *E. coli* sequence). In addition, two cysteines (position 93/96) close to the oligo-His motif are missing in *P. putida* (Figure 2.1). Cysteines are often partners in the binding of divalent cations such as zinc and iron. However, the function of these four cysteine residues in *E. coli* Fur is unknown. Evidence from spin label binding studies indicates that

---

Figure 2.1 Comparison of 18 Fur and Fur-like sequences from Gram-negative and Gram-positive bacteria. Conserved identical amino acids are indicated by a star and similar ones by a caret. In the upper part, the closely related Fur sequences from Gram-negative organisms are compared; in the lower part, the more distantly related Fur and Fur-like sequences from Gram-positive organisms are included. ΛΛΛΛΛ indicates helical regions and →→→ indicates β-sheet structure as predicted by Sanders et al. (1994) for the *E. coli* Fur. The sequences can be found by a BLAST search against the *E. coli* Fur in the nonredundant GenBank protein sequences. Abbreviations and publications: YERPE, *Yersinia pestis* (Staggs and Perry 1992); VIBAN, *Vibrio anguillarum* (Tomalsky et al. 1994); VIBVU, *V. vulnificus* (Litwin and Calderwood 1993a); VIBCH, *V. cholerae* (Litwin et al. 1992); HAEIN *Haemophilus influenzae* (Fleischmann et al. 1995); LEGPN, *Legionella pneumophila* (Hickey and Cianciotto 1994); PSEAE, *Pseudomonas aeruginosa* (Prince et al. 1993); PSEPU, *P. putida* (Venturi et al. 1995); BORPE, *Bordetella pertussis* (Beall and Sanden 1995); NEIGO, *Neisseria gonorrhoeae* (Berish et al. 1993); NEIME, *N. meningitidis* (Thomas and Sparling 1994); CAMJE, *Campylobacter jejuni* (Wooldridge et al. 1994); CAMUP, *C. uppsaliensis;* BACSU, *Bacillus subtilis;* SYNEC, *Synechococcus* sp. (Ghassemian and Straus 1996); STAEP, *Staphylococcus epidermidis* (Heidrich et al. 1996). The sequences were compared using the computer program CLUSTAL from the PC/GENE package.

```
                                   ΛΛΛΛΛΛΛΛΛΛ        ΛΛΛΛΛΛ
FUR_ECOLI    M-----TD-NN-TA-----LKKAGLKVTLPRLKILEVLQE-PDNHHVSAEDLYKR    42
FUR_YERPE    M-----TD-NN-KA-----LKNAGLKVTLPRLKILEVLQN-PACHHVSAEDLYKI    42
FUR_VIBAN    M-----SD-NN-QA-----LKDAGLKVTLPRLKILEVLQQ-PECQHISAEELYKK    42
FUR_VIBVU    M-----SD-NN-QA-----LKDAGLKVTLPRLKILEVLQQ-PDCQHISAEDLYKK    42
FUR_VIBCH    M-----SD-NN-QA-----LKDAGLKVTLPRLKILEVLQQ-PECQHISAEELYKK    42
FUR_HAEIN    M-----SE-GNIKL-----LKKVGLKITEPRLTILALMQN-HKNEHFSAEDVYKI    43
FUR_LEGPN    M-----EE-S--QQ-----LKDAGLKITLPRIKVLQILEQ-SRNHHLSAEAVYKA    41
FUR_PSEAE    M-----VE-N--SE-----LRKAGLKVTLPRVKILQMLDS-AEQRHMSAEDVYKA    41
FUR_PSEPU    M-----VE-N--SE-----LRKAGLKVTLPRVKILQMLDS-TEQRHMSAEECIKA    41
FUR_BORPE    M--------SDQSE-----LKNMGLKATFPRLKILDIFRK-SDLRHLSAEDVYRA    41
FUR_NEIGO    M-----EKFSNIAQ-----LKDSGLKVTGPRLKILDLFEK-HAEEHLSAEDVYRI    44
FUR_NEIME    M-----EKFSNIAQ-----LKDSGLKVTGPRLKILDLFET-HAEEHLSAEDVYRI    44
             *         ^         * ^ ^  *** *  ** ^^^ ^^^     *  *** ^   ^

FUR_CAMJE    MLIENNVEYDVLLERFKKILRQGGLKYTKQREVLLKTLYH-SD-THYTPESLYME    53
FUR_CAMUP    MLMEN-LEYDVLLEKFKKILREGGLKYTKQREVLLKTLYH-SD-THYTPESLYME    52
FUR1_BACSU   M--------ENRIDRIKKQLHSSSYKLTPQREATVRVLLE-NEEDHLSAEDVY--    44
FUR_SYNEC    M------TYT--AASLKAELNERGWRLTPQREEILRVFQNLPAGEHLSAEDLYNH    47
FUR_STAEP    MNTNDAI----------KILKDNGLKYTDKRKDMLDIFVE--EDKYLNAKHIQQK    43
FUR2_BACSU   MNVQEAL----------NLLKGNGYKYTNKREDMLQLFAD--SDRYLTAKNVLSA    43
             *            *^   ^  ^ ^  *   *     ^  ^         ^  ^^^

                     ΛΛΛΛΛ   TURN  ΛΛΛΛΛ      ⇒⇒⇒        ⇐⇐⇐
FUR_ECOLI    LIDMGEE--IGLATVYRVLNQFDDAGIVTRHNFEGGKSVFEL---TQQHHHDHLI    92
FUR_YERPE    LIDIGEE--IGLATVYRCSEQFDDAGIVTRHNFEGGKSVFEL---TQQHHHDHLI    92
FUR_VIBAN    LIDLGEE--IGLATVYRVLNQFDDAGIVTRHHFEGGKSVFEL---STQHHHDHLV    92
FUR_VIBVU    LIDLGEE--IGLATVYRVLNQFDDAGIVTRHHFEGGKSVFEL---STQHHHDHLV    92
FUR_VIBCH    LIDLSEE--IGLATVYRVLNQFDDAGIVTRHHFEGGKSVFEL---STQHHHDHLV    92
FUR_HAEIN    FLEQGCE--IGLATVYRVLNQFDEAHIVIRHNFEGNKSVFEL---APTEHHDHII    93
FUR_LEGPN    LLESGED--VGLATVYRVLTQFEAAGLVSRHNFEGGHSVFEL---SQGEHHDHLV    91
FUR_PSEAE    LMEAGED--VGLATVYRVLTQFEAAGLVVRHNFDGGHAVFEL---ADSGHHDHMV    91
FUR_PSEPU    LMEAGED--VGLATVYRVLTQFEAAGLVVRHNFDGGHAVFEL---ADGGHHDHMV    91
FUR_BORPE    LIAENVE--IGLATVYRVLTQFEQAGILTRSQFDTGKAVFEL---NDGDHHDHLI    91
FUR_NEIGO    LLEEGVE--IGVATIYRVLTQFEQAGILQRHHFETGKAVYEL---DKGDHHDHIV    94
FUR_NEIME    LLEEGVE--IGVATIYRVLTQFEQAGILQRHHFETGKAVYEL---DKGDHHDHIV    94
             ^^^  ^  ^   ^*^*^^ ^*    *^* ^  ^^ ^       * ****^^

FUR_CAMJE    IKQAEPDLNVGIATVIVLLNLLEEAEMVTSISFGSAGKKYEL---ANKPHHDHMI    105
FUR_CAMUP    IKQAEPDSNVGIATVYRTLNLLEEAEMVTSLSLDSAGKKYEL---SNKPHHDHMI    104
FUR1_BACSU   LLVKEKSPEIGLATVYRTLELLTELKVVDKINFGDGVSRYDLRKEGAAHFHHHLV    99
FUR_SYNEC    LL--SRNSPISLSTIYRTLKELNQP--------------------LKHHHHLI    78
FUR_STAEP    M--DKDYPGISFDTVYRNLHLFKDLGIIESTELEGEM-KFRIAC-T-NHHHHFI    93
FUR2_BACSU   L--NDDYPGLSFDTIYRNLSLYEELGILETTELSGEK-LFRFKC-SFTHHHHFI    94
             ^          ^^  ^*^        ^^         ^          *^*^^

FUR_ECOLI    CLDCGKVIEFSDDSIEARQREIAAKHGIRLTNHSLYLYGHCA-EGDCREDEHAH-EGK    148
FUR_YERPE    CLDCGKVIEFSNESIESLQREIAKQHGIKLTNHSLYLYGHCE-TGNCREDESAH-SKR    148
FUR_VIBAN    CLDCGEVIEFSDEVIEQRQREIAEQYNVQLTNHSLYLYGHCA-DGSCKQNPNAHKSKR    149
FUR_VIBVU    CLDCGEVIEFSDDIIEERQKEIAAAYNVQLTNHSLYLYGKCG-DGSCKGNPDAHKRKS    149
FUR_VIBCH    CLDCGEVIEFSDDVIEQRQKEIAAKYNVQLTNHSLYLYGKCGSDGSCKDNPNAHKPKK    150
FUR_HAEIN    CEDCGKVFEFTDNIIEQRQREISEKYGIKLKTHNVYLYGKCSDINHCDEN-----NSK    146
FUR_LEGPN    CVKCGRVEEFVDIIEQRQKAIAERAHFKMTDHALNIYGIC------------PQCQ    136
FUR_PSEAE    CVDTGEVIEFMDAEIEKRQKEIVRERGFELVDHNLVLYV--------------RKKK    134
FUR_PSEPU    NVETSEVIEFMDAEIEKRQREIVAEHGFELVDHNLVLYV--------------RKKK    134
FUR_BORPE    CTNCGTVFEFSDPDIEKRQYKVAKDNGFVLESHAMVLYGIC---GNCQ------KG-R    139
FUR_NEIGO    CVKCGEVTEFHNPEIEALQDKIAEENGYRIVDHALYMYGVC---SDCQA-----KGKR    144
FUR_NEIME    CVKCGEVTEFHNPEIEALQDKIAEENGYRIVDHALYMYGVC---SDCQA-----KGKR    144
             ^ ^  *  ** ^     **    * ^          ^  ^*^^^  ^*

FUR_CAMJE    CKNCGKIIEFENPIIERQQALIAKEHGFKLTGHLMQLYGVC---GDCNNQKA--KVKI    158
FUR_CAMUP    CKVCGKIIEFENPIIERQQSLIANEHHFKLTGHLMQLYGIC---SDCN-HKT--KVKI    156
FUR1_BACSU   CMEFGAVDEIEGDLLEDVEEIIERDWKFKIKDHRLTFHGICHR---CNG-----KETE    149
FUR_SYNEC    CVSCSKTIEFKSDSVLKIGAKTSEKEGYHLLDCQLTIHGVCP---TC------QRSLV    127
FUR_STAEP    CENCGDTKVIDFCPIEQIKQYLP---NVTIHTHKLEVYGVCE---SC------QKNA-    138
FUR2_BACSU   CLACGKTKEIESCPMDKLCDDLD---GYQVSGHKFEIYGTCP---DCTAENQENTTA-    145
             ^  ^  ^   ^*^         ^   ^ ^!     ^
```

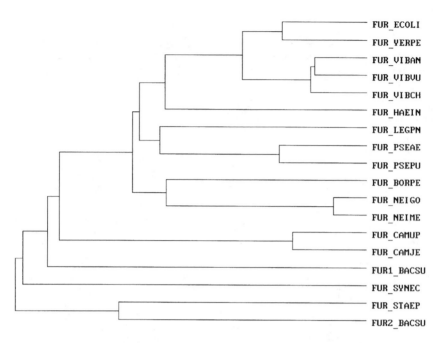

Figure 2.2 Hypothetical phylogenetic tree of Fur sequences derived from the comparison of the 18 sequences shown in Figure 2.1. For abbreviations, see Figure 2.1.

all four unpaired cysteines of *E. coli* Fur are not involved in Fe^{2+} or Mn^{2+} binding (Saito et al. 1991; Hamed and Neilands 1994). Lysine and arginine residues provide many Fur proteins with a very basic C-terminal end, which may facilitate binding to DNA.

2.3 Are There Fur-like Proteins in Gram-positive Bacteria?

The long-known requirement for iron limitation for diphtheria toxin formation raised the interest in iron regulation in *Corynebacterium diphtheriae*. The *dtxR* gene responsible for the iron regulation of toxin production was cloned and sequenced (Boyd et al. 1990). The deduced amino acid sequence shows virtually no similarity to the Fur proteins. Using gel retardation assays and footprinting, the binding site for DtxR was identified in the promoter region of the diphtheria toxin gene (Tao et al. 1992; Schmitt and Holmes 1993). Binding of DtxR to two additional iron-regulated promoters from *C. diphtheriae* was studied (Schmitt and Holmes 1994).

Investigations on the biosynthesis of the therapeutically important ferrioxamine B of *Streptomyces pilosus* identified a promoter site that is similar to the DtxR binding site (Günter et al. 1993). This promoter exhibits iron-regulated expression of a reporter gene in *Streptomyces lividans,* and mutations within a 19-bp dyad symmetric region from the promoter result in an iron-independent expression of the reporter gene. The region of dyad symmetry includes the transcriptional start site, indicating that the postulated DtxR-like repressor would cover this site during repression. The DtxR-like genes have been cloned and sequenced recently from *Streptomyces lividans. Streptomyces pilosus* (Günter-Seeboth and Schupp 1995), and *Brevibacterium lactofermentum* (Oguiza et al. 1995). Only the N-terminal 140 aa residues from about 230 aa residues are well conserved in this protein family.

The genes for the ferrichrome-uptake system of *Bacillus subtilis* were cloned and sequenced (Schneider and Hantke 1993, 1996). The common promoter region of these genes contains sequences very similar to a Fur binding site; these sequences are recognized by the *E. coli* Fur protein *in vivo* (Stojiljkovic et al. 1994). In addition, genes that complement *entD* and *entABCE* genes were cloned from *B. subtilis,* which is known to produce the siderophore dihydroxybenzoyl glycine (Grossman et al. 1993). The gene complementing *entD* was sequenced; the *entD* homolog (*sfp*) encodes the Sfp protein, which is involved in secretion of surfactin (a cyclic lipopeptide surfactant). The *sfp* gene is also required for growth under iron-restricting conditions (Grossman et al. 1993). A Fur box with unknown function is found between an uncharacterized open reading frame (ORF) and the *sfp* gene. Furthermore, from *B. subtilis,* a manganese-regulated gene and an iron-regulated gene fused to *lacZ* were sequenced; both genes contain Fur box sequences in the putative promoter regions (Chen et al. 1993b; see also Chapter 4).

From these observations, it is inferred that *B. subtilis* and the other Gram-positive organisms with a low G+C content utilize a Fur-like regulatory protein, whereas the high-G+C-containing Gram-positive organisms such as *Streptomyces* and *Corynebacteria* utilize DtxR-like proteins for iron-dependent regulation. Further support for this suggestion comes from the recent cloning of a *fur*-like gene from *Staphylococcus epidermidis;* the gene clearly is homologous to Fur and not to DtxR (Heidrich et al. 1996). Two Fur-like sequences were found in the sequence of the genome of *B. subtilis.* One shows some similarity to the *S. epidermidis* Fur-like sequence (Figures 2.1 and 2.2). The second is more similar to the Fur sequences from Gram-negative bacteria. The *Synechococcus* Fur sequence (Ghassemian and Straus 1996) is more similar to the Fur sequences of Gram-positive organisms than to those of Gram-negative bacteria.

2.4 Structural Considerations on the Interaction of Fur, Fe²⁺,
    and DNA

The Fur protein lacks the classical helix-turn-helix motif that is found in many DNA binding proteins. Binding studies with fragments of Fur indicate that the

C-terminal part of the protein is responsible for metal binding, whereas the N-terminal domain has some affinity for DNA (Coy and Neilands 1991). Binding of Mn^{2+} to Fur enhances degradation by trypsin (Braun et al. 1990; Coy and Neilands 1991). The N-terminal domain of Fur is first degraded, leaving an intermediate C-terminal fragment of Fur, which is more slowly degraded. A similar enhanced protease sensitivity is also observed with catabolite activator protein (CAP) upon binding cAMP, which is correlated with DNA binding (Heyduk and Lee 1989). These results indicate that the domain necessary for DNA binding in both Fur and CAP is induced by exposure to a ligand (Coy and Neilands 1991).

Conflicting results were presented for the binding site of Mn^{2+}. The four reduced cysteines of Fur are not directly involved in the binding of Fe^{2+} or Mn^{2+} (Coy and Neilands 1991; Saito et al. 1991; Hamed and Neilands 1994). The metal is only bound to histidines and to carboxylates of aspartate and glutamate. Saito et al. (1991) proposed a Mn^{2+}-binding site between H32, H131, and possibly H31 which would connect the N- and C-terminal domains of the protein, whereas Hamed and Neilands (1994) suggested only one binding site in the C-terminal part of Fur.

The dyad symmetry of the Fur box and negative complementation in *fur⁺fur⁻* merodiploid strains (Braun et al. 1990) suggests that Fur acts as a multimer when it binds to DNA. One of the Fur derivatives used for negative complementation, Fur 11/12 (A11T, G12D), shows upon addition of Mn^{2+} no enhanced trypsin sensitivity, indicating that the necessary conformational change for DNA binding is not possible (Braun et al. 1990). Trypsin sensitivity of another mutant, Fur90 (H90Y), is lowered by Mn^{2+}, indicating that Mn^{2+} binds without any conformational change in the N-terminal domain (Braun et al. 1990). To prove the two-domain structure of Fur, hybrid repressor proteins were made by exchanging the C-terminal domains between Fur and the λ repressor cI. According to structural predictions, the region between Fur residues 74–80 is highly flexible, which suggests that this is the hinge region between the two domains. In accordance with these considerations, the N-terminus of the trypsin cleavage product is S77 (Coy and Neilands 1991), which is predicted to reside in a turn of the polypeptide chain (Saito et al. 1991). The N-terminal domain of Fur (77 residues) and the C-terminal domain of λ cI (122 residues) were fused; the fusion protein inhibits an Fe^{2+}-Fur-regulated *lacZ* fusion, confirming the N-terminal location of the DNA binding domain. However, the fusion protein was not able to complement the growth of *fur* mutants on succinate (Hantke 1987). The C-terminal domain of Fur fused to the N-terminus of cI is able to repress a λP_{R-} regulated *lacZ* fusion, indicating that the C-terminal part is necessary for oligomerization of Fur (Stojilj-kovic and Hantke 1995). It remains to be determined whether binding of one metal per monomer (Hamed and Neilands 1994) leads to dimerization of Fur or whether iron-free Fur already has a dimeric structure. No iron-dependent activation of cI-Fur has been observed, which is an argument in favor of an iron-independent dimeric structure of Fur (Stojiljkovic and Hantke 1995).

The X-ray structure of the repressor LexA (regulator of *recA* and the SOS system) shows that it belongs to the superfamily of CAP-like DNA binding proteins. The predicted structure of Fur also shows similarities in the N-terminal domain to the CAP superfamily. A DNA-interacting helix is predicted between residues 36 and 49 and between 57 and 62 (Holm et al. 1994). *Synechococcus* Fur is interesting because it exactly lacks the domain of a two-stranded β- sheet, a common structural element in the CAP-type regulator proteins (Holm et al. 1994; Figure 2.2).

The region covered by Fur in DNase footprinting studies is relatively large (30–80 bp) for a dimer of 34 kDa (de Lorenzo et al. 1987). Also, in electron microscopic studies, it seems that more than one dimer covers the operator region (Frechon and Le Cam 1994). It is not known whether this is an artifact generated by the high concentrations of Fur used in the binding assays or whether it reflects the situation *in vivo.*

2.5 Phenotype of Fur Mutants

In *Salmonella* and *E. coli,* the first *fur* mutants were recognized by their phenotype to overproduce enterochelin (Ernst et al. 1978), to be derepressed for iron uptake, and to synthesize constitutively the siderophore receptor proteins in the outer membrane (Hantke 1981). Later, it was shown that *fur* mutants are unable to grow on a minimal medium with nonfermentable carbon sources such as succinate, fumarate, or acetate (Hantke 1987). *E. coli* Δ*fur recA* double mutants are not viable under oxic conditions (Touati et al. 1995), which shows that Fur is also important for iron metabolism in the cell and for detoxification of iron(II), which generates cell-damaging radicals in the Fenton reaction with oxygen.

In many bacteria, it was possible to select *fur* mutants with manganese (Hantke 1987). Manganese may substitute for ferrous iron in the Fur protein and repress iron-regulated genes as observed *in vitro.* If the sole source of iron requires derepression of a specific uptake system, the bacterium will be starved for iron unless it acquires a mutation that at least partially derepresses the iron-uptake system. Mutants in *fur* are able to grow in the presence of manganese. However, the critical manganese concentration must be determined for every organism. Deletions of *fur* have been so far isolated only in enteric bacteria and in vibrios. In many other organisms, *fur* seems to have a vital function because it has not been possible to obtain deletions of *fur* in *N. gonorrhoeae* (Thomas and Sparling 1996), *Pseudomonas* (Hasset et al. 1996), and *Synechococcus* (Ghassemian and Straus 1996). *P. aeruginosa* Fur point mutants show a phenotype different than that expected from *E. coli fur* mutants. In particular, they are partly deficient in the uptake of their siderophores ferripyochelin and ferripyoverdin, they have a longer lag phase when growing aerobically, and they are more sensitive to paraquat and H_2O_2 than the parent strain. *Synechococcus* sp. PCC 7942 has a polyploid genome. After insertional inactivation of *fur,* it was selected for a pure *fur* mutant by segregation; however, no mutants that had lost their wild-type copy of *fur* were obtained. These heteroallelic mutants showed constitutive production of siderophore. In addition, flavodoxin synthesis was derepressed (Ghassemian and Straus 1996). Flavodoxin synthesis is derepressed in cyanobacteria and many other organisms under low-iron growth conditions, but not in *E. coli* (Mayhew and Ludwig 1975).

3. Fur-dependent Iron-uptake Systems

3.1 Iron(III) Uptake by Outer Membrane Receptors

All proteins in the outer membrane of *E. coli* that are derepressed in *fur* mutants are receptors for certain siderophores. Six receptors and their substrates were identified in *E. coli* K-12: FhuA, ferrichrome (Coulton et al. 1986); FhuE, coprogen (Sauer et al. 1990); FecA, iron dicitrate (Pressler et al. 1988); FepA, enterochelin (enterobactin) (Lundrigan and Kadner 1986); Cir, dihydroxybenzoyl serine (catecholate) (Nau and Konisky 1989); and Fiu, dihydroxybenzoyl serine (catecholate) (Hantke 1990). Some *E. coli* strains have the receptors for aerobactin (Iut; mostly plasmid encoded; Krone et al. 1985), and for yersiniabactin (FyuA; Rakin et al. 1995; Haag et al. 1994). It is interesting to note that *E. coli* produces at least three receptors (FhuA, FecA, and FhuE) for utilization of siderophores that it does not produce. A similar saprophytic mechanism of iron acquisition has been observed in *Serratia marcescens* (Angerer et al. 1992), *Yersinia enterocolitica* (Bäumler et al. 1993), *Pseudomonas aeruginosa* (Dean and Poole 1993), and many other bacteria.

3.2 Iron(III) Uptake by Outer Membrane Receptors is
TonB dependent

The activity of all iron-regulated outer membrane receptors is dependent on the TonB protein. TonB is anchored in the cytoplasmic membrane, extends into the periplasm, and is thought to function as an energy transducer that drives siderophore uptake through the outer membrane receptors (Braun et al. 1991; Postle 1990a,b). Once inside the periplasm, siderophores interact with a binding-protein-dependent transport system. TonB may gain its energy by interacting with the complex of the two cytoplasmic membrane proteins ExbB and ExbD (Kampfenkel and Braun 1992, 1993). The expression of *tonB* is regulated by Fur, as was shown in the first description of the *fur* gene in *E. coli* (Hantke 1981).

Because of the general iron(III) transport defect, *tonB* mutants are derepressed for iron-transport systems in normal growth medium. No difference in derepression between a *tonB-lacZ* mutant and a *fur tonB-lacZ* mutant has been observed. A threefold Fur-dependent repression of the *tonB-lacZ* fusion is obtained when the cells are supplied with iron(II) generated by the addition of ascorbate. Iron(II) uptake is independent of TonB (Kammler et al. 1993). Also, under anoxic conditions, a Fur-dependent regulation of *tonB-lacZ* is observed (Hantke 1981). These results were later confirmed with a *tonB-lacZ* fusion in a *tonB*+ background (Postle 1990a,b). In addition, it was shown that the previously reported growth-phase-dependent regulation of *tonB,* which is correlated with supercoiling of DNA (Dorman et al. 1988), is most probably the result of gradual iron starvation when the cells reach the stationary phase (Postle 1990a,b). The genes *exbBD* are Fur regulated (Hantke and Zimmermann 1981), probably through a Fur box in the *exbB* promoter (Eick-Helmerich and Braun 1989). In *P. aeruginosa,* the three genes *tonB, exbB,* and *exbD* are found in an operon (Bitter et al. 1993).

TonB-dependent receptors are found in many other Gram-negative bacteria, and more than 40 sequences are now known. An alignment of sequences shows that even the more closely related proteins may have very different substrate specificities (Figure 2.3). A broad range of iron-carrying substrates can be recognized by these proteins. For example, the transferrin receptor TbpA (Cornelissen et al. 1992; Legrain et al. 1993) and the lactoferrin receptor LbpA (Petterson et al. 1994) in gonococci belong to this protein family. A second outer membrane transferrin-binding lipoprotein, TbpB, encoded upstream of *tbpA,* has no similarities to TonB-dependent receptors. TbpB interacts with TbpA for efficient iron uptake from transferrin as mutations in either of these genes lead to a defect in transferrin iron uptake (Irwin et al. 1993; Anderson et al. 1994). Also, hemin transport is TonB dependent (Stojiljkovic and Hantke 1992; Jarosik et al. 1994), and two receptor proteins have been characterized (Henderson and Payne 1994; Stojiljkovic and Hantke 1992).

The expression of all these TonB-dependent receptor proteins (with the exception of BtuB, the receptor for vitamin B_{12}) is regulated by iron and, in many cases, the involvement of Fur has been demonstrated. In some cases, there are additional regulatory circuits that specifically induce expression of the transport system only in the presence of their substrate (see Section 4).

3.3 Iron(III) Uptake: Binding-protein-dependent Transport Systems

After the TonB-dependent receptor has delivered its substrate into the periplasm, a binding-protein-dependent system transports the iron-containing siderophore across the cytoplasmic membrane. Often these systems are encoded in an Fe^{2+}-Fur-regulated operon, together with their cognate outer membrane receptor, and sequence alignments have shown that they constitute a subfamily in the

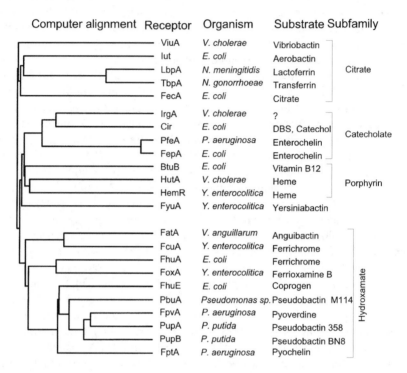

Figure 2.3 Hypothetical phylogenetic tree of 23 members of the TonB-dependent receptor proteins. Subfamilies were named according to a prominent member of the group; however, substrate specificity within the groups is often very diverse. The description of the different receptors can be found: ViuA (Butterton et al. 1992); Iut (Krone et al. 1985); LbpA (Petterson et al. 1994); TbpA (Cornelissen et al. 1992; Legrain et al. 1993); FecA (Pressler et al. 1988); IrgA (Goldber et al. 1992); Cir (Nau and Konisky 1989); PfeA (Dean and Poole 1993); FepA (Lundrigan and Kadner 1986); BtuB (Heller and Kadner 1985); HutA (Henderson and Payne 1994); HemR (Stojiljkovic and Hantke 1992); FyuA (Rakin et al. 1994); FatA (Actis et al. 1988); FcuA (Koebnik et al. 1993); FhuA (Coulton et al. 1986); FoxA (Bäumler and Hantke 1992); FhuE (Sauer et al. 1990); PbuA (Morris et al. 1994); FpvA (Poole et al. 1993); PupA (Bitter et al. 1991); PupB (Koster et al. 1993); FptA (Ankenbauer and Quan 1994).

binding-protein-dependent transport systems (Köster et al. 1991; Saurin et al. 1994; Tam and Saier 1993). However, significantly fewer sequences are known for the binding-protein-dependent transport systems (Table 2.1) than for receptor proteins. One reason is that several receptors deliver their substrates to one binding-protein-dependent transport system. For example, three outer membrane receptors, FhuA, FhuE, and Iut, bind the hydroxamates ferrichrome, coprogen, and aerobactin, respectively, which in *E. coli* are all transported into the cytoplasm via FhuCDB (Table 2.1; Braun and Hantke 1991). There are only two additional binding-protein-dependent transport systems in *E. coli* known to transport sidero-phores: FepBDGC for catecholates (Chenault and Earhart 1991; Chenault and Earhart 1992; Shea and McIntosh 1991) and FecBCDE for ferric dicitrate (Staudenmaier et al. 1989). From other organisms, only the system for anguibactin uptake into *V. anguillarum* (Köster et al. 1991), and for hemin uptake into *Y. enterocolitica* (Stojiljkovic and Hantke 1994) are known (Table 2.1). In the Gram-positive organism *B. subtilis,* one binding-protein-dependent transport system for ferrichrome has been cloned (Schneider and Hantke 1993). Interestingly, the binding protein FhuD$_{BS}$ is anchored as a lipoprotein in the cytoplasmic membrane; this seems to be a general feature of binding proteins in Gram-positive bacteria.

The iron(III) transport system Sfu (Table 2.1), originally characterized in *Serratia marcescens* (Angerer et al. 1990), belongs to another subfamily of binding-protein-dependent transport systems (Saurin et al. 1994; Tam and Saier 1993). It is not known which iron(III) source is used in nature by this uptake system. Homologous iron binding proteins have also been observed in *Neisseria* spp. and *Haemophilus influenzae.* In *Neisseria* spp., the Fbp system transports the iron, which somehow has been removed from transferrin or lactoferrin by the proteins TbpAB or LbpAB, respectively, in the outer membrane (Chen et al. 1993a). The same function is expected for HitABC in *Haemophilus,* which possesses transferrin binding proteins in the outer membrane.

3.4 Iron(II) Uptake

Ferrous iron transport systems are certainly far more distributed in the microbial world than one would expect from the very few publications in this field. Recently, the first ferrous iron transport system was characterized in *E. coli* (Kammler et al. 1993). It consists of two proteins, a small cytoplasmic protein of 8.4 kDa with unknown function and an 84-kDa protein with similarity to ATPases in the cytoplasmic membrane. This led to the assumption that ferrous iron transport is mediated by an ATP-driven pump. Expression of this system is regulated by Fur. In addition, uptake is stimulated under anoxic conditions by Fnr, the fumarate-nitrate-reductase regulator (Kammler et al. 1993).

Table 2.1 Iron-binding-protein-dependent transport systems

Organism	Binding Protein	Membrane Component	ATP Binding Component	Substrates	References
E. coli	FhuD	FhuB	FhuC	Ferrichrome, aerobactin, coprogen	Köster and Braun 1986; Burkhardt and Braun 1987; Braun and Hantke 1991
	FecB	FecC, FecD	FecE	Ferric di-citrate	Staudenmaier et al. 1989
	FepB	FepD, FepG	FepC	Enterochelin, dihydroxy-benzoylserine	Chenault and Earhart 1992; Shea and McIntosh 1991
Y. enterocolitica	HemT	HemU	HemV	Heme	Stojiljkovic and Hantke 1994
V. anguillarum	FatB	FatC, FatD	?	Anguibactin	Köster et al. 1991
Campylobacter coli	CeuE	CeuB, CeuC	CeuD	Enterochelin	Richardson and Park 1995
Erwinia chrysanthemi	CbrA	CbrB, CbrC	CbrD	?	Mahe et al. 1995
B. subtilis	FhuD	FhuB, FhuG	FhuC	Ferrichrome	Schneider and Hantke 1993, 1996
S. marcescens	SfuA	SfuB	SfuC	Iron(III)	Angerer et al. 1990
H. influenzae	HitA	HitB	HitC	Iron(III)	Sanders et al. 1994
Neisseria spp.	FbpA	FbpB	FbpC	Iron(III)	Adhikari et al. 1996

4. Induction by Fe^{3+} Siderophores

*4.1 Transmembrane Transcription Induction from the Cell Surface
and Fe^{2+}-Fur Repression of the Ferric Citrate Transport System
of E. coli*

Under oxic conditions, *E. coli* cannot grow on citrate as a carbon source because citrate is not taken up. However, citrate serves as a siderophore that, at a 20-fold surplus of citrate over iron, forms a Fe^{3+}-dicitrate complex (Hussein et al. 1981). The transport system is inducible by citrate (Frost and Rosenberg 1973) and requires low amounts of iron (Hussein et al. 1981). Higher amounts of iron repress synthesis of the transport system via Fe^{2+}-Fur.

The Fec transport system maps at 97.3 minutes on the *E. coli* K-12 chromosome (Veitinger and Braun 1992), where five transport genes (*fecABCDE*) and two regulatory genes (*fecIR*) are encoded (Figure 2.4). *fecI* codes for a protein that activates transcription of the *fec* transport genes, and *fecR* determines a protein that responds to ferric citrate in the growth medium (Van Hove et al. 1990). *fecA* codes for an outer membrane receptor (Pressler et al. 1988), *fecB* for a periplasmic binding protein, *fecC* and *fecD* for highly hydrophobic integral cytoplasmic membrane proteins, and *fecE* for a hydrophilic protein that is associated with the cytoplasmic membrane and contains the two motifs of ATP-binding proteins (Staudenmaier et al. 1989). FecE can be specifically labeled with 8-azido-ATP, suggesting that ATP is the principal energy source of citrate-mediated iron transport (Schultz-Hauser et al. 1992). The Fec transport system displays the typical properties of a periplasmic binding-protein-dependent transport system. It is unclear how far ferric citrate enters the cell and where Fe^{3+} dissociates from citrate. We have shown that ferric citrate protects the FecB protein from degradation by proteinase K (Baun and Herrmann, unpublished results), which implies that ferric citrate is taken up into the periplasm, where it binds to FecB. Studies of mutants revealed that transport of Fe^{3+} across the outer membrane requires the FecA, TonB, ExbB, and ExbD proteins (Zimmermann et al. 1984).

Regulation of the ferric citrate transport system has been studied by various means. Growth is determined on nutrient agar plates containing 1 mM ferric citrate as sole iron source, and 0.2 mM dipyridyl for depletion of the available iron. Determination of the $^{55}Fe^{3+}$ transport rates in the presence of 1 mM citrate provides quantitative data on the level of induction and the function of the various *fec* genes. The amount of FecA protein is determined by SDS-PAGE of outer membrane proteins. FecA is the only transport protein that is made in sufficient amounts to allow detection by staining. The β-galactosidase activities of a chromosomal *fecB*::Mud1 (Ap*lac*) operon fusion and a plasmid-encoded *fecA-lacZ* gene fusion yield quantitative data on the transcription of the *fec* transport genes (Zimmermann et al. 1984).

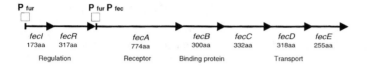

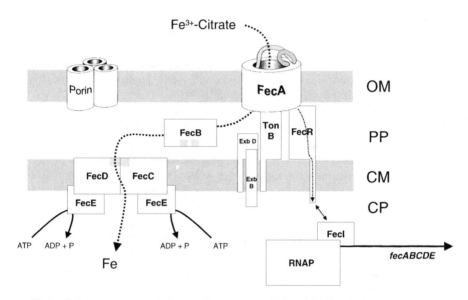

Figure 2.4 Arrangement of the regulatory genes *fecI* and *fecR* and the transport genes *fecABCDE* of the ferric citrate transport system on the chromosome of *E. coli* (upper panel). P. denotes the promoters upstream of *fecI* and *fecA,* fur indicates potential binding sites for Fe^{2+}-Fur, fec indicates the binding site for the FecI RNA polymerase (RNAP) complex in response to ferric citrate, and aa denotes amino acid residues. The lower panel shows the location of the gene products in the outer membrane (OM), the periplasm (PP), the cytoplasmic membrane (CM), and the cytoplasm (CP) of cells. The dotted line on the left indicates the transport route into the cells; the dotted line on the right indicates the signaling pathway, starting from FecA and proceeding through FecR to FecI. It is assumed that the permeability of FecA is determined by a loop that closes the channel unless it is actively opened by an energy-consuming process via the action of the TonB, ExbB, and ExbD proteins, which are involved in transport and signaling across the outer membrane. The functional relationships of the proteins are demonstrated. The stoichiometry of the proteins has not been determined. Porin trimers, through which hydrophilic substances not larger than 600 Daltons diffuse, are shown for comparison. The diagram was drawn by S. Plantör, this Institute.

Earlier studies have shown that fluorocitrate and phosphocitrate induce the system but transport iron poorly. A 10-fold increase in the intracellular citrate concentration does not induce the Fec system. A *fec* transport mutant, the mutation of which maps outside the *fec*A gene, produces increased amounts of FecA in the presence of ferric citrate (Hussein et al. 1981). Later, it was shown with genetically defined mutants that, indeed, *fecB* (Zimmermann et al. 1984) and *fecBCDE* (Härle et al. 1995) are not required for *fec* induction by ferric citrate. These data suggested that ferric citrate does not have to enter the cytoplasm, but enters the periplasm to induce the system. As will be shown later in this section, for induction, ferric citrate does not even have to be taken up into the periplasm.

The scenario of having two regulatory genes and induction from outside the cytoplasm is reminiscent of the two-component regulatory systems in which a transmembrane protein serves as sensor of the inducer and transmits the signal to the receiver, which in turn activates transcription (Stock et al. 1990). Signal transduction involves phosphorylation of the sensor, followed by phosphorylation of the receiver. The histidine phosphorylation site of the sensor and the aspartate phosphorylation site of the receiver, both contained in conserved sequences among the various two-component systems, are not present in FecR and FecI (Van Hove et al. 1990). Fec transmembrane regulation follows other rules.

Chromosomal Mud1-insertion mutants in *fecI* are unable to grow with ferric citrate as the iron source, have no FecA protein, and do not transcribe *fecB-lacZ*, indicating an essential role of FecI in *fec* induction (Van Hove et al. 1990). Complementation of a *fecI* mutant by plasmid-encoded *fecI* yields a constitutive expression of chromosomal *fecA* and *fecB*, at a level about 10% of that obtained by induction of a *fecIR* wild type with ferric citrate. The FecI sequence reveals a putative DNA binding helix-turn-helix motif (Van Hove et al. 1990). Point mutations in the helix motifs either increase or decrease induction of *fecB* transcription (Ochs et al. 1996).

For studying *fec* transcription *in vitro*, FecI was overexpressed, isolated as inclusion bodies, and renatured. Mobility of a 75-bp DNA fragment upstream of *fecA* is retarded on an agarose gel by FecI and RNA polymerase core enzyme devoid of σ^{70}. Cell lysates shift the same DNA fragment when the cells are grown in the presence of ferric citrate and overexpress FecIRA. A mutational analysis of the putative promoter DNA revealed a cluster of point mutations in the 75-bp fragment that showed a reduced FecI-RNA polymerase binding (Angerer et al. 1995). The 75-bp fragment contains the −35 and −10 regions of the transcription initiation site of the *fec* transport genes upstream of *fecA* (Enz et al. 1995). Northern blot hybridizations revealed two major transcripts, one comprising *fecI* and *fecR,* and the other comprising *fecA*. mRNA from the *fecBCDE* genes downstream of *fecA* is only detected after amplification of *fec* mRNA by reverse transcription PCR. A hairpin structure is located between *fecA* and *fecB,* which may reduce transcription of the *fecBCDE* genes and may stabilize the *fecA* transcript if the *fecABCDE* transcript is degraded from the 3'-end.

Sequence comparison revealed that FecI belongs to a new class of sigma factors (Lonetto et al. 1994). The predicted sigma factors regulate carotinoid biosynthesis in *Myxococcus xanthus* (CarQ), alginate biosynthesis in *Pseudomonas aeruginosa* (AlgU), synthesis of a periplasmic protease in *E. coli* (sigma^E), Ni^{2+} and Co^{2+} resistance in *Alcaligenes eutrophus* (CnrH)), and agar degradation by *S. coelicolor* (sigma^e). All these regulatory proteins receive their signals from extracytoplasmic compartments and have been designated ECF (extra cytoplasmic function) sigma factors (Lonetto et al. 1994). FecI appears to be an unusually specific sigma factor that activates only transcription of the *fec* transport genes. Sequencing of the *Bacillus subtilis* 168 genome also revealed a DNA sequence, designated OrfX20, that has similarity to *fecI* and for which no function has been determined (Sorokin et al. 1993). FecI of *B. subtilis* complements a *fecI-fecR* mutant of *E. coli,* resulting in a constitutive phenotype higher than that with FecI, but that does not confer induction by ferric citrate via *E. coli* FecR (S. Brutsche et al., unpublished results).

E. coli overexpressing FecI transcribe *fec* transport genes even in the absence of FecR, but do not respond to ferric citrate. FecR is required for ferric citrate regulation. Point mutations in *fecR* do not respond to ferric citrate, and some express *fecA-lacZ* constitutively (Wriedt et al. 1995). All of the mutated FecR derivatives except one are no longer cleaved after residue 181 by a cellular protease, as is observed with wild-type FecR, indicating a conformational change in the mutated FecR derivatives. An *E. coli* chromosomal *fecR* mutant (WA176), containing an amber stop codon in the 19th triplet, does not grow on ferric citrate and transcribes *fec* transport genes only poorly when *fecI* is encoded on the chromosome. Two other mutants that synthesize C-terminally truncated FecR proteins of 81 and 201 residues (length of FecR wild-type 317 residues) express *fecA* and *fecB* constitutively and independently of ferric citrate. To identify the minimal size of FecR that leads to the constitutive phenotype, plasmid-encoded C-terminally truncated *fecR* derivatives were constructed; polypeptides with more than 59 N-terminal residues complement the chromosomal *fecR* mutant and display a constitutive phenotype (Ochs et al. 1995). These data suggest that the N-terminal portion of FecR is sufficient for induction, but that the C-terminal portion is required for the response to ferric citrate. Chromosomally encoded FecI requires FecR and ferric citrate to induce the *fec* transport genes. Therefore, FecI activates transcription in the presence of FecR, which responds to ferric citrate. Overexpressed FecI induces transcription without the help of FecR and ferric citrate, but to a lower level (10%) than with FecR and ferric citrate (Ochs et al. 1995).

Because ferric citrate does not have to enter the cytoplasm to induce the *fec* transport genes, FecR likely transduces the signal from the periplasm into the cytoplasm. FecR contains a single hydrophobic sequence (GLLLLLGAGGGW-QLWQ) between residues 85 and 100. Fusions between FecR, starting at residues 61 and 81, and β-lactamase render cells sensitive to 5 µg/ml of ampicillin, whereas a fusion to residue 107 is resistant to 100 µg/ml of ampicillin (Ochs et al. 1995). These results are consistent with a model that localizes residues 1–84 of FecR in the cytoplasm and residues 100–317 in the periplasm. The smallest FecR derivatives, with 59, 65, and 68 residues, that display a constitutive phenotype, would interact with FecI in the cytoplasm even in the absence of ferric citrate. These derivatives and the larger truncated FecR polypeptides do not respond to ferric citrate, but still interact with FecI, causing a constitutive phenotype.

To determine the genes required for induction of the *fec* transport system, an *E. coli* strain was constructed (AA93) that lacked the entire *fec* region. Strain AA93 was transformed with plasmids carrying various *fec* regulatory and transport genes. The transformant that fully induced the *fec* transport system contained minimally *fecI, fecR,* and *fecA,* and *tonB, exbB,* and *exbD* (Ochs et al. 1995, 1996).

Because the components required for ferric citrate translocation through the outer membrane are required for *fec* induction by ferric citrate, the obvious assumption is that ferric citrate enters the periplasm to induce the *fec* transport genes. Once the transport system is induced, the minimal concentration of ferric citrate to maintain the system should then be lower than the concentration required to induce the system. In fact, the amount of FecA greatly increases after induction, and synthesis of TonB, ExbB, and ExbD increases during iron deprivation caused by citrate under conditions (*fecBCDE* mutants) in which iron is not taken up through the cytoplasmic membrane. However, the same minimal concentration of ferric citrate (0.1 mM) to induce the system is required to maintain the system; this argues against transport of ferric citrate through the outer membrane to be essential for induction.

To determine whether *fec* induction by ferric citrate requires FecA, conditions were used that bypassed FecA. Ferric citrate is small enough (ferric dicitrate: 434 Da) to diffuse through the porins. At ferric citrate concentrations that stimulate growth on iron-deficient nutrient agar plates, *fecA-lacZ* and *fecB-lacZ* transcription is not enhanced in a *fecA* deletion mutant. This result indicates that ferric citrate in the periplasm does not induce the system. Rather, FecA in the outer membrane is essential; therefore, ferric citrate may induce the *fec* system from the cell surface. The unlikely possibility that the inducing concentration of ferric citrate is higher than the growth-promoting concentration is ruled out by the following finding. Overproduced FecR, in the presence of chromosomally encoded FecI, induces *fec* transcription in nutrient broth that contains trace amounts of ferric citrate. As this ferric citrate concentration does not stimulate growth, a lower ferric citrate concentration is required for induction than for growth. No induction occurs in M9 minimal salt medium without supplementation of ferric citrate, showing that overproduction of FecR per se does not induce the system.

The direct participation of FecA in induction was proven with *fecA* point mutants. *E. coli* AA93 Δ*fec* constitutively expresses plasmid-encoded *fecA-lacZ* in the absence of ferric citrate (Härle et al. 1995). One mutant displays a TonB-independent induction, but a TonB-dependent transport. Uncoupling of Fec-A-mediated induction from FecA-catalyzed transport is also achieved by deletion of residues 14–68 of the mature FecA protein. Strains synthesizing FecA Δ14–68 are no longer inducible by ferric citrate, but show wild-type ferric citrate transport activity (Kim et al., 1997). The N-terminal FecA portion shows no similarity to any of the mature iron receptors. About 70 residues of FecA extend from the homologous region of the ferric siderophore receptors so that FecA is longer than the other *E. coli* receptors. This becomes especially apparent when the sequence of FecA is compared to that of other iron receptors. Of all the iron receptors, FecA shares the greatest similarity with RumA of *Morganella morganii* (Kühn et al. 1995), except for the first N-terminal 73 residues, which are absent in RumA. RumA is the receptor for rhizoferrin, which consists of two citrate moieties linked by 1,4-diaminobutane (Drechsel et al. 1991); RumA is not inducible by ferric rhizoferrin.

The current model proposes a conformational change of FecA induced by ferric citrate. The TonB-ExbB-ExbD complex is involved in the conformational change. The signal is then transmitted from FecA to FecR by direct interaction. FecR transduces the signal across the cytoplasmic membrane to FecI, which assumes a conformation that facilitates interaction with RNA polymerase and binding to the promoter upstream of *fecA*. The constitutive phenotype of the short, truncated FecR derivatives located in the cytoplasm possibly arises from the lack of interaction with FecA in the outer membrane. The longer derivatives possibly also do not interact with FecA or are unable to respond to the conformational change in FecA. This mechanism also ascribes a new function to the TonB, ExbB, and ExbD proteins in that they are not only required for uptake of substrates through the outer membrane, but are also required for signal transduction across the outer membrane. The finding of a *tonB*-independent *fecA* mutant in signal transduction that retains the TonB requirement for transport suggests that the assumed conformational changes in FecA for induction and transport are mechanistically not identical.

As will be described in Section 4.2, induction systems similar to that of Fec are present in *Pseudomonas putida*. It is likely that these systems will not remain the only examples of transcription initiation from the cell surface. It is conceivable that gene expression observed in bacteria as a response to interaction with human, animal, and plant cells involves mechanisms that are similar to the Fec induction prototype.

Transcription of *fecI* and *fecR* are not autoregulated (Ochs et al. 1996). Rather, *fecI* and *fecR* transcription requires σ^{70} and is subject to repression by iron via the Fur protein (Van Hove et al. 1990). Upstream of *fecA* and *fecI* are typical Fur binding sequences. Lack of the Fur repressor in a *fur* deletion derivative does not alter the low expression of FecA in strain WA176 *fecR*, indicating that the FecR phenotype is not affected by Fe^{2+}-Fur.

4.2 Induction of PupB Receptor Gene Expression of Pseudomonas putida *WCS358 by Pseudobactin BN8*

Synthesis of the PupB receptor of *Pseudomonas putida* WCS358 is induced by its cognate ferric siderophores, pseudobactin BN8 and pseudobactin BN7. Two regulatory genes, *pupI* and *pupR,* are located upstream of *pupB.* PupI is 43% identical to FecI and displays a helix-turn-helix motif at the same position as FecI. PupR is 43% identical to FecR, contains a potential transmembrane segment between residues 85 and 103 similar to FecR, but in contrast to FecR, and exhibits a second hydrophobic sequence from residues 238 to 263, which reads QAVAAVAPAWSQGMLVAQGQPLAAFI (Venturi et al. 1993).

Convincing evidence for the involvement of PupB in the induction process has been shown by an elegant approach (Koster et al. 1993). A *pupB* mutant shows 50% transport activity, but is no longer inducible, suggesting that pseudobactin BN8 in the periplasm does not induce *pubB* expression (Venturi 1994) and that PubB might be directly involved in signal transduction. This notion was examined by replacing the 86 N-terminal residues of PupB with the corresponding region of PupA, another receptor of *P. putida* WCS358, which recognizes pseudobactin 358. The chimeric PupAB protein transports pseudobactin BN8, but is incapable of inducing *pupB* expression. The converse hybrid protein, PupBA, induces *pupB* expression in response to pseudobactin 358, showing that the ferric siderophore is recognized by regions of PupA downstream of the PupB fragment, and that the N-terminal PupB fragment confers the induction.

Synthesis of pseudobactin 358 seems to be under the control of the PfrI protein, which displays 38% identity to FecI and 27% identity to PupI. Interestingly, no analogs of PupR and FecR have been found, suggesting another mechanism for PfrI activity control (Venturi et al. 1993).

In *Pseudomonas aeruginosa,* pyoverdin synthesis is under the positive control of the *pvdS* gene product, and PvdS shows significant homologies to FecI and PupI and other members of the ECF family (Cunliffe et al. 1995; Miyazaki et al. 1995). In *Pseudomonas fluorescens* M114, a *fecI*-like positive regulatory gene, *pbrA,* controls pseudobactin M114 synthesis (Sexton et al. 1996), *pvdS* and *pbrA* expression are, like *fecI* expression, not autoregulated, but are iron regulated and, in *E. coli,* repressed by Fe^{2+}-Fur. In *P. aeruginosa* and *P. fluorescens,* no FecR homologs have been identified, and it has not been shown that ferric siderophores and outer membrane receptors are involved in transcription initiation (Venturi et al. 1995). There might be a number of mechanisms to explain how bacteria react via ECF sigma factors to environmental signals.

4.3 Induction of the PfeA Receptor Gene Expression of Pseudomonas aeruginosa by Ferric Enterobactin

The ferric enterobactin receptor PfeA of *P. aeruginosa* is induced by ferric enterobactin (Dean and Poole 1993). Two genes upstream of *pfeA* encode proteins that are highly similar to response regulator and sensor components of the so-called two-component regulatory systems. Analysis of the amino acid sequences derived from the nucleotide sequences suggests a membrane association of the PfeS protein and a cytoplasmic location of the PfeR protein. These structural features suggest a transmembrane signaling device across the cytoplasmic membrane through a phosphorylation/dephosphorylation cascade, which is mechanistically entirely different from the transport systems induced by ferric citrate and the pseudobactins.

4.4 Induction of Pyochelin and Ferric Pyochelin Receptor Protein Synthesis by Ferric Pyochelin

Pyochelin activates expression of the pyochelin receptor gene (Heinrichs and Poole 1993). Pyochelin synthesis and ferric pyochelin receptor synthesis are induced by the *pchR* gene product, which displays in its C-terminal portion a strong similarity to the C-terminus of the AraC protein. AraC serves in a variety of enterobacteria as a transcriptional repressor in the absence of arabinose and as an activator in the presence of arabinose. Expression of *pchR* is probably iron controlled because a typical Fe^{2+} Fur binding sequence is present upstream of *pchR*. These data indicate a third mechanism of ferric-siderophore-controlled transcription of siderophore synthesis and transport genes.

4.5 Positive and Negative Regulation of the Ferric Anguibactin Transport System of Vibrio anguillarum

A sufficient iron supply is essential for the virulence of the fish pathogen *V. anguillarum.* The 65-kb plasmid pJM1 encodes an iron-uptake system that consists of the siderophore anguibactin and a transport system that provides the bacterium with the essential iron and is thus directly responsible for the pathogenicity of the strain (Crosa 1980). Two plasmid-encoded synergistically acting transcriptional activators, AngR and Taf, regulate synthesis of the anguibactin biosynthetic genes, and possibly the gene for the outer membrane transport protein FatA (Tomalsky et al. 1988). The 110-kDa protein AngR possesses two helix-turn-helix motifs with corresponding leucine zippers that may provide interactive domains with other coregulators, such as Taf (Tomalsky et al. 1993). In addition to its transcriptional activating function, AngR also complements a mutation in the *E. coli entE* gene, which encodes the enterobactin biosynthetic enzyme 2,3-dihydroxybenzoate-AMP ligase. Therefore, AngR may function in *V. anguillarum* as an EntE-like enzyme for the biosynthesis of anguibactin. A similar regulatory device operates in the expression of the iron-regulated high-molecular-weight protein 2 (Irp2) of *Yersinia enterocolitica,* which exhibits 44.5% identity to AngR (Guilvout et al. 1993).

A Fur-type protein has been found in *V. anguillarum* that, under high-iron conditions, represses *fatA* transcription as well as anguibactin synthesis (Tomalsky et al. 1994). A further mechanism of control in this bacterium is provided by an antisense RNA, RNAa. RNAa is complementary to the 5'-untranslated end of *fatA* mRNA and to the 3'-translated end of *fatB,* including the intergenic region between *fatA* and *fatB*. RNAa synthesis is maximal under iron-rich conditions. RNAa fine-tunes *fatA* and *fatB* transcripts that escape Fe^{2+}-Fur repression and increases the rate of turnover of *fatB* mRNA (Waldbeser et al. 1993; Salinas et al. 1993).

5. Fur-independent Iron Regulation

The systems described contain *fur* genes or promoter regions with putative Fur-binding sites, suggesting regulation of ferric siderophore transport and/or synthesis of genes or their regulatory genes by Fe^{2+}-Fur. However, evidence exists that iron displays additional regulatory activities that are not mediated by Fur. In an *E. coli fur* deletion mutant, supplementation of iron in the growth medium reduces to some extent the amount of iron-regulated outer membrane proteins (Ochs et al. 1995). This points to an iron regulatory mechanism that acts independently of Fur. In a study on iron regulation of anaerobically expressed respiratory genes of *E. coli, narG, dmsA,* and *frdA* were expressed 15-fold less when iron was limited; this also occurred in a *fur* mutant (Cotter et al. 1992). Similar observations have been made in other organisms. In *Vibrio cholerae,* 17 proteins are repressed by iron, independently of Fur (Litwin and Calderwood 1994). Another class of proteins synthesized in the *fur* mutant at a low iron concentration is strongly repressed by iron, but is not expressed in the *fur⁺* strain grown in low- or high-iron medium. In addition, induction by iron of proteins that are Fur dependent or Fur independent is observed (Litwin and Calderwood 1994). In a *fur* mutant of *S. typhimurium,* two proteins are induced in dipyridyl-containing medium, suggesting induction by iron limitation, which is not caused by conversion of Fe^{2+}-Fur into unloaded Fur (Foster and Hall 1992). These results indicate complex iron regulatory mechanisms that presumably also affect genes that have nothing to do with the iron supply.

Acknowledgments

We thank Russell Bishop and Karen Brune for critical reading of the manuscript. The authors' work was supported by the Deutsche Forschungsgemeinschaft (SFB323, Ha1186, BR330/14) and Fonds der Chemischen Industrie.

References

Actis, L. A., M. E. Tolmasky, D. H. Farrel, and J. H. Crosa. 1988. Genetics and molecular characterization of essential components of the *Vibrio anguillarum* plasmid-mediated iron-transport system. *J. Biol. Chem.* 263:2853–2860.

Adhikari, P., S. A. Berish, A. J. Nowalk, K. L. Veraldi, S. A. Morse, and T. A. Mietzner. 1996. The *fbpABC* locus of *Neisseria gonorrhoeae* functions in the periplasm-to-cytosol transport of iron. *J. Bacteriol.* 178:2145–2149.

Anderson, J. E., P. F. Sparling, and C. N. Cornelissen. 1994. Gonococcal transferrin-binding protein 2 facilitates but is not essential for transferrin utilization. *J. Bacteriol.* 176:3162–3170.

Angerer, A., B. Klupp, and V. Braun. 1992. Iron transport systems of *Serratia marcescens*. *J. Bacteriol.* 174:1378–1387.

Angerer, A., S. Enz, M. Ochs, and V. Braun. 1995. Transcriptional regulation of ferric citrate transport in *Escherichia coli:* FecI belongs to a new subfamily of σ^{70}-type factors that respond to extracytoplasmic stimuli. *Mol. Microbiol.* 18:163–174.

Angerer, A. M., S. Gaisser, and V. Braun. 1990. Nucleotide sequence of the *sfuA, sfuB, sfuC* genes of *Serratia marcescens* suggests a periplasmic binding protein-dependent iron transport system. *J. Bacteriol.* 172:572–578.

Ankenbauer, R. G., and H. N. Quan. 1994. FptA, the Fe(III)-pyochelin receptor of *Pseudomonas aeruginosa:* A phenolate siderophore receptor homologous to hydroxamate siderophore receptors. *J. Bacteriol.* 176:307–319.

Bäumler, A., and K. Hantke. 1992. Ferrioxamine uptake in *Yersinia enterocolitica:* Characterization of the receptor protein FoxA. *Mol. Microbiol.* 6:1309–1321.

Bäumler, A., R. Koebnik, I. Stojiljkovic, J. Heesemann, V. Braun, and K. Hantke. 1993. Survey on newly characterized iron uptake systems of *Yersinia enterocolitica*. Zbl. Bakt. 278:416–424.

Beall, B. W., and G. N. Sanden. 1995. Cloning and initial characterization of the *Bordetella pertussis fur* gene. *Current Microbiol.* 30:223–226.

Berish, S. A., S. Subbarao, C.-Y. Chen, D. L. Trees, and S. A. Morse. 1993. Identification and cloning of a *fur* homolog from *Neisseria gonorrhoeae*. *Infect. Imm.* 61:4599–4606.

Bitter, W., J. D. Marugg, L. A. deWeger, J. Thomassen, and P. J. Weisbeek. 1991. The ferric-pseudobactin receptor PupA of *Pseudomonas putida* WCS538: Homology to

TonB-dependent *Escherichia coli* receptors and specificity of the protein. *Mol. Microbiol.* 5:647–655.

Bitter, W., J. Tommassen, and P. J. Weisbeek. 1993. Identification and characterization of the *exbB, exbD,* and *tonB* genes of *Pseudomonas putida* WCS358. *Mol. Microbiol.* 7:117–130.

Boyd, J. M., O. N. Manish, and J. R. Murphy. 1990. Molecular cloning and DNA sequence analysis of a diphtheria *tox* iron-dependent regulatory element from *Corynebacterium diphtheriae. Proc. Natl. Acad. Sci. U.S.A.* 87:5968–5972.

Braun, V., K. Günter, and K. Hantke. 1991. Transport of iron across the outer membrane. *Biol. Metals* 4:14–22.

Braun, V., R. E. W. Hancock, K. Hantke, and A. Hartmann. 1976. Functional organization of the outer membrane of *Escherichia coli:* Phage and colicin receptors as components of iron uptake systems. *J. Supramol. Struct.* 5:37–58.

Braun, V., and K. Hantke. 1991. Genetics of bacterial iron transport. In *Handbook of Microbial Iron Chelates,* ed. G. Winkelmann. pp. 107–138. CRC Press, Boca Raton, FL.

Braun, V., and C. Herrmann. Unpublished results.

Braun, V., S. Schäffer, K. Hantke, and W. Tröger. 1990. Regulation of gene expression by iron. *Coll. Mosbach* 41:164–179.

Braun, V., and H. Wolff. 1973. Characterization of the receptor protein for phage T5 and Colicin M in the outer membrane of *E. coli* B. *FEBS Lett.* 34:77–80.

Brickman, T. J., B. A. Ozenberger, and M. A. McIntosh. 1990. Regulation of divergent transcription from the iron-responsive *fepB-entC* promoter-operator regions in *Escherichia coli. J. Mol. Biol.* 212:669–682.

Brutsche, S., K. Hantke, and V. Braun. Unpublished results.

Burkhardt, R., and V. Braun. 1987. Nucleotide sequence of the *fhuC* and *fhuD* genes involved in iron(III) hydroxamate transport: Domains in FhuC homologous to ATP-binding proteins. *Mol. Gen. Genet.* 209:49–55.

Butterton, J. R., J. A. Stoebner, S. M. Payne, and S. B. Calderwood. 1992. Cloning, sequencing, and transcriptional regulation of *viuA,* the gene encoding the ferric vibriobactin receptor of *Vibrio cholerae. J. Bacteriol.* 174:3729–3738.

Casadaban, M. J., and S. N. Cohen. 1979. Lactose genes fused to exogenous promoters in one step using a Mu*lac* bacteriophage: In vitro probe for transcriptional control sequences. *Proc. Natl. Acad. Sci. U.S.A.* 76:4530–4533.

Chen, C.-Y., S. A. Berish, S. A. Morse, and T. A. Mietzner. 1993a. The ferric iron-binding protein of *Neisseria* ssp. functions as a periplasmic transport protein in iron acquisition from human transferrin. *Mol. Microbiol.* 10:311–318.

Chen, L., L. P. James, and J. D. Helmann. 1993b. Metalloregulation in *Bacillus subtilis:* Isolation and characterization of two genes differentially repressed by metal ions. *J. Bacteriol.* 175:5428–5437.

Chenault, S. S., and C. F. Earhart. 1991. Organization of genes encoding membrane proteins of the *Escherichia coli* ferrienterobactin permease. *Mol. Microbiol.* 5:1405–1413.

Chenault, S. S., and C. F. Earhart. 1992. Identification of hydrophobic proteins FepD and FepG of *Escherichia coli* ferrienterobactin permease. *J. Gen. Microbiol.* 138:2167–2171.

Cornelissen, C. N., G. D. Biswas, J. Tsai, D. K. Paruchuri, S. A. Thompson, and P. F. Sparling. 1992. Gonococcal transferrin binding protein 1 is required for transferrin utilization and is homologous to TonB-dependent outer membrane receptors. *J. Bacteriol.* 174:5788–5797.

Cotter, P. A., S. Darie, and R. P. Gunsalus. 1992. The effect of iron limitation on expression of the aerobic and anaerobic electron transport pathway genes in *Escherichia coli*. *FEMS Microbiol. Lett.* 100:227–232.

Coulton, J. W., P. Mason, D. R. Cameron, G. Carmel, R. Jean, and H. N. Rode. 1986. Protein fusions of β-galactosidase to the ferrichrome-iron receptor of *Escherichia coli*. *J. Bacteriol.* 165:181–192.

Coy, M., and J. B. Neilands. 1991. Structural dynamics and functional domains of the Fur protein. *Biochem.* 30:8201–8210.

Crosa, J. H. 1980. A plasmid associated with virulence in the marine fish pathogen *Vibrio anguillarum* specifies an iron-sequestering system. *Nature (London)* 284:566–568.

Cunliffe, H. E., T. R. Merriman, and J. L. Lamont. 1995. Cloning and characterization of *pvdS*, a gene required for pyoverdine synthesis in *Pseudomonas aeruginosa*: PvdS is probably an alternative sigma factor. *J. Bacteriol.* 177:2744–2750.

Dean, C. R., and K. Poole. 1993. Expression of the ferric enterobactin receptor (PfeA) of *Pseudomonas aeruginosa*: Involvement of a two-component regulatory system. *Mol. Microbiol.* 8:1095–1103.

de Lorenzo, V., M. Herrero, F. Giovannini, and J. B. Neilands. 1988. Fur (ferric uptake regulation) protein and CAP (catabolite-activator protein) modulate transcription of *fur* gene in *Escherichia coli*. *Eur. J. Biochem.* 173:537–546.

de Lorenzo, V., S. Wee, M. Herrero, and J. B. Neilands. 1987. Operator sequences of the aerobactin operon of plasmid ColV-K30 binding the ferric uptake regulation (*fur*) repressor. *J. Bacteriol.* 169:2624–2630.

Desai, P. J., A. Angerer, and C. A. Genco. 1996. Analysis of Fur binding to operator sequences within the *Neisseria gonorrhoeae fbpA* promoter. *J. Bacteriol.* 178: 5020–5023.

Dorman, C. J., G. C. Barr, N. N. Bhriain, and C. F. Higgins. 1988. DNA supercoiling and the anaerobic and growth phase regulation of *tonB* gene expression. *J. Bacteriol.* 170:2816–2826.

Drechsel, H., J. Metzger, S. Freund, G. Jung, J. R. Boelaert, and G. Winkelmann. 1991. Rhizoferrin—A novel siderophore from the fungus *Rhizopus microsprus* var. *rhizopodiformis*. *Biol. Metals* 4:238–243.

Eick-Helmerich, K., and V. Braun. 1989. Import of biopolymers into *Escherichia coli*: Nucleotide sequences of the *exbB* and *exbD* genes are homologous to those of the *tolQ* and *tolR* genes, respectively. *J. Bacteriol.* 171:5117–5126.

Enz, S., V. Braun, and J. H. Crosa. 1995. Transcription of the region encoding the ferric dicitrate-transport system in *Escherichia coli:* Similarity between promoters for *fecA* and for extracytoplasmic function sigma factors. *Gene* 163:13–18.

Ernst, J. F., R. L. Bennet, and L. I. Rothfield. 1978. Constitutive expression of the iron-enterochelin and ferrichrome uptake systems in a mutant strain of *Salmonella typhimurium. J. Bacteriol.* 135:928–934.

Fleischmann, R. D., et al. 1995. Whole-genome random sequencing and assembly of *Haemophilus influenzae* Rd. *Science* 269:496–512.

Foster, J. W., and H. K. Hall. 1992. Effect of *Salmonella typhimurium* ferric-uptake regulator (*fur*) mutations on iron and pH-regulated protein synthesis. *J. Bacteriol.* 174:4317–4323.

Frechon, D., and E. Le Cam. 1994. Fur (ferric uptake regulation) protein interaction with target DNA: Comparison of gel retardation, footprinting and electron microscopy analyses. *Biochem. Biophys. Res. Comm.* 201:346–355.

Frost, G. E., and H. Rosenberg. 1973. The inducible citrate-dependent iron transport system in *Escherichia coli* K12. *Biochim. Biophys. Acta* 330:90–101.

Ghassemian, M., and N. A. Straus. 1996. Fur regulates the expression of iron-stress genes in the cyanobacterium *Synechococcus* sp. strain PCC 7942. *Microbiology* 142:1469–1476.

Goldber, M. B., S. A. Boyko, J. R. Butterton, J. A. Stoebner, S. M. Payne, and S. B. Calderwood. 1992. Characterization of a *Vibrio cholerae* virulence factor homologous to the family of tonB-dependent proteins. *Mol. Microbiol.* 6:2407–2418.

Griggs, D., and J. Konisky. 1989. Mechanism for iron-regulated transcription of the *Escherichia coli cir* gene: Metal-dependent binding of Fur protein to the promoters. *J. Bacteriol.* 171:1048–1054.

Grossman, T. H., M. Tuckman, S. Ellestad, and M. S. Osburne. 1993. Isolation and characterization of *Bacillus subtilis* genes involved in siderophore biosynthesis: Relationship between *B. subtilis sfp°* and *E. coli entD* gene. *J. Bacteriol.* 175:6203–6211.

Guilvout, I., O. Mercereau-Puijalon, S. Bonnefroy, A. P. Pugsley, and E. Carniel. 1993. High-Molecular-weight protein 2 of *Yersinia enterocolitica* is homologous to AngR of *Vibrio anguillarum* and belongs to a family of proteins involved in nonribosomal peptide synthesis. *J. Bacteriol.* 175:5488–5504.

Günter, K., C. Toupet, and T. Schupp. 1993. Characterization of an iron-regulated promoter involved in desferrioxamine B synthesis in *Streptomyces pilosus:* Repressor-binding site and homology to the diphtheria toxin gene promoter. *J. Bateriol.* 175:3295–3302.

Günter-Seeboth, K., and T. Schupp. 1995. Cloning and sequence analysis of the *Corynebacterium diphtheriae dtxR* homologue from *Streptomyces lividans* and *S. pilosus* encoding a putative iron repressor protein. *Gene* 166:117–119.

Haag, H., K. Hantke, H. Drechsel, I. Stojiljkovic, G. Jung, and H. Zähner. 1994. Purification of yersiniabactin: A possible virulence factor of *Yersinia enterocolitica. J. Gen. Microbiol.* 139:2159–2165.

Halliwell, B., and J.M.C. Gutteridge. 1984. Oxygen toxicity, oxygen radicals, transition metals and disease. *Biochem. J.* 219:1–14.

Hamed, M. Y., and J. B. Neilands. 1994. An electronic spin resonance study of the Mn(II) and Cu(II) complexes of the Fur repressor protein. *J. Inorg. Biochem.* 53:235–248.

Hantke, K. 1981. Regulation of ferric iron transport in *Escherichia coli* K12: Isolation of a constitutive mutant. *Mol. Gen. Genet.* 182:288–292.

Hantke, K. 1983. Identification of an iron uptake system specific for coprogen and rhodotorulic acid in *Escherichia coli* K12. *Mol. Gen. Genet.* 191:301–306.

Hantke, K. 1984. Cloning of the repressor protein gene of iron-regulated systems in *Escherichia coli* K12. *Mol. Gen. Genet.* 197:337–341.

Hantke, K. 1987. Selection procedure for deregulated iron transport mutants in *Escherichia coli* K-12: *fur* not only affects iron metabolism. *Mol. Gen. Genet.* 210:135–139.

Hantke, K. 1990. Dihydroxybenzoylserine—A siderophore for *E. coli*. *FEMS Microbiol. Lett.* 67:5–8.

Hantke, K., and Zimmermann, L. 1981. The importance of the *exbB* gene for vitamin B12 and ferric iron transport. *FEMS Microbiol. Lett.* 12:31–35.

Härle, C., I. Kim, A. Angerer, and V. Braun. 1995. Signal transfer through three compartments. Transcription initiation of the *Escherichia coli* ferric citrate transport system from the cell surface. *EMBO J.* 14:1430–1438.

Hassett, D. J., P. A. Sokol, M. L. Howell, J.-F. Ma, H. T. Schweizer, U. Ochsner, and M. L. Vasil. 1996. Ferric uptake regulator (Fur) mutants of *Pseudomonas aeruginosa* demonstrate defective siderophore-mediated iron uptake, altered aerobic growth, and decreased superoxide dismutase and catalase activities. *J. Bacteriol.* 178:3996–4003.

Heidrich, C., K. Hantke, G. Bierbaum, and H. G. Sahl. 1996. Identification and analysis of a gene encoding a Fur-like protein of *Staphylococcus epidermidis*. *FEMS Microbiol. Lett.* 140:253–259.

Heinrichs, D. E., and K. Poole. 1993. Cloning and sequence analysis of a gene (*pchR*) encoding an AraC family activator of pyochelin and ferripyrochelin receptor synthesis in *Pseudomonas aeruginosa*. *J. Bacteriol.* 175:5882–5889.

Heller, K., and R. J. Kadner. 1985. Nucleotide sequence of the gene for the vitamin B12 receptor protein in the outer membrane of *Escherichia coli*. *J. Bacteriol.* 161:904–908.

Henderson, D. P., and S. M. Payne. 1994. Characterization of the *Vibrio cholerae* outer membrane heme transport protein HutA: Sequence of the gene, regulation of expression, and homology to the family of TonB-dependent proteins. *J. Bacteriol.* 176:3269–3277.

Heyduk, T., and J. C. Lee. 1989. *Escherichia coli* cAMP receptor protein: Evidence for three protein conformational states with different promoter binding affinities. *Biochem.* 28:6914–6924.

Hickey, E. K., and N. P. Cianciotto. 1994. Cloning and sequencing of the *Legionella pneumophila fur* gene. *Gene* 143:117–121.

Holm, L., C. Sander, H. R. Rüterjahns, M. Schnarr, R. Fogh, R. Boelens, and R. Kaplan. 1994. LexA repressor and iron uptake regulator from *Escherichia coli:* New members of the CAP-like DNA binding domain superfamily. *Protein Engin.* 7:1449–1453.

Hunt, M. D., G. S. Pettis, and M. A. McIntosh. 1994. Promoter and operator determinants for Fur-mediated iron regulation in the bidirectional *fep*A-*fes* control region of the *Escherichia coli* enterobactin gene system. *J. Bacteriol.* 176:3944–3955.

Hussein, S., K. Hantke, and V. Braun. 1981. Citrate-dependent iron transport system in *Escherichia coli* K-12. *Eur. J. Biochem.* 117:431–437.

Irwin, S. W., N. Averil, C. Y. Cheng, and A. B. Schryvers. 1993. Preparation and analysis of isogenic mutants in the transferrin receptor protein genes, *tbpA* and *tbpB*, from *Neisseria meningitidis. Mol. Microbiol.* 8:1125–1133.

Jarosik, G. P., J. D. Sanders, L. D. Cope, U. Muller-Eberhard, and E. J. Hansen. 1994. A functional *tonB* gene is required for both utilization of heme and virulence expression by *Haemophilus influenzae* type b. *Infect. Imm.* 62:2470–2477.

Kammler, M., C. Schön, and K. Hantke. 1993. Characterization of the ferrous iron uptake system of *Escherichia coli. J. Bacteriol.* 175:6216–6219.

Kampfenkel, K., and V. Braun. 1992. Membrane topology of the *Escherichia coli* ExbD protein. *J. Bacteriol.* 174:5485–5487.

Kampfenkel, K., and V. Braun. 1993. Topology of the ExbB protein in the cytoplasmic membrane of *Escherichia coli. J. Biol. Chem.* 268:6050–6057.

Kim, I., S. Plantör, A. Stiefel, A. Angerer, and V. Braun. 1997. Transcription induction of the ferric citrate transport genes via the N-terminus of the FecA outer membrane protein, the Ton system and the electrochemical potential of the cytoplasmic membrane. *Mol. Microbiol.* 23:333–344.

Koster, M., J. van de Vossenberg, J. Leong, and P. J. Weisbeek. 1993. Identification and characterization of the *pupB* gene encoding the indudible ferric pseudobactin receptor of *Pseudomonas putida* WCS358. *Mol. Microbiol.* 8:591–601.

Koster, M., W. van Klompenburg, W. Bitter, and P. Weisbeek. 1994. Role for the outer membrane ferric-siderophore receptor PupB in signal transduction across the bacterial cell envelope. *EMBO J.* 13:2805–2813.

Köster, W., and V. Braun. 1986. Iron hydroxamate transport of *Escherichia coli:* Nucleotide sequence of the *fhuB* gene and identification of the protein. *Mol. Gen. Genet.* 204:435–442.

Köster, W. L., L. A. Actis, L. S. Waldbeser, M. E. Tolmasky, and J. H. Crosa. 1991. Molecular characterization of the iron transport system mediated by the pJM1-plasmid in *Vibrio anguillarum* 775. *J. Biol. Chem.* 266:23829–23833.

Krone, W. J. A., F. Steghuis, G. Koningstein, C. van Doorn, B. Roosendaal, F. K. deGraaf, and B. Oudega. 1985. Characterization of the pColV-K30 encoded colicin DF13/aerobactin outer membrane receptor protein of *Escherichia coli;* isolation and purification of the protein and analysis of its nucleotide sequence and primary structure. *FEMS Microbiol. Lett.* 26:153–161.

Kühn, S., Braun, V., and W. Köster. 1995. Ferric rhizoferrin uptake into *Morganella morganii:* characterization of genes involved in the uptake of a polyhydroxycarboxylate siderophore. *J. Bacteriol.* 178:496–504.

Legrain, M., V. Mazarin, S. W. Irwin, B. Bouchon, M.-J. Quentin-Millet, E. Jacobs, and

A. B. Schryvers. 1993. Cloning and characterization of *Neisseria meningitidis* genes encoding the transferrin-binding proteins Tbp1 and Tbp2. *Gene* 130:73–80.

Litwin, C. M., S. A. Boyko, and S. B. Calderwood. 1992. Cloning, sequencing and transcriptional regulation of the *Vibrio cholerae fur* gene. *J. Bacteriol.* 174:1897–1903.

Litwin, C. M., and S. B. Calderwood. 1993a. Cloning and genetic analysis of the *Vibrio vulnificus fur* gene and construction of a *fur* mutant by in vivo marker exchange. *J. Bacteriol.* 175:706–715.

Litwin, C. M., and S. B. Calderwood. 1993b. Role of iron in regulation of virulence genes. *Clin. Microbiol. Rev.* 6:137–149.

Litwin, C. M., and S. B. Calderwood. 1994. Analysis of the complexity of gene regulation by *fur* in *Vibrio cholerae. J. Bacteriol.* 176:240–248.

Lonetto, M., K. L. Brown, K. Rudd, and M. J. Buttner. 1994. Analysis of the *Streptomyces coelicolor sigE* gene reveals a new sub-family of eubacterial RNA polymerase σ factors involved in the regulation of extracytoplasmic functions. *Proc. Natl. Acad. Sci. U.S.A.* 91:7573–7577.

Lundrigan, M. D., and R. J. Kadner. 1986. Nucleotide sequence of the gene for the ferrienterochelin receptor FepA in *Escherichia coli. J. Biol. Chem.* 261:10797–10801.

Mahe, B., C. Masclaux, L. Rauscher, C. Enard, and D. Expert. 1995. Differential expression of two siderophore-dependent iron-acquisition pathways in *Erwinia chrysanthemi* 3937: Characterization of a novel ferrisiderophore permease of the ABC transporter family. *Mol. Microbiol.* 18:33–43.

Martinez, J. L., M. Herrero, and V. de Lorenzo. 1994. The organization of intercistronic regions of the aerobactin operon of pColV-K30 may account for the differential expression of the *iucABCD iutA* genes. *J. Mol. Biol.* 238:288–293.

Mayhew, S. G., and M. L. Ludwig. 1975. Flavodoxins and electron-transferring flavoproteins. In *The Enzymes* 3rd edition, ed. P. D. Boyer. Vol. XIIB, pp. 59–118. Academic Press, New York.

Miyazaki, H., H. Kato, T. Nakazawa, and M. Tsuda. 1995. A positive regulatory gene, *pvdS,* for expression of pyoverdin biosynthesis genes in *Pseudomonas aeruginosa. Mol. Gen. Genet.* 248:17–24.

Morris, J., D. F. Donelly, E. O'Neill, F. McConnell, and F. O'Gara. 1994. Nucleotide sequence analysis and potential environmental distribution of a ferric pseudobactin receptor gene of *Pseudomonas* sp. strain M114. *Mol. Gen. Genet.* 242:9–16.

Nau, C. D., and J. Konisky. 1989. Evolutionary relationship between the *tonB*-dependent outer membrane transport proteins: Nucleotide and amino acid sequences of the *Escherichia coli* colicin I receptor gene. *J. Bacteriol.* 171:1041–1047.

Ochs, M., A. Angerer, S. Enz, and V. Braun. 1996. Surface signaling in transcriptional regulation of the ferric citrate transport system of *Escherichia coli:* Mutational analysis of the alternate sigma factor FecI supports its essential role in *fec* transport gene transcription. *Mol. Gen. Genet.* 250:455–465.

Ochs, M., S. Veitinger, I. Kim, D. Welz, A. Angerer, and V. Braun. 1995. Regulation of citrate-dependent iron transport of *Escherichia coli:* FecR is required for transcription activation by FecI. *Mol. Microbiol.* 15:119–132.

Oguiza, J. A., X. Tao, A. T. Marcos, J. F. Martin, and J. R. Murphy. 1995. Molecular cloning, DNA sequence analysis and characterization of the *Corynebacterium diphtheriae dtxR* homolog from *Brevibacterium lactofermentum. J. Bacteriol.* 177:465–467.

O'Sullivan, D. J., D. N. Dowling, V. de Lorenzo, and F. O'Gara. 1994. *Escherirchia coli* ferric uptake regulator (Fur) can mediate regulation of a pseudomonad iron-regulated promoter. *FEMS Microbiol. Lett.* 117:327–332.

Pettersson, A., A. Maas, and J. Thomassen. 1994. Identification of the *iroA* gene product of *Neisseria meningitidis* as a lactoferrin receptor. *J. Bacteriol.* 176:1764–1766.

Poole, K., S. Neshat, K. Krebes, and D. E. Heinrichs. 1993. Cloning and nucleotide sequence analysis of the ferripyoverdine receptor gene *fpvA* of *Pseudomonas aeruginosa. J. Bacteriol.* 175:4597–4604.

Postle, K. 1990a. Aerobic regulation of the *Escherichia coli tonB* gene by changes in iron availability and the *fur* locus. *J. Bacteriol.* 172:2287–2293.

Postle, K. 1990b. TonB and the Gram-negative dilemma. *Mol. Microbiol.* 4:2019–2025.

Pressler, U., H. Staudenmaier, L. Zimmermann, and V. Braun. 1988. Genetics of the iron dicitrate transport system of *Escherichia coli. J. Bacteriol.* 170:2716–2724.

Prince, R. W., C. D. Cox, and M. L. Vasil. 1993. Coordinate regulation of siderophore and exotoxin A production: Molecular cloning and sequencing of the *Pseudomonas aeruginosa fur* gene. *J. Bacteriol.* 175:2589–2598.

Ptashne, M. 1986. Gene regulation by proteins acting nearby and at a distance. *Nature* 322:697–701.

Rakin, A., E. Saken, D. Harmsen, and J. Heesemann. 1994. The pesticin receptor of *Yersinia enterocolitica:* A novel virulence factor with dual function. *Mol. Microbiol.* 13:253–263.

Rakin, A., P. Urbitsch, and J. Heesemann. 1995. Evidence for two evolutionary lineages of highly pathogenic *Yersinia* species. *J. Bacteriol.* 177:2292–2298.

Richardson, P. T., and S. F. Park. 1995. Enterochelin acquisition in *Campylobacter coli:* Characterization of components of a binding-protein-dependent transport system. *Microbiology* 141:3181–3191.

Saito, T., M. R. Wormald, and R. J. P. Williams. 1991. Some structural features of the iron-uptake regulation protein. *Eur. J. Biochem.* 197:29–38.

Salinas, P. C., L. S. Waldbeser, and J. H. Crosa. 1993. Regulation of the expression of bacterial iron transport genes: Possible role of an antisense RNA. *Gene* 123:33–38.

Sanders, J. D., D. Cope, and E. J. Hansen. 1994. Identification of a locus involved in the utilization of iron by *Haemophilus influenzae. Infect. Imm.* 62:4515–4525.

Sauer, M., K. Hantke, and V. Braun. 1990. Sequence of the *fhuE* outer membrane receptor gene of *Escherichia coli* K-12 and properties of mutants. *Mol. Microbiol.* 4:427–437.

Saurin, W., W. Köster, and E. Dassa. 1994. Bacterial binding protein-dependent permeases: Characterization of distinctive signatures for functionally related integral cytoplasmic membrane proteins. *Mol. Microbiol.* 12:993–1004.

Schäffer, S., K. Hantke, and V. Braun. 1985. Nucleotide sequence of the iron regulatory gene *fur. Mol. Gen. Genet.* 201:204–212.

Schmitt, M. P., and R. K. Holmes. 1993. Analysis of diphtheria toxin repressor-operator interactions and characterization of a mutant repressor with decreased binding activity for divalent metals. *Mol. Microbiol.* 9:173–181.

Schmitt, M. P., and R. K. Holmes. 1994. Cloning, sequence, and footprint analysis of two promoter/operators of *Corynebacterium diphtheriae* that are regulated by the diphtheria toxin repressor (DtxR) and iron. *J. Bacteriol.* 176:1141–1149.

Schneider, R., and K. Hantke. 1993. Iron-hydroxamate uptake systems in *Bacillus subtilis:* Identification of a lipoprotein as part of a binding protein dependent transport system. *Mol. Microbiol.* 8:111–121.

Schneider, R., and K. Hantke. 1996. Unpublished results; for the *fhuBGC* sequence see EMBL Nucleotide Sequence Database, accession number X93092.

Schultz-Hauser, G., B. Van Hove, and V. Braun. 1992. 8-Azido-ATP labelling of the FecE protein of the *Escherichia coli* iron citrate transport system. *FEMS Microbiol. Lett.* 95:231–234.

Sexton, R., P. R. Gill Jr., D. N. Dowling, and F. O'Gara. 1996. Transcriptional regulation of the iron-responsive sigma factor gene *pbrA. Mol. Gen. Genet.* 250:50–58.

Shea, C. M., and M. A. McIntosh. 1991. Nucleotide sequence and genetic organization of the ferric enterobactin system—Homology to other periplasmic binding protein-dependent transport systems in *Escherichia coli. Mol. Microbiol.* 5:1415–1428.

Sorokin, A., E. Zumstein, V. Azevedo, S. D. Ehrlich, and P. Serror. 1993. The organization of the *Bacillus subtilis* 168 chromosome region between the *spoVA* and *serA* genetic loci, based on sequence data. *Mol. Microbiol.* 10:385–395.

Staggs, T. M., and R. D. Perry. 1992. Fur regulon in *Yersinia* species. *Mol. Microbiol.* 6:2507–2516.

Staudenmaier, H., B. Van Hove, Z. Yaraghi, and V. Braun. 1989. Nucleotide sequences of the *fecBCDE* genes and locations of the proteins suggest a periplasmic-binding-protein-dependent transport mechanism for iron(III) dicitrate in *Escherichia coli. J. Bacteriol.* 171:2626–2633.

Stock, J. B., A. M. Stock, and J. M. Mottonen. 1990. Signal transduction in bacteria. *Nature* 344:395–400.

Stojilijkovic, I., A. J. Baeumler, and K. Hantke. 1994. Fur regulon in Gram-negative bacteria, identification and characterization of new iron-regulated *Escherichia coli* genes by a Fur titration assay. *J. Mol. Biol.* 236:531–545.

Stojilijkovic, I., and K. Hantke. 1992. Hemin uptake system of *Yersinia enterocolitica:* Similarity with other TonB-dependent systems in gram-negative bacteria. *EMBO J.* 11:4359–4367.

Stojiljkovic, I., and K. Hantke 1994. Transport of haemin across the cytoplasmic membrane through a haemin-specific periplasmic binding-protein-dependent transport system in *Yersinia enterocolitica. Mol. Microbiol.* 13:719–732.

Stojiljkovic, I., and K. Hantke. 1995. Functional domains of the *Escherichia coli* ferric uptake regulator protein (Fur). *Mol. Gen. Genet.* 247:199–205.

Tam, R., and M. H. Saier. 1993. Structural, functional, and evolutionary relationships among extracellular solute-binding receptors of bacteria. *Microbiol. Rev.* 57:320–346.

Tao, X., J. Boyd, and J. R. Murphy. 1992. Specific binding of the diphtheria *tox* regulatory element DtxR to the *tox* operator requires divalent heavy metal ions and a 9-base-pair interrupted palindromic sequence. *Proc. Natl. Acad. Sci. U.S.A.* 89:5897–5901.

Tardat, B., and D. Touati. 1993. Iron and oxygen regulation of *Escherichia coli* MnSOD expression: Competition between the global regulators Fur and ArcA for binding to DNA. *Mol. Microbiol.* 9:53–63.

Thomas, C. E., and P. F. Sparling. 1994. Identification and cloning of a *fur* homologue from *Neisseria meningitidis. Mol. Microbiol.* 11:725–737.

Thomas, C. E., and P. F. Sparling. 1996. Isolation and analysis of a *fur* mutant of *Neisseria gonorrhoeae. J. Bacteriol.* 178:4224–4232.

Tomalsky, M. E., L. A. Actis, and J. H. Crosa. 1988. Genetic analysis of the iron uptake region of the *Vibrio anguillarum* plasmid pJM1: Molecular cloning of genetic determinants encoding a novel *trans* factor of siderophore biosynthesis. *J. Bacteriol.* 170:1913–1919.

Tomalsky, M. E., L. A. Actis, and J. H. Crosa. 1993. A single amino acid change in AngR, a protein encoded by pJM1-like virulence plasmids, results in hyperproduction of anguibactin. *Infect. Imm.* 61:3228–3233.

Tomalsky, M. E., L. S. Wertheimer, L. A. Actis, and J. H. Crosa. 1994. Characterization of the *Vibrio anguillarum fur* gene. Role in regulation of expression of the FatA outer membrane protein and catechols. *J. Bacteriol.* 176:213–220.

Touati, D., M. Jaques, B. Tardat, L. Bouchard, and S. Despied. 1995. Lethal oxidative damage and mutagenesis are generated by iron in Δ*fur* mutants of *Escherichia coli:* Protective role of superoxide dismutase. *J. Bacteriol.* 177:2305–2314.

Van Hove, B., H. Staudenmaier, and V. Braun. 1990. Novel two-component transmembrane transcription control: Regulation of iron dicitrate transport in *Escherichia coli* K-12. *J. Bacteriol.* 172:6749–6758.

Veitinger, S., and V. Braun. 1992. Localization of the entire *fec* region at 97.3 minutes on the *Escherichia coli* K-12 chromosome. *J. Bacteriol.* 174:3838–3839.

Venturi, V. 1994. Molecular genetics of siderophore mediated iron acquisition in *Pseudomonas putida.* Thesis, University of Utrecht, The Netherlands.

Venturi, V., C. Ottevanger, J. Leong, and P. J. Weisbeek 1993. Identification and characterization of a siderophore regulatory gene (pfrA) of *Pseudomonas putida* WCS358: Homology to the alginate regulatory gene *algQ* of *Pseudomonas aeruginosa. Mol. Microbiol.* 10:63–73.

Venturi, V., C. Ottevanger, M. Bracke, and P. Weisbeek. 1995. Iron regulation of siderophore biosynthesis and transport in *Pseudomonas putida* WCS358: Involvement of a transcriptional activator and of the Fur protein. *Mol. Microbiol.* 15:1081–1093.

Venturi, V., P. Weisbeek, and M. Koster. 1995. Gene regulation of siderophore-mediated iron acquisition in *Pseudomonas:* Not only the Fur repressor. *Mol. Microbiol.* 17:603–610.

Wächtershäuser, G. 1990. Evolution of the first metabolic cycles. *Proc. Natl. Acad. Sci. U.S.A.* 87:200–204.

Waldbeser, L. S., M. E. Tolmasky, L. A. Actis, and J. H. Crosa. 1993. Mechanisms for negative regulation by iron of the *fatA* outer membrane protein gene expression in *Vibrio anguillarum* 775. *J. Biol. Chem.* 268:10433–10439.

Wee, S., J. B. Neilands, M. L. Bittner, B. C. Hemming, B. L. Haymore, and R. Seetharam. 1988. Expression, isolation and properties of Fur (ferric iron uptake regulation) protein of *Escherichia coli* K12. *Biol. Metals* 1:62–68.

Wooldridge, K. G., P. H. Williams, and J. M. Ketley. 1994. Iron-responsive genetic regulation in *Campylobacter jejuni:* Cloning and characterization of a fur homolog. *J. Bacteriol.* 176:5852–5856.

Wriedt, K., A. Angerer, and V. Braun. 1995. Transcriptional regulation from the cell surface: Conformational changes in the transmembrane protein FecR lead to altered transcription of the ferric citrate transport genes in *Escherichia coli. J. Bacteriol.* 177:3320–3322.

Zimmermann, L., K. Hantke, and V. Braun. 1984. Exogenous induction of the iron dicitrate transport system of *Escherichia coli* K-12. *J. Bacteriol.* 159:271–277.

3

Metal Cation Regulation in Gram-Positive Bacteria

John D. Helmann

1. Introduction

Metal ions have a profound influence on gene regulation in bacterial cells. First, metal ions affect the production of numerous secondary metabolites of industrial and medical importance. For example, the role of iron in the regulation of diphtheria toxin production has been appreciated for over 60 years (Pappenheimer 1936, 1977). Similarly, optimization of metal ion levels is often essential for the high-level production of secreted enzymes from bacilli, actinomycetes, and related organisms and for the synthesis of secondary metabolites such as antibiotics and surfactants (Weinberg 1990). Second, several essential metals regulate their own uptake, and toxic metals often induce specific detoxification or export machinery (Silver and Ji 1994; Silver and Walderhaug 1992). Lessons learned from analysis of these bacterial metal ion transport systems have shed light on human diseases arising from disorders of metal ion homeostasis (DiDonato and Sarkar 1997). Third, metal ion uptake systems, particularly those for iron, are important for the virulence of numerous pathogens (Cornelissen and Sparling 1994; Wooldridge and Williams 1993). Fourth, it is now appreciated that metal ions can alter the expression of proteins important in oxidative stress responses and therefore they have the capacity to affect the susceptibility of bacteria to host immune defenses (Chen et al. 1995).

In this review I summarize molecular studies of metalloregulation in Gram-positive bacteria. My emphasis is on the molecular biology of the regulatory systems and, where known, the mechanisms of metal ion selectivity. Ultimately, work in this field promises to illuminate the chemical basis of selective and reversible metal ion binding by metalloregulatory proteins and aid in the understanding of how metal ions affect diverse aspects of cellular physiology.

2. Mercury

Genetic elements that confer an ability to grow in the presence of high levels of mercuric ion are widespread among both Gram-negative and Gram-positive bacteria (Nakamura and Silver 1994; Smith 1967; Summers 1986). These mercury resistance (*mer*) genes are found on transposons, such as Tn*21* and Tn*501*, on plasmids, and in some cases on the chromosome. The *mer* genes are generally found in a single operon inducible by mercuric ion (Figure 3.1). Induction requires the presence of the regulatory protein, MerR (O'Halloran and Walsh 1987). In addition, some of the broad-spectrum resistance systems can be induced by organomercurials, which are then degraded to mercuric ion and a hydrocarbon component by organomercurial lyase, the product of the *merB* gene (Misra 1992; Walsh et al. 1988). Interestingly, the ability of organomercurials to induce the broad-spectrum resistance operons is a property of the MerR protein and appears to reflect an ability of MerR to be activated by either organomercurials or inorganic mercuric ion (Yu et al. 1994). For the purposes of this chapter, I will emphasize studies of *mer* resistance systems from *Staphylococcus aureus* plasmid pI258 (Laddaga et al. 1987) and *Bacillus* sp. RC607 (Wang et al. 1989). Although

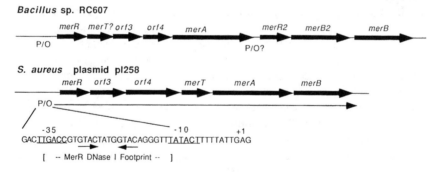

Figure 3.1 Organization of *mer* operons in *Bacillus* sp. RC607 and *S. aureus* plasmid pI258. Transcription initiates within the promoter/operator (P/O) region. The operons encode regulatory (MerR), transport (MerT), and detoxification (MerA and MerB) functions. The recently completed sequence of the *Bacillus merA-merB* intergenic region reveals the presence of an additional putative regulator (*merR2*) and lyase (*merB2*), although the function of these genes has not yet been demonstrated (S. Silver, personal communication). The *S. aureus* MerR protein has been purified as a glutathione transferase fusion protein and the DNAase I footprint determined (Chu et al. 1992).

Bacillus sp. RC607 was originally isolated from Boston Harbor, the corresponding resistance determinant is broadly distributed. A recent study of mercury-resistant isolates from Minamata Bay, Japan, revealed a high degree of similarity between their resistance genes and the *Bacillus* sp. RC607 *mer* operon (Nakamura and Silver 1994). A mercury resistance operon has also been sequenced from *Streptomyces lividans,* but little is known about the regulation of that operon (Sedlmeier and Altenbuchner 1992).

The Gram-positive mercury resistance systems are clearly homologous to the Gram-negative systems, in the mechanisms of both resistance and regulation (Helmann et al. 1989). The regulation of two of these systems has been documented: one from an *S. aureus* plasmid, and the second a chromosomal resistance element from a marine *Bacillus* (Wang et al. 1989; Chu et al. 1992; Skinner et al. 1991). Each resistance operon consists of genes involved in mercuric ion transport (*merT*), reduction (*merA*), organomercurial cleavage (*merB*), and regulation (*merR*) together with other, unassigned, open reading frames (Helmann et al. 1989; Laddaga et al. 1987; Wang et al. 1989). The mechanism of resistance is the reduction of mercuric ion to metallic mercury by the NADPH-dependent flavoprotein, MerA (Walsh et al. 1988). The structure of the MerA protein from *Bacillus* sp. RC607 has been solved by X-ray crystallography (Schiering et al. 1991). In the sequenced Gram-negative resistance elements the regulatory gene, *merR,* is transcribed divergently from the resistance genes with both transcripts initiating in a common promoter/regulatory region (Silver and Walderhaug 1992; Summers 1986, 1992). In contrast, there is apparently a single *mer* promoter in the Gram-positive resistance systems and *merR* is the first gene of the resistance operon (Helmann et al. 1989; Skinner et al. 1991).

2.1 The MerR Protein

The MerR proteins are among the best-characterized metalloregulatory proteins (Summers 1992). MerR is a small (ca. 14-kDa monomer mass) protein that binds DNA as a dimer at a single operator site overlapping the *mer* promoter (Chu et al. 1992; Helmann et al. 1989; O'Halloran et al. 1989). Each MerR protein contains an amino-terminal domain with a helix-turn-helix type DNA-binding motif (Helmann et al. 1989). The second helix of this motif, thought to be responsible for site-specific DNA-binding, is the most highly conserved region of MerR proteins (Figure 3.2). In the *Bacillus* sp. RC607 MerR protein, the amino-terminal DNA-binding region is linked to a carboxyl-terminal metal-binding domain by a protease-sensitive linker region (Figures 3.2 and 3.3). Addition of Hg(II) stabilizes the carboxyl-terminal metal binding domain against protease attack, suggesting that metal ion coordination decreases the flexibility or accessibility of protease-sensitive sites in this domain (Figure 3.3). Sequence comparisons suggest that MerR may be a prototype for a family of helix-turn-helix

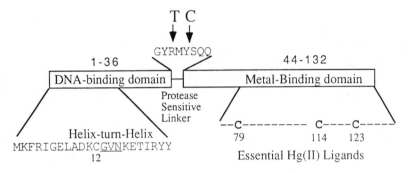

Figure 3.2 Structure-function map of the *Bacillus* sp. RC607 MerR protein. Protease digestion separates MerR into a relatively stable carboxyl-terminal domain (see Figure 3.3) and an amino-terminal fragment containing the determinants of DNA-binding. The three cysteine residues crucial for interaction with Hg(II) are in the C-terminal domain; the cysteine proximal to the turn of the helix-turn-helix DNA-binding motif is not important for Hg(II) binding (Helmann et al. 1990). T and C indicate the primary cleavage sites of trypsin and α-chymotrypsin, respectively, in partial proteolysis experiments (Figure 3.3).

containing regulatory proteins. Proteins related in sequence to MerR include the *Escherichia coli* SoxR regulator of the oxidative stress response (Demple and Amábile-Cuevas 1991), the *B. subtilis* GlnR regulator of glutamine synthetase (Schreier et al. 1989), and the *Bradyrhizobium japonicum* NolA protein involved in formation of root nodules (Sadowsky et al. 1991). Interestingly, SoxR contains a [2Fe-2S] cluster that allows this protein to sense superoxide (Ding et al. 1996; Gaudu et al. 1997).

The MerR proteins from Gram-positive organisms are about 40% identical with their Gram-negative counterparts. In addition, all *mer* operators share a conserved GTAC N_4 GTAC motif. Indeed, the MerR protein from *Bacillus* sp. RC607 binds to both its own operator and to the regulatory region of the Tn*501* transposon (Helmann et al. 1989). Conversely, the Tn*501* MerR protein binds with high affinity to the *Bacillus* regulatory region (Helmann et al. 1989). Genetic analysis clearly indicates that the conserved base pairs of the *mer* operator are essential for MerR function in vivo (Lee et al. 1993; Park et al. 1992).

MerR acts as a repressor of the *mer* genes, in the absence of mercury, and as a transcriptional activator, when Hg(II) is present. This dual role is clear from genetic studies: inactivation of the *merR* gene leads to constitutive, low-level expression of the *mer* operon, whereas high-level expression of *mer* genes requires both MerR and the inducer, Hg(II) (Ni Bhriain et al. 1983; Ross et al. 1989). Mutant MerR proteins that no longer bind Hg(II) but still bind DNA act as repressors even in the presence of the inducer (Ross et al. 1989; Shewchuk et al. 1989b). Conversely, constitutive mutants of MerR activate transcription even

-Hg(II)   +Hg(II)
0 2 4 8 16 32 64   0 2 4 8 16 32 64

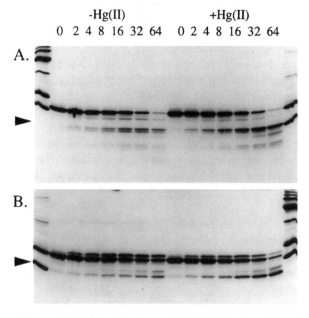

Figure 3.3 Partial proteolysis of the *Bacillus* sp. RC607 MerR Protein. MerR protein (0.5 mg/ml) was treated with 5 μg/ml trypsin (panel A) or α-chymotrypsin (panel B) in buffer P (10 mM TrisHCl, pH 7.5, 200 mM NaCl, 2 mM 2-mercaptoethanol, 5% glycerol) in the absence or presence of 40 μM HgCl₂ as indicated. Incubations were at 30°C for the indicated time (minutes), reactions were terminated by boiling in SDS-PAGE sample buffer, resolved by electrophoresis on SDS-PAGE (20%), and staining with coomassie blue. The presence of Hg(II) does not appear to alter the initial rate of cleavage of MerR with trypsin, as judged by the rate of disappearance of full-length protein, but the persistence of the ca. 12-kD partial proteolysis product is greatly enhanced. Amino-terminal sequencing identifies this product as the results of cleavage at Arg39 (Figure 3.2). Similar effects are seen in the presence of α-chymotrypsin.

in the absence of Hg(II) (Comess et al. 1994; Parkhill et al. 1993). MerR is rather unusual in that it binds to its operator in both the presence and absence of inducer (O'Halloran and Walsh 1987). In fact, it appears that the repressed *mer* promoter contains both MerR and a bound RNA polymerase poised for a rapid response should the cell be challenged with Hg(II) (Heltzel et al. 1990; Lee et al. 1993).

The MerR:Hg(II) complex binds to the *mer* promoter in the spacer region between the −35 and −10 recognition elements and activates transcription initiation (Summers 1992). The mechanism of activation involves, at least in part, a distortion of the spacer DNA (Ansari et al. 1992, 1995). The *mer* promoter regions are characterized by a long spacer region (19 bp in Gram-negative systems and 20 bp in Gram-positive systems compared to a consensus length of 17 bp). Genetic studies indicate that the long spacer is essential for inducibility: deletion of one base-pair from the Tn501 spacer element results in a constitutive promoter (Lund and Brown 1989). The MerR:Hg(II) complex apparently compensates for this long spacer by unwinding and bending the spacer DNA (Ansari et al. 1995). Mutant MerR proteins that activate transcription in the absence of Hg(II) distort the spacer region even in the absence of inducer (Comess et al. 1994; Parkhill et al. 1993). Thus, there is a good correlation between the ability of MerR to distort the spacer region and to activate transcription. Whether or not DNA distortion is sufficient for activation is not yet established. It is possible that specific protein-protein contacts are also needed to facilitate activation (Caslake et al. 1997).

2.2 Metal-binding Motifs of MerR

The metal-binding domain of MerR is characterized by three invariant cysteine residues (Helmann et al. 1989). These are present at positions 79, 114, and 123 in the numbering of the *Bacillus* protein (Figure 3.2). Mutations that alter any of these three cysteines (or their positional equivalents in related MerR proteins) eliminate Hg(II) binding and prevent transcriptional activation of the *mer* resistance genes (although MerR still functions as a repressor) (Ross et al. 1989; Shewchuk et al. 1989a; Helmann et al. 1990). Biochemical studies indicate that both the Tn501 MerR and the *Bacillus* sp. RC607 MerR bind a single Hg(II) per dimer (Helmann et al. 1990, Shewchuk et al. 1989a). As Hg(II) binds with high affinity to thiolates, the three invariant cysteine residues of MerR were obvious candidates for Hg(II) binding ligands. The geometry of the metal binding site was difficult to envision, however, because Hg(II) normally accepts between two and four ligands and there are six thiolates, three from each MerR monomer, that could potentially interact with the single bound Hg(II).

A detailed model for the MerR:Hg(II) complex was developed from biochemical studies using mutant derivatives of the *Bacillus* sp. RC607 MerR protein (Helmann et al. 1990) (Figure 3.4). Mutation of any of the three candidate cysteine ligands (C79, C114, or C123) eliminates high-affinity Hg(II) binding. High-affinity binding in this case refers to binding of Hg(II) in the presence of 10 mM 2-mercaptoethanol, which mimics the physiological levels of thiol in the cytoplasm of the bacterial cell. Significantly, mixing of inactive MerR proteins allows the formation of MerR heterodimers that bind Hg(II) with wild-type affinity. This complementation was only observed with certain combinations of mutant proteins; the Cys 114 and Cys 123 thiolates had to be present on the same monomer in the heterodimer. This led to the proposal that Cys 114 and Cys 123 from one monomer could function together with Cys 79 from the second

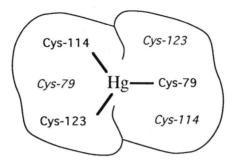

Figure 3.4 Model of Hg(II) coordination to the *Bacillus* MerR protein. Heterodimer reconstitution experiments (see text) led to the proposed asymmetric, tricoordinate binding site for Hg(II) in MerR (Helmann et al. 1990). The three residues indicated in italics can presumably form a second (symmetry-related) site that, possibly for steric reasons, is not occupied.

monomer to form a high-affinity Hg(II) binding pocket. This is supported by the observation that a multiply mutant MerR protein, in which all cysteines except Cys 79 were changed to alanine, can complement a multiply mutant protein that retains only Cys 114 and Cys 123 for *in vitro* Hg(II) binding (Helmann et al. 1990). In the resulting heterodimer only three cysteines are present, yet this protein binds Hg(II) with wild-type affinity. These studies argue that the MerR:Hg(II) activator contains an asymmetrically bound Hg(II) coordinated to three cysteine residues (Helmann et al. 1990). Although there are two symmetrically related binding sites for Hg(II) in the wild-type protein, it is proposed that the two sites physically overlap such that only one site can be occupied at a time. The Tn*501* MerR:Hg(II) complex is also tricoordinate as supported by physical analyses including extended X-ray fine structure analysis (EXAFS), [199]Hg NMR, and UV spectroscopy (Utschig et al. 1995; Wright et al. 1990). It is interesting to note that Cd(II) also binds to MerR, although at higher concentrations than Hg(II), and the MerR:Cd(II) complex activates transcription both *in vivo* and *in vitro* (Ralston and O'Halloran 1990; Summers 1992). However, the *mer* gene products do not detoxify cadmium. Therefore, Cd(II) is a gratuitous inducer of *mer* genes.

3. Cadmium

Cadmium, like mercury, is a toxic heavy metal with no known physiological role. Several cadmium resistance systems are regulated specifically by cadmium (Endo and Silver 1995, Silver and Walderhaug 1992). One family of cadmium resistance systems acts to efflux cadmium via an ATP-dependent transporter encoded by the *cadA* gene. Intriguingly, this ATP-dependent transporter is quite similar in sequence to the candidate gene for Menkes' syndrome, a lethal X-linked trait in humans resulting from improper copper transport (Silver et al. 1993; DiDonato and Sarkar 1997). The second protein encoded by these systems, designated CadC, is similar in sequence to the ArsR regulatory protein for inducible arsenate resistance (Shi et al. 1994, 1996) and is required for high-level resistance to cadmium. The CadAC-type efflux system was originally identified on *S. aureus* plasmid pI258 (Nucifora et al. 1989; Yoon et al. 1991) and has subsequently been identified in the chromosome of *B. firmus* (Ivey et al. 1992) and on a *Listeria monocytogenes* plasmid (Lebrun et al. 1994a,b). The *L. monocytogenes* cadmium resistance genes are found on a transposable element, Tn*5422,* which is a member of the Tn*917* family of elements (Lebrun et al. 1994b). This element may account, in part, for the observation that nearly 36% of isolated *L. monocytogenes* strains are cadmium resistant (Lebrun et al. 1994b).

The *S. aureus* pI258 CadC protein has been purified and shown to directly regulate transcription in vitro (Endo and Silver 1995). The purified protein binds specifically to an inverted repeat element overlapping the promoter for the *cadCA* operon, and this binding is relieved by addition of Cd(II), Pb(II), and Bi(III) (Figure 3.5). This therefore replicates the metal ion selectivity of regulation as documented in vivo (Yoon et al. 1991). The CadC protein is a member of a growing family of metalloregulatory proteins related to ArsR (Figure 3.6), the regulator of arsenic resistance operons from both Gram-positive and Gram-negative organisms (Shi et al. 1994). It is likely that all of these proteins sense metal ions by direct coordination to conserved cysteine thiolates (Shi et al. 1994; Bairoch 1993). In the case of ArsR, two or possibly three cysteines participate in arsenic binding (Shi et al. 1996).

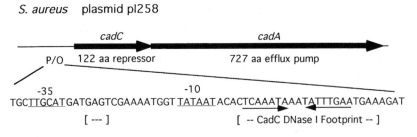

Figure 3.5 Regulation of the *cadCA* operon from plasmid pI258. The CadC repressor protein binds to an inverted repeat element overlapping the transcription start site. Regions of DNase I protection are indicated (Endo and Silver 1995). In the presence of inducing metal ions DNA binding is relieved and the operon is induced.

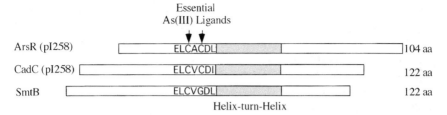

Figure 3.6 The ArsR/CadC family of metalloregulatory proteins. The ArsR and CadC proteins have a conserved structure with two highly conserved cysteine residues proximal to a DNA-binding helix-turn-helix motif. In the case of ArsR encoded by the *E. coli* plasmid R773 the two indicated cysteines have been shown to be sufficient for arsenic binding and consequent dissociation of the repressor from DNA (Shi et al. 1996). It is not yet established whether these same two cysteines are necessary or sufficient for the Cd(II)-responsive CadC proteins. In the case of SmtB, a zinc responsive repressor of cyanobacterial metallothionein expression (Erbe et al. 1995), the second cysteine is not present and other amino acids are presumably involved in the Zn(II)-induced dissociation of the repressor (Bairoch 1993).

4. Iron

Iron is an essential nutrient for nearly all living organisms (Guerinot, 1994). Its low solubility in aerobic environments is often limiting for growth, and, consequently, competition for iron can be a decisive factor in determining survival. Most microbes, as well as many fungi, synthesize and secrete low-molecular-weight, iron-specific chelators known as siderophores. Siderophores are structurally diverse, but are often either catechol- or hydroxamate-containing compounds (Neilands 1995). The ferri-siderophore complexes are recognized by specific receptors and internalized prior to the release of bound iron by reduction to ferrous iron and, in some cases, by hydrolysis of the siderophore.

When intracellular iron levels are low, genes for the biosynthesis and uptake of siderophores are induced (Bagg and Neilands 1987a,b; Silver and Walderhaug 1992). Mechanisms of iron regulation are best understood in *E. coli* (Fur regulon, see Chapter 2) and *Corynebacterium diphtheriae* (DtxR regulon) (Tao et al. 1994). Fur and DtxR both bind metal ions as corepressors and they control similar sets of genes, yet they are not obviously homologous (however, see Section 3.6). To date, Fur has been best characterized from Gram-negative bacteria, whereas DtxR has been studied in several genera of Gram-positive bacteria. This has led some to suggest that Fur regulates iron uptake in Gram-negative bacteria whereas DtxR regulates iron uptake in Gram-positive bacteria. It is now clear that this is a gross oversimplification. As a result of several microbial genome sequencing projects, it is now clear that there are numerous Fur homologs in Gram-positive bacteria and, conversely, there are DtxR-like proteins in several Gram-negative organisms (including *E. coli*) as well as in *Methacococcus jannaschii*, a member of the domain Archae. The regulatory roles of these various putative metal-binding repressors proteins is not yet clear. Some indication of the complexity of this problem can be gleaned from the observation that in *B. subtilis* there are no fewer than three distinct Fur homologs and a DtxR homolog (YqhN) and all four proteins appear to have distinct DNA-binding selectivity and regulatory roles.

4.1 The E. coli Fur Regulon

In *E. coli*, a metalloregulatory protein designated Fur (ferric uptake repressor) binds to a conserved operator sequence (fur box) to repress the genes for sidero-phore biosynthesis and uptake (Schäffer et al 1985; Bagg and Neilands 1987a,b). Fur requires a divalent metal ion as a corepressor (Bagg and Neilands 1987a). Fe(II) is thought to be the physiological corepressor, but Fur also binds Mn(II), Co(II), and Cd(II) in vitro, Mn(II) also binds Fur in vivo and represses the expression of some Fur-regulated genes (Privalle and Fridovich 1993). The ability of manganese to prevent iron uptake may explain the observation that many Mn(II)-resistant mutants are altered in the *fur* gene (Hantke 1987). Indeed, this has provided a convenient selection for *fur* mutants in *E. coli*, *Serratia marcescens*, *Vibrio* spp., *Yersinia enterocolitica*, and *Bordetella bronchiseptica* (Brickman and Armstrong 1995; Hantke 1987; Lam et al. 1994; Tomalsky et al. 1994). However, we have found that *B. subtilis* mutants selected for Mn(II) resistance are not derepressed for siderophore biosynthesis. As iron-regulated genes in *B. subtilis* are generally not repressed by Mn(II), we suggest that the *B. subtilis* Fur protein has a more stringent metal ion recognition than *E. coli* Fur (Chen et al. 1993).

E. coli Fur is a dimeric DNA-binding protein related to the CAP family of proteins (Holm et al. 1994). Fur contains an amino-terminal DNA-binding motif and a carboxyl-terminal metal binding domain (Figure 3.7) (Coy and Neilands 1991; Stojilkovic and Hantke 1995). The identity of the metal binding ligands in Fur remains unknown, although there is evidence for involvement of both histidines and cysteines in binding to various cations (Carayre and Neilands 1991; Coy et al. 1994; Lam et al. 1994; Saito et al. 1991a,b).

Genes for Fur-like proteins have been sequenced from many genera of Gram-negative bacteria. Most of these proteins are very closely related, having >50% amino acid identity. Indeed, *fur* genes from *Neisseria gonorrhoeae*, *Y. pestis*, *Legionella pneumophila*, *V. cholerae*, and *V. anguillarum* all complement an *E. coli fur* mutation (Brickman and Armstrong 1995, and references therein). The corresponding proteins are therefore likely to be very similar to *E. coli* Fur in both structure and function. In contrast the Fur-like proteins of *B. subtilis* are comparatively diverged members of this family (Figure 3.8).

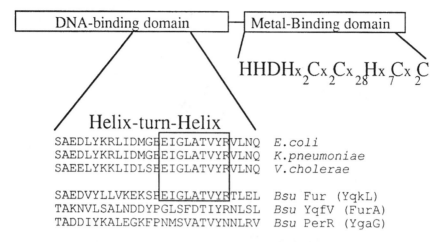

$$HHDHx_2Cx_2Cx_{28}Hx_7Cx_2C$$

Helix-turn-Helix

```
SAEDLYKRLIDMGHEIGLATVYRVLNQ   E.coli
SAEDLYKRLIDMGHEIGLATVYRVLNQ   K.pneumoniae
SAEELYKKLIDLSHEIGLATVYRVLNQ   V.cholerae

SAEDVYLLVKEKSHEIGLATVYRTLEL   Bsu Fur (YqkL)
TAKNVLSALNDDYPGLSFDTIYRNLSL   Bsu YqfV (FurA)
TADDIYKALEGKFPNMSVATVYNNLRV   Bsu PerR (YgaG)
```

Figure 3.7 Structure-function map of Fur proteins. Fur proteins contain two independently functioning domains (Coy and Neilands 1991; Stojilkovic and Hantke 1995): an amino-terminal DNA-binding domain and a carboxyl-terminal metal-binding domain. The proposed helix-turn-helix DNA-binding motif from various Fur proteins is illustrated. Note that although *B. subtilis* has genes for three predicted Fur homologs, only Fur contains a recognition helix highly similar to those of Gram-negative Fur proteins. This is consistent with the observation that FurB encodes the iron uptake repressor in *B. subtilis,* and siderophore biosynthesis and uptake genes have fur boxes highly similar to those recognized by *E. coli* Fur. The targets of the YqfA (FurA) protein and the role of this protein in sensing iron or other metal ions are not currently known. The carboxyl-terminal metal binding domain is typically very histidine and cysteine rich. A typical sequence for this region is shown.

4.2 *The* B. subtilis *Fur Regulon*

In contrast with the extensive characterization of iron uptake systems in Gram-negative bacteria (Guerinot 1994), there is less information available for *B. subtilis.* This is despite the fact that the first described bacterial siderophore, itoic acid, was identified nearly 40 years ago in iron-starved *B. subtilis* cultures (Ito and Neilands 1958). Synthesis of itoic acid [a mixture of 2,3-dihydroxybenzoic acid (DHB) and 2,3-dihydroxylbenzoyl glycine (DHBG)] is induced by iron starvation (Peters and Warren 1968; Tai et al. 1990; Walsh et al. 1971). Early experiments indicated that this induction was likely mediated by an increase in transcription of the corresponding biosynthetic genes. However, the molecular basis of this metalloregulation has only recently been determined.

The hypothesis that *B. subtilis* contains a Fur-like regulatory protein arose from the observation that iron-regulated promoters contain *cis*-acting sequence

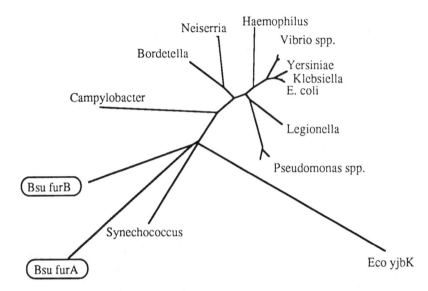

Figure 3.8 The Fur family of bacterial repressor proteins. Fur proteins were aligned using CLUSTAL W (Thompson et al. 1994), and the resulting multiple sequence alignment was analyzed using the ProtDist and Fitch programs of the PHYLIP suite of phylogenetic inference programs (Felsenstein 1989). The resulting unrooted phylogenetic tree was plotted using TreePlot. With the exception of the two *B. subtilis* proteins (circled), all the known Fur-like proteins are from Gram-negative bacteria. FurA is the same as YqfV and FurB is now called Fur.

elements similar to the fur box recognized by *E. coli* Fur (Chen et al. 1993; Schneider and Hanke 1993). This hypothesis is strongly supported by recent analysis of the genes for dihydroxybenzoic acid siderophore biosynthesis (*dhb*). A perfect (19/19) match to the fur box consensus overlaps the *dhb* promoter and mutation of this element leads to constitutive expression (Rowland and Taber 1996). Regulation of the *dhb* operon by iron has also been demonstrated by direct enzymatic assay of the encoded products (Walsh et al. 1971) and by siderophore assays (Chen et al. 1993; Ito and Neilands 1958). Other operons postulated to be part of the Fur regulon are compiled in Table 3.1. Intriguingly, the predicted YwbL protein is very similar to an iron-uptake protein of *Saccharomyces cerevisiae* (Stearman et al. 1996), and YwcC may encode a DNA binding regulator. The *feuABC* operon was identified as a locus affecting resistance to protonophores (Quirk et al. 1994). The FeuABC proteins are postulated to form a transport system for the uptake of DHB or other catecholate siderophores. The iron regulation of *mrgC* (Chen et al. 1993), *fhuD* (our unpublished data), and *dhb* (Rowland and Taber 1996) is known to be Fur dependent; molecular studies of the other putative

Table 3.1 Compilation of the *Bacillus subtilis* Fe uptake (Fur) regulon

Gene(s)	Fur Box	Fur regulated	Function
dhbACEBF	GATAATGATAATCATTATC	yes	DHB siderophore synthesis
fhuBGC	GATAATGATTATCAATTGC	n.d.[a]	Ferric hydroxamate uptake
fhuD	GATAATGATTCTCTTTTTC	yes	Ferric hydroxamate uptake
feuABC (a)	CCAATTGATAATAGTTATC	n.d.	Putative uptake system for DHB siderophore
(b)	GATAATAGTTATCAATTGA		
mrgC	GCGATTGAAAATCATTCTC	yes	Unknown
ywbLMN	TACAATGATAATCATTTTC	n.d.	Similar to yeast FTS3
ywcCD	GATAAGGGTAATCATTACC	n.d.	Unknown
yxeB	GATAATGATAATCATTACT	n.d.	FhuD homolog
citH	CATAAAGACAATCATCATC	n.d.	Citrate transporter
yclN	GATAATGATAATCAATTAC	n.d.	Homolog of ferric anguibactin transport system
fliD (a)	GATACTCAAATACATAATC	yes	Flagellar-based motility:
(b)	GATAATGATAACGAGGTTC		operon encodes hook protein and regulators
ahpCF	TATAATTAGAATTATTATT	yes	Alkylhydroperoxide reductase

[a]Not determined.

members of the Fur regulon of *B. subtilis* have only just begun. Interestingly, the Fur regulon appears to include the genes for alkylhydroperoxide reductase (*ahpCF*), an important component of the oxidative stress response of *B. subtilis* (see Section 6.1). This operon is preceded by a fur box sequence as well as multiple sequences resembling the operators of peroxide-inducible genes (per box). In general, *ahpCF* expression is much more strongly affected by manganese than by iron. The iron repression, but not the manganese repression, requires the *B. subtilis* Fur protein (our unpublished data).

4.3 The B. subtilis *Fur Protein*

The gene encoding the *B. subtilis* Fur protein was recently identified as part of the effort to sequence the genome. Indeed, *B. subtilis* contains three genes encoding Fur homologs, designated *yqfV*, *fur* (formerly *yqkL*), and *perR* (formerly *ygaG*). The *B. subtilis* Fur-like proteins are among the most distantly related members of the Fur family of proteins (Figure 3.8) and are also highly diverged from each other with only 23 to 29% identity between pairs of proteins. Gene disruption experiments reveal that *yqkL* encodes the Fur protein in *B. subtilis* which is consistent with the observation the it is most similar to Gram-negative

Fur proteins in its DNA recognition helix. The function of YqfV is still obscure while PerR represses genes involved in the peroxide stress response (see Section 6.1). It is interesting to note that *E. coli* also contains a second, distantly related member of this protein family (YjbK) of unknown function (Holm et al. 1994). Recently, Fur-like proteins have been described in *Streptococcus pyogenes* and *Staphylococcus aureus,* although the role of these proteins is not yet determined. Whereas the *S. pyogenes* protein is most similar to PerR, particularly in the DNA recognition region, the *S. aureus* protein matches most closely to YqfV. Whether these organisms contain multiple Fur-like proteins, as in *B. subtilis,* or whether these are the only Fur-like proteins is not yet clear.

4.4 The Corynebacterium diphtheriae *DtxR Regulon*

Regulation of gene expression by iron in the Gram-positive pathogen, *C. diphtheriae,* requires a distinct metalloregulatory protein, DtxR (Tao et al. 1994). DtxR was originally defined as a regulator required for the iron-mediated repression of diphtheria toxin. Like Fur, this small, dimeric DNA binding protein requires a metal ion to activate its DNA binding function and thereby allow the repression of transcription. In vitro, DtxR binds DNA in the presence of Fe(II), Cd(II), Co(II), Mn(II), or Ni(II) (Schmitt and Holmes 1993; Tao et al. 1994). The DtxR regulon includes genes for iron assimilation, one of which is related to the *B. subtilis fhuD* hydroxamate siderophore uptake gene (Schmitt and Holmes 1994) and another that allows acquisition of iron from heme (Schmitt 1997). Therefore, DtxR is functionally analogous to Fur (Table 3.2). This view is supported by the finding that a *Streptomyces pilosus* gene for biosynthesis of desferrioxamine B siderophore contains a *dtxR*-like operator site (Günter et al. 1993).

DtxR homologs have been identified in numerous genera of bacteria and recent results suggest that this is a very broadly distributed family of regulatory proteins (Table 3.2). In *Mycobacterium* spp., a protein with 60% identity to DtxR has been described and designated IdeR (iron-dependent regulator) (Schmitt et al. 1995). In *M. smegmatis,* an *ideR* mutant strain was unable to repress siderophore in the presence of iron (Dussurget et al. 1996), emphasizing again the parallels between IdeR and Fur regulation. The role of the DtxR-like protein *B. subtilis* is not currently known, but all known iron uptake functions in this organism are regulated by a Fur homolog (see Section 3.3).

4.5 Metal Ion Recognition by DtxR

DtxR is unique among bacterial metalloregulatory proteins in that an X-ray crystal structure is available of the protein both with and without associated metal ions (Qiu et al. 1995; Schiering et al. 1995). The overall structure of this protein

Table 3.2 DtxR family of regulatory proteins

Organism	Regulator	Target Gene(s)	Size (aa)	Operator	Reference
				(consensus) AGGTTAG-CTAACCT	
Corynebacterium diphtheriae	DtxR	*tox*	226	TTAGGATAGCTTTACCTAA	Tao and Murphy 1994
		irp1		TTAGGTTAGCCAAACCTTT	Tao et al. 1994
		irp2		GCAGGGTAGCCTAACCTAA	Schmitt and Holmes 1994
					Schmitt and Holmes 1994
Streptomyces spp.	"DtxR"	*desA*	~230	TTAGGTTAGGCTCACCTAA	Günter et al. 1993
					Günter and Schupp 1995
Mycobacterium spp.	IdeR	n.d.[a]	~230	n.d.	Schmitt et al. 1995
Brevibacterium sp.	DmdR	n.d.	228	n.d.	Oguiza et al. 1995
Staphylococcus epidermidis	SirR	n.d.	214	n.d.	SESIRR_1; X99128
Bacillus subtilis	YqhN	n.d.	142	n.d.	Que and Helmann, unpub.; D84432
Treponema pallidum	TroR	n.d.	153	n.d.	TPU55214_6
Methanococcus jannaschii	MJ0568	n.d.	125	n.d.	H64370
Escherichia coli	o155	n.d.	155	n.d.	ECAE000184_2

[a] "n.d." indicates not determined.

reveals three folded domains in each monomer: an amino-terminal DNA binding domain, a central metal binding domain, and a carboxyl terminal domain of unknown function. Despite the solution of several structures at atomic resolution, the molecular basis of metal ion recognition by DtxR has been controversial. The crystal structures reveal two metal ion binding sites: site 1 was occupied in all the crystal forms, whereas site 2 was only occupied in the structure determined with Cd(II). However, it is site 2 which contains most of the ligands identified genetically as essential for DtxR function. As the crystals revealed a variable degree of oxidation of Cys 102 to the sulfinyl group ($R-SO_{2-}$), the ability of the protein to bind metal in site 2 may have been adversely affected (Qiu et al. 1995). It is also possible that site 1 is a "structural" site and site 2 is the "metal-sensing" site. Finally, the crystal structure fails to account for the observation that DtxR can readily form disulfide cross-linked dimers in solution: in the crystal structure the two Cys-102 residues are separated by 23 Å. Significantly, the crystal structure in the absence of metal ions shows a different arrangement of the putative ligands (Tao et al. 1994).

A resolution of these apparently inconsistent results has recently been proposed based on X-ray analysis of a Ni(II) complex of a mutant derivative of DtxR in which Cys 102 is replaced by Asp (Ding et al. 1996). This mutation still allows metalloregulation but eliminates problems associated with the oxidation of the Cys 102 side chain. Based on this work, it is proposed that the metal-sensing site in DtxR contains the following ligands (Figure 3.9): Asp (or Cys) 102, Glu 105, His 106, Met 10, the main chain carbonyl of residue 102, and a water molecule in an octahedral arrangement. It is postulated that metal occupancy of this site affects the orientation of the DNA binding motif, which is near the Met

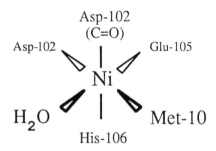

Figure 3.9 The Metal-sensing site of DtxR (Ding et al. 1996). The X-ray crystal structure of a mutant (Cys 102 → Asp 102) DtxR protein complexed with Ni(II) was determined. The identity and significance of the various metal ligands was further verified by site-directed mutagenesis and biochemical characterization. Although in this structure the carbonyl group of Asp 102 is a ligand, it is likely that the thiolate side chain of Cys 102 can serve this function in the wild-type protein *in vivo*.

10 residue. This model is consistent with the refined structure reported for the wild-type DtxR protein in the presence of either Mn(II) or Co(II) (Qiu et al. 1996). In this structure the putative metal sensing site was again only partially occupied and the Cys 102 carbonyl, but not the thiol side chain, was metal ligand. This is unexpected since it is known that Cys 102 can only be replaced by Asp *in vivo* and still allow metal responsive repressor function. It is possible that the formation of a mixed disulfide, or other modification of the Cys 102 side chain, may have prevented a direct participation in coordination to Mn(II) (Qiu et al. 1996). In this latter study, it was also possible for the first time to visualize the relatively disordered third domain of DtxR which has an SH3-like fold. Intriguingly, this third domain is absent in several of the recently identified DtxR homologs which contain between 125 and 155 amino acids (Table 3.2).

4.6 Analysis of the Fur Family of Proteins: Structural Link to the DtxR Family

The relationship between DtxR and Fur is unclear: there is no statistically significant amino acid identity, they do not complement each other *in vivo,* and there is little similarity between the *dtxR* and *fur* operators. Nevertheless, it is quite possible that DtxR and Fur share a similar three-dimensional structure, at least in their DNA-binding domains. Multiple sequence alignment and profile analysis suggests that Fur contains an amino-terminal DNA-binding domain with a predicted topology similar to CAP (Holm et al. 1994) linked to a carboxyl-terminal metal binding domain (Figure 3.7). The model is supported both by proteolysis experiments (Coy and Neilands 1991) and the construction of hybrid proteins with DNA-binding or dimerization domains of the λ repressor (Stojil-kovic and Hantke 1995). DxtR, like Fur, has an amino-terminal DNA-binding domain upstream of a metal-binding domain. The topology of the DtxR DNA binding domain is also very similar to the CAP/LexA family of proteins (Qiu et al. 1995; Schiering et al. 1995). Thus, DtxR may be related to Fur in both structure and function.

5. Manganese

Manganese is an important nutrient for bacteria and is accumulated from the environment via a high-affinity transport system (Silver and Walderhaug 1992). In *B. subtilis,* the activity of the manganous ion transport system is regulated by manganese. This regulation appears to be at the level of gene expression, suggesting that there is specific Mn(II) metalloregulatory mechanism (Eisenstadt et al. 1973; Fisher et al. 1973). Manganese is clearly an important regulatory ion in *B. subtilis* as high manganese levels are required for sporulation and manganese regulates the production of numerous secondary metabolites. These observations led Eugene Weinberg to propose that manganous ion may play a central regulatory role in Gram-positive organisms, akin to the role assigned to iron in Gram-negative bacteria and zinc in the regulation of secondary metabolite production in fungi (Weinberg 1990). This hypothesis provides ample motivation for the identification of proteins mediating selective gene regulation by manganese. The protein(s) regulating the expression of Mn(II)-uptake systems are still uncharacterized, however.

Although the putative Mn(II)-specific regulator of manganese uptake genes has not been identified, Mn(II) has been found to be a potent corepressor of a family of oxidative stress genes in *B. subtilis* (Chen et al. 1995). In section 6, I review the evidence suggesting that oxidative stress responses are regulated by Mn(II) and other transition metals.

6. Metal Ion Homestasis and Oxidative Stress

During aerobic growth, the incomplete reduction of molecular oxygen leads to the formation of reactive oxygen intermediates such as superoxide anion, hydrogen peroxide (H_2O_2), and hydroxyl radical (Cadenas 1989). These reactive oxygen species modify proteins, lipids, and DNA and contribute to processes as diverse as aging, cancer, and cardiovascular disease (Ames et al. 1993; Shigenaga et al. 1994; Stadtman 1992).

The ability of reactive oxygen intermediates to inflict damage is exacerbated by the presence of free iron (Imlay and Linn 1988). Iron reacts with H_2O_2 by the well-characterized Haber-Weiss-Fenton reactions, to generate highly reactive hydroxyl radicals, which cleave DNA and proteins (Imlay et al. 1988; Stadtman and Berlett 1991). This chemistry plays a role in the killing of *E. coli* by low concentrations of H_2O_2 (mode I killing) since addition of an iron chelator (dipyridyl) prevents toxicity. The toxicity of superoxide is also related to iron levels: superoxide attacks [4Fe-4S] clusters and releases free iron, which can then promote DNA damage (Keyer et al. 1995).

The regulation of metal ion homeostasis and oxidative stress responses are intertwined in many organisms. For example, iron uptake sensitizes cells to oxidative damage (Keyer and Imlay 1996). As a result, *E. coli fur* mutants, which constitutively import iron, are extremely sensitive to H_2O_2. Mutants lacking both Fur and RecA (essential for recombinational repair) cannot grow aerobically (Touati et al. 1995). Conversely, cells grown anaerobically are sensitive to peroxide-mediated killing because of the induction of the Fnr-dependent ferrous iron uptake (*feo*) system (Keyer et al. 1995). In addition, some metal ions, notably Mn(II) and Cu(II), may protect against oxidative stress. For example, *Lactobacillus plantarum* can be grown in the absence of any added iron, but under that condition accumulates millimolar levels of Mn(II), which acts as an inorganic superoxide dismutase (SOD) (Archibald and Fridovich 1981; Fridovich 1995). In *S. cerevisiae,* strains lacking cytosolic Cu/Zn SOD are auxotrophic for lysine and methionine when grown in air. This defect can be bypassed genetically by (a) mutations in PMR1 (a P-type ATPase), which increases cellular levels of Mn, (b) mutations in BSD2, which increase Cu accumulation (Liu and Culotta 1994), or (c) overexpression of ATX1, a copper binding protein (similar to MerP mercury transporters), which scavenges both superoxide and H_2O_2 (Lin and Culotta 1995). Indeed, the *S. cerevisiae* Cu/Zn SOD functions both as an antioxidant and in protection against copper toxicity. Consistent with this role in copper homeostasis, it is transcriptionally induced by copper (Culotta et al. 1995). As these examples demonstrate, there is ample precedent for thinking that oxidative stress genes might be regulated by metal ion availability. Recent studies in *B. subtilis* and *C. diphtheriae* demonstrate that metal ions do in fact affect the expression of peroxide stress genes in these organisms (Chen et al. 1995; Tai and Zhu 1995).

6.1 Regulation of the B. subtilis Peroxide Stress Response by Mn(II)

We were led to analyzing the peroxide stress response in *B. subtilis* as a result of studies of genes induced by metal ion chelators (Chen et al. 1993). In the course of our work we identified one gene, *mrgA,* expressed in stationary phase cells only if both manganese and iron are limiting. Repression is most responsive to added manganese (half-maximal repression at 0.2 µM), suggesting that this may be the primary regulatory ion in vivo. MrgA is a member of the Dps(PexB) family of DNA-binding proteins produced in stationary phase in several bacteria (Chen and Helmann 1995). Like *E. coli* Dps (Altuvia et al. 1994), MrgA is part of a peroxide stimulon comprising MrgA, catalase (KatA), and alkylhydroperoxide reductase (AhpCF) (Chen et al. 1995; Dowds 1994; Hartford and Dowds 1994). Interestingly, all of these genes are repressed by manganese, iron, or both. Subsequent analysis demonstrated that MrgA is part of the peroxide stimulon: synthesis of MrgA is stimulated by treatment with H_2O_2 and *mrgA* mutant strains are more sensitive to peroxide killing. The repression of peroxide stress genes by Mn(II) may have important physiological implications: when grown in environments rich in Mn(II), cells may be much more susceptible to peroxide killing (Chen et al. 1995).

To determine the mechanisms responsible for regulating *mrgA* transcription in response to metal ions, we fused the *mrgA* promoter to the *cat* (chloramphenicol resistance) and *lacZ* genes and introduced this operon into an SPβ prophage. As noted above, the postexponential induction of MrgA is blocked by added Mn(II) or iron salts. By selecting for chloramphenicol resistance in the presence of metal ions, we obtained both *cis-* and *trans-*acting regulatory mutations. At least two classes of *trans*-acting mutations were obtained. *Trans*-acting mutations in *katA* result in reduced catalase activity and presumably elevated H_2O_2 levels. The latter accounts for elevated *mrgA* expression in these mutants. We also obtained *trans*-acting mutations, which derepress several members of the peroxide regulon and define *perR* (Chen et al. 1995).

Analysis of *cis*-acting mutations derepressed for *mrgA* synthesis identified an inverted repeat element required for response to both metal ions and H_2O_2 (Chen et al. 1995). We designated this operator a *per box* to denote its role in regulation by hydrogen peroxide (Table 3.3). When *B. subtilis* cells are treated with H_2O_2 the most strongly induced proteins are MrgA, KatA, AhpC, and AhpF (Dowds 1994). Remarkably, *mrgA* and the gene for the major vegetative catalase (*katA*) both have nearly identical operators located immediately upstream of the −35 element. Similar operator sequences also overlap the promoter for the major heme biosynthesis operon, *hemAXCDBL*, and the *ahpCF* operon encoding alkyl-hydroperoxide reductase. Indeed, *katA, hemA, ahpC,* and *mrgA* are all coordinately regulated: all four promoters are induced by entry into stationary phase, and this induction can be blocked by metal ions (Chen et al. 1995 and our unpublished data). The picture that is emerging from these studies is that Mn(II), and to a lesser extent Fe(III), acts as a corepressor of peroxide stress genes in *B. subtilis.* Metal ions are also likely to repress peroxide stress genes in other organisms: an iron-repressible gene in *C. diphtheriae* encodes the small subunit of alkyl hydroperoxide reductase (AhpC). This iron-mediated regulation is independent of DtxR (Tai and Zhu 1995).

The *aphCF* operon appears to be an unusual case in that it is part of both the PerR and Fur regulons. Originally, we identified the *ahpCF* operon during a screen for Tn*10* insertions that derepress *mrgA-lacZ* in the presence of Mn(II) (Bsat et al. 1996). Like catalase mutants, the increased accumulation of endogenously produced reactive oxygen species in the *ahpC*::Tn*10* mutants is thought

Table 3.3 Compilation of the *Bacillus subtilis* PerR regulon

Gene(s)	Per Box	Gene Function
	TTATAAT-ATTATAA	consensus
mrgA	CTAAATTATAATTATTATAATTTAG	DNA binding protein
katA	CTATTTTATAATAATTATAAAATAA	Catalase
hemAXCDBL (a)	CTATGTTATAATTATTATAAATAAT	Heme biosynthesis
(b)	CTATGTTAGAATGATTATAAATTAA	
perR (a)	AAGAGTTACACTAATTATAAACATT	Repressor of per regulon
(b)	ACTAATTATAAACATTACAATGTAA	
furB	TCAGTTTATAATAATTATAGTTGGA	Iron uptake repressor
ahpCF (a)	TATAATTAGAATTATTATTGAAACG	Alkylhydroperoxide
(b)	AAAATATATATTAATTAATAATTCA	reductase

to derepress the peroxide regulon. Sequence analysis of the promoter region of *ahpCF* revealed sequences with similarity to the per box and, in addition, a fur box (Tables 3.1 and 3.3). In fact, a strain lacking Fur can no longer repress *ahpCF* in response to added iron, but the much more dramatic repression in response to Mn(II) is unaffected.

The metalloregulatory protein controlling the peroxide stress response, PerR, has recently been identified as the product of the *ygaG* gene, one of the three genes encoding Fur-like proteins in *B. subtilis* (Herbig and Helmann, unpublished studies). Our current working model for iron and manganese regulation in *B. subtilis* postulates (at least) two separate regulatory systems: an iron-specific homeostasis system mediated by one Fur homolog (YqkL), and a second Fur homolog (PerR) that represses peroxide stress genes in response to either manganese or iron. Intriguingly, Fur itself seems to be part of the PerR regulon (Bsat and Helmann, unpublished data). We anticipate that this model will be further refined as target genes, and corresponding regulatory metal ions, are identified for the third *B. subtilis* Fur homolog (YqfV) and the DtxR-like repressor (product of the *yqhN* gene).

7. Copper

The regulation of gene expression by copper ion has been an active area of research (see Chapter 4). In *Enterococcus hirae*, copper resistance is mediated by two P-type ATPases. Expression of their genes is regulated in a complex manner by copper: repression is maximal at 10 µM Cu(II) with induction observed at either lower or higher levels. This metalloregulation appears to depend on two linked regulatory loci, *copY* and *copZ*. CopY is a "metal-fist" type repressor and contains the metal-binding motif, $CxCx_4CxC$. In the presence of copper, CopY is released from its DNA binding sites (Strausak and Solioz 1997). In contrast, CopZ acts as an activator, possibly by opposing the action of CopY (Odermatt and Solioz 1995). Interestingly, CopZ has a region of amino acid similarity with the MerP and MerA proteins in a region containing a conserved GMxCxxC motif postulated to bind metal ions. The detailed interactions between these proteins responsible for copper regulation are not yet understood (Strausak and Solioz 1997).

8. Zinc

Zinc, like iron and manganese, is an important cofactor in numerous enzymes. The mechanism of zinc uptake and its regulation has not been well studied in bacteria (Silver and Walderhaug 1992). It seems quite likely, however, that zinc will regulate several intracellular processes. Zinc is widely appreciated for its role as an essential cofactor in nucleic acid polymerases and eukaryotic DNA binding and gene regulatory proteins (Berg and Shi 1996). It should be stressed, however, that these appear to be metalloproteins rather than metalloregulatory proteins. Zinc appears to act as a structural scaffold for the folding of small protein domains (fingers) rather than as a signal of intracellular zinc availability.

One system for which specific regulation by zinc has been demonstrated is the production of esterase from *Streptomyces scabies* (Babcock et al. 1992). This soil bacterium produces an extracellular esterase that may be important in invasion of susceptible plant tissues. The cloning of the gene for a zinc-inducible esterase has allowed the identification of relevant *cis*-acting sequences. A 23-bp region of the esterase promoter is protected against DNaseI digestion when incubated with cell extracts. Deletion of this sequence abolished high-level expression of esterase, suggesting that a zinc-dependent transcriptional activator may bind to this sequence. Further work on this interesting system promises to shed light on a novel zinc-metalloregulatory protein. Regulation of gene expression by zinc will undoubtedly turn out to be widespread: for example, SmtB regulation of metallothionein expression in cyanobacteria is zinc responsive (Erbe et al. 1995).

9. Alternative Sigma Factors and Metal-activated Gene Expression

A powerful mechanism of gene regulation exploited by many bacteria is the modification of the promoter selectivity of RNA polymerase by the elaboration of new sigma subunits. Alternative sigma factors promise to play an important role in metalloregulation in both Gram-positive and Gram-negative bacteria. A divergent group of sigma factors has recently been described which appear to regulate a variety of extracytoplasmic functions, and are therefore designated the ECF sigma factors (Lonetto et al. 1994). The ECF sigma factors include a second heat shock sigma in *E. coli*, an *E. coli* regulator of ferric citrate transport (FecI), positive regulators of siderophore biosynthesis in various Pseudomonads (PupI, PbrA, PvdS, and PfrI), and positive regulators of metal ion resistance operons in *Alcaligenes* spp. (CnrH and NccH) (Table 3.4). The activity of these positive regulatory factors is often regulated by corresponding antisigma factors, which in at least some cases, can sense metal ions. ECF sigma factors are also well represented in Gram-positive bacteria: *B. subtilis* harbors five such loci although the role of these proteins, if any, in metal ion metabolism or resistance is not yet known.

Table 3.4 ECF sigma factors implicated in metal homeostasis

Organism	Sigma	Metal(s)	Target Gene(s)	Reference
Escherichia coli	FecI	Fe(III)	Ferric citrate uptake	Angerer et al. 1995
Pseudomonas putida	PupI	Fe(III)	Iron uptake genes	Koster et al. 1994
P. fluorescens	PbrA	Fe(III)	Iron uptake genes	Sexton et al. 1995
P. aeruginosa	PvdS	Fe(III)	Iron uptake genes	Cunliffe et al. 1995
P. putida	PfrI	Fe(III)	Iron uptake genes	Venturi et al. 1995
Alcaligines eutrophus	CnrH	Ni, Co	Resistance genes	Liesegang et al. 1993
A. xylosidans	NccH	Ni, Cd, Co	Resistance genes	Schmidt and Schlegel 1994

10. Conclusions

It has been appreciated for decades that the chemical composition of culture medium can influence the production of secondary metabolites of bacteria. In particular, metal ions often activate or repress the synthesis of extracellular enzymes, antibiotics, or the entry into developmental programs such as endospore formation. The required metal ion concentrations for these processes are often much narrower than those necessary for growth. In some cases, this may reflect an essential role of a metal as a cofactor for a key biosynthetic enzyme, but in other cases the metal ions are influencing metabolic processes through global control of gene expression. Examples include iron regulation by proteins such as Fur and DtxR. Similar proteins may influence gene expression in response to changing levels of ions such as zinc and manganese.

Another class of metal-responsive systems encodes resistance determinants. MerR was discovered in the course of defining *mer* resistance systems. Specific regulation of copper resistance systems has been described in several Gram-negative systems (see Chapter 4, this volume) and similar systems exist in Gram-positive organisms (Odermatt and Solioz 1995). Several systems encode multiple metal ion resistance, such as those identified in *Alcaligenes* spp. (Nies and Silver 1995), and we can therefore anticipate the characterization of additional broad-specificity metalloregulatory proteins.

The next several years promise to see the identification and characterization of many new metalloregulatory proteins and the elucidation of their mechanisms of highly selective and reversible metal ion recognition. The lessons learned from these systems will aid those aspiring to engineer metal-binding sites in proteins and contribute to a better understanding of the complex integration of signals controlling bacterial gene expression.

Acknowledgment
The work in my laboratory on metalloregulation is supported by the National Science Foundation (MCB-9630411).

References

Altuvia, S., M. Almirón, G. Huisman, R. Kolter, and G. Storz. 1994. The *dps* promoter is activated by OxyR during growth and by IHF and σ^S in stationary phase. *Mol. Microbiol.* 13:265–272.

Ames, B. N., M. K. Shigenaga, and T. M. Hagen. 1993. Oxidants, antioxidants, and the degenerative diseases of aging. *Proc. Natl. Acad. Sci. U.S.A.* 90:7915–7922.

Angerer, A., S. Enz, M. Ochs, and V. Braun. 1995. Transcriptional regulation of ferric citrate transport in *Escherichia coli* K-12. FecI belongs to a new subfamily of σ^{70}-type factors that respond to extracytoplasmic stimuli. *Mol. Microbiol.* 18:163–174.

Ansari, A. Z., J. E. Bradner, and T. V. O'Halloran. 1995. DNA-bend modulation in a repressor-to activator switching mechanism. *Nature* 374:371–375.

Ansari, A. Z., M. L. Chael, and T. V. O'Halloran. 1992. Allosteric underwinding of DNA is a critical step in positive control by Hg-MerR. *Nature* 355:87–89.

Archibald, F. S. and I. Fridovich. 1981. Manganese, superoxide dismutase, and oxygen tolerance in some lactic acid bacteria. *J. Bacteriol.* 146:928–936.

Babcock, M. J., M. McGrew, and J. L. Schottel. 1992. Identification of a protein-binding sequence involved in expression of an esterase gene from *Streptomyces scabies*. *J. Bacteriol.* 174:4287–4293

Bagg, A. and J. B. Neilands. 1987a. Ferric uptake regulation protein acts as a repressor, employing iron(II) as a cofactor to bind the operator of an iron transport operon in *Escherichia coli*. *Biochemistry* 26:5471–5477.

Bagg, A. and J. B. Neilands. 1987b. Molecular mechanism of regulation of siderophore-mediated iron assimilation. *Microbiol. Rev.* 51:509–518.

Bairoch, A. 1993. A possible mechanism for metal-ion induced DNA-protein dissociation in a family of prokaryotic transcriptional regulators. *Nucl. Acids Res.* 21:2515.

Berg, J. M. and Y. Shi. 1996. The galvanization of biology: A growing appreciation for the roles of zinc. *Science* 271:1081–1085.

Brickman, T. J. and S. K. Armstrong. 1995. *Bordetalla pertusis fur* gene restores iron repressibility of siderophore and protein expression to deregulated *Bordetella bronchiseptica* mutants. *J. Bacteriol.* 177:268–270.

Bsat, N., L. Chen, and J. D. Helmann. 1996. Mutation of the *Bacillus subtilis* alkyl hydroperoxide reductase (*ahpCF*) operon reveals compensatory interactions among hydrogen peroxide stress genes. *J. Bacteriol.* 178:6579–6586.

Cadenas, E. 1989. Biochemistry of oxygen toxicity. *Ann. Rev. Biochem.* 58:79–110.

Carayre, S. D. and J. B. Neilands. 1991. Structure-activity correlations for the ferric uptake repressor (Fur) protein for *Escherichia coli* K12, in *Iron Biominerals,* eds. R. B. Frankel and R. P. Blakemore p. 387–396. Plenum Press; New York.

Caslake, L. F., S. I. Ashraf, and A. O. Summers. 1997. Mutations in the alpha and sigma-70 subunits of RNA polymerase affect expression of the *mer* operon. *J. Bacteriol.* 179:1787–1795.

Chen, L. and J. D. Helmann. 1995. *Bacillus subtilis* MrgA is a Dps(PexB) homologue: Evidence for metalloregulation of an oxidative stress gene. *Mol. Microbiol.* 18:295–300.

Chen, L., L. P. James, and J. D. Helmann. 1993. Metalloregulation in *Bacillus subtilis:* Isolation and characterization of two genes differentially repressed by metal ions. *J. Bacteriol.* 175:5428–5437.

Chen, L., L. Keramati, and J. D. Helmann. 1995. Coordinate regulation of *Bacillus subtilis* peroxide stress genes by hydrogen peroxide and metal ions. *Proc. Natl. Acad. Sci. U.S.A.* 92:8190–8194.

Chu, L., D. Mukhopadhyay, H. Yu, K.-S. Kim, and T. K. Misra. 1992. Regulation of the *Staphylococcus aureus* plasmid pI258 mercury resistance operon. *J. Bacteriol.* 174:7044–7047.

Comess, K. M., L. M. Shewchuk, K. Ivanetich, and C. T. Walsh. 1994. Construction of a synthetic gene for the metalloregulatory protein MerR and analysis of regionally mutated proteins for transcriptional regulation. *Biochemistry* 33:4175–4186.

Cornelissen, C. N., and P. F. Sparling. 1994. Iron piracy: Acquisition of transferrin-bound iron by bacterial pathogens. *Mol. Microbiol.* 14:843–850.

Coy, M. and J. B. Neilands. 1991. Structural dynamics and functional domains of the Fur protein. *Biochem.* 30:8201–8210.

Coy, M., C. Doyle, J. Besser, and J. B. Neilands. 1994. Site-directed mutagenesis of the ferric uptake regulation gene of *Escherichia coli*. *BioMetals* 7:292–298.

Culotta, V. C., H.-D. Joh, S.-J. Lin, K. H. Slekar, and J. Strain. 1995. A physiological role for *Saccharomyces cerevisiae* copper/zinc superoxide dismutase in copper buffering. *J. Biol. Chem.* 270:29991–29997.

Cunliffe, H. E., T. R. Merriman, and I. L. Lamont. 1995. Cloning and characterization of *pvdS,* a gene required for pyoverdine synthesis in *Pseudomonas aeruginosa:* PvdS is probably an alternative sigma factor. *J. Bacteriol.* 177:2744–2750.

Demple, B., and C. F. Amábile-Cuevas. 1991. Redox-redux: The control of oxidative stress responses. *Cell* 67:837–839.

DiDonato, M., and B. Sarkar. 1997. Copper transport and its alterations in Menkes and Wilson diseases. *Biochem. Biopys. Acta* 1360:3–16.

Ding, H., E. Hidalgo, and B. Demple. 1996. The redox state of the [2Fe-2S] clusters in SoxR protein regulates its activity as a transcription factor. *J. Biol. Chem.* 271:33173–33175.

Ding, X., H. Zeng, N. Schiering, D. Ringe, and J. R. Murphy. 1996. Identification of the primary metal ion-activation sites of the diphtheria *tox* repressor by X-ray crystallography and site-directed mutagenesis. *Nature Struct. Biol.* 3:382–387.

Dowds, B. C. A. 1994. The oxidative stress response in *Bacillus subtilis*. *FEMS Microbiol. Lett.* 124:255–264.

Dussurget, O., M. Rodriguez, and I. Smith. An *ideR* mutant of *Mycobacterium smegmatis* has derepressed siderophore production and an altered oxidative stress response. *Mol. Microbiol.* 22:535–544.

Eisenstadt, E., S. Fisher, C-L. Der, and S. Silver. 1973. Manganese transport in *Bacillus subtilis* W23 during growth and sporulation. *J. Bacteriol.* 113:1363–1372.

Endo, G., and S. Silver. 1995. CadC, the transcriptional regulatory protein of the cadmium resistance system of *Staphylococcus aureus* plasmid pI258. *J. Bacteriol.* 177:4437–4441.

Erbe, J. L., K. B. Taylor, and L. M. Hall. 1995. Metalloregulation of the cyanobacterial *smt* locus: Identification of SmtB binding sites and direct interaction with metals. *Nucl. Acids Res.* 23:2472–2478.

Felsenstein, J. 1989. PHYLIP—Phylogeny inference package (Version 3.2). *Cladistics* 5:164–166.

Fisher, S., L. Buxbaum, K. Toth, E. Eisenstadt, and S. Silver. 1973. Regulation of manganese accumulation and exchange in *Bacillus subtilis* W23. *J. Bacteriol.* 113:1373–1380.

Fridovich, I. 1995. Superoxide radical and superoxide dismutase. *Ann. Rev. Biochem.* 64:97–112.

Gaudu, P., N. Moon, and B. Weiss. 1997. Regulation of the *soxRS* oxidative stress regulon. Reversible oxidation of the Fe-S centers of SoxR in vivo. *J. Biol. Chem.* 272:5082–5086.

Guerinot, M. L. 1994. Microbial iron transport. *Ann. Rev. Microbiol.* 48:743–772.

Günter, K., C. Toupet, and T. Schupp. 1993. Characterization of an iron-regulated promoter involved in desferrioxamine B synthesis in *Streptomyces pilosus*: Repressor-binding site and homology to the diphtheria toxin gene promoter. *J. Bacteriol.* 175:3295–3302.

Günter-Seeboth, K., and T. Schupp. 1995. Cloning and sequence analysis of the *Corynebacterium diphtheriae dtxR* homologue from *Streptomyces lividans* and *Streptomyces pilosus* encoding a putative iron repressor protein. *Gene* 166:177–199.

Hantke, K. 1987. Selection procedure for deregulated iron transport mutants (*fur*) in *Escherichia coli* K 12: *fur* not only affects iron metabolism. *Mol. Gen. Genet.* 210:135–139.

Hartford, O. M., and B. C. A. Dowds. 1994. Isolation and characterization of a hydrogen peroxide resistant mutant of *Bacillus subtilis*. *Microbiology* 140:297–304.

Helmann, J. D., B. T. Ballard, and C. T. Walsh. 1990. The *Bacillus* species RC607 MerR protein binds Hg(II) as a tricoordinate, metal-bridged dimer. *Science* 247:946–948.

Helmann, J. D., Y. Wang, I. Mahler, and C. T. Walsh. 1989. Homologous metalloregulatory proteins from both Gram-positive and Gram-negative bacteria control transcription of mercury resistance operons. *J. Bacteriol.* 171:222–229.

Heltzel, A., I. W. Lee, P. A. Toti, and A. O. Summers. 1990. Activator-dependent preinduction binding of σ-70 RNA polymerase at the metal-regulated *mer* promoter. *Biochem.* 29:9572–9584.

Holm, L., C. Sander, H. Rüterjans, M. Schnarr, R. Fogh, R. Boelens, and R. Kaptein. 1994. LexA repressor and iron uptake regulator from *Escherichia coli:* New members of the CAP-like DNA binding domain superfamily. *Protein Engin.* 7:1449–1453.

Imlay, J. A., S. M. Chin, and S. Linn. 1988. Toxic DNA damage by hydrogen peroxide through the Fenton reaction in vivo and in vitro. *Science* 240:640–642.

Imlay, J. A., and S. Linn. 1988. DNA damage and oxygen radical toxicity. *Science* 240:1302–1309.

Ito, T., and J. B. Neilands. 1958. Products of "low-iron fermentation" with *Bacillus subtilis:* Isolation, characterization and synthesis of 2,3-dihydroxybenzoylglycine. *J. Am. Chem. Soc.* 80:4645–4647.

Ivey, D. M., A. A. Guffanti, Z. Shen, N. Kudyan, and T. A. Krulwich. 1992. The *cadC* gene product of alkalophilic *Bacillus firmus* OF4 partially restores Na⁺ resistance to an *Escherichia coli* strain lacking an Na⁺/H⁺ antiporter (NhaA). *J. Bacteriol.* 174:4878–4884.

Keyer, K., and J. A. Imlay. 1996. Superoxide accelerates DNA damage by elecating free-iron levels. *Proc. Natl. Acad. Sci. U.S.A.* 93:13635–13640.

Keyer, K., A. S. Gort, and J. A. Imlay. 1995. Superoxide and the production of oxidative DNA damage. *J. Bacteriol.* 177:6782–6790.

Koster, M., K. van Klompemburg, W. Bitter, J. Leong, and P. J. Weisbeek. 1994. Role of the outer membrane ferric siderophore receptor PupB in signal transduction across the bacterial cell envelope. *EMBO J.* 13:2805–2813.

Laddaga, R. A., L. Chu, T. Misra, and S. Silver. 1987. Nucleotide sequence and expression of the mercurial-resistance operon from *Staphylococcus aureus* plasmid pI258. *Proc. Natl. Acad. Sci. U.S.A.* 84:5106–5110.

Lam, M. S., C. M. Litwin, P. A. Carroll, and S. B. Calderwood. 1994. *Vibrio cholerae fur* mutations associated with loss of repressor activity: Implications for the structural-functional relationships of Fur. *J. Bacteriol.* 176:5108–5115.

Lebrun, M., A. Audurier, and P. Cossart. 1994a. Plasmid-borne cadmium resistance genes in *Listeria monocytogenes* are similar to *cadA* and *cadC* of *Staphylococcus aureus* and are induced by cadmium. *J. Bacteriol.* 176:3040–3048.

Lebrun, M., A. Audurier, and P. Cossart. 1994b. Plasmid-borne cadmium resistance genes in *Listeria monocytogenes* are present on Tn5442, a novel transposon closely related to Tn917. *J. Bacteriol.* 176:3049–3061.

Lee, I. W., V. Livrelli, S. J. Park, P. A. Totis, and A. O. Summers. 1993. *In vivo* DNA-protein interactions at the divergent mercury resistant *mer* promoters II. Repressor-activator MerR-RNA polymerase interaction with *merOP* Mutants. *J. Biol. Chem.* 268:2632–2639.

Liesegang, H., K. Lemke, R. A. Siddiqui, and H.-G. Schlegel. 1993. Characterization of the inducible nickel and cobalt determinant *cnr* from pMOL28 of *Alcaligines eutrophus* CH34. *J. Bacteriol.* 175:767–778.

Lin, S.-J., and V. C. Culotta. 1995. The ATX1 gene of *Saccharomyces cerevisiae* encodes a small metal homeostasis factor that protects cells against reactive oxygen toxicity. *Proc. Natl. Acad. Sci. U.S.A.* 92:3784–3788.

Liu, X.-F., and V. C. Culotta. 1994. The requirement for yeast superoxide dismutase is bypassed through mutations in BSD2, a novel metal homeostasis gene. *Mol. Cell. Biol.* 14:7037–7045.

Lonetto, M. A., K. L. Brown, K. E. Rudd, and M. J. Buttner. 1994. Analysis of the *Streptomyces coelicolor sigE* gene reveals the existence of a subfamily of eubacterial σ factors involved in the regulation of extracytoplasmic functions. *Proc. Natl. Acad. Sci. U.S.A.* 91:7573–7577.

Lund, P., and N. Brown. 1989. Up-promoter, utations in the positively-regulated *mer* promoter of Tn*501. Nucl. Acids Res.* 17:5517–5527.

Misra, T. K. 1992. Bacterial resistances to inorganic mercury salts and organomercurials. *Plasmid* 27:4–16.

Nakamura, K. and S. Silver. 1994. Molecular analysis of mercury-resistant bacillus isolates from dediment of Minimata Bay, Japan. *Appl. Environ. Microbiol.* 60:4596–4599.

Neilands, J. B. 1995. Siderophores: structure and function of microbial iron transport compounds. *J. Biol. Chem.* 270:26723–26726.

Ni Bhriain, N. A. M., S. Silver, and T. J. Foster. 1983. Tn5 insertion mutations in the mercuric resistance gene derived from plasmid R100. *J. Bacteriol.* 155:690–703.

Nies, D. H., and S. Silver. 1995. Ion efflux systems involved in bacterial metal resistance. *J. Ind. Microbiol.* 14:186–189.

Nucifora, G., L. Chu, T. K. Misra, and S. Silver. 1989. Cadmium resistance from *Staphylococcus aureus* plasmid pI258 results from a cadmium-Efflux ATPase. *Proc. Natl. Acad. Sci. U.S.A.* 86:3544–3548.

Odermatt, A., and M. Solioz. 1995. Two *trans*-acting metalloregulatory proteins controlling expression of the copper-ATPases of *Enterococcus hirae. J. Biol. Chem.* 270:4349–4354.

Oguiza, J. A., X. Tao, A. T. Marcos, J. F. Martin, and J. R. Murphy. 1995. Molecular cloning, DNA sequence analysis, and characterization of the *Corynebacterium diphtheriae dtxR* homolog from *Brevibacterium lactofermentum. J. Bacteriol.* 177:465–467.

O'Halloran, T., and C. T. Walsh. 1987. Metalloregulatory DNA-binding protein encoded by the *merR* gene: Isolation and characterization. *Science* 235:211–214.

O'Halloran, T. V. 1993. Transition metals in control of gene rxpression. *Science* 261:715–725.

O'Halloran, T. V., B. Frantz, M. K. Shin, D. M. Ralston, and J. C. Wright. 1989. The MerR heavy metal receptor mediates positive activation in a topologically novel transcription complex. *Cell* 56:119–129.

Pappenheimer, A. M., Jr. 1936. Studies in diphtheria toxin production. I. The effect of iron and copper. *Br. J. Exp. Pathol.* 17:335–341.

Pappenheimer, A. M., Jr. 1977. Diphtheria toxin. *Ann. Rev. Biochem.* 46:69–94.

Park, S.-J., J. Wireman, and A. O. Summers. 1992. Genetic analysis of the Tn*21 mer* operator-promoter. *J. Bacteriol.* 174:2160–2171.

Parkhill, J., A. Z. Ansari. J. G. Wright, N. L. Brown, and T. V. O'Halloran. 1993. Construction and characterization of a mercury-independent MerR activator (MerR[AC]): Transcriptional activation in the absence of Hg(II) is accompanied by DNA distortion. *EMBO J.* 12:413–421.

Peters, W. J., and R. A. J. Warren. 1968. Itoic acid biosynthesis in *Bacillus subtilis. J. Bacteriol.* 95:360–366.

Privalle, C. T. and I. Fridovich. 1993. Iron specificity of the Fur-dependent regulation of the biosynthesis of the manganese-containing superoxide dismutase in *Escherichia coli. J. Biol. Chem.* 268:5178–5181.

Qiu, X., C. L. M. J. Verlinde, S. Zhang, M. P. Schmitt, R. K. Holmes, and W. G. J. Hol. 1995. Three-dimensional structure of the diphtheria toxin repressor in complex with divalent cation co-repressors. *Structure* 3:87–100.

Qiu, X., E. Pohl, R. K. Holmes, and W. G. J. Hol. 1996. High-resolution structure of the diphtheria toxin repressor complexed with cobalt and manganese reveals an SH3-like third domain and suggest a possible role of phosphate as co-corepressor. *Biochem.* 35:12292–12302.

Quirk, P. G., A. A. Guffanti, S. Clejan, J. Cheng, and T. A. Krulwich. 1994. Isolation of Tn*917* insertional mutants of *Bacillus subtilis* that are resistant to the protonophore carbonyl cyanide *m*-chlorophenylhydrazone. *Biochim. et Biophys. Acta* 1186:27–34.

Ralston, D. M., and T. V. O'Halloran. 1990. Ultrasensitivity and heavy-metal selectivity of the allosterically modulated MerR transcription complex. *Proc. Natl. Acad. Sci. U.S.A.* 87:3846–3850.

Ross, W., S.-J. Park, and A. O. Summers. 1989. Genetic analysis of transcriptional activation and repression in the Tn*21 mer* operon. *J. Bacteriol.* 171:4009–4018.

Rowland, B. M., and H. W. Taber. 1996. Duplicate isochorismate synthase genes of *Bacillus subtilis:* Regulation and involvement in the biosynthesis of menaquinone and 2,3-dihydroxybenzoate. *J. Bacteriol.* 178:854–861.

Sadowsky, M. J., P. B. Cregan, M. Gottfert, A. Sharma, D. Gerhold, F. Rodriguez-Quinones, H. H. Keyser, H. Hennecke, and G. Stacey. 1991. The *Bradyrhizobium japonicum nolA* gene and its involvement in the genotype-specific nodulation of soybeans. *Proc. Natl. Acad. Sci. U.S.A.* 88:637–641.

Saito, T., M. R. Wormald, and R. J. P. Williams. 1991a. Some structural features of the iron-uptake regulation protein. *Eur. J. Biochem.* 197:29–38.

Saito, T., D. Duly, and R. J. P. Williams. 1991b. The histidines of the iron-uptake regulation protein, Fur. *Eur. J. Biochem.* 197:39–42.

Schäffer, S., K. Hantke, and V. Braun. 1985. Nucleotide sequence of the iron regulatory gene *fur. Mol. Gen. Genet.* 200:110–113.

Schiering, N., W. Kabsch, M. J. Moore, M. D. Distefano, C. T. Walsh, and E. F. Pai. 1991. Structure of the detoxification catalyst mercuric ion reductase from *Bacillus* sp. strain RC607. *Nature* 352:168–172.

Schiering, N., X. Tao, H. Zeng, J. R. Murphy, G. A. Petsko, and D. Ringe. 1995. Structures of the apo- and the metal ion-activated forms of the diphtheria *tox* repressor from *Corynebacterium diphtheriae*. *Proc. Natl. Acad. Sci. U.S.A.* 92:9843–9850.

Schmidt, T. and H. G. Schlegel. 1994. Combined nickel-cobalt-cadmium resistance encoded by the *ncc* locus of *Alcaligenes xylosoxidans* 31A. *J. Bacteriol.* 176:7045–7054.

Schmitt, M. P. 1997. Utilization of host iron sources by *Corynebacterium diphtheriae:* Identification of a gene whose product is homologous to eukaryotic heme oxygenases and is required for acquisition of iron from heme and hemoglobin. *J. Bacteriol.* 179:838–845.

Schmitt, M. P., and R. K. Holmes. 1993. Analysis of diphtheria toxin repressor-operator interactions and characterization of a mutant repressor with decreased binding activity for divalent metals. *Mol. Microbiol.* 9:173–181.

Scmitt, M. P. and R. K. Holmes. 1994. Cloning, sequence, and footprint analysis of two promoter/operators from *Corynebacterium diphtheriae* that are regulated by the diphtheria toxin repressor (DtxR) and iron. *J. Bacteriol.* 176:1141–1149.

Schmitt, M. P., M. Predich, L. Doukhan, I. Smith, and R. K. Holmes. 1995. Characterization of an iron-dependent regulatory protein (IdeR) of *Mycobacterium tuberculosis* as a functional homolog of the diphtheria toxin repressor (DtxR) from *Corynebacterium diphtheriae*. *Infect. Imm.* 63:4284–4289.

Schneider, R., and K. Hantke. 1993. Iron-hydroxamate uptake systems in *Bacillus subtilis:* Identification of a lipoprotein as part of a binding protein-dependent transport system. *Mol. Microbiol.* 8:111–121.

Schreier, H. J., S. W. Brown, K. D. Hirschi, J. F. Nomellini, and A. L. Sonenshein. 1989. Regulation of *Bacillus subtilis* glutamine synthetase gene expression by the product of the *glnR* gene. *J. Mol. Biol.* 210:51–63.

Sedlmeier, R., and J. Altenbuchner. 1992. Cloning and DNA sequence analysis of the mercury resistance genes of *Streptomyces lividans*. *Mol. Gen. Genet.* 236:76–85.

Sexton, R., P. R. Gill, Jr., M. J. Callanan, D. J. O'Sullivan, D. N. Dowling, and F. O'Gara. 1995. Iron-responsive gene expression in *Pseudomonas fluorescens* M114: Cloning and characterization of a transcription activating factor. *Mol. Microbiol.* 15:297–306.

Shewchuk, L. M., J. D. Helmann, W. Ross, S. J. Park, A. O. Summers, and C. T. Walsh. 1989b. Transcriptional switching by the MerR protein: Activation and repression mutants implicate distinct DNA and mercury(II) binding domains. *Biochem.* 28:2340–2344.

Shewchuk, L. M., G. L. Verdine, and C. T. Walsh. 1989a. Transcriptional switching by the metalloregulatory MerR protein: Initial characterization of DNA and mercury(II)-binding activities. *Biochem.* 28:2331–2339.

Shi, W., J. Dong, R. A. Scott, M. Y. Ksenzenko, and B. P. Rosen. 1996. The role of arsenic-thiol interactions in metalloregulation of the *ars* operon. *J. Biol. Chem.* 271:9291–9297.

Shi, W., J. Wu, and B. P. Rosen. 1994. Identification of a putative metal binding site in a new family of metalloregulatory proteins. *J. Biol. Chem.* 269:19826–19829.

Shigenaga, M. K., T. M. Hagen, and B. N. Ames. 1994. Oxidative damage and mitochondrial decay in aging. *Proc. Natl. Acad. Sci. U.S.A.* 91:10771–10778.

Silver, S., and G. Ji. 1994. Newer systems for bacterial resistance to toxic heavy metals. *Environment. Health Perspect.* 102 suppl. 3:107–113.

Silver, S., G. Nucifora, and L. T. Phung. 1993. Human Menkes' X-chromosome disease and the staphylococcal cadmium-resistant ATPase: A remarkable similarity in protein sequences. *Mol. Microbiol.* 10:7–12.

Silver, S. and M. Walderhaug. 1992. Gene regulation of plasmid-and chromosome-determined inorganic ion transport in bacteria. *Microbiol. Rev.* 56:195–228.

Skinner, J. S., E. Ribot, and R. A. Laddaga. 1991. Transcriptional analysis of the *Staphylococcal aureus* plasmid PI258 mercury resistance determinant. *J. Bacteriol.* 173:5234–5238.

Smith, D. H. 1967. R factors mediate resistance to mercury, nickel, and cobalt. *Science* 156:1114–1116.

Stadtman, E. R. 1992. Protein oxidation and aging. *Science* 257:1220–1224.

Stadtman, E. R. and B. S. Berlett. 1991. Fenton chemistry: amino acid oxidation. *J. Biol. Chem.* 266:17201–17211.

Staggs, T. M., and R. D. Perry. 1992. Fur regulation in *Yersinia* species. *Mol. Microbiol.* 6:2507–2516.

Stearman, R., D. S. Yuan, Y. Yamaguchi-Iwai, R. D. Klausner, and A. Dancis. 1996. A permease-oxidase complex involved in high-affinity iron uptake in yeast. *Science* 271:1552–1557.

Stojilkovic, I. and K. Hantke. 1995. Functional domains of the *Escherichia coli* ferric uptake regulator protein (Fur). *Mol. Gen. Genet.* 247:199–205.

Strausak, D., and M. Solioz. 1997. CopY is a copper-inducible repressor of the *Enterococcus hirae* copper ATPases. *J. Biol. Chem.* 272:8932–8936.

Summers, A. O. 1986. Organization, expression, and evolution of genes for mercury resistance. *Ann. Rev. Microbiol.* 40:607–634.

Summers, A. O. 1992. Untwist and shout: A heavy metal-responsive transcriptional regulator. *J. Bacteriol.* 174:3097–3101.

Tai, S. P. S., A. E. Krafft, P. Nootheti, and R. K. Holmes. 1990. Coordinate regulation of siderophore and diphtheria toxin production by iron in *Corynebacterium diphtheriae*. *Microb. Pathog.* 9:267–273.

Tai, S. S. and Y. Y. Zhu. 1995. Cloning of a *Corynebacterium diphtheriae* iron-repressible gene that shares sequence homology with the AphC subunit of alkyl hydroperoxide reductase of *Salmonella typhimurium*. *J. Bacteriol.* 177:3512–3517.

Tao, X. and J. R. Murphy. 1994. Determination of the minimal essential nucleotide sequence for diphtheria *tox* repressor binding by in vitro affinity selection. *Proc. Natl. Acad. Sci. U.S.A.* 91:9646–9650.

Tao, X., N. Schiering, H. Zeng, D. Ringe, and J. Murphy. 1994. Iron, DtxR, and the regulation of diphtheria toxin expression. *Mol. Microbiol.* 14:191–197.

Thompson, J. D., D. G. Higgins, and T. J. Gibson. 1994. CLUSTAL W: Improving the sensitivity of progressive multiple sequence alignment through sequence weighting, positions-specific gap penalties and weight matrix choice. *Nucl. Acids Res.* 22:4673–4680.

Tomalsky, M. E., A. M. Wertheimer, L. A. Actis, and J. H. Crosa. 1994. Characterization of the *Vibrio anguillarum fur* gene: Role in the regulation of expression of the FatA outer membrane protein and catechols. *J. Bacteriol.* 176:213–220.

Touati, D., M. Jacques, B. Tardat, L. Bouchard, and S. Despied. 1995. Lethal oxidative damage and mutagenesis are generated by iron in Δ*fur* mutants of *Escherichia coli:* Protective role of superoxide dismutase. *J. Bacteriol.* 177:2305–2314.

Utschig, L. M., J. W. Bryson, and T. V. O'Halloran. 1995. Mercury-199 NMR of the metal receptor site in MerR and its protein-DNA complex. *Science* 268:380–385.

Venturi, V., C. Ottevanger, M. Bracke, and P. Weisbeek. 1995. Iron regulation of siderophore biosynthesis and transport in *Pseudomonas putida* WCS358: Involvement of a transcriptional activator and of the Fur protein. *Mol. Microbiol.* 15:1081–1093.

Walsh, B. L., W. J. Peters, and R. A. J. Warren. 1971. The regulation of phenolic acid synthesis in *Bacillus subtilis. Can. J. Micro.* 17:53–59.

Walsh, C. T., M. D. Distefano, M. J. Moore, L. M. Shewchuk, and G. L. Verdine. 1988. Molecular basis of bacterial resistance to organomercurial and inorganic mercury salts. *FASEB J.* 2:124–130.

Wang, Y., M. Moore, H. S. Levinson, S. Silver, C. Walsh, and I. Mahler. 1989. Nucleotide sequence of a chromosomal mercury resistance determinant from a *Bacillus* sp. with broad-spectrum mercury resistance. *J. Bacteriol.* 171:83–92.

Weinberg, E. D. 1990. Roles of trace metals in transcriptional control of microbial secondary metabolism. *Biol. Metals* 2:191–196.

Wooldridge, K. G., and P. H. Williams. 1993. Iron uptake mechanisms of pathogenic bacteria. *FEMS Microbiol. Rev.* 12:325–348.

Wright, J. G., H.-T. Tsang, J. E. Penner-Hahn, and T. V. O'Halloran. 1990. Coordination chemistry of the Hg-MerR metalloregulatory protein: evidence for a novel tridentate Hg-cysteine receptor site. *J. Am. Chem. Soc.* 112:2434–2435.

Yoon, K. P., T. K. Misra, and S. Silver. 1991. Regulation of the *cadA* cadmium resistance determinant of *Staphylococcus aureus* plasmid pI258. *J. Bacteriol.* 173:7643–7649.

Yu, H., D. Mukhopadhyay, and T. K. Misra. 1994. Purification and characterization of a novel organometallic receptor protein regulating the expression of the broad spectrum mercury-resistance operon of plasmid pDU1358. *J. Biol. Chem.* 269:15697–15702.

4

Two-Component Systems in the Regulation of Heavy Metal Resistance

Dietrich H. Nies and Nigel L. Brown

1. Regulation of Bacterial Heavy Metal Resistances: An Introduction

Microbial resistance to heavy metals is widespread. This is not surprising, as early in evolutionary history microorganisms would have been in contact with toxic concentrations of heavy metals. Geochemical events caused the release of heavy metals from the earth's crust into the biosphere, and still do. The change from an anoxic to an oxic biosphere altered the redox state and biological availability of a number of heavy metals. Resistances to a wide range of toxic metal ions have been reported in bacteria. These include resistances to metals that are purely toxic, with no ascribed biological function (such as mercury and cadmium), and to metals that are toxic in excess but are required in small amounts for the correct functioning of the bacterial cell (such as copper and zinc) (Silver and Ji 1994; Silver and Walderhaug 1992). Bacteria, and other unicellular organisms, must have mechanisms which allow them to avoid heavy metal toxicity, yet still accumulate enough of the essential heavy metals to allow normal cell growth.

As the bacterial cell does not have discrete cytoplasmic compartments, an efficient resistance system must be specific for one, or a small group of, heavy metals. Otherwise, the mechanisms conferring resistance might deplete the cell of an essential metal. For toxic, but essential metals, expression of a resistance system must be not only specific, but carefully regulated, such that it does not deplete the cell of the target metal. All heavy metal resistance systems that have been studied are carefully regulated by the heavy metal at the level of transcription.

There are a lot of definitions of "heavy metal", but the best relates to the density and chemical properties. Heavy metals have a density of more than 5 g cm^{-3}, and with the exception of non metals and the group IA and IIA metals, many elements fit this definition. Most heavy metals are transition elements with incompletely filled d-orbitals. These di-orbitals give heavy metal ions their unique ability to form complexes that are (a) redox active, (b) Lewis acids, (c) or both. This makes heavy metal ions useful catalysts in biochemical pathways, and life is not possible without them. On the other hand, if such heavy metal-catalyzed reactions occur at the wrong place or time, cells can be severely damaged. Thus, many heavy metals are essential but also toxic at higher concentrations.

Of all the heavy metals, only a few have any biological importance. There are three reasons for this. First, the occurrence of any element decreases sharply with increasing atomic mass (Schaifers 1984). Thus, elements with high atomic masses are rare, and high concentrations of these elements usually do not occur in ecosystems. There are a few exceptions to this general rule, such as lithium, beryllium, and boron, which are rare although they have a low atomic mass. These are not of ecological importance. Iron and its neighbors in the Periodic Table are wide spread and are biologically important. Second, because life depends on water, the solubility of a metal ion determines its biological availability. Some divalent and all tri- or tetravalent cations are nearly insoluble at neutral pH values and are not biologically accessible. The third factor is the toxicity. Heavy metal cations with high atomic masses tend to bind strongly to sulfide groups. There is a linear relationship between the minimal inhibitory concentration for *Escherichia coli* of a given heavy metal and the logarithm of the complex binding constant of its sulfide (unpublished results). Thus, "sulfur-lovers" are always very toxic; the higher their affinity for sulfur, the more toxic they are. Elements like mercury, cadmium, and silver are thus too toxic to fulfill physiological functions, and are solely toxic. Arsenate and chromate are also toxic with no beneficial functions, because the oxyanions AsO_4^{3-} and CrO_4^{2-} strongly interfere with the phosphate (PO_4^{3-}) (Silver et al. 1981) or sulfate (SO_4^{2-}) metabolism (Nies and Silver 1995).

The divalent cations of cobalt, nickel, copper, and zinc, however, are medium "sulfur-lovers" and these metals (known as trace elements) have essential functions at low concentrations but are toxic at high concentrations. The intracellular concentration of these metals must be finely adjusted to avoid either metal deprivation or metal toxicity, and careful homeostasis is necessary. In contrast, homeostasis of the purely toxic metals is simple: the cell must quickly eliminate them.

1.1 Mercury, a Purely Toxic Metal

One of the most toxic metal cations is Hg(II). The unique chemical features of mercury are matched by its unique detoxification mechanism (Silver and Walderhaug 1995) which involves uptake of the toxic ion instead of efflux, which is required for most metal resistances (Silver and Walderhaug 1992). Uptake is necessary because cytoplasmic NADPH is required for the reduction of Hg(II) to elemental Hg(0). Uptake into the cell also prevents Hg(II) inhibition of periplasmic or surface proteins.

Bacterial mercury resistance determinants have been described in some detail (Hobman and Brown 1997; Silver and Phung 1996). The proteins for uptake of Hg(II) (MerT, MerP, and MerC) and reduction to Hg(0) (mercuric reductase, MerA) are usually encoded by a single *mer* operon. In broad-spectrum mercury resistance determinants the organomercury lyase enzyme may be encoded in the same *mer* operon. Since Hg(II) is very toxic, induction of the *mer* genes must be very fast, and a tightly controlled strong promoter is needed. The main regulator, MerR, and a co-regulator, MerD, appear to work in concert to control expression of the *mer* genes.

The *mer* promoter has −35 and −10 sequences, which resemble the optimum for a sigma-70 promoter. However, these are spaced too far apart for strong constitutive activity (Parkhill et al. 1993; Summers 1992). The main regulator of *mer* is the MerR protein, which acts as both repressor and activator. MerR binds to an operator within the *mer* promoter and represses transcription. Activation is due to the MerR dimer binding a single Hg(II) ion on three cysteine residues, two on one subunit and the third on the second subunit (Helmann et al. 1990), leading to a MerR-dependent change in conformation of the DNA-(MerR)$_2$-complex (O'Halloran et al. 1989). The DNA is bent and twisted and the −35 and −10 sites become accessible for binding by RNA polymerase holoenzyme (Ansari et al. 1995; Ansari et al. 1992; Summers 1992). The preassembly of DNA, RNA polymerase, and MerR at the promoter before activation by the binding of a single mercuric ion allows a rapid (hypersensitive) response of the *mer* promoter to increases in mercuric ion concentration (Ralston and O'Halloran 1990; Summers 1992). This ensures that the resistance genes are fully switched on before the toxic effects of mercuric salts are fully realized, without having to switch on the operon at very low Hg(II) concentrations.

Due to the high affinity to Hg(II) and sulfide, Hg(II) is bound tightly to MerR dimers. Deactivation of the *mer* promoter may in part be due to binding of MerR which is not bound to Hg(II) and has a higher binding constant for the *mer* operator (Parkhill et al. 1993); it may also be due in part to the MerD protein, a coregulator of *mer*. MerD apparently competes with MerR for the operator site, but is a repressor only, and may prevent further expression of the *mer* determinant (Mukhopadhyay et al. 1991).

1.2 Arsenate and the Phosphate Problem

Resistance to arsenate also involves a coregulator that limits activity of the main regulator (Rosen et al. 1995). Arsenate is structurally similar to phosphate and is taken up by phosphate transport systems. As in many other cases, phosphate and arsenate are transported into the cell by a fast, relatively nonspecific constitutive uptake system. Under conditions where phosphate is limiting for growth, a specific uptake system is induced. High-affinity systems usually consume more energy than the fast unspecific systems. In *E. coli*, the Pit-system is the fast (high V_{max}) nonspecific uptake system for phosphate, and the inducible high-affinity system is Pst. Pst shows high discrimination between phosphate and arsenate. As with other ions, accumulation via the constitutive nonspecific system cannot be shut down if the concentration of arsenate becomes high, and accumulation of the (now toxic) arsenate ion continues (Nies and Silver 1995).

Within the cell, the problem of the discrimination of arsenate and phosphate continues, because of their similar size, shape, and charge. How can a regulator or a transporter bind arsenate with high specificity in a high-phosphate background? These oxyanions can readily be differentiated by a chemical reaction: arsenate As(V) is reduced to arsenite As(III), catalyzed by the ArsC arsenate reductase. The cofactor for this reaction is glutaredoxin in Gram-negative or thioredoxin in Gram-positive bacteria (Gladysheva et al. 1994; Ji and Silver 1992).

Arsenite has no structural homolog in the cell. It serves as an inducer for the ArsR repressor and as the substrate for the ArsB membrane transport protein, which exports the toxic anion. The export of any anion decreases the charge gradient across the cytoplasmic membrane, so the movement of arsenite by ArsB is driven by the chemiosmotic gradient. In some cases, an alternative energy source is used to drive out the toxic arsenite: besides the chemiosmotic gradient, ATP is used by the ArsA-ATPase-plus-ArsB-complex for arsenite efflux. This sophisticated ArsAB transport system is more energy expensive than ArsB alone. The sophisticated Ars systems are not only controlled by the ArsR transcriptional repressor protein, but also by a second protein, the ArsD corepressor, which limits the upper level response of the system (Rosen et al. 1995).

1.3 Other Systems

There are other solely toxic heavy metal ions of environmental importance: Cd(II), Ag(I), Cr(VI), and Pb(II). Cadmium resistance in Gram-positive bacteria is governed by the CadA protein, which is a Cd(II) [and Zn(II)]-effluxing P-type ATPase (Silver and Walderhaug 1995; Solioz and Vulpe 1996). Expression of CadA is regulated by CadC (Endo and Silver 1995), a repressor related in sequence to ArsR (Rosen et al. 1995). A third member of the small ArsR family is SmtB, the regulator for metallothionein synthesis in the cyanobacterium *Synechococcus* (Morby et al. 1993; Silver and Phung 1996) (see chapter by Robinson et al. in this volume).

There is little information available about the molecular biology of bacterial lead resistance, and this was also true for silver resistance until recently. Appropriately, silver resistance has been cloned in Simon Silver's group, and DNA sequencing is complete but not yet published. The *sil* resistance system seems similar to the cobalt-zinc-cadmium resistance system Czc (see below, S. Silver, personal communication). The Czc, copper, and silver resistance systems are regulated by two-component sensor/responder complexes (see next section).

Chromate is reduced by bacterial cells, but there is also decreased accumulation of chromate which is the bases of plasmid chromate resistance (Nies and Silver 1995). In *Alcaligenes eutrophus,* both reduction and resistance occur (N. Peitzsch, G. Eberz, and D. H. Nies, in preparation). However, they are not linked as is the case for arsenate resistance. Chromate is reduced with highest efficiency when the cells are starved for sulfur. Under these conditions, the *chr* resistance system is not induced (N. Peitzsch, G. Eberz, and D. H. Nies, in preparation). The *chr* system is based on reduced accumulation (probably by efflux) of chromate catalyzed by the ChrA transporter (Cervates and Silver 1992; Nies and Silver 1995).

Regulation of *chr* is still unclear. The *chr* determinant of *A. eutrophus* is located on a megaplasmid adjacent to another heavy metal resistance determinant, *cnr,* which encodes resistance to cobalt and nickel. The *cnr* and *chr* determinants are transcribed from opposite DNA-strands (that is in different directions), and the genes regulating these determinants may be the first genes located in the regions upstream of the genes for membrane transport.

There are three possible proteins involved in regulation of *cnr:* CnrH, which is a sigma factor (accessory component of RNA polymerase) of the extracellular function family (Lonetto et al. 1994), CnrX, which might be an antisigma factor (Liesegang et al. 1993), and the product of Orf150, which is located upstream of the *cnrXH* genes. The putative Orf150 product is a hydrophilic protein with a molecular weight of ca. 15,000. Orf150 product may be another DNA-binding protein.

2. Two-Component Regulatory Systems

Although some of the preceding resistance determinants are regulated by two proteins, e.g., MerR/D or ArsR/D, these are not two-component regulatory systems in the usual sense. This name is normally used to describe systems in which regulation of transcription occurs through a histidine kinase "sensor" and a "response regulator" (Hoch and Silhavy 1995; Nixon et al. 1986). Two-component systems have properties that other regulatory systems lack. First, sensing and transcriptional control are spatially decoupled. The sensor is often located at the cytoplasmic membrane and receives periplasmic and/or cytoplasmic signals. Signal input leads to phosphorylation of a histidine residue of the sensor protein; the phosphate residue is then transferred to an aspartate residue of the response regulator protein (Stock et al. 1995). At the moment, more than 150 two-component regulatory systems are known; and they were also found in yeasts and plants (Brown et al. 1993a; Hoch and Silhavy 1995; Maeda et al. 1994).

This is a real signal-processing pathway. A given signal intensity leads to a corresponding phosphorylation of the response regulator. Many independent processes can interfere with the level of phosphorylation. Histidine kinases of other two-component systems may transfer a phosphate group to the response regulator (cross talk), low-molecular-weight phosphorylated compounds such as acetylphosphate are able to phosphorylate a response regulator directly, which may give informational input about the general physiology of the cell (Stock et al. 1995).

Dephosphorylation of the response regulator can also be controlled. Response regulators can autodephosphorylate with time. The histidine kinase can also stimulate dephosphorylation, and this can depend on a specific signal input. In addition, other proteins may be required for dephosphorylation (Stock et al. 1995). Thus, the phosphorylation level of the response regulator is the product of a multisignal input.

There are four classes of response regulators: the CheY, NtrC, FixJ, and OmpR families (Pao et al. 1994; Volz 1995). CheY-like proteins are short and consist only of the phosphorylation domain. The other three families have at least an additional DNA-binding domain. In the DNA-binding response regulators, the phosphorylation level determines the affinity of the regulator to the DNA and the activity of the response regulator as repressor or activator. The signal output of these DNA-binding response regulators is the frequency of initiation of transcription.

In the following sections, we concentrate on copper and zinc resistance determinants in Gram-negative bacteria and describe the functions of two-component systems in their regulation. Bearing in mind that Cu(II) and Zn(II) are cations which are both toxic and essential, it is clear that bacterial cells use two-component systems to regulate homeostasis of these metal cations. Other metal homeostasis systems which use two-component regulators will undoubtedly be found. One has already been identified: the ferric enterobactin receptor in *Pseudomonas aeruginosa* (Dean et al. 1996; Dean and Poole 1993). In this case the PfeS sensor phosphorylates the PfeR regulator and induces production of the enterobactin receptor PfeA. As mentioned above, another system regulated by sensor/receptor is plasmid/borne silver resistance in plasmids of *E. coli*. Here the *silR* and *silS* genes are thought to regulate expression of the resistance genes (S. Silver, personal communication).

3. Copper Resistance

Plasmid-encoded resistance to copper has been documented in several species, the best studied of which are *Escherichia coli* (Brown et al. 1995; 1993b; Tetaz and Luke 1983), *Pseudomonas syringae* pv. *tomato* (Bender and Cooksey 1986; Cooksey 1993; Cooksey 1994) and *Xanthamonas campestris* pv. *juglandis* (Lee et al. 1994). The mechanism of plasmid-borne resistance to copper has been reviewed, and is best understood for the inducible systems from *P. syringae* and *E. coli.* Recently, information on the regulation of resistance became available in both systems (Mills et al. 1993; 1994; Rouch and Brown 1997). Chromosomal resistance to copper has been described in *Enterococcus hirae* (Odermatt et al. 1992; 1993) and its regulation has also been studied (Odermatt and Solioz 1995; Strausak and Solioz 1997).

The original sequencing of the *cop* resistance determinant of *P. syringae* (Mellano and Cooksey 1988) detected four genes (*copABCD*) responsible for copper resistance, and expressed from a single promoter. Resistance was inducible by copper (Mellano and Cooksey 1988). Of the four structural gene products, CopA and CopB are essential to resistance, whereas CopC and CopD are supplementary for full resistance. CopA and CopC are periplasmic copper binding proteins, and CopB is an outer membrane protein that also binds copper. The mechanism of resistance appears to be binding of excess copper by the CopA and CopB proteins (Figure 4.1). Copper-resistant strains of *P. syringae* on plates containing Cu(II) salts turn blue, from the accumulated copper salts (Cooksey 1993). The CopC (periplasmic) and CopD (inner membrane) proteins are thought to be responsible for continued uptake of copper when the resistance operon is expressed. If CopC and CopD are expressed in the absence of CopA and CopB, then the cells show a copper-hypersensitive phenotype, indicating that additional copper is being transported into the cells (Cha and Cooksey 1993).

Copper resistance in *E. coli* was first identified in enteric bacteria isolated from Australian pig farms. The plasmid-borne copper resistance genes (*pco*) were found on a conjugative plasmid pRJ1004 (Tetaz and Luke 1983), and resistance was inducible by cupric salts (Rouch et al. 1985). The mechanism of copper resistance conferred by the *pco* determinant (Figure 4.1) appears to be somewhat different from that of *cop* in *P. syringae.* In *E. coli,* the colonies turn brown, and resistance involves copper export (Brown et al. 1995). Following induction of *pco* with subtoxic concentrations of copper, exponentially growing copper-resistant *E. coli* accumulate less copper than sensitive cells. The Pco system includes four structural gene products: PcoA, PcoB, PcoC, and PcoD (Figure 4.2), which are homologous to the *P. syringae* CopA, CopB, CopC, and CopD proteins, respectively. All four Pco products are required for copper resistance in *E. coli* and for reduced accumulation of copper (S. R. Barrett and N. L. Brown unpublished). A fifth gene product, PcoE, is a periplasmic copper-binding protein.

Since the predicted gene products are closely related (CopA-PcoA, 76%; CopB-PcoB, 55%; CopC-PcoC, 60%; and CopD-PcoD, 38% amino acid identities), it is not understood why the mechanisms of resistance appear different. Interestingly, there is a major difference in the number of putative copper-binding octapeptide

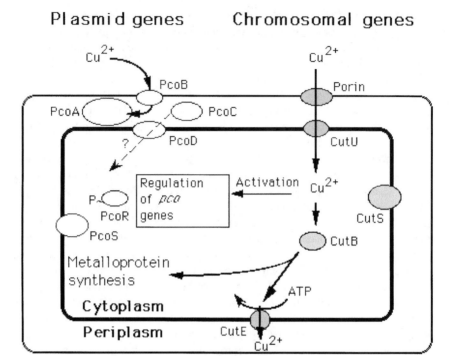

Figure 4.1 Genes and functions for copper resistance and transport in *E. coli;* modified from Silver and Phung (1996) with permission. CutU (uptake) and CutE (efflux) are hypothesized copper transporters, based on the chromosomal products of *Enterococcus hirae* (Odermatt et al. 1993). CutB is a proposed intracellular copper binding protein (Brown et al. 1993b); PcoA, PcoB, PcoC, and PcoE are the copper resistance proteins (Brown et al. 1995). PcoS and PcoR are the plasmid regulatory proteins and CutS is the chromosomal sensor responsible for additional regulation of the plasmid-borne genes (Rouch and Brown 1997), and presumed to regulate the chromosomal genes for copper homeostasis.

sites (consensus DHSXMXGM) observed in CopB and PcoB: CopB has five repeats of this sequence showing good matches to the consensus, whereas PcoB has only one, poorly matching copy. Similar motifs occur in CopA, PcoA, and PcoE. A product equivalent to PcoE has not been found in the *cop* determinant, yet PcoE is produced in large amounts in the periplasmic compartment of *E. coli* (C. J. D. Yates and N. L. Brown unpublished). One can speculate that differences in the chromosomal background between the two species leads to the apparent differences in the phenotype of resistant cells.

The copper resistant determinant of *A. eutrophus* CH34 (Dressler et al. 1991)

Figure 4.2 Genetic structure of the *pco* copper resistance determinant of *E. coli* showing the positions of the genes, the promoters (p), and the rho-independent terminator (t) (Brown et al. 1995; Rouch and Brown 1997). The numbers at the bottom are the positions in kilobasepairs. Modified from (Rouch and Brown 1997).

has recently been cloned and sequenced (T. Borremans and D. van der Lelie, personal communication). The orientation of the operon is *copSR-copABCD-copF* with *copF* encoding a copper ATPase similar to PacS of *Synechococcus* and to the Cu-ATPase involved in Menkes' disease (T. Borremans and D. van der Lelie, personal commun.). This is the first example of a *pco-* or *cop*-like resistance system containing a copper P-type ATPase.

Copper resistance in *Enterococcus hirae* is very different. In this case a chromosomally encoded membrane ATPase is thought to confer resistance by pumping copper out of the cell (Odermatt et al. 1992; 1993; Solioz and Odermatt 1995; Solioz and Vulpe 1996). This system also pumps out silver ions (Solioz and Odermatt 1995). The genes encoding these proteins, also called *copA* and *copB*, are probably better considered as part of the normal mechanisms of copper homeostasis in *E. hirae*, rather than as resistance genes *per se*.

3.1 Regulation of Copper Resistance

The products of the *copR* and *copS* genes of *P. syringae* (Mills et al. 1993) and the comparable *pcoR* and *pcoS* genes of *E. coli* (Brown et al. 1995; 1993b) are by homology members of the two-component regulatory systems (Hoch and Silhavy 1995). CopS/PcoS is the membrane-bound sensor and CopR/PcoR is the aspartophosphate-containing DNA-binding protein. There is a difference, however, in that the *cop* genes of *P. syringae* were expressed from a single promoter (P_{copA}) (Mellano and Cooksey 1988) whereare the *pco* genes of *E. coli* are regulated by copper from two promoters (P_{pcoA} and P_{pcoE}) (Rouch and Brown 1997).

The CopR protein was partially purified (Mills et al. 1994). CopR bound to the P_{copA} promoter, and in addition to a chromosomal homolog promoter (P_{copH}) in DNA mobility shift assays. DNase I footprinting of CopR on P_{copA} and P_{copH}

showed a conserved sequence, a "cop box" (Figure 4.3), centered at −43.5 bp relative to the promoters.

By subcloning fragments of *pco* and by primer extension, two copper-inducible promoters, P_{pcoA} and P_{pcoE}, were identified (Rouch and Brown 1997). These promoters are start points for transcribing the structural genes *pcoABCDRS* and *pcoE*, respectively (Figure 4.2). Evidence that at least the *pcoR* gene as well as *pcoABCD* are transcribed from P_{pcoA} was obtained by using a reporter gene fusion within *pcoR*, however, low level constitutive expression of the *pcoR* reporter gene fusion was seen when P_{pcoA} was deleted (Rouch and Brown 1997).

The PcoR and CopR gene products (26 kDa) are OmpR-like (Pao et al. 1994), and there is a conserved aspartate (Asp$_{52}$ in PcoR) that is the putative phosphorylation site. This group of similar regulators includes OmpR, PhoB, VirG, and CutR, a copper-responsive regulatory locus from *Streptomyces lividans,* which has recently been shown to be involved in the regulation of antibiotic synthesis (Chang et al. 1996). The PcoS and CopS gene products (53 kDa) contain two hydrophobic potential membrane-spanning regions near their N-termini (amino acids 11 to 34 and 169 to 192) and there is the conserved histidine (His257 in PcoS) equivalent to that which is known to be the autophosphorylation site in EnvZ (Forst et al. 1989).

Studies of expression of *pco* in a "minicell" system (Rouch 1986) showed a soluble protein of 25 kDa which was induced in the presence of added copper. Transposon insertion mutants lacking this inducible protein mapped to the region near the 3′ end of *pcoS.* Insertions further upstream in the *pcoS* region still produced the protein, and could complement transposon insertion mutations in the 3′ end of *pcoS.* One explanation may be, that *pcoS* encodes two gene products, one (PcoS) being the full-length protein, and the other (PcoS*) being the C-terminal part of the protein expressed from an internal translational start point. An ATG methionine codon and a good putative ribosome binding sequence were identified in the sequence (Brown et al. 1995) at a location compatible with the synthesis of the PcoS* protein. Such a protein would lack the two hydrophobic regions of PcoS and would not insert into the membrane. It would retain the autophosphorylated histidine site at position 21 in the PcoS* protein. One can speculate that the *pcoS* gene may encode both membrane-bound and soluble cytoplasmic sensors.

Upstream of the P_{pcoA} and P_{pcoE} transcriptional start points were sequences showing similarity to the *cop* boxes identified in *P. syringae*. The consensus between these sequences upstream of P_{copA}, P_{copH}, P_{pcoA}, and P_{pcoE} is shown in Figure 4.3. The dyad symmetrical sequence is centered 43.5 bp upstream of the transcriptional start point of P_{pcoA}, a similar separation to that in P_{copA} and P_{copH}. That in P_{pcoE} is centered 54.5 bp upstream of the transcript start, and is unlikely therefore to interact with the regulatory protein and transcriptional complex in the same way as in the outer promoters. Binding of PcoR at −54.5 may allow

```
                                                                          -35
                                                                  /////////////
pcoA    125   G C T A T G T A C A T T A C A C G A T T G T A A A T G A A T T T G T T T   -35
                          |_____|  /////////////
pcoE   6485   A C C T G G A A G G T G A C A A A A T T G T C A T T C A G T C A C G C G A T A
              ___
copA     54   G T T T T C A A G C T T A C A G A A A A T G T A A A T C G C G C C G C T A
              _____
copH     54   G T C G C C C A G C T T A C G G A A A A T G T A A T T A C C T C G T T G

Consensus      A g - T t A C a - a A - T G T a A A T - a - - - - G
                                    < >
```

Figure 4.3 Alignment of "copper box" sequences from P$_{pcoA}$ and P$_{pcoE}$ from *E. coli*, P$_{copA}$ from the *P. syringae cop* determinant and P$_{copH}$ from the *P. syringae* chromosome. The vertical bars show the extent of DNase footprinting in P$_{copA}$ and P$_{copH}$, and the copper boxes of all four sequences are boxed. The consensus sequence is given underneath, and the −35 sequences of the two *pco* promoters are marked. The common dyad center is shown by arrowheads (<>). Modified from (Rouch and Brown 1997).

direct contact of the regulator with RNA polymerase. OmpR also binds some distance upstream of the *ompF* promoter, although there are multiple binding sites and the exact mode of OmpR-mediated regulation is not clear (Rampersaud et al. 1994).

P_{pcoE} is about 10 times as active as P_{pcoA} in transcriptional assays (Rouch and Brown 1997). This may be reflected in production of the respective gene products, as PcoE is produced in large amounts in the periplasm when expressed from its own promoter (C. J. D. Yates and N. L. Brown, unpublished). PcoE is proposed to function in *E. coli* as the primary protection against slightly elevated external copper concentrations. When the copper concentrations are very high, the PcoA, PcoB, PcoC, and PcoD products are also recruited.

PCR fragments containing progressive deletions of the copper boxes, −35 and −10 sequences, and the transcriptional start points were used to express the *cat* reporter gene (Rouch and Brown 1997). These data showed that copper inducibility was dependent on the copper box. Sequences upstream of this were not apparently required, but once the copper box was deleted, copper-inducible transcription was lost.

Interestingly, the presence of the *pcoRS* genes was not essential to copper-inducible transcription. This suggests that chromosomal genes may function in copper-inducible regulation, and cross-react with the *pcoRS* gene functions. Whether these are similar to the chromosomal homolog *copH* in *P. syringae* is not known, but one can speculate that such genes exist.

In *E. hirae* two metalloregulatory proteins have been identified, CopY and CopZ, which are required for induction of the *copAB* operon by copper. CopY apparently acts as a "metallist" repressor, and CopZ acts as an activator (Odermatt and Solioz 1995). It is not known if such regulators also occur in *E. coli* or *P. syringae*. CopY binds as a homodimer to sequences containing TTACA in inverted repeats at two places in the *copAB* promoter (Strausak and Solioz 1997). This sequence also occurs in the 'copper boxes' of the *pco* and *P. syringae cop* promoters (Rouch and Brown 1997). However, CopY and CopZ are not two-component response regulators of the histidine kinase family, but work in a different way.

The *pco* promoters in high copy number also respond to zinc. Whether this is a direct response or a secondary response due to added zinc affecting the intracellular concentration of copper is not yet known. There is no obvious reason why the *pco* genes should be required in the presence of excess zinc. The response to 0.6 mM added zinc is about 25% of that to 0.6 mM added copper.

The dose-response of the *pco* promoters to copper differs from that of the better studied dose-response, for the *mer* promoter (O'Halloran 1993; Rouch et al. 1995). Whereas the *mer* promoter shows a hypersensitive response both *in vivo* and *in vitro* with an apparent Hill coefficient (a measure of the deviation from normal Michaelis-Menten kinetics) of about 2.6 (Ralston and O'Halloran

1990; Summers 1992), the *pco* promoters have a *hypo*sensitive response. The apparent Hill coefficient is 0.67 (Rouch and Brown 1997; Rouch et al. 1995), and the response is not fully sigmoidal. This makes physiological sense, as induction of the copper resistance genes must be very carefully regulated to prevent both excess copper causing toxicity and the cells becoming copper depleted due to copper being exported from the cell.

At low copper concentrations the normal mechanisms of copper homeostasis in *E. coli* prevent intracellular copper concentrations becoming sufficiently high to cause toxicity, but at high concentrations the *pco* resistance mechanism is induced to export excess copper from the cell. If the *pco* promoters were induced with normal Michaelis-Menten kinetics, the resistance genes might be expressed at too low a copper concentration and cause the cells to become copper deficient. Alternatively, they might not be expressed until toxic effects were already occurring. Having a low and gradually increasing rate of induction with increasing copper concentration supports homeaostasis.

4. The Czc Metal Resistance System of *Alcaligenes eutrophus* CH34

Bacteria related to A. *eutrophus* CH34 have been isolated from various metal-contaminated regions in the world (Diels and Mergeay 1990; Dressler et al. 1991; Kunito et al. 1996; Mergeay et al. 1985). The species "*A. eutrophus*" was recently reclassified as *Ralstonia eutropha* (Yabuuchi et al. 1995). However, since CH34 is very different from the "normal" *R. eutropha* strains, classification of all the metal-resistant CH34-like organisms is still open (M. Mergeay, personal communication). Strain CH34 will remain as A. *eutrophus* until this work is done.

One major difference between CH34 and *R. eutropha* strains is the plasmid content. *R. eutropha* strains have 450 kb megaplasmids harboring genes that enable the bacterium to grow autotrophically on molecular hydrogen, oxygen, and carbon dioxide (Andersen et al. 1981). Some CH34-like bacteria cannot grow autotrophically. When CH34-like strains are chemoautotrophs, all required genes are located on the chromosome (Mergeay et al. 1985). These bacteria carry smaller plasmids that are involved in heavy metal resistance. CH34 contains two plasmids: pMOL30, harboring resistance to the divalent cations of cobalt, zinc, and cadmium (*czc* determinant), and pMOL28, carrying resistance to nickel and cobalt (*cnr* determinant) and to chromate (*chr* determinant) (Mergeay et al. 1985; Nies and Silver 1989).

Both cation resistance systems are based on inducible ion export (Nies and Silver 1989). In the case of the Czc system, the membrane-bound cation efflux complex CzcABC is a cation-proton-antiporter (Nies 1995). The substrate affinity of the efflux complex is an important factor for cation homeostasis in the bacterium. Zn(II) is transported with a Hill coefficient of 2. Therefore, Zn(II) will be transported more slowly at low (trace element) concentrations, preventing zinc starvation. At high (toxic) zinc concentrations, however, the system reaches the maximum transport velocity, and Zn(II) is efficiently removed. Cd(II) and Co(II) are effluxed following Michaelis-Menten substrate saturation, toxic Cd(II) with high and essential Co(II) with lower affinity (Nies 1995).

4.1 Structure and Transcription of the czc Determinant

The three structural genes *czcCBA* are transcribed as a tricistronic mRNA of 6200 nucleotides (van der Lelie et al. 1997). These three genes are encoded in the structural gene region located in the center of the *czc* determinant (Figure 4.4). Two regulatory regions flank the structural genes, the upstream regulatory region, and the downstream regulatory region. Nine *czc* genes are transcribed in the same direction (Figure 4.4) (van der Lelie et at. 1997). The downstream regulatory region is followed on the same DNA strand by an open reading frame (Orf131) with similarities to a fungal β-lactamase, and by the 3′-end of a shikimate kinase gene on the other strand. Thus, the *czc* determinant ends with the downstream regulatory region. Sequencing of the upstream regulatory region is continuing, however, there are no open reading frames upstream of *czc* which are transcribed in the same direction as the *czc* genes (C. Große, G. Grass and D. H. Nies, in preparation).

Each of the two regulatory regions is thought to contain three genes, all of which are induced by the three substrates of the *czc* system. The upstream regulatory region encodes a predicted membrane-bound protein, CzcN, showing similarity to the NccN gene product of the nickel-cobalt-cadmium resistance of *Alcaligenes xylosoxidans* (Schmidt and Schlegel 1994). The function of both N-proteins is unclear. The *czcN* gene is followed by the small open reading frame Orf69a encoding a potential product of 69 amino acids, which may have one transmembrane alpha helix and is thus similar to the small "glue" proteins that are required for the stability of multicomponent membrane-bound protein complexes (Altendorf and Epstein 1996). If Orf69a does encode a protein, it may be translationally coupled to *czcN*. The third gene of the upstream region, *czcI*, is located 105 base pairs away from Orf69a (Diels et al. 1995), and *czcI* is followed by a rho-independent terminator. The *czcI* gene ends about 40 base pairs from the start of *czcC*. The CzcI gene product has a strong potential leader-peptidase signal sequence and CzcI may be located in the periplasm. None of these putative upstream regulatory products appears to be located in the cytoplasm.

The structural gene region is followed by a likely rho-independent terminator which also shows consensus motifs of a sigma-70 promoter (van der Lelie et al. 1997). The −35 sequence ATGACG is located five base pairs upstream of the stem-loop and contains the stop codon TGA of *czcA*. The −10 sequence CATCGT

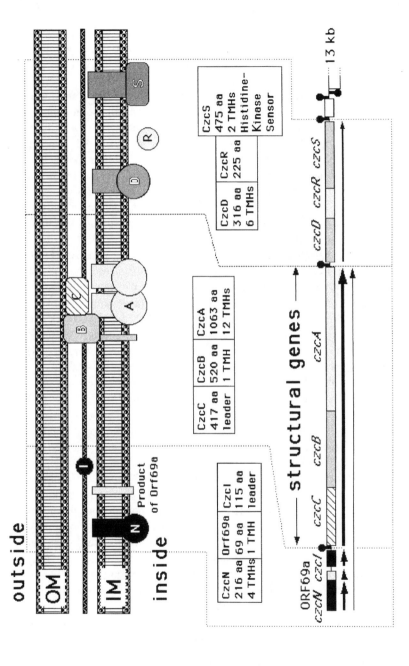

Figure 4.4 The *czc* determinant of *A. eutrophus* CH34. The map of genes is shown below and the subcellular locations of the gene products above. Upstream and downstream regulatory gene regions are indicated. Stem-loop structures in the gene map are also shown as lollipops. Arrows indicate mRNA transcripts. Abbreviations are OM, outer membrane; IM, inner membrane; TMH, transmembrane helix; and aa, amino acids.

is in the loop. The T-stretch of the terminator is immediately followed by the ribosome binding site AGGAG of *czcD* (Nies et al. 1989). The CzcD protein is a membrane-bound protein of the cation diffusion facilator family (Nies and Silver 1995).

The three genes of the downstream regulatory region are probably translationally coupled. The *czcR* gene starts with a GTG with a strong ribosome binding site that is located within the last codons of *czcD* (van der Lelie et al. 1997). CzcR is a response regulator by homology (Hoch and Silhavy 1995; van der Lelie et al. 1997). The *czcR* gene is followed by *czcS* encoding a membrane-bound histidine-kinase sensor. Only four bases are between *czcR* and *czcS*, and therefore it is probable that a tricistronic transcript of the *czcDRS* genes is made. Evidence for this large transcript comes from reverse transcriptase analysis (C. Große, G. Grass and D. H. Nies, in preparation). The physiological functions of the *czcDRS* products, therefore, may be tightly linked.

Transcription of *czc* has been studied in detail (C. Große, G. Grass, and D. H. Nies, in preparation). Besides the 6200 nucleotide *czcCBA* message, the sizes of the *czcN* and the *czcl* messages were about 500 nt and 400 nt, respectively. All *czc* transcripts followed the same induction pattern: they were barely visible in uninduced cells, and cadmium was a weak inducer. Cobalt was a stronger and zinc the strongest inducer. With RT-PCR, evidence for a long minor transcript from *czcN* to at least *czcC* was obtained. However, *czcA* and *czcD* (Figure 4.4) clearly occur on separate transcripts.

4.2 New Insights from New and Old Data

When a DNA fragment missing the start of *czcN* and the last part of *czcD*, and also *czcRS*, was transferred into *A. eutrophus,* inducible metal resistance still occured (Nies 1992). Therefore, there is control of *czcCBA* expression independent of CzcR and CzcS. Unknown chromosomally encoded proteins, and/or *cis*-acting sites in the *czcC* promoter region might be responsible for this action.

The transcriptional start sites of *czcCBA* region is not homologous to known RNA polymerase sigma factor binding sites. There is a weak homology to a newly recognized extracellular functions (ECF) sigma factor consensus site (Lonetto et al. 1994). Moreover, there is a gene for a specific ECF sigma factor, *cnrH*, as part of the regulatory region of the *czc*-related resistance system *cnr* (cobalt and nickel resistance) which is located on a different plasmid in *A. eutrophus* CH34 (Liesegang et al. 1993). Because Czc function does not require this other plasmid, any ECF sigma factor used in *czc* expression must be encoded by the bacterial chromosome. This specificity would explain the narrow bacterial cell host range for *czc* expression (Nies et al. 1987).

Non-specific induction of an ECF sigma factor required for *czc* transcription also explains the difference in the induction pattern of *czc* mRNAs and of β-galactosidase activity from a *czcC::lacZ* translational reporter gene fusion, lacking *czcRS*. Induction of the β-galactosidase activity was slow and low, and was obtained with Zn(II), Cd(II), Co(II), Ni(II), Cu(II), Hg(II), Mn(II), and even Al(III) (Nies 1992). In contrast, induction of the *czc* mRNAs was rapid, strong, and specific for Zn(II)—with lower activity with Co(II) or Cd(II) (van der Lelie et al. 1997) (C. Große, G. Grass, and D. H. Nies, in preparation).

When the structural gene region *czcCBA* was constitutively expressed from a second plasmid *in trans* to the *czcC::lacZ* reporter gene, expression of the β-galactosidase activity was constitutive at a low level (Nies 1992). When *czcD* was added to *czcCBA*, inducibility was restored, albeit weakly. These data fit the hypothesis above, of nonspecific induction of *czc* by an ECF sigma factor. Constitutive expression of CzcCBA would decrease the intracellular cation concentration essential to trigger the nonspecific ECF response. Than, CzcD would catalyze uptake of the inducing cations.

4.3 The Sensory Histidine Kinase CzcS: Relationship to Other Sensors and Mechanism of Action

The CzcS protein shows high sequence similarity (27% amino acids identities) with a newly sequenced histidine kinase sensor (Itoh et al. 1996) (GenBank 1736637), the gene for which is located at about 43 min at the *E. coli* chromosome. It is premature to hypothesize the cation specificity of this histidine kinase. However, it belongs to a six-member subbranch of probably heavy metal cation sensors (Figure 4.5). The other members of this group are the product of a gene at about 11 min on the *E. coli* chromosome (GenBank1778485), a newly sequenced histidine kinase sensor from a plasmid determinant of resistance to silver cations (K. Matsui and S. Silver, personal communication), and the two copper sensing kinases PcoS from *E. coli* (Brown et al. 1995) and CopS from *P. syringae* (Mills et al. 1993). These six proteins are clearly separated from other sensory histidine kinases (Figure 4.5).

The proteins most homologous to CzcS all contain two predicted transmembrane alpha helices, one close to the amino terminus (starting between position 10 and position 16), and the other starting at about position 170. This allows for an extracellular domain of over 100 amino acids in length, which is hypothesized to contain histidines and other amino acids that sense and provide specificity for particular cations. Alternatively, since this hypothesis has not been tested by site-directed mutagenesis, additional metal-binding amino acid residues may occur after the second transmembrane span (and therefore in the cytoplasm). Having both, extracellular and intracellular sensing was hypothesized above for the PcoS and PcoS* protein pair.

At about position 250 in all six of these proteins is the absolutely-conserved histidine, which is the site of autophosphorylation of the protein by Mg-ATP.

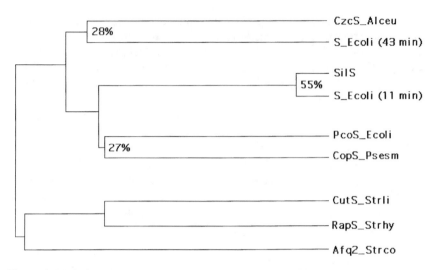

Figure 4.5 Relationship of CzcS to the most-closely related sequences. The top six sequences are listed in the text, and the remaining sequences are best seen by doing a BLASTP search, which will also display newer results. For the three top pairs, the percent amino acid identities are given.

Following the histidine are three key motifs (the so-called N-, D/F-, and G-boxes) containing conserved residues that are involved in ATP-binding to the protein (Stock et al. 1995). The N-boxes, G-boxes and the F-boxes are nicely conserved in the six potential heavy metal cation sensor proteins, but the usually conserved aspartate of the D/F-box is replaced in five of the six sequences by an asparagine. In four of the six proteins, histidine and/or cysteine residues occur immediately in front of the D-box, suggesting another possible location for metal sensing.

The specificity of the six heavy metal cation sensor kinases for specific target kinases are different. The residues providing this specificity and even the general location of the residues providing the specificity appear to differ amongst the six.

4.4 The Response Regulator CzcR

The sequence similarities of the CzcR subgroup of response regulators that are transphosphorylated from the CzcS subgroup of metal-responding sensor kinases is greater yet (Figure 4.6). CzcR is 54% identical with the response regulator encoded at 43 min on the *E. coli* chromosome, whereas the comparable pair of sensor kinases are only 27% identical. CzcR and the other five heavy

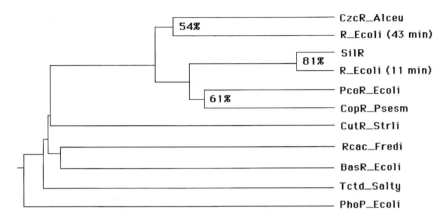

Figure 4.6 Relationship of CzcR to the most-closely related sequences. The codes are similar to Figure 5.5. For the three top pairs, the percent amino acid identities are given.

metal response regulators belong to the OmpR subfamily of response regulators (Volz 1995).

Of the approximately 150 response regulators (R) protein sequences available (Hoch and Silhavy 1995), those in the OmpR subfamily all appear to respond to extracellular signals, although the signals are as varied as heavy metal cations, oxyanions, carbon sources, and even osmotic pressure. OmpR regulates synthesis of two outer membrane proteins. The signal is osmolarity, and the membrane sensor is the EnvZ histidine kinase. This two-component system is remarkable in that it represses synthesis of OmpC and activates synthesis of OmpF at low osmolarity (Pratt and Silhavy 1995). At high osmolarity, the opposite happens: OmpC synthesis is stimulated and OmpF synthesis is repressed. With EnvZ missing, *ompC* transcription is completely repressed and *ompF* transcription is derepressed. Thus, the unphosphorylated OmpR appears to act as a repressor for *ompC* and as an activator for *ompF*. Phosphorylation increases the DNA-binding ability of OmpR to both promoters tenfold (Aiba et al. 1989). Thus, binding alone does not determine if the regulation is positive or negative.

That the response regulator R proteins are more closely clustered than the sensor kinase S protein sequences was somewhat of a surprise. The S proteins are functioning in parallel fashion in response to extracellular signals. There might have been less constraint on the intracellular R proteins. The close clustering of the response regulators indicates very similar biochemical interactions, both with the sensor kinase proteins, and with the DNA promoter sequence.

The EnvZ/OmpR system responds incrementally to signal input; increasing or decreasing osmolarity leads to a gradual increase or decrease of *ompC* and *ompF* expression (Pratt and Silhavy 1995). Phosphorylation of OmpR results in increased multimer-binding to both promoters (Rampersaud et al. 1994). There should be no absolute on/off regulation where homoeostasis is required, and the size of the response must reflect the size of the signal. In case of CzcR, CopR and PcoR, such a gradual response is also important, because zinc and copper are toxic as well as essential. The resistance system needs to detoxify the cation quickly, but not let the cells be depleted of metal. Like OmpR, differences in the oligomerization of CzcR in response to the phosphorylation level may affect transcription.

Like other response regulators of the OmpR family (Pratt and Silhavy 1995; Stock et al. 1995), CzcR is composed of two domains. The first domain has homology to CheY (Volz 1995) and all other response regulators. There are no surprises in this region, all amino acid residues e.g. required for phosphorylation (D_{51}) or Mg(II)-binding (E_7, D_8) are present. The same holds true for the other members of the CzcR branch of heavy metal cation responding regulators. The first domain has a size of about 120 amino acid residues and is followed by a predicted helix-turn-helix structure which might be the DNA-binding site.

4.5 The Novel Sensory Transporter CzcD is a Member of the Cation Diffusion Facilitator Family

In the ten years since CzcD was described (Nies et al. 1989), about two dozen additional related proteins have been identified from DNA sequencing. This ever-growing family has been designated CDF for cation diffusion facilitators (Nies and Silver 1995). They come from all forms of life, bacteria, eucaryotes and archaea. There are already subfamilies within the CDF family tree.

CzcD is probably a sensor, in addition to CzcS, of the extracellular cation concentration (Nies 1992). The length is now known to be 316 amino acids (van der Lelie et al. 1997). There are six predicted transmembrane alpha helices and some potential metal binding sites located in the hydrophilic amino- and carboxyl-termini. The subcluster of CzcD-related proteins includes three sequences from bacterial genome analysis, and these are of unknown function (Figure 4.7). The branches of the tree contains from eukaryotes, yeast and animals. There are additional bacterial members of CzcD-related proteins also.

Some yeast and animal members of the CzcD family have been implicated in metal cation transport. These include ZnT-1 and ZnT-2 (Palmiter et al. 1996; Palmiter and Findley 1995), which confer transport of and resistance to zinc. The two S. cerevisiae proteins ZRC1 (Kamizomo et al. 1989) and COT1 (Conklin et al. 1992) are located in the mitochondrial membrane, and thought to be involved in transport of cadmium, zinc, and cobalt. ZRC1 and COT1 complement the other (Conklin et al. 1994).

ZRC1 was independently cloned and sequenced (Inoue et al. 1993) as a gene involved in oxygen stress response. It was shown to regulate yeast glutathione

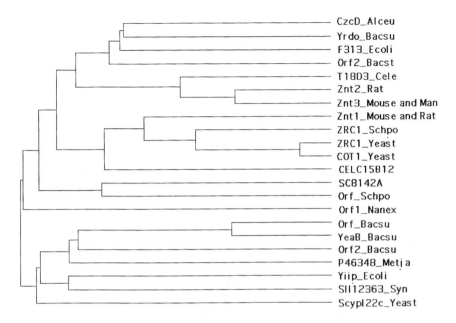

Figure 4.7 Relationship of CzcD to the new CzcD-related family of cation diffusion facilitator proteins (CDF). Individual sequences are listed in the text or should be sought in NCBI.

biosynthesis (Inoue et al. 1993). Thus, ZRC1 is also a co-regulator for heavy metal homeostasis in yeast as CzcD is in *A. eutrophus.*

Acknowledgments

We are grateful to Simon Silver for his harrassing discussion on metal resistance, to Duncan Rouch and Barry Lee for models and speculations, and to Niels van der Lelie for discussion and communication of unpublished results. Work from N. L. Brown's laboratory has been supported by the Medical Research Council, the Agriculture and Food Research Council and the Royal Society; that from D. H. Nies' laboratory by the DFG (Nie 262/1-1 and SFB363), Forschungsmittel des Landes Sachsen-Anhalt, and Fonds der Chemischen Industrie. The collaboration between our laboratories is part of the European Environmental Research Organisation Network on "Molecular Adaptation of Microorganisms to Pollution by Heavy Metals".

References

Aiba, H., F. Nakasai, S. Mizushima, and T. Mizuno. 1989. Phosphorylation of a bacterial activator protein, OmpR, by a protein kinase, EnvZ, results in stimulation of its DNA-binding ability. *J. Biochem. (Tokyo)* 106:5–7.

Altendorf, K., and W. Epstein. 1996. The Kdp-ATPase of *Escherichia coli*. *In* Advances in Cell and Molecular Biology of Membranes and Organelles, Ed. R. E. Dalbey. pp. 401–418. JAI Press, Greenwich, CT.

Andersen, K., R. C. Tait, and W. R. King. 1981. Plasmids required for utilization of molecular hydrogen by *Alcaligenes eutrophus*. *Arch. Microbiol.* 129:384–390.

Ansari, A. Z., J. E. Bradner, and T. V. O'Halloran. 1995. DNA-bend modulation in a repressor-to-activator switching mechanism. *Nature* 374:371–375.

Ansari, A. Z., M. L. Chael, and T. V. O'Halloran. 1992. Allosteric underwinding of DNA is a critical step in positive control of transcription by Hg-MerR. *Nature* 355:87–89.

Bender, C. L., and D. A. Cooksey. 1986. Indigenous plasmids in *Pseudomonas syringae* pv. *tomato:* Conjugative transfer and role in copper resistance. *J. Bacteriol.* 165:534–541.

Brown, J. L., S. North, and H. Bussey. 1993a. SKN7, a yeast multicopy suppressor of a mutation affecting cell wall β-glucan assembly, encodes a product with domains homologous to prokaryotic two-component regulators and to heat shock transcription factors. *J. Bacteriol.* 175:6908–6915.

Brown, N. L., S. R. Barrett, J. Camakaris, B.T.O. Lee, and D. A. Rouch. 1995. Molecular genetics and transport analysis of the copper-resistance determinant (*pco*) from *Escherichia coli plasmid* pRJ1004. *Mol. Microbiol.* 17:1153–1166.

Brown, N. L., B. T. O. Lee, and S. Silver. 1993b. Bacterial transport of and resistance to copper. *In* Metal Ions in Biological Systems, Eds. H. Sigel and A. Sigel. pp. 405–430. Marcel Dekker, New York.

Cervantes, C., and S. Silver. 1992. Plasmid chromate resistance and chromium reduction. *Plasmid* 27:65–71.

Cha, J. S., and D. A. Cooksey. 1993. Copper hypersensitivity and uptake in *Pseudomonas syringae* containing cloned components of the copper resistance operon. *Appl. Environ. Microbiol.* 59:1671–1674.

Chang, H., M. Chen, Y. Shieh, M. J. Bibb, and C. W. Chen. 1996. The *cutRS* signal transduction system of *Streptomyces lividans* represses the biosynthesis of the polyketids antibiotic actinorhodin. *Mol. Microbiol.* 21:1075–1085.

Conklin, D. S., M. R. Culbertson, and C. Kung. 1994. Interactions between gene products involved in divalent cation transport in *Sacchromyces cerevisiae*. *Mol. Gen. Genet.* 244:303–311.

Conklin, D. S., J. A. McMaster, M. R. Culbertson, and C. Kung. 1992. COT1, a gene involved in cobalt accumulation in *Saccharomyces cerevisiae*. *Mol. Cell. Biol.* 12:3678–3688.

Cooksey, D. A. 1993. Copper uptake and resistance in bacteria. *Mol. Microbiol.* 7:1–5.

Cooksey, D. A. 1994. Molecular mechanisms of copper resistance and accumulation in bacteria. *FEMS Microbiol. Rev.* 14:381–386.

Dean, C. R., S. Neshat, and K. Poole. 1996. PfeR, an enterobactin-responsive activator of ferric enterobactin receptor gene expression in *Pseudomonas aeruginosa*. *J. Bacteriol.* 178:5361–5369.

Dean, C. R., and K. Poole. 1993. Expression of the ferric enterobactin receptor (PfeA) of *Pseudomonas aeruginosa:* involvement of a two-component regulatory system. *Mol. Microbiol.* 8:1095–1103.

Diels, L., Q. Dong, D. van der Lelie, W. Baeyens, and M. Mergeay. 1995. The *czc* operon of *Alcaligenes eutrophus* CH34: from resistance mechanism to the removal of heavy metals. *J. Indust. Microbiol.* 14:142–153.

Diels, L., and M. Mergeay. 1990. DNA probe-mediated detection of resistant bacteria from soils highly polluted by heavy metals. *Appl. Environ. Microbiol.* 56:1485–1491.

Dressler, C., U. Kues, D. H. Nies, and B. Friedrich. 1991. Determinants encoding multiple metal resistance in newly isolated copper-resistant bacteria. *Appl. Environ. Microbiol.* 57:3079–3085.

Endo, G., and S. Silver. 1995. CadC, the transcriptional regulatory protein of the cadmium resistance system of *Staphylococcus aureus* plasmid pl258. *J. Bacteriol.* 177:4437–4441.

Forst, S., J. Delgado, and M. Inouye. 1989. Phosphorylation of OmpR by the osmosensor EnvZ modulates expression of the *ompF* and *ompC* genes in *Escherichia coli. Proc. Natl. Acad. Sci. U.S.A.* 86:6052–6056.

Gladysheva, T. B., K. L. Oden, and B. P. Rosen. 1994. Properties of the arsenate reductase of plasmid R773. *Biochemistry* 33:7288–7293.

Helmann, J. D., B. T. Ballard, and C. T. Walsh. 1990. The MerR metalloregulatory protein binds mercuric ion as a tricoordinate, metal-bridged dimer. *Science* 247:946–948.

Hobman, J. L., and N. L. Brown. 1997. Bacterial mercury resistance genes. *In* Metal Ions in Biological Systems, Eds. H. Sigel and A. Sigel. pp. 527–568. Marcel Dekker, New York.

Hoch, J. A., and T. J. Silhavy. 1995. Two Component Signal Transduction. ASM Press, Washington, 488 pp.

Inoue, Y., S. Kobayashi, and A. Kimura. 1993. Cloning and phenotypic characterization of a gene enhancing resistance against oxidative stress in *Saccharomyces cerevisiae. J. Ferment. Bioeng.* 75:327–331.

Ji, G., and S. Silver. 1992. Reduction of arsenate to arsenite by the ArsC protein of the arsenic resistance operon of *Staphylococcus aureus* plasmid pl258. *Proc. Natl. Acad. Sci. U.S.A.* 89:9474–9478.

Kamizomo, A., M. Nishizawa, A. Teranishi, K. Murata, and A. Kimura. 1989. Identification of a gene conferring resistance to zinc and cadmium ions in the yeast *Saccharomyces cerevisiae. Mol. Gen. Genet.* 219:161–167.

Kunito, T., T. Kusano, H. Oyaizu, K. Senoo, S. Kanazawa, and S. Matsumoto. 1996. Cloning and sequence analysis of *czc* genes in *Alcaligenes* sp. strain CT14. *Biosci. Biotech. Biochem.* 60:699–704.

Lee, Y. A., M. Hendson, N. J. Panopoulos, and M. N. Schroth. 1994. Molecular cloning, chromosomal mapping, and sequence analysis of copper resistance genes from *Xanthomonas campestris* pv. *juglandis:* Homology with small blue copper proteins and multi-copper oxidase. *J. Bacteriol.* 176:173–188.

Liesegang, H., K. Lemke, R. A. Siddiqui, and H.-G. Schlegel. 1993. Characterization of the inducible nickel and cobalt resistance determinant *cnr* from pMOL28 of *Alcaligenes eutrophus* CH34. *J. Bacteriol.* 175:767–778.

Lonetto, M. A., K. L. Brown, K. E. Rudd, and M. J. Buttner. 1994. Analysis of the *Streptomyces coelicolor sigF* gene reveals the existence of a subfamily of eubacterial RNA polymerase σ factors involved in the regulation of extracytoplasmic functions. *Proc. Natl. Acad. Sci. U.S.A.* 91:7573–7577.

Maeda, T., S. M. Wurgler-Murphy, and H. Saito. 1994. A two-component system that regulates an osmosensing MAP kinase cascade in yeast. *Nature* 369:242–245.

Mellano, M. A., and D. A. Cooksey. 1988. Induction of the copper resistance operon from *Pseudomonas syringae. J. Bacteriol.* 170:4399–4401.

Mergeay, M., D. Nies, H. G. Schlegel, J. Gerits, P. Charles, and F. van Gijsegem. 1985. *Alcaligenes eutrophus* CH34 is a facultative chemolithotroph with plasmid-bound resistance to heavy metals. *J. Bacteriol.* 162:328–334.

Mills, S. D., C. A. Jasalavich, and D. A. Cooksey. 1993. A two-component regulatory system for copper-inducible expression of the copper resistance operon of *Pseudomonas syringae. J. Bacteriol.* 175:1656–1664.

Mills, S. D., C. K. Lim, and D. A. Cooksey. 1994. Purification and characterization of CopR, a transcriptional activator protein that binds to a conserved domain (*cop* box) in copper-inducible promoters of *Pseudomonas syringae. Mol. Gen. Genet.* 244:341–351.

Morby, A. P., J. S. Turner, J. W. Huckle, and N. J. Robinson. 1993. SmtB is a metal-dependent repressor of the cyanobacterial metallothionein gene *smt*A: Identification of a Zn inhibited DNA-protein complex. *Nucleic Acids Res.* 21:921–925.

Mukhopadhyay, D., H. Yu, G. Nucifora, and T. K. Misra. 1991. Purification and functional characterization of MerD. A coregulator of the mercury resistance operon in Gram-negative bacteria. *J. Biol. Chem.* 266:18538–18542.

Nies, D., M. Mergeay, B. Friedrich, and H. G. Schlegel. 1987. Cloning of plasmid genes encoding resistance to cadmium, zinc, and cobalt in *Alcaligenes eutrophus* CH34. *J. Bacteriol.* 169:4865–4868.

Nies, D. H. 1995. The cobalt, zinc, and cadmium efflux system CzcABC from *Alcaligenes eutrophus* functions as a cation-proton-antiporter in *Escherichia coli. J. Bacteriol.* 177:2707–2712.

Nies, D. H. 1992. CzcR and CzcD, gene products affecting regulation of cobalt, zinc and cadmium resistance (*czc*) in *Alcaligenes eutrophus. J. Bacteriol.* 174:8102–8110.

Nies, D. H., A. Nies, L. Chu, and S. Silver. 1989. Expression and nucleotide sequence of a plasmid-determined divalent cation efflux system from *Alcaligenes eutrophus. Proc. Natl. Acad. Sci. U.S.A.* 86:7351–7355.

Nies, D. H., and S. Silver. 1995. Ion efflux systems involved in bacterial metal resistances. *J. Indust. Microbiol.* 14:186–199.

Nies, D. H., and S. Silver. 1989. Plasmid-determined inducible efflux is responsible for resistance to cadmium, zinc, and cobalt in *Alcaligenes eutrophus. J. Bacteriol.* 171:896–900.

Nixon, B. T., C. W. Ronson, and F. M. Ausubel. 1986. Two-component regulatory systems responsive to environmental stimuli share strongly conserved domains with the nitrogen assimilation regulatory genes *ntrB* and *ntrC*. *Proc. Natl. Acad. Sci. U.S.A.* 83:7850–7854.

Odermatt, A., and M. Solioz. 1995. Two trans-acting metalloregulatory proteins controlling expression of the copper-ATPases of *Enterococcus hirae*. *J. Biol. Chem.* 270:4349–4354.

Odermatt, A., H. Suter, R. Krapf, and M. Solioz. 1992. An ATPase operon involved in copper resistance by *Enterococcus hirae*. *Ann. N.Y. Acad. Sci.* 671:484–486.

Odermatt, A., H. Suter, R. Krapf, and M. Solioz. 1993. Primary structure of two P-type ATPases involved in copper homeostasis in *Enterococcus hirae*. *J. Biol. Chem.* 268:12775–12779.

O'Halloran, T. V. 1993. Transition metals in control of gene expression. *Science* 261:715–725.

O'Halloran, T. V., B. Frantz, M. K. Shin, D. M. Ralston, and J. G. Wright. 1989. The MerR heavy metal receptor mediates positive activation in a topologically novel transcription complex. *Cell* 56:119–129.

Palmiter, R. D., T. B. Cole, and S. D. Findley. 1996. ZnT-2, a mammalian protein that confers resistance to zinc by facilitating vesicular sequestration. *EMBO J.* 15:1784–1791.

Palmiter, R. D., and S. D. Findley. 1995. Cloning and functional characterization of a mammalian zinc transporter that confers resistance to zinc. *EMBO J.* 14:639–649.

Pao, G. M., R. Tam, L. S. Lipschitz, and M. H. Saier, Jr. 1994. Response regulators: Structure, function and evolution. *Res. Microbiol.* 145:356–362.

Parkhill, J., A. Z. Ansari, J. G. Wright, N. L. Brown, and T. V. O'Halloran. 1993. Construction and characterization of a mercury-independent MerR activator (MerRAC): transcriptional activation in the absence of Hg(II) is accompanied by DNA distortion. *EMBO J.* 12:413–421.

Pratt, L. A., and T. J. Silhavy. 1995. Porin regulon of *Escherichia coli*. *In* Two-Component Signal Transduction, Eds. J. A. Hoch and T. J. Silhavy. pp. 105–127. ASM Press, Washington.

Ralston, D., and T. V. O'Halloran. 1990. Ultrasensitivity and heavy-metal selectivity of the allosterically modulated MerR transcription complex. *Proc. Natl. Acad. Sci. U.S.A.* 87:3846–3850.

Rampersaud, A., S. L. Harlocker, and M. Inouye. 1994. The OmpR protein of *Escherichia coli* binds to sites in the *ompF* promoter region in a hierarchical manner determined by its degree of phosphorylation. *J. Biol. Chem.* 269:12559–12566.

Rosen, B. P., H. Bhattacharjee, and W. P. Shi. 1995. Mechanisms of metalloregulation of an anion-translocating ATPase. *J. Bioenerget. Biomemb.* 27:85–91.

Rouch, D., J. Camakaris, B. T. O. Lee, and R. K. J. Luke. 1985. Inducible plasmid mediated copper resistance in Escherichia coli. *J. Gen. Microbiol.* 131:939–943.

Rouch, D. A. 1986. Plasmid-mediated copper resistance in *E. coli*. Ph.D. Thesis. The University of Melbourne.

Rouch, D. A., and N. L. Brown. 1997. Copper-inducible transcriptional regulation at two promoters in the *Escherichia coli* copper resistance determinant *pco*. *Microbiology* 143:1191–1202.

Rouch, D. A., J. Parkhill, and N. L. Brown. 1995. Induction of bacterial mercury- and copper-responsive promoters: functional differences between inducible systems and implications for their use in gene-fusions for in vivo metal biosensors. *J. Indust. Microbiol.* 14:249–253.

Schaifers, K. T., G. 1984. Meyers Handbuch Weltall. Bibliographisches Institut, Mannheim, pp.

Schmidt, T., and H. G. Schlegel. 1994. Combined nickel-cobalt-cadmium resistance encoded by the *ncc* locus of *Alcaligenes xylosoxidans* 31A. *J. Bacteriol.* 176:7045–7054.

Silver, S., K. Budd, K. M. Leahy, W. V. Shaw, D. Hammond, R. P. Novick, G. R. Willsky, M. H. Malamy, and H. Rosenberg. 1981. Inducible plasmid-determined resistance to arsenate, arsenite, and antimony(III) in *Escherichia coli* and *Staphylococcus aureus*. *J. Bacteriol.* 146:983–996.

Silver, S., and G. Ji. 1994. Newer systems for bacterial resistances to toxic heavy metals. *Environ. Health Perspect.* 102:107–113.

Silver, S., and L. T. Phung. 1996. Bacterial heavy metals resistance: New surprises. *Annu. Rev. Microbiol.* 50:753–789.

Silver, S., and M. Walderhaug. 1995. Bacterial plasmid-mediated resistances to mercury, cadmium and copper. *In* Toxicology of Metals. Biochemical Aspects. Eds. R. A. Goyer and M. G. Cherian. pp. 435–458. Springer Verlag, Berlin.

Silver, S., and M. Walderhaug. 1992. Gene regulation of plasmid- and chromosome-determined inorganic ion transport in bacteria. *Microbiol. Rev.* 56:195–228.

Solioz, M., and A. Odermatt. 1995. Copper and silver transport by CopB-ATPase in membrane vesicles of *Enterococcus hirae*. *J. Biol. Chem.* 270:9217–9221.

Solioz, M., and C. Vulpe. 1996. Cpx-type ATPases: a class of P-type ATPases that pump heavy metals. *Trends Biochem. Sci.* 21:237–41.

Stock, J. B., M. G. Surette, M. Levit, and P. Park. 1995. Two-component signal transduction systems: Structure-function relationships and mechanisms of catalysis. *In* Two-Component Signal Transduction, Eds. J. A. Hoch and T. J. Silhavy. pp. 25–51. ASM Press, Washington.

Strausak, D., and M. Solioz. 1997. CopY is a copper-inducible repressor of the *Enterococcus hirae* copper ATPases. *J. Biol. Chem:* In the Press.

Summers, A. O. 1992. Untwist and shout: a heavy metal-responsive transcriptional regulator. *J. Bacteriol.* 174:3097–3101.

Tetaz, T. J., and R.K.J. Luke. 1983. Plasmid controlled resistance to copper in *Escherichia coli*. *J. Bacteriol* 154:1263–1268.

van der Lelie, D., T. Schwuchow, U. Schwidetzky, S. Wuertz, W. Baeyens, M. Mergeay, and D. H. Nies. 1997. Two component regulatory system involved in transcriptional control of heavy metal homeostasis in *Alcaligenes eutrophus. Mol. Microbiol.* 23:493–503.

Volz, K. 1995. Structural and functional conservation in response regulators. *In* Two-Component Signal Transduction, Eds. J. A. Hoch and T. J. Silhavy, pp. 53–64. ASM Press, Washington.

Yabuuchi, E., Y. Kosako, I. Yano, H. Hotta, and Y. Nishiuchi. 1995. Transfer of two *Burkholderia* and an *Alcaligenes* species to *Ralstonia* gen. nov.: proposal of *Ralstonia pickettii* (Ralston, Palleroni and Doudoroff 1973) comb. nov., *Ralstonia solanaceaum* (Smith, 1896) comb. nov. and *Ralstonia eutropha* (Davis 1969) comb. nov. *Microbiol. Immunol.* 39:897–904.

5

Phosphate Signaling and the Control of Gene Expression in *Escherichia coli*

Barry L. Wanner

1. Introduction

Phosphorus (P) exists in several chemical forms usable in the biological world. These include three forms of inorganic phosphates [orthophosphates, pyrophosphates, and metaphosphates (also called polyphosphates)], innumerable organophosphates, and phosphonates [compounds with a direct carbon-phosphorus (C–P) bond]. As an essential nutrient, P plays a key role in numerous biological processes. It is a component of membrane lipids, complex carbohydrates such as lipopolysaccharides, nucleic acids, and many proteins (where P is incorporated posttranslationally); P occupies also a central role in energy metabolism. Hence, P assimilation and metabolism are of critical importance for growth and survival in all cells from *Escherichia coli* to humans. Regardless of the cell type, inorganic orthophosphate (P_i) is the preferred P source; alternative P compounds are used primarily only when P_i is unavailable (Wanner 1996).

In spite of the abundance of P_i in nature, cells frequently encounter P_i limitation because much of its natural supply occurs as insoluble salts. No doubt because of the scarcity of usable P_i, microorganisms (including bacteria, yeast, and *Neurospora*) regulate the synthesis of many proteins in response to the levels of environmental (extracellular) P_i; many of these proteins are involved in scavenging small amounts of P_i or alternative P sources. A global analysis of proteins made in *E. coli* has revealed that the synthesis of about 25% of the proteins detectable by two-dimensional gel analysis responds to P_i limitation (VanBogelen et al. 1996). P_i also has major regulatory effects in *Neurospora crassa* (Mann et al. 1989), *Saccharomyces cerevisiae* (Ogawa et al. 1995), *Schizosaccharomyces pombe* (Fankhauser et al. 1995), and other lower eukaryotes. In mammalian cells, P_i limitation induces synthesis of a P_i transporter (Murer and Biber 1996). Furthermore, P_i homeostasis is essential for the normal physiological functioning of many human cell types, with the kidney playing a major role in P_i homeostasis. At least five human genetic diseases are now attributed to defects in P_i homeostasis (Rowe et al. 1996; Econs 1996).

This chapter concerns how cells detect environmental P_i levels and regulate gene expression in response to these levels. How cells respond to environmental (extracellular) signals is of fundamental importance in biology. As the processes of P_i signaling and gene regulation are best understood in *E. coli,* I have focused my attention on our knowledge of P_i-regulated gene expression in this model organism.

2. P_i-regulated Genes and the Pho Regulon

It has been known for a long time that a large number of *E. coli* genes are regulated by environmental P_i levels. It is also known that many of these P_i-regulated genes are subject to a common molecular control; those P_i-regulated genes that are controlled by the PhoB-PhoR two-component regulatory system are collectively called the phosphate (Pho) regulon (Wanner 1996). Altogether, 38 different Pho regulon genes have now been identified in *E. coli* or closely related bacteria, including seven new genes (*phnR* to *phnX*) from *Salmonella typhimurium* (Jiang et al. 1995; W. W. Metcalf, W. Jiang, and B. L. Wanner, manuscript in preparation). Of these, 31 are present in *E. coli* and 7 are absent. Only 21 Pho regulon genes are known to be present in *S. typhimurium,* 16 are known to be absent, and the status of 2 others is unknown (Figure 5.1). Many of the corresponding gene products have also been well characterized. All probably have a role in the use of various P compounds as the sole P source for growth. These include four different ABC (ATP-binding cassette) transporters for the uptake of P-containing compounds (Table 5.1), as well as a number of proteins with roles in degradation of various P compounds. The *phoA-psiF* operon encodes bacterial alkaline phosphatase (Bap; the *phoA* gene product) and a downstream open reading frame (Orf; the *psiF* gene product) of unknown function. Bap is a nonspecific phosphomonoesterase that is capable of hydrolyzing many nontransportable organophosphates in the periplasm; the released P_i is then taken up by a P_i transporter. The *phoBR* operon encodes two primary regulatory proteins of the Rho regulon: the response regulator PhoB, a transcriptional activator, and its partner protein, the P_i sensor histidine kinase PhoR. Both PhoB and PhoR are required for transcriptional activation of Pho regulon genes in response to P_i limitation. These proteins are described in Section 3. PhoE is an outer membrane porin with specificity favoring anions; PhoE may allow entry of complex organophosphate anions into the periplasm for breakdown by a variety

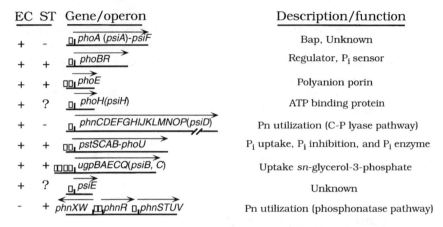

EC	ST	Gene/operon	Description/function
+	–	*phoA (psiA)-psiF*	Bap, Unknown
+	+	*phoBR*	Regulator, P_i sensor
+	+	*phoE*	Polyanion porin
+	?	*phoH(psiH)*	ATP binding protein
+	–	*phnCDEFGHIJKLMNOP(psiD)*	Pn utilization (C-P lyase pathway)
+	+	*pstSCAB-phoU*	P_i uptake, P_i inhibition, and P_i enzyme
+	+	*ugpBAECQ(psiB, C)*	Uptake *sn*-glycerol-3-phosphate
+	?	*psiE*	Unknown
–	+	*phnXW phnR phnSTUV*	Pn utilization (phosphonatase pathway)

Figure 5.1 Pho regulon genes of *E. coli* (EC) and *S. typhimurium* (ST). Several of these genes were identified as phosphate-starvation-inducible (*psi*) genes many years ago. Corresponding *psi* designations are given in parentheses. + = present; – = absent; ? = uncertain. Rectangles, arrows, and vertical bars indicate Pho boxes, promoters, and transcripts, respectively. Bap is bacterial alkaline phosphatase; Pn is phosphonate. Adapted from Fig. 5 in Wanner (1996). Sources are given in Wanner (1996), except for the *phnR* to *phnX* gene cluster (W. W. Metcalf, W. Jiang, and B. L. Wanner, manuscript in preparation).

Table 5.1 ABC transporters of the Pho regulon

Transporter[a]	Periplasmic Binding Protein	ABC Family Traffic ATPase	Membrane Channel Proteins
PhnCDE[b]	PhnD	PhnC	PhnE
PhnSTUV[b]	PhnS	PhnT	PhnU, PhnV
PstSCAB	PstS	PstB	PstA, PstC
UgpBAEC	UgpB	UgpC	UgpA, UgpE

[a]All except PhnSTUV are present in *E. coli,* and all except PhnCDE are present in *S. typhimurium.* Citations are given in Wanner (1996).

[b]Transport functions of the PhnCDE and PhnSTUV transporters are inferred on the basis of sequence similarities and growth phenotypes of mutants. Transport assays have not been done because of the unavailability of suitable radiolabeled substrates from commercial sources.

of degradative enzymes including Bap. PhoH is an ATP binding protein (with sequence similarity to ABC family traffic ATPases) of unknown function. The function of PsiE is also unknown.

Two gene clusters of the Pho regulon encode systems for uptake and breakdown of phosphonates for use as a sole P source (Wanner 1996). These systems differ in regard to both their substrate specificity and catalytic mechanism of the C–P bond cleavage. One, the 14-gene *phnC*-to-*phnP* operon, is present in *E. coli,* but absent in *S. typhimurium;* the other, the complex *phnR*-to-*phnX* gene cluster, is present in *S. typhimurium,* but absent in *E. coli.* The *phnC*-to-*phnP* operon encodes an ABC transporter (comprised of PhnC, PhnD, and PhnE; Table 5.1) and a C-P lyase (comprised of PhnG to PhnN and PhnP). The C-P lyase pathway allows for growth of *E. coli* on a broad range of phosphonates, including ones with or without substitutions on the 2-carbon adjacent to the C-P bond, like methylphosphonate, ethylphosphonate, 2-aminoethylphosphonate, and many others as a sole P source. Two additional gene products (PhnF and PhnO) have no obligatory role in phosphonate utilization. They have signature motifs of gene regulatory proteins. PhnF belongs to the GntR family of gene regulatory proteins; PhnO has a helix-turn-helix motif that is commonly found in DNA binding proteins. Hence, PhnF and PhnO are likely involved in gene regulation at an unknown site(s). PhoB and PhoR are required for transcriptional activation of the P_{phnC} promoter in response to P_i limitation; PhnF and PhnO may be involved in specific control of a gene(s) within the *phnC*-to-*phnP* operon. No evidence of a specific control of the C-P lyase pathway exists, however. In spite of the wealth of information on the molecular genetics and molecular biology of the genes for the C-P lyase pathway, its mechanism of C–P bond fission is poorly understood because of the inability to detect C-P lyase activity in a cell-free system (Wanner 1994).

The *phnR*-to-*phnX* gene cluster encodes an ABC transporter (comprised of PhnS, PhnT, PhnU, and PhnV; Table 5.1) and enzymes (the *phnW* and *phnX* gene products) for C-P bond cleavage by the "phosphonatase" pathway (Jiang et al. 1995; W. W. Metcalf, W. Jiang, and B. L. Wanner, manuscript in preparation). The phosphonate transport system of the C-P lyase and phosphonatase pathways share no more sequence similarity than other ABC transporters for a variety of substrates. Based on mutational effects on growth, the ABC transporter of the C-P lyase pathway has a broad substrate specificity. In contrast, the ABC transporter of the phosphonatase pathway, like the enzymes of this pathway, has a narrow substrate specificity. The phosphonatase pathway acts primarily on the major naturally occurring and abundant phosphonate, 2-aminoethylphosphonate. C–P bond cleavage by the phosphonatase pathway involves two steps. First, 2-aminoethylphosphonate is converted to phosphonoacetaldehyde by a transaminase (the *phnW* gene product); phosphonacetaldehyde is subsequently converted to acetaldehyde and P_i by phosphonacetaldehyde hydrolase (trivial name phosphonatase; the *phnX* gene product). A seventh gene product (PhnR) has no obligatory role in phosphonate utilization. Like PhnF of the C-P lyase gene cluster, PhnR belongs to the GntR family of regulatory proteins. Hence, PhnR is probably involved in gene regulation at an unknown site. PhnF and PhnR are no more similar than other members of the GntR family of regulatory proteins, however. Interestingly, early studies on the phosphonatase pathway in *Bacillus cereus* revealed that induction of the uptake system is subject to a general (repression) control by P_i and a specific (induction) control by 2-aminoethylphosphonate (or the gratuitous inducer aminomethylphosphonate; Rosenberg and La Nauze 1967). In *B. cereus,* repression occurred when P_i was in excess (regardless of the presence of a phosphonate), induction occurred under conditions of P_i limitation, and induction was maximal under conditions of P_i limitation in the presence of 2-aminoethylphosphonate (or aminomethylphosphonate). Accordingly, the PhoB-PhoR system may be required for activation of the *phnR-to-phnX* gene cluster by P_i limitation, and PhnR may be involved in specific control of a particular *phn* gene(s) by 2-aminoethylphosphonate or a related compound. New studies are needed to determine whether PhnR is involved in a specific control of an *S. typhimurium phn* gene(s). Anyway, phosphonates are apparently a common P source in nature because many diverse bacteria are capable of phosphonate uptake and breakdown. Curiously, particular bacteria (for example, *Enterobacter aerogenes*) encode both the C-P lyase and phosphonatase pathways (Lee et al. 1992), even though substrates of the phosphonatase pathway are probably also substrates of the C-P lyase pathway.

An ABC transporter for P_i-specific transport, the Pst system (Table 5.1), is encoded by the *pstSCAB-phoU* operon along with a protein called PhoU. Although it is clear that PhoU has no role in P_i uptake, it may have an auxiliary role in the overall process of P_i assimilation via the Pst system (Steed and Wanner 1993). Once P_i is taken up, P_i is incorporated into ATP, which then serves as the primary phosphoryl donor in cellular metabolism. Accordingly, PhoU may be an enzyme for incorporation of P_i entering via the Pst system into ATP. Evidence favoring this hypothesis resulted from examining the consequences of *phoU* mutations. A $\Delta phoU$ mutation was shown to result in an extremely deleterious effect on growth that was largely abolished by a *pst* mutation, as well as mutations that reduce synthesis of the Pst transporter. That is, the severe growth defect results from the presence of a functional Pst transporter in the absence of PhoU. Furthermore, the growth defect appeared to be due to P_i sensitivity, suggesting that excess intracellular P_i entering via the Pst system may accumulate in the absence of PhoU function. This could occur if PhoU were an enzyme in P_i metabolism. Direct evidence that PhoU is an enzyme is lacking, however. Both the Pst transporter and PhoU are also required for P_i inhibition of Pho regulon gene expression. Their role in P_i control is discussed in Section 4.

The fourth ABC transporter of the Pho regulon, the Ugp system (Table 5.1), for uptake of *sn*-glycerol-3-phosphate and glycerophosphoryl diesters, is encoded by the *ugpBAECQ* operon together with a glycerophosphoryl phosphodiesterase (the *ugpQ* gene product). The expression of the *ugpBAECQ* operon is regulated by both P_i via the PhoR-PhoB two-component regulatory system and the carbon source via cyclic AMP (cAMP) and the cAMP activator protein acting on separate promoters (Su et al. 1991; Kasahara et al. 1991). Paradoxically, the Ugp system transports *sn*-glycerol-3-phosphate solely for use as a P source; the Ugp system cannot transport *sn*-glycerol-3-phosphate for use as a carbon source. This is because intracellular P_i accumulates during growth on *sn*-glycerol-3-phosphate as a carbon source, and excess P_i interferes with *sn*-glycerol-3-phosphate uptake by the Ugp system (Brzoska et al. 1994).

These Pho regulon genes and operons are arranged in 11 transcriptional units, each of which is preceded by a promoter containing one or more "Pho box" sequences (Makino et al. 1996) required for regulation (Figure 5-1). The typical Pho box has two 7-bp direct repeats with the well-conserved consensus CTGTCAT separated by a 4-bp segment that corresponds to a poorly conserved −35 region. Pho box sequences preceding 7 Pho regulon promoters are depicted in Figure 5.2. As shown, the P_{pstS} and P_{ugpB} promoters contain two and five additional copies, respectively, of the 7-bp direct repeat at the same spacing farther upstream. The occurrence of a 7-bp unit every 11 bp may allow binding of multiple PhoB molecules to the same face of the helix. Like the promoters shown in Figure 5.2, the P_{phnR} and P_{phnS} promoters also have typical Pho box sequences overlapping the respective −35 regions; in contrast, the P_{phnW} promoter is unique for it appears to have a Pho box sequence overlapping its −10 region (S.-K. Kim and B. L. Wanner, unpublished data).

3. The PhoB-PhoR Two-component Regulatory System

P_i control of the Pho regulon is a paradigm of a signal transduction pathway in which occupancy of a cell surface receptor(s) regulates gene expression in the cytoplasm. P_i signaling in *E. coli* involves seven proteins, all of which probably interact in a membrane-associated signaling complex. These P_i signaling proteins include: (1) PhoB and PhoR, which are required for transcriptional activation in response to P_i limitation; (2) all four components of the Pst transporter that are required for inhibition when P_i is in excess; and (3) the negative regulator of an unknown function called PhoU.

PhoB and PhoR are members of the large family of two-component regulatory systems that are crucial in a variety of signal transduction pathways in bacteria (Parkinson and Kofoid 1992). Most of these systems share a common biochemical signaling mechanism in which a sensor protein (usually membrane-associated, like PhoR) autophosphorylates on a conserved histidine residue in response to

```
                                                                      +1
phnC (-41)   CTGTTAGTCACTTTTAATTAACCAAATCGTCACAATAATCCG
                                                                      +1
phoA (-40)   CTGTCATAAAGTTGTCACGGCCGAGACTTATAGTCGCTTTG
                                                                      +1
phoB (-40)   TTTTCATAAATCTGTCATAAATCTGACGCATAATGACGTCG
                                                                      +1
phoE (-41)   CTGTAATATATCTTTAACAATCTCAGGTTAAAAACTTTCCTG
                                                                      +1
phoH (-37)   CTGTCATCACTCTGTCATCTTTCCAGTAGAAACTAATG

pstS (-63)   CTGTCATAAAACTGTCATATTC-
                                                                      +1
             CTTACATATAACTGTCACCTGTTTGTCCTATTTTGCTTCTCG
ugpB (-95)   CCGTCACCGCC-

             TTGTCATCTTTCTGACACCTTA-

             CTATCTTACAAATGTAACAAAA-
                                                                      +1
             AAGTTATTTTTCTGTAATTCGAGCATGTCATGTTACCCCG
```

Consensus: CTGTCATA(AT)A(TA)CTGT(CA)A(CT)

Information Content

18-bp Consensus	7-bp Direct Repeat

A	1	1	1	1	1	9	0	6	4	7	3	1	0	0	1	4	10	0	A	2	1	1	2	5	20	0
C	7	0	0	0	7	0	0	2	2	1	1	7	0	0	0	6	0	5	C	15	1	0	0	14	0	6
G	0	0	7	0	0	0	1	0	0	0	1	0	0	8	0	0	0	0	G	0	0	16	0	0	0	1
T	2	9	2	9	2	1	9	2	4	2	5	2	10	2	9	0	0	5	T	4	19	4	19	2	1	14

Figure 5.2 Pho regulon promoters. Sequences known to be required for transcriptional activation are shown. Adapted from Fig. 6 in Wanner (1996). Sources are given in Wanner (1996). Individual Pho box sequences are overlined, −10 regions are underlined, and mRNA start sites are marked as +1. The −35 segment of the "18-bp consensus" is underlined. Preliminary studies indicate that similar sequences preceding P_{phnR}, P_{phnS}, and P_{phnW} are also required for P_i regulation of those promoters (S.-K. Kim and B. L. Wanner, unpublished data).

an environmental stimulus and then transfers the phosphoryl group to a conserved aspartate residue on its partner response regulator (commonly a transcriptional activator, like PhoB). As many as 50 pairs of these sensors and regulators probably exist in a single bacterium such as *E. coli*. Thus, interactions among partner proteins such as PhoR and PhoB must be highly specific in order to prevent undesirable interactions, termed cross talk (noise), between nonpartner proteins.

3.1 The Response Regulator PhoB

PhoB is the central regulator of the Pho regulon; it is known to govern at least 10 different Pho regulon promoters, and in all of these cases PhoB is a transcriptional activator. How PhoB regulates the P_{phnW} promoter is unknown. Like response regulators of other two-component regulatory systems, PhoB activity is controlled by differential phosphorylation. Activation occurs in response to P_i

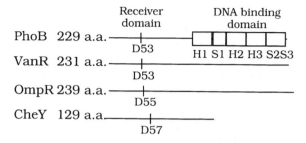

Figure 5.3 Simple diagram of PhoB and related response regulators. Conserved aspartate residues that are phosphorylated are indicated. Boxes labeled H1, H2, H3, S1, S2, and S3 correspond to α-helices (H) and β-strands (S) in the histone 5 structure (Suzuki and Makino 1995).

limitation and is mediated by PhoR specifically phosphorylating D53 in the PhoB N-terminal receiver domain. Inhibition is reestablished upon growth shift to P_i excess conditions and is probably brought about by PhoR and an excess of PhoU (or PstB) dephosphorylating phospho-PhoB. Transcriptional activation requires interaction of the PhoB C-terminal histone 5-like DNA binding domain with DNA.

PhoB belongs to the "OmpR family" of response regulators (including ArcA, CreB, KdpE, PhoP, VanR, VirG, and others; (Figure 5.3). These proteins are composed of two domains: a highly conserved N-terminal receiver (phosphoryla-tion) domain of ca. 100 amino acids and a C-terminal DNA-binding domain. Structural conservation probably exists among the N-terminal domains of these proteins and other members of the CheY superfamily (Volz 1993). Structural similarity probably also exists between the C-terminal domains of the OmpR family members (Martinez-Hackert et al. 1996; Kondo et al. 1994). The C-terminal domains of PhoB and OmpR may share structural similarity with histone 5. In accordance with the structure of histone 5, the helix 2/helix 3 (H2/H3) segments of PhoB and OmpR are predicted to interact with the σ^{70} (Makino et al. 1996) and α (Pratt and Silhavy 1994) subunits of RNA polymerase, respec-tively. Yet it is poorly understood how phosphorylation of these proteins leads to transcriptional activation. Although phosphorylation has been proposed to result in activation by oligomerization of OmpR at the P_{ompF} promoter (Harlocker et al. 1995), spatial considerations make this unlikely to account for regulation at other promoters. The spacing between the divergently transcribed P_{phnW} and P_{phnR} promoters is insufficient to allow formation of highly ordered oligomers.

3.2 The P_i Sensor Histidine Kinase PhoR

PhoR acts as the P_i sensor by analogy to signaling kinases of many other two-component regulatory systems. PhoR activates PhoB by phosphorylation under conditions of P_i limitation; PhoR, probably together with PhoU or PstB, inactivates phospho-PhoB by dephosphorylation upon growth shift to P_i excess conditions; and PhoR together with PhoU and an intact Pst system prevent phosphorylation of PhoB when P_i is in excess. PhoR is a cytoplasmic integral membrane protein of 431 amino acids in length and appears to be composed of four domains (Figure 5.4). These include: (1) an N-terminal "sensor domain" of about 50 residues; (2) a cytoplasmic "linker domain" of about 150 residues; (3) a phosphotransfer domain with an invariant histidyl residue (H213, the autophosphorylation site) that is sometimes called the H box subdomain; and (4) a C-terminal catalytic domain containing N, G1, F, and G2 motifs, which are also sometimes subdivided into the N box, D/F box, and G box subdomains (Parkinson 1995; Parkinson and Kofoid 1992). The highly conserved transmitter sequences among PhoR and other signaling histidine kinases include the C-terminal domain(s) beginning near residue 159 of PhoR.

Many altered function PhoR mutants exist, including ones with single amino acid substitutions in the linker, phosphotransfer, and catalytic domains. Based on *in vivo* effects of these and other *phoR* mutations [reviewed in Wanner (1996)], both the N-terminal sensor and linker domains are probably essential for signal input, and the phosphotransfer and catalytic domains are required for autophosphorylation, phosphotransfer, and dephosphorylation reactions. Both the linker domain and a separate C-terminal segment may be involved in protein-protein interactions, PhoR dimer formation, or PhoR interaction with another protein, as dominant mutations alter residues in these segments.

On the basis of its "membrane-binding domain," PhoR represents a unique class of signaling kinases. The vast majority of membrane-associated histidine

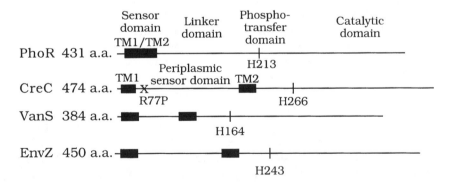

Figure 5.4 Simple diagram of PhoR and related signaling sensor kinases. Filled boxes are hydrophobic segments predicted to be transmembrane domains TM1 and TM2. Conserved histidine residues that are phosphorylated are indicated. The *creC510* (R77P) mutation (Wanner 1996) that results in constitutive signaling is also shown.

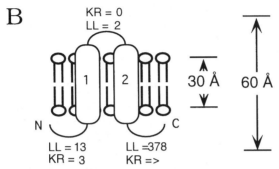

A   MLERLSWKRLVLE<u>LLLCCLPAFILGAFFGYLPWF</u>

<u>TM2</u>   TM1

LLASVTGLLIWHFWNLLRLSWWLWVDRSMTPP-

B   KR = 0
LL = 2

1   2

30 Å   60 Å

N   C

LL = 13   LL =378
KR = 3   KR =>

Figure 5.5 PhoR N-terminus. A. Sequence showing transmembrane segments TM1 and TM2. B. Segments predicted to span the membrane based on TopPredII (Claros and Von Heijne 1994). A fluid lipid bilayer is shown to be ca. 60 Å across with a hydrophobic core of ca. 30 Å and two interfaces of 15 Å each; a typical α-helical peptide may traverse the membrane (perpendicularly, as shown) or it may lie on the surface (horizontally) completely embedded within the interfacial layer (White and Wimley 1994). Top, periplasmic face; bottom, cytoplasmic face. LL, loop length; KR, lysine and arginine residues.

kinases contain two predicted α-helical transmembrane (TM1 and TM2) segments separated by a large "periplasmic loop" that is often ca. 100 residues or more in length (like CreC, EnvZ, and VanS in Figure 5.4). In contrast, TM1 and TM2 in PhoR are nearly adjacent and therefore capable of forming a helical hairpin (Figure 5.5). Furthermore, of the more than 200 histidine kinases now in GenBank (including ca. 130 unique ones), only four are predicted to have similar N-terminal structures. These include three PhoR proteins from diverse bacteria (PhoR$_{EC}$, PhoR$_{PA}$, and PhoR$_{HI}$; Table 5.2) and one other histidine kinase (EnvZ from *Xenorhabdus nematophilus,* XN; Tabatabai and Forst 1995). These PhoRs are predicted to have two or fewer hydrophilic residues in the periplasm, whereas EnvZ$_{XN}$ is predicted to have four or five. Other atypical histidine kinases of *E. coli* include some with multiple membrane-spanning segments (e.g., BaeS, EvgS, KdpD, UhpB, and YehU), two exceptional ones with short periplasmic domains (ArcB and PmrB, with 14 and 30 predicted periplasmic residues, respectively), and some that are cytoplasmic (e.g., BasS, CheA, NtrB, and StnORF).

The only other signaling protein predicted to have an N-terminal structure similar to PhoR is the methyl-accepting transducer HtrI of the archaeon *Halobacterium salinarium* (Bogomolni et al. 1994; Marwan et al. 1995). Interestingly,

Table 5.2 P_i signaling proteins of diverse bacteria

Organism (ID)[a]	Proteins
Caulobacter crescentus (CC)	PhoB, PhoU, PstB
E. coli (EC), *S. typhimurium* (ST), and closely related bacteria	PhoB, PhoR, PhoU, PstA, PstB, PstC, PstS
Haemophilus influenzae (HI)	PhoB, PhoR, PstA, PstB, PstC, PstS
Pseudomonas aeruginosa (PA)	PhoB, PhoR, PhoU, PstA, PstB, PstC, PstS
Mycobacterium leprae (ML)	PhoU, PstA, PstB, PstC, PstS
Rhizobium meliloti (RM)	PhoB, PhoU
Mycoplasma genitalium (MG)	PhoU, PstA, PstB, PstS

[a]ID, identification acronym. Sources of information are: *C. crescentus* (Y. Brun, personal communication); *E. coli* and several other members of the Enterobacteriaceae, for which all sequenced Pho genes are, as expected, greater than 97% identical [as cited in Wanner (1996)]; *H. influenzae* (Fleischmann et al. 1995), *P. aeruginosa* (Nikata et al. 1996; Kato et al. 1994; Anba et al. 1990); *M. leprae* (unpublished entry, GenBank); *R. meliloti* (unpublished entry, GenBank); and *M. genitalium* (Fraser et al. 1995).

HtrI is functionally associated with the photoreceptor sensory rhodopsin I (SRI). Therefore, the HtrI-SRI membrane complex may provide a precedent for how PhoR acts in the control of the Pho regulon. HtrI is required for phototaxis and is thought to act in an analogous manner to the methyl-accepting chemotaxis proteins (MCPs) Tar, Tsr, Trg, and Tap of *E. coli* and *S. typhimurium* (Amsler and Matsumura 1995). However, unlike those MCPs, which have large periplasmic domains for signal detection, HtrI receives signal input by its interaction with SRI and then provides signal output to CheA, leading to CheA autophosphorylation and a consequent signaling response (Rudolph et al. 1995). Accordingly, *E. coli* PhoR may be a multifunctional protein with an N-terminal HtrI-like domain for receiving signal input via an interaction with the Pst system and PhoU joined to a C-terminal signaling kinase domain.

4. P_i Inhibition by the Pst Transporter and PhoU

The Pst system has been studied extensively as a transporter. This system is highly specific for P_i; it recognizes the inorganic oxyanions arsenate and sulfate poorly or not at all (Ledvina et al. 1996). Specificity residues in PstS (originally named PhoS), a 34.4-kDa protein with a single P_i binding site (K_d ca. 1 µM). Its structure has now been refined to 1.05 Å and shows 12 strong hydrogen bonds between the four oxygens of P_i and polar side chains within the binding pocket. Involvement of PstS in P_i control is supported by the finding that missense changes altering the P_i binding pocket abolish P_i inhibition of gene expression. PstS probably contacts the channel proteins PstA and PstC on the periplasmic face of the inner membrane. Residues that may define the P_i binding site or be important for translocation of P_i have been identified in PstA and PstC (Webb et al. 1992). Generally, mutations that abolish P_i transport also abolish P_i inhibition of gene expression. An exceptional one (an R220E change in PstA) abolishes only transport (Cox et al. 1988), indicating that these processes can be uncoupled. This residue may lie on the periplasmic side of the membrane, where it can interact with PstS. Or, as the topology of PstA is unknown, the R220 residue may instead lie on the cytoplasmic face of the membrane, where it may instead interact with PhoR or PhoU. It has also been shown that, as expected, the ABC transporter component PstB is an ATPase (Chan and Torriani 1996). Importantly, the Pst system is nonessential for growth with P_i as a P source provided that P_i is in excess (greater than ca. 10 μM); thus it is possible to study mutational effects of the Pst transporter on P_i signaling. In the absence of the Pst system, a low-affinity P_i transporter (Pit) allows for uptake. A *pit* mutation has no effect on P_i control (Wanner 1996).

The negative regulator PhoU is 241 residues in length and is peripherally associated with the membrane, even though it lacks features of a typical membrane protein. Its membrane association may be due to an interaction with a Pst component, PhoR, or both. Until recently, studies on PhoU used a sole point mutation, the *phoU35* allele, which abolishes signaling (inhibition) without affecting P_i uptake. PhoU was purified to homogeneity in the lab of the late Harry Rosenberg (Surin et al. 1986). However, those studies were carried out prior to the discovery of two-component regulatory systems in bacteria and of protein phosphorylation in these systems, so no PhoU assay was available at the time. The *phoU35* lesion results in an A147E change in a relatively long predicted α-helical region that may define a site of interaction with a Pst component or PhoR. Accordingly, PhoU may act together with PhoR in the dephosphorylation of phospho-PhoB. It was recently discovered that a $\Delta phoU$ mutation, unlike the *phoU35* allele, simultaneously abolishes P_i signaling and causes a severe growth defect (Steed and Wanner 1993). On the basis of a "P_i sensitivity" phenotype, it was proposed that PhoU has two functions: one as a negative regulator and the other as an enzyme in P_i metabolism. If true, this may have broad implications in cellular metabolism. PhoU homologs have now been identified in six diverse bacteria (Table 5.2); all are encoded by structural genes adjacent to a *pst* or *phoB* gene. Curiously, *H. influenzae* lacks a PhoU protein, although it encodes the six other P_i signaling proteins in a single gene cluster.

5. Transmembrane Signal Transduction by Environmental P_i: A Model

On the basis of numerous observations, P_i signaling appears to involve three processes: (1) activation, (2) deactivation, and (3) inhibition (Wanner 1986;

1996); each of these involves highly specific protein-protein interactions. These processes are depicted in the model shown in Figure 5.6. Activation occurs under conditions of P_i limitation and requires the interaction of PhoR and PhoB. In response to P_i limitation, PhoR is autophosphorylated by ATP, phospho-PhoR transfers the phosphoryl group to PhoB, and phospho-PhoB in turn activates transcription at Pho regulon promoters (the P_i signaling response). Deactivation occurs upon a growth shift from P_i-limiting (activation) to P_i-excess (inhibition) conditions. Deactivation is a distinct intermediate step-down process, requiring both PhoR and an excess of PhoU (or PstB) to reestablish inhibition following a period of P_i limitation. Current evidence does not allow distinguishing whether deactivation requires PhoU, PstB, or both. In this process, PhoR and PhoU (or PstB) dephosphorylate phospho-PhoB. Inhibition occurs when P_i is in excess and requires the interaction of all seven P_i signaling proteins (PhoB, PhoR, PhoU, PstA, PstB, PstC, and PstS) in an "inhibition complex." The inhibition complex prevents phosphorylation of PhoB when P_i is in excess.

Activation, deactivation, and inhibition are all dynamic processes, requiring

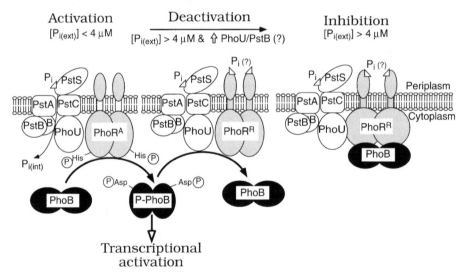

Figure 5.6 Model for transmembrane signal transduction by environmental P_i. Adapted from Fig. 4 in Wanner (1996). Small triangles mark a P_i binding site on PstS and a hypothetical regulatory site on PhoR. RhoB, response regulator; P-PhoB, phospho-PhoB; PhoR^A, activation (autophosphorylated) form of PhoR; PhoR^R, inhibition (repression) form of PhoR; PhoU, *phoU* gene product; Pst components; PstA and PstC, integral membrane channel proteins; PstB, traffic ATPase (permease); PstS, periplasmic P_i binding protein. $P_{i(ext)}$, external P_i, $P_{i(int)}$, internal P_i. The model is adopted from one that was originally illustrated in Wanner (1990) and modified in Wanner (1995, 1996).

PhoR. In the absence of PhoR, two other signaling pathways—involving CreC (formerly called PhoM) or acetyl phosphate—result in high activation of PhoB in response to carbon sources (Wanner 1992; Wanner and Wilmes-Riesenberg, 1992). To explain its dual roles, PhoR has been proposed to exist in two forms; these were called PhoR^R (PhoR repressor form) and PhoR^A (PhoR activator form) long before there was a mechanistic understanding of how they may act (Wanner and Latterell 1980). *In vitro* experiments have since shown that PhoR is a histidine autokinase that transfers phosphate to PhoB; phospho-PhoB (but not free PhoB) is a transcriptional activator (Makino et al. 1989). It is now believed that PhoR^R and PhoR^A determine the amount of phospho-PhoB, which in turn determines the level of Pho regulon gene expression.

In normal cells, activation occurs rapidly upon exhaustion of environmental P_i; activation requires only PhoR and PhoB because high activation still occurs in mutants lacking a Pst component or PhoU. As illustrated in Figure 5.6, a conformational change in PhoR resulting in the formation of PhoR^A may occur when extracellular P_i drops to a low level. PhoR^A corresponds to phospho-PhoR. It transfers phosphate to PhoB; phospho-PhoB then activates transcription of Pho regulon genes including the *pstSCAB-phoU* and the *phoBR* operons encoding the seven P_i signaling proteins. One purpose of this activation is to increase synthesis of the Pst system so that P_i can be more efficiently scavenged when scarce. More PhoU, PstB, or both is also required to reestablish inhibition (deactivation) following a period of P_i limitation (Wanner 1986, 1996). PhoR^R and PhoU (or PstB) may act together to dephosphorylate phospho-PhoB in order to turn off gene expression when cells are shifted from conditions of P_i limitation (activation) to P_i excess (inhibition). Eventually, the amounts of all seven proteins are lowered to inhibition levels, presumably because of simple dilution during growth. When P_i is in excess (> ca. 4 μM), all seven P_i signaling proteins are required to prevent phosphorylation of PhoB (inhibition).

5.1 Mechanism of Detecting Environmental P_i

It is unknown how the sensor PhoR detects P_i. It may detect P_i directly via a P_i regulatory site or it may detect it indirectly via an interaction(s) with PhoU or a Pst component(s). These alternatives are depicted in Figure 5.6, where P_i is shown to bind two proteins in the periplasm: PstS and an extracellular (putative P_i regulatory site) domain of PhoR. PstS is the high-affinity P_i binding protein (K_d ca. 1 μM) of the Pst system that binds P_i with exquisite specificity (Ledvina et al. 1996). P_i binding to PhoR is hypothetical. In order to account for a paradox in Pho regulon control, two alternative mechanisms were proposed: a PhoR P_i regulatory site model and a stoichiometric model (Wanner 1995). The paradox concerns the role(s) of PstS in P_i signaling and P_i uptake. If P_i were to bind only PstS, then P_i would be expected to bind PstS, causing inhibition when in excess, and to bind PstS for the purpose of transport without causing inhibition under conditions of P_i limitation. The PhoR regulatory site model would permit P_i binding to PstS under both conditions and P_i binding to PhoR only when in excess (i.e., when PstS is more saturated with P_i). If this is true, the primary role of the Pst system in P_i signaling may be to increase the local P_i concentration to facilitate P_i binding to the PhoR regulatory site. Alternatively, a stoichiometric model in which different ratios of PhoB, PhoR, PhoU, and the Pst components exist under conditions of P_i excess and limitation can account for the paradox. In this model, altering the levels of these signaling proteins would be expected to have dramatic signaling consequences, which is known to occur (Wanner 1996).

The hypothesized inhibition complex contains one molecule of PstS, two of PstB, one each of PstA and PstC, and two of PhoR. Stoichiometries of the Pst components are based on analogous periplasmic binding protein-dependent transporters (Doige and Ames 1993). PhoR is shown as a dimer to allow for its autophosphorylation (as PhoR^A) by a transphosphorylation mechanism (Yang and Inouye 1991). No particular stoichiometry of PhoU is intended; whether PhoU and PhoR interact with each other or with a Pst component is wholly speculative. Alternatively, PhoU (or a Pst component) may interact with PhoB. It is uncertain whether PhoB should be considered as a component of the inhibition complex. PhoB and PhoR are shown to interact as PhoR^A and PhoR^R have direct roles in the phosphorylation of PhoB and (probably) in the dephosphorylation of phospho-PhoB, respectively. These interactions are likely to be transient, however. More detailed descriptions of how PhoR, PhoU, and the Pst transporter may act in P_i signaling have been described elsewhere (Wanner 1995).

6. ABC Transporters in Other Signaling Systems

The Pst transporter belongs to the superfamily of ABC transporters that are present in a wide variety of cells, including mammalian cells (Higgins 1995). Analogous transporters (also called traffic ATPases) have been implicated in three other two-component signaling pathways: the control of sporulation and the control of competence in *Bacillus subtilis* by the *spoOK* locus (Hoch 1995; Solomon et al. 1995) and the control of multicellular fruiting body formation in *Myxococcus xanthus* by the *sasA* locus (Kaplan and Plamann 1996). Yet, these signaling systems are quite different from P_i signaling in *E. coli*. In the *B. subtilis* and *M. xanthus* examples, the transporter has an indirect role; in these cases, the transporter is required for transport of a small signaling molecule (a peptide) and not for signaling per se (Hoch 1995). In contrast, P_i inhibition by the Pst transporter is independent of P_i uptake. In the maltose regulon of *E. coli,* which does not involve a two-component regulatory system, the ABC protein MalK has been directly implicated in signaling by maltose (Covitz et al. 1994). P_i signaling by the Pst transporter may involve a fundamentally similar process.

7. P_i Signaling Proteins in Diverse Bacteria

Regulatory networks similar to the *E. coli* Pho regulon now appear to exist in seven phylogenically distant bacteria (Table 5.2). In all of these cases, the protein sequences show high degrees of sequence identities, and the respective *pho* or *pst* genes are often adjacent. With one exception, P_i signaling in these bacteria may also involve regulatory coupling between a sensor histidine kinase and Pst transporter. *M. genitalium* is exceptional because it encodes no two-component regulatory protein (Fraser et al. 1995), although genes for PhoU and Pst homologs are adjacent (A. Haldimann and B. L. Wanner, unpublished data).

In contrast, partner proteins of a number of other two-component "Pho systems" are quite different (Table 5.3), including the PhoQ-PhoP systems of *E. coli* (EC; Kasahara et al. 1992; Groisman et al. 1992) and *S. typhimurium* (ST; Miller et al. 1989; Groisman et al. 1989), the ChvG-ChvI system of *Agrobacterium tumefaciens* (AT; Mantis and Winans 1993; Charles and Nester 1993), the PhoR-PhoP and ResE-ResD systems of *B. subtilis* (BS; Sun et al. 1996; Lee and Hulett 1992; Seki et al. 1987), the PhoR-PhoP system of *M. leprae* (ML; GenBank), and the SphS-SphR system of *Synechococcus* sp. (Aiba and Mizuno 1994; Nagaya et al. 1994). On the basis of sequence similarities, the response regulators $ChvI_{AT}$, $PhoP_{EC/ST}$, $PhoP_{BS}$, $PhoP_{ML}$, $ResD_{BS}$, and SphR are no more related to $PhoB_{EC}$

Table 5.3 Regulatory proteins belonging to other Pho systems

Organism (ID)[a]	Proteins (RR, HK)[b]
E. coli (EC) and *S. typhimurium* (ST)	PhoP, PhoQ
Agrobacterium tumefaciens (AT)	ChvI, ChvG
Bacillus subtilis (BS)	PhoP, PhoR
Bacillus subtilis (BS)	ResD, ResE
Mycobacterium leprae (ML)	PhoP, PhoR
Synechococcus sp. (SP)	SphR, SphS

[a]ID, identification acronym.
[b]RR, response regulator; HK, histidine kinase.
Sources of information are: *E. coli* PhoP and PhoQ (Kasahara et al. 1992; Groisman et al. 1992), *S. typhimurium* PhoP and PhoQ (Miller et al. 1989; Groisman et al. 1989), *A. tumefaciens* ChvI and ChvG (Mantis and Winans 1993; Charles and Nester 1993), *B. subtilis* PhoP, PhoR, ResD, and ResE (Sun et al. 1996; Lee and Hulett 1992; Seki et al. 1987), *Mycobacterium leprae* PhoP and PhoR (unpublished entry, GenBank), and *Synechococcus* sp. SphR and SphS (Aiba and Mizuno 1994; Nagaya et al. 1994).

than to many other response regulators. Likewise, the histidine kinases $ChvG_{AT}$, $PhoQ_{EC/ST}$, $PhoR_{BS}$, $PhoR_{ML}$, $ResE_{BS}$, and SphS are no more related to $PhoR_{EC}$ than to many other histidine kinases. None of these two-component regulatory systems appears to be coupled to a transporter. Accordingly, the mechanism(s) by which these systems detect environmental signals is probably fundamentally different. It is even unclear whether they respond to extracellular P_i per se. In this regard, the PhoQ-PhoP two-component regulatory system was thought to be regulated by P_i (Kier et al., 1979; Groisman et al. 1989); it was recently shown to be regulated by magnesium instead (García Véscovi et al. 1996).

8. Additional Controls of the Pho Regulon

In addition to its control by environmental P_i, the Pho regulon is controlled by two other signaling pathways (at least in mutants). As described above, the primary pathway for control of the Pho regulon responds to environmental P_i and involves regulation (via phosphorylation and dephosphorylation reactions) of PhoB activity by PhoR, PhoU, and the Pst transporter. In contrast, both of the other pathways are P_i independent. One of these P_i-independent pathways responds to an unknown catabolite and involves the catabolite regulatory sensor kinase CreC (originally called PhoM); the other P_i-independent pathway responds to acetyl phosphate levels and leads to activation by a process requiring acetyl phosphate (Wanner and Wilmes-Riesenberg 1992). Because large regulatory effects due to CreC and acetyl phosphate are seen only in the absence of PhoR, it is uncertain whether these P_i-independent controls also have a role in Pho regulon control in normal cells. Yet dramatic effects due to CreC and acetyl phosphate are observed in the presence of normal (physiological) amounts of the regulatory proteins (PhoB and CreC) and enzymes for acetyl phosphate synthesis (phosphotransacetylase and acetate kinase). Furthermore, activation of PhoB by CreC or acetyl phosphate is quite substantial; under certain conditions activation of PhoB by CreC or acetyl phosphate can even exceed activation of PhoB by PhoR. For these and other reasons, the controls of the Pho regulon by these P_i-independent pathways may be forms of "cross-regulation," which may be important for the overall global control of cell growth and metabolism (Wanner 1992).

9. Cross-Regulation and Acetyl Phosphate

Cross-regulation refers to the control of a response regulator of one two-component regulatory system by a different regulatory system. All three signaling pathways for activation of the Pho regulon lead to phosphorylation of PhoB

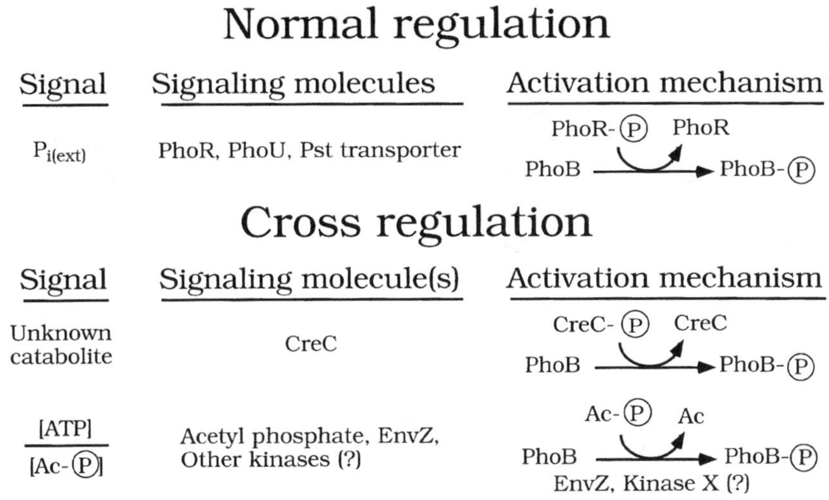

Normal regulation

Signal	Signaling molecules	Activation mechanism
$P_{i(ext)}$	PhoR, PhoU, Pst transporter	PhoR-Ⓟ PhoR PhoB ⟶ PhoB-Ⓟ

Cross regulation

Signal	Signaling molecule(s)	Activation mechanism
Unknown catabolite	CreC	CreC-Ⓟ CreC PhoB ⟶ PhoB-Ⓟ
$\dfrac{[ATP]}{[Ac\text{-}Ⓟ]}$	Acetyl phosphate, EnvZ, Other kinases (?)	Ac-Ⓟ Ac PhoB ⟶ PhoB-Ⓟ EnvZ, Kinase X (?)

Figure 5.7 Normal regulation and cross-regulation of the Pho regulon. Circled P refers to the activated phosphoryl group on PhoR, CreC, or acetyl phosphate (Ac-P). $P_{i(ext)}$, extracellular P_i, Kinase X, unknown histidine kinase; Ac, acetate.

(Figure 5.7). Normal regulation responds to environmental P_i as the signal and requires as signaling molecules the sensor kinase PhoR, PhoU, and the Pst transporter. Under conditions of P_i limitation, PhoR is autophosphorylated and phospho-PhoR transfers the phosphoryl group to PhoB. Cross-regulation by acetyl phosphate may respond to the ATP-to-acetyl phosphate ratio [see Wanner (1992)] and leads to phosphorylation of PhoB by a process requiring acetyl phosphate (Wanner and Wilmes-Riesenberg 1992). Activation of PhoB by acetyl phosphate clearly also requires the sensor histidine kinase EnvZ under certain conditions (Kim et al. 1996). Accordingly, activation of PhoB by acetyl phosphate (and perhaps other two-component response regulators as well) probably always requires a certain kinase(s) that can vary depending upon the growth condition. The requirement for a sensor kinase(s) for activation of PhoB by acetyl phosphate *in vivo* is consistent with the inefficient phosphorylation of PhoB by acetyl phosphate *in vitro*. As the *in vivo* concentration of acetyl phosphate seldom exceeds 1 mM and the K_M of PhoB for acetyl phosphate is ca. 7 mM, it seems unlikely that acetyl phosphate contributes significantly to the direct activation of PhoB *in vivo* (McCleary 1996).

It is likely that cells coordinate the expression of genes for many diverse branches of metabolism in order to maximize growth yield or rate. This coordination is likely to involve many signal inputs and transcriptional controls, as well as other controls. Different genes, operons, or regulons may also share controls to form an overlapping regulatory network to coordinate particular metabolic processes and eventually for global regulation. Cross-regulation may be an important means of global control of genes for separate but related metabolic processes. Activation of PhoB both by CreC and acetyl phosphate appears to be coupled to carbon and energy metabolism. As pathways of carbon and energy metabolism are formally also pathways of P_i metabolism, cross-regulation of the Pho regulon by the CreC and acetyl phosphate signaling pathways may provide a regulatory coupling(s) between the uptake of environmental P_i and intracellular P_i metabolism. Regulatory interactions of this sort may be a form of cross-regulation between different two-component regulatory systems, as well as other regulatory networks (Wanner 1992; Wanner et al. 1995).

10. Concluding Comments

A vast amount of information is now known about many P_i-regulated genes and the respective gene products. Much is also known about the biochemical mechanism of gene activation. Yet many mysteries remain. For example, in regard to phosphonate metabolism, seven to nine gene products may be required as part of a C-P lyase enzyme complex. The catalytic mechanism of C-P bond fission by the lyase is poorly understood. All well-characterized Pho regulon promoters are activated by a phosphorylated transcription factor (phospho-PhoB) but not by unphosphorylated PhoB. How a phosphorylated, but not an unphosphorylated, regulatory protein leads to transcriptional activation is unknown. Three additional proteins (PhnF, PhnO, and PhnR) may also regulate specific Pho regulon genes. Where and how do these proteins act? Do they act together with PhoB (in a regulatory complex) or act independently? P_i signaling involves both transmembrane signal transduction and highly specific protein-protein interactions. Yet it is unknown how P_i is detected. Does the sensor kinase PhoR detect P_i directly? Or, does PhoR detect P_i indirectly via an interaction with a Pst component, or PhoU? How is the signal for P_i limitation transduced across the membrane? The PhoB-PhoR two-component regulatory system is one of many (perhaps 50 different) pairs of these regulatory proteins in a single bacterium. Yet each of these systems appears to be highly specific (Haldimann et al. 1996). What are the specificity determinants of protein-protein interactions between various homologous (partner) and heterologous (nonpartner) two-component regulatory proteins? In particular cases, interactions suggestive of cross-regulation occur between certain nonpartner proteins. Are interactions of this sort important? Have they evolved for a specific purpose(s)? Acetyl phosphate appears to be a key signaling molecule. How important is acetyl phosphate for cross-regulation between heterologous sensor kinase and response regulator proteins? Searching for answers to these and other questions is the goal of new research in the area of P_i signaling and the control of P_i-regulated genes in *E. coli,* and other cells.

Acknowledgments

I thank Andreas Haldimann and Irwin Tessman for carefully reading this manuscript and for helpful comments. Research in this laboratory has been supported by NIH grant GM35392 and NSF grant MCB 9405929.

References

Aiba, H., and T. Mizuno. 1994. A novel gene whose expression is regulated by the response-regulator, SphR, in response to phosphate limitation in *Synechococcus* species PCC7942. *Mol. Microbiol.* 13:25–34.

Amsler, C. D., and P. Matsumura. 1995. Chemotactic signal transduction in *Escherichia coli* and *Salmonella typhimurium.* In *Two-Component Signal Transduction,* eds. J. A. Hoch and T. J. Silhavy. pp. 89–103. American Society for Microbiology, Washington, D.C.

Anba, J., M. Bidaud, M. L. Vasil, and A. Lazdunski. 1990. Nucleotide sequence of the *Pseudomonas aeruginosa phoB* gene, the regulatory gene for the phosphate regulon. *J. Bacteriol.* 172:4685–4689.

Bogomolni, R. A., W. Stoeckenius, I. Szundi, E. Perozo, K. D. Olson, and J. L. Spudich. 1994. Removal of transducer HtrI allows electrogenic proton translocation by sensory rhodopsin I. *Proc. Natl. Acad. Sci. U.S.A.* 91:10188–10192.

Brzoska, P., M. Rimmele, K. Brzostek, and W. Boos. 1994. The *pho* regulon-dependent Ugp uptake system for glycerol-3-phosphate in *Escherichia coli* is *trans* inhibited by P_i. *J. Bacteriol.* 176:15–20.

Chan, F. Y., and A. Torriani. 1996. PstB protein of the phosphate-specific transport system of *Escherichia coli* is an ATPase. *J. Bacteriol.* 178:3974–3977.

Charles, T. C., and E. W. Nester. 1993. A chromosomally encoded two-component sensory transduction system is required for virulence of *Agrobacterium tumefaciens. J. Bacteriol.* 175:6614–6625.

Claros, M. G., and G. Von Heijne. 1994. TopPredII: An improved software for membrane protein structure predictions. *Comput. Appl. Biosci.* 10:685–686.

Covitz, K.-M.Y., C. H. Panagiotidis, L.-I. Hor, M. Reyes, N. A. Treptow, and H. A. Shuman. 1994. Mutations that alter the transmembrane signalling pathway in an ATP binding cassette (ABC) transporter. *EMBO J.* 13:1752–1759.

Cox, G. B., D. Webb, J. Godovac-Zimmermann, and H. Rosenberg. 1988. Arg-220 of the PstA protein is required for phosphate transport through the phosphate-specific transport system in *Escherichia coli* but not for alkaline phosphatase repression. *J. Bacteriol.* 170:2283–2286.

Doige, C. A., and G.F.-L. Ames. 1993. ATP-dependent transport systems in bacteria and humans: Relevance to cystic fibrosis and multidrug resistance. *Ann. Rev. Microbiol.* 47:291–319.

Econs, M. J. 1996. Positional cloning of the HYP gene: A review. *Kidney Int.* 49:1033–1037.

Fankhauser, H., A. M. Schweingruber, E. Edenharter, and M. E. Schweingruber. 1995. Growth of a mutant defective in a putative phosphoinositide-specific phospholipase C of *Schizosaccharomyces pombe* is restored by low concentrations of phosphate and inositol. *Current Genet.* 28:199–203.

Fleischmann, R. D., M. D. Adams, O. White, R. A. Clayton, E. F. Kirkness, A. R. Kerlavage, C. J. Bult, J.-F. Tomb, B. A. Dougherty, J. M. Merrick, K. McKenney, G. Sutton, W. FitzHugh, C. Fields, J. D. Gocayne, J. Scott, R. Shirley, L.-I. Liu, A. Glodek, J. M. Kelley, J. F. Weidman, C. A. Phillips, T. Spriggs, E. Hedblom, M. D. Cotton, T. R. Utterback, M. C. Hanna, D. T. Nguyen, D. M. Saudek, R. C. Brandon, L. D. Fine, J. L. Fritchman, J. L. Fuhrmann, N.S.M. Geoghagen, C. L. Gnehm, L. A. McDonald, K. V. Small, C. M. Fraser, H. O. Smith, and J. C. Venter. 1995. Whole-genome random sequencing and assembly of *Haemophilus influenzae* Rd. *Science* 269:496–512.

Fraser, C. M., J. D. Gocayne, O. White, M. D. Adams, R. A. Clayton, R. D. Fleischmann, C. J. Bult, A. R. Kerlavage, G. Sutton, J. M. Kelley, J. L. Fritchman, J. F. Weidman, K. V. Small, M. Sandusky, J. Fuhrmann, D. Nguyen, T. R. Utterback, D. M. Saudek, C. A. Phillips, J. M. Merrick, J.-F. Tomb, B. A. Dougherty, K. F. Bott, P.-C. Hu, T. S. Lucier, S. N. Peterson, H. O. Smith, C. A. Hutchison III, and J. C. Venter. 1995. The minimal gene complement of *Mycoplasma gentitalium*. *Science* 270:397–403.

García Véscovi, E., F. C. Soncini, and E. A. Groisman. 1996. Magnesium as an extracellular signal: Environmental regulation of *Salmonella* virulence. *Cell* 84:165–174.

Groisman, E. A., E. Chiao, C. J. Lipps, and F. Heffron. 1989. *Salmonella typhimurium phoP* virulence gene is a transcriptional regulator. *Proc. Natl. Acad. Sci. U.S.A.* 86:7077–7081.

Groisman, E. A., F. Heffron, and F. Solomon. 1992. Molecular genetic analysis of the *Escherichia coli phoP* locus. *J. Bacteriol.* 174:486–491.

Haldimann, A., M. K. Prahalad, S. L. Fisher, S.-K. Kim, C. T. Walsh, and B. L. Wanner. 1996. Altered recognition mutants of the response regulator PhoB: a new genetic strategy for studying protein-protein interactions. Proc. Natl. Acad. Sci. USA 93:14361–14366.

Harlocker, S. L., L. Bergstrom, and M. Inouye. 1995. Tandem binding of six OmpR proteins to the *ompF* upstream regulatory sequence of *Escherichia coli*. *J. Biol. Chem.* 270:26849–26856.

Higgins, C. F. 1995. The ABC of channel regulation. *Cell* 82:693–696.

Hoch, J. A. 1995. Control of cellular development in sporulating bacteria by the phospho-relay two-component signal transduction system. In *Two-Component Signal Transduction*, eds. J. A. Hoch and T. J. Silhavy. pp. 129–144. American Society for Microbiology, Washington, D.C.

Jiang, W., W. W. Metcalf, K.-S. Lee, and B. L. Wanner. 1995. Molecular cloning, mapping, and regulation of Pho regulon genes for phosphonate breakdown by the phosphonatase pathway of *Salmonella typhimurium* LT2. *J. Bacteriol.* 177:6411–6421.

Kaplan, H.B., and L. Plamann. 1996. A *Myxococcus xanthus* cell density-sensing system required for multicellular development. *FEMS Microbiol. Lett.* 139:89–95.

Kasahara, M., K. Makino, M. Amemura, A. Nakata, and H. Shinagawa. 1991. Dual regulation of the *ugp* operon by phosphate and carbon starvation at two interspaced promoters. *J. Bacteriol.* 173:549–558.

Kasahara, M., A. Nakata, and H. Shinagawa. 1992. Molecular analysis of the *Escherichia coli phoP-phoQ* operon. *J. Bacteriol.* 174:492–498.

Kato, J., Y. Sakai, T. Nikata, and H. Ohtake. 1994. Cloning and characterization of a *Pseudomonas aeruginosa* gene involved in the negative regulation of phosphate taxis. *J. Bacteriol.* 176:5874–5877.

Kier, L. D., R. M. Weppelman, and B. N. Ames. 1979. Regulation of nonspecific acid phosphatase in *Salmonella: phoN* and *phoP* genes. *J. Bacteriol.* 138:155–161.

Kim, S.-K., M. R. Wilmes-Riesenberg, and B. L. Wanner. 1996. Involvement of the sensor kinase EnvZ in the *in vivo* activation of the response regulator PhoB by acetyl phosphate. *Mol. Microbiol.* (in press).

Kondo, H., T. Miyaji, M. Suzuki, S. Tate, T. Mizuno, Y. Nishimura, and I. Tanaka. 1994. Crystallization and X-ray studies of the DNA-binding domain of OmpR protein, a positive regulator involved in activation of osmoregulatory genes in *Escherichia coli*. *J. Mol. Biol.* 235:780–782.

Ledvina, P. S., N. H. Yao, A. Choudhary, and F. A. Quiocho. 1996. Negative electrostatic surface potential of protein sites specific for anionic ligands. *Proc. Natl. Acad. Sci. U.S.A.* 93:6786–6791.

Lee, J. W., and F. M. Hulett. 1992. Nucleotide sequence of the *phoP* gene encoding PhoP, the response regulator of the phosphate regulon of *Bacillus subtilis*. *Nucl. Acids Res.* 20:5848.

Lee, K.-S., W. W. Metcalf, and B. L. Wanner. 1992. Evidence for two phosphonate degradative pathways in *Enterobacter aerogenes*. *J. Bacteriol.* 174:2501–2510.

Makino, K., M. Amemura, T. Kawamoto, S. Kimura, H. Shinagawa, A. Nakata, and M. Suzuki. 1996. DNA binding of PhoB and its interaction with RNA polymerase. *J. Mol. Biol.* 259:15–26.

Makino, K., H. Shinagawa, M. Amemura, T. Kawamoto, M. Yamada, and A. Nakata. 1989. Signal transduction in the phosphate regulon of *Escherichia coli* involves phosphotransfer between PhoR and PhoB proteins. *J. Mol. Biol.* 210:551–559.

Mann, B. J., B. J. Bowman, J. Grotelueschen, and R. L. Metzenberg. 1989. Nucleotide sequence of *pho-4+*, encoding a phosphate-repressible phosphate permease of *Neurospora crassa. Gene* 83:281–289.

Mantis, N. J., and S. C. Winans. 1993. The chromosomal response regulatory gene *chvI* of *Agrobacterium tumefaciens* complements an *Escherichia coli phoB* mutation and is required for virulence. *J. Bacteriol.* 175:6626–6636.

Martínez-Hackert, E., S. Harlocker, M. Inouye, H. M. Berman, and A. M. Stock. 1996. Crystallization, X-ray studies, and site-directed cysteine mutagenesis of the DNA-binding domain of OmpR. *Protein Sci.* 5:1429–1433.

Marwan, W., S. I. Bibikov, M. Montrone, and D. Oesterhelt. 1995. Mechanism of photosensory adaptation in *Halobacterium salinarium. J. Mol. Biol.* 246:493–499.

McCleary, W. R. 1996. The activation of PhoB by acetylphosphate. *Mol. Microbiol.* 20:1155–1163.

Miller, S. I., A. M. Kukral, and J. J. Mekalanos. 1989. A two-component regulatory system (*phoP phoQ*) controls *Salmonella typhimurium* virulence. *Proc. Natl. Acad. Sci. U.S.A.* 86:5054–5058.

Murer, H., and J. Biber. 1996. Molecular mechanisms of renal apical Na phosphate cotransport. *Ann. Rev. Physiol.* 58:607–618.

Nagaya, M., H. Aiba, and T. Mizuno. 1994. The *sphR* product, a two-component system response regulator protein, regulates phosphate assimilation in *Synechococcus* sp. strain PCC 7942 by binding to two sites upstream from the *phoA* promoter. *J. Bacteriol.* 176:2210–2215.

Nikata, T., Y. Sakai, K. Shibata, J. Kato, A. Kuroda, and H. Ohtake. 1996. Molecular analysis of the phosphate specific transport (*pst*) operon of *Pseudomonas aeruginosa. Mol. Gen. Genet.* 250:692–698.

Ogawa, N., K. Noguchi, H. Sawai, Y. Yamashita, C. Yompakdee, and Y. Oshima. 1995. Functional domains of Pho81p, an inhibitor of Pho85p protein kinase, in the transduction pathway of P_i signals in *Saccharomyces cerevisiae. Mol. Cell. Biol.* 15:997–1004.

Parkinson, J. S. 1995. Genetic approaches for signaling pathways and proteins. In *Two-Component Signal Transduction,* eds. J. A. Hoch and T. J. Silhavy. pp. 9–23. American Society for Microbiology, Washington, D.C.

Parkinson, J. S., and E. C. Kofoid. 1992. Communication modules in bacterial signaling proteins. *Ann. Rev. Genet.* 26:71–112.

Pratt, L. A., and T. J. Silhavy. 1994. OmpR mutants specifically defective for transcriptional activation. *J. Mol. Biol.* 243:579–594.

Rosenberg, H., and J. M. LaNauze. 1967. The metabolism of phosphonates by microorganisms. The transport of aminoethylphosphonic acid in *Bacillus cereus. Biochim. Biophys. Acta* 141:79–90.

Rowe, P.S.N., J. N. Goulding, F. Francis, C. Oudet, M. J. Econs, A. Hanauer, H. Lehrach, A. P. Read, R. C. Mountford, T. Summerfield, J. Weissenbach, W. Fraser, M. K. Drezner, K. E. Davies, and J.L.H. O'Riordan. 1996. The gene for X-linked hypophosphataemic rickets maps to a 200–300 kb region in Xp22.1, and is located on a single YAC containing a putative vitamin D response element (VDRE). *Hum. Genet.* 97:345–352.

Rudolph, J., N. Tolliday, C. Schmitt, S. C. Schuster, and D. Oesterhelt. 1995. Phosphorylation in halobacterial signal transduction. *EMBO J.* 14:4249–4257.

Seki, T., H. Yoshikawa, H. Takahashi, and H. Saito. 1987. Cloning and nucleotide sequence of *phoP*, the regulatory gene for alkaline phosphatase and phosphodiesterase in *Bacillus subtilis. J. Bacteriol.* 169:2913–2916.

Solomon, J. M., R. Magnuson, A. Srivastava, and A. D. Grossman. 1995. Convergent sensing pathways mediate response to two extracellular competence factors in *Bacillus subtilis. Genes Dev.* 9:547–558.

Steed, P. M., and B. L. Wanner. 1993. Use of the *rep* technique for allele replacement to construct mutants with deletions of the *pstSCAB-phoU* operon: Evidence of a new role for the PhoU protein in the phosphate regulon. *J. Bacteriol.* 175:6797–6809.

Su, T.-Z., H. P. Schweizer, and D. L. Oxender. 1991. Carbon-starvation induction of the *ugp* operon, encoding the binding protein-dependent *sn*-glycerol-3-phosphate transport system in *Escherichia coli. Mol. Gen. Genet.* 230:28–32.

Sun, G. F., E. Sharkova, R. Chesnut, S. Birkey, M. F. Duggan, A. Sorokin, P. Pujic, S. D. Ehrlich, and F. M. Hulett. 1996. Regulators of aerobic and anaerobic respiration in *Bacillus subtilis. J. Bacteriol.* 178:1374–1385.

Surin, B. P., N. E. Dixon, and H. Rosenberg. 1986. Purification of the PhoU protein, a negative regulator of the *pho* regulon of *Escherichia coli* K-12. *J. Bacteriol.* 168:631–635.

Suzuki, M., and K. Makino. 1995. The DNA binding domains of transcription factors, PhoB and OmpR, adopt the same folding as that of histone H5 and bind to RNA polymerase using the ends of two α-helices. *Proc. Japan Acad.* 71:132–137.

Tabatabai, N., and S. Forst. 1995. Molecular analysis of the two-component genes, *ompR* and *envZ*, in the symbiotic bacterium *Xenorhabdus nematophilus. Mol. Microbiol.* 17:643–652.

Van Bogelen, R. A., E. R. Olson, B. L. Wanner, and F. C. Neidhardt. 1996. Global analysis of proteins synthesized during phosphorus restriction in *Escherichia coli. J. Bacteriol.* 178.

Volz, K. 1993. Structural conservation in the CheY superfamily. *Biochem.* 32:11741–11753.

Wanner, B. L. 1986. Novel regulatory mutants of the phosphate regulon in *Escherichia coli* K-12. *J. Mol. Biol.* 191:39–58.

Wanner, B. L. 1990. Phosphorus assimilation and its control of gene expression in *Escherichia coli.* In *The Molecular Basis of Bacterial Metabolism,* pp. 152–163. eds. G. Hauska and R. Thauer, Springer-Verlag, Heidelberg.

Wanner, B. L. 1992. Minireview. Is cross regulation by phosphorylation of two-component response regulator proteins important in bacteria? *J. Bacteriol.* 174:2053–2058.

Wanner, B. L. 1994. Molecular genetics of carbon-phosphorus bond cleavage in bacteria. *Biodegradation* 5:175–184.

Wanner, B. L. 1995. Signal transduction and cross regulation in the *Escherichia coli* phosphate regulon by PhoR, CreC, and acetyl phosphate. In *Two-Component Signal Transduction,* eds. J. A. Hoch and T. J. Silhavy. pp. 203–221. American Society for Microbiology, Washington, D.C.

Wanner, B. L. 1996. Phosphorus assimilation and control of the phosphate regulon. In *Escherichia coli* and *Salmonella typhimurium Cellular and Molecular Biology,* eds. F. C. Neidhardt, R. Curtiss III, J. L. Ingraham, E.C.C. Lin, K. B. Low, Jr., B. Magasanik,

W. Reznikoff, M. Riley, M. Schaechter, and H. E. Umbarger. pp. 1357–1381. American Society for Microbiology, Washington, D.C.

Wanner, B. L., W. Jiang, S.-K. Kim, S. Yamagata, A. Haldimann, and L. L. Daniels. 1995. Are the multiple signal transduction pathways of the Pho regulon due to cross talk or cross regulation? In *Regulation of Gene Expression in Escherichia coli,* eds. E.C.C. Lin and A. S. Lynch. pp. 297–315. R. G. Landes Co., Austin, TX.

Wanner, B. L., and P. Latterell. 1980. Mutants affected in alkaline phosphatase expression: Evidence for multiple positive regulators of the phosphate regulon in *Escherichia coli. Genetics* 96:242–266.

Wanner, B. L., and M. R. Wilmes-Riesenberg. 1992. Involvement of phosphotransacety-lase, acetate kinase, and acetyl phosphate synthesis in the control of the phosphate regulon in *Escherichia coli. J. Bacteriol.* 174:2124–2130.

Webb, D. C., H. Rosenberg, and G. B. Cox. 1992. Mutational analysis of the *Escherichia coli* phosphate-specific transport system, a member of the traffic ATPase (or ABC) family of membrane transporters. A role for proline residues in transmembrane helices. *J. Biol. Chem.* 267:24661–24668.

White, S. H., and W. C. Wimley. 1994. Peptides in bilayers: Structural and thermodynamic basis for partitioning and folding. *Curr. Opin. Struct. Biol.* 4:79–86.

Yang, Y., and M. Inouye. 1991. Intermolecular complementation between two defective mutant signal-transducing receptors of *Escherichia coli. Proc. Natl. Acad. Sci. U.S.A.* 88:11057–11061.

II

ANIMAL METAL METABOLISM

6

Translational Regulation of Bioiron

Elizabeth C. Theil

1. The Problem: Iron Need vs Iron Insolubility

Iron is central to respiration, photosynthesis, nitrogen fixation, and, in most organisms, the reduction of ribose to deoxyribose, rate limiting in DNA synthesis. Dioxygen, which allows high-efficiency bioenergetics, at the same time converts soluble ferrous ions to insoluble ferric ions; iron concentrations in cells are almost a trillion times the solubility of the free ferric ion under physiological conditions. Ferritin is the protein that concentrates ferric iron in all known organisms [reviewed in Theil (1987, 1990) Waldo and Theil (1996) Harrison and Lilley (1990)]. Induction of ferritin synthesis by iron also protects cells from oxidant stress (Balla et al. 1992). Gene regulation of ferritin synthesis is precise and complex and, in animals, is coordinately regulated with transferrin receptor (TfR) synthesis [reviewed in Theil (1990, 1993, 1994, 1997) Hentze and Kuhn (1996) Rouault and Klausner (1996) Klausner et al. (1993) Munro (1990)]. The problem of acquiring and concentrating iron is solved by the use of environmental iron to regulate expression of both iron-storage (ferritin) and iron-uptake (TfR) genes; changes in expression of ferritin and TfR are also regulated by growth factors and hormones and during cell differentiation.

Iron, when readily available, represses TfR synthesis expression and derepresses ferritin synthesis. Conversely, when iron is limiting, ferritin synthesis is repressed and TfR synthesis is derepressed. Whether the same coordination of expression of genes controlling iron uptake and storage occurs in bacteria, with ferritin/siderophore expression or in plants with ferritin/mugeneic acid synthesis and proton secretion, has yet to be explored. Available evidence suggests that in the case of ferritin at least, the observed conservation of protein sequence and regulatory signals does not extend to genetic regulatory mechanisms (Kimata and Theil 1994; Lescure et al. 1991) or gene organization (Proudhon et al. 1996).

mRNA is a major target for iron-induced changes in expression of ferritin and TfR in animals. mRNA regulation depends on a family of noncoding sequences called IREs (iron responsive elements); IREs are among the longest-known conserved RNA sequences (n=28–30) (Figure 6.1). The phylogenetic conservation of each IRE family member is high (>96–99% sequence identity), but variation in sequence is only 39% between translation IREs (ferritin and erythroid aminolevulinate synthase) and stability IREs (TfR) (Table 6.1).

IREs that control translation have relatively G/C-rich sequences and are near the 5' cap of mRNA. In contrast, IREs that control mRNA instability are AU rich and are in the 3' UTR (untranslated region) and can include the canonical AUUUA found in many mRNA rapid turnover elements. All IREs have the secondary structure of a hairpin loop (Figure 6.1 and 6.2) sharing the terminal loop sequence: CAGUG/C (Figure 6.1). Tertiary structure is indicated by the complex reactivity with protein and chemical nucleases [Wang et al. (1990) Harrell et al. (1991) Dix et al. (1993) Thorpe et al. (1996)]. IREs can be thought of as composites of generic RNA regulatory elements (Figure 6.3) (stable, G/C-

---

Figure 6.1 Primary and secondary structure of iron regulatory elements. A. Comparison of ▶ the regulatory elements for translation (ferritin, Fr and erythroid aminolevulinate synthase, eALAS) and mRNA stability (transferrin receptor, Tfr). Seven phylogenetically conserved members of the families are shown. All of the members of the ferritin IRE are displayed; however, the very high phylogenetic conservation for each family member (Fr, eALAS, Tfr-a, Tfr-b, Trf-c, Tfr-d, and Tfr-e) is similar. Differences in sequence identity between family members are shown in Table 6.1. The variation is 39–76% with the exception of Tfr-b and Tfr-c, where the sequence identity is 90% because of duplication of the palindromic sequence AUUAU/AUAAU. Sequences predicted to be paired are shown in capital letters; the internal bulge/loop is boxed. Note that the 28 nucleotides used in the original comparisons of IREs (Theil 1990; Klausner et al. 1993; Munro 1990) can be extended to 30 for ferritin (Theil 1993, 1994), and to 29 for eALAS and Tfr-a, which reveals, among the Tfr-IREs, the canonical AUUUA sequence of mRNA stability elements. (Figure reproduced from Theil 1994 with permission). B. Predicted secondary structure of the ferritin IRE. Shown are the hairpin loop mutation (G/A), which alters structural stability and the conserved triplet of base pairs in the flanking region (FL) specific to the ferritin IRE. Functional specificity of the ferritin IRE includes quantitatively greater effects on translation upon IRP binding (Bhasker et al. 1993), modulation of the IRP effect by the FL (Harrell et al. 1991), and enhanced rates of translation in the absence of the IRP or the presence of high cellular iron (Zahringer et al. 1976; Schaefer and Theil 1981; Shull and Theil 1983).

The Iron Regulatory Element (IRE) Family

Iron Storage	1		10	IRE	20		28
Human Fr-H	GUUUCC	ugc	UUCAA	cagugc	UUGGA	c	GGAAC
Mouse Fr-H	GUUUCC	ugc	UUCAA	cagugc	UUGAA	c	GGAAC
Rat Fr-H	GUUUCC	ugc	UUCAA	cagugc	UUGAA	c	GGAAC
Chicken Fr-H	GU-UCC	ugc	gUCAA	cagugc	UUGGa	c	GGAAC
Frog Fr-H	GU-UCU	ugc	UUCAA	cagugu	UUGAA	c	GGAAC
Frog Fr-H′ (M)	GU-UCU	ugc	UUCAA	cagugu	UUGAA	c	GGAAC
Human Fr-L	GUCUCU	ugc	uUCAA	cagugu	UUGA	c	GAACA
Rabbit Fr-L	GUGUCU	ugc	UUCAA	cagugu	UUGAA	c	GGAAC
Rat Fr-L	GUAUCU	ugc	UUCAA	cagugu	UUGGA	c	GGAAC
Iron Uptake							
Human/Rat/Chick Tfr-a	AUUUAU	c	AGUGA	cagagu	UCACU	·	AUAAAU
Human/Rat/Chick Tfr-b	AUUAU	c	GGAAG	cagugc	CUUCC	·	AUAAU
Human/Rat/Chick Tfr-c	AUUAU	c	GGGAG	cagugu	CUUCC	·	AUAAU
Human/Rat/Chick Tfr-d	UAU	c	GGAGA	caguga	UCUCC	·	AUA
Human/Rat/Chick Tfr-e	AUUAU	c	GGGAA	cagugu	UUCCC	·	AUAAU
Heme Synthesis							
Human/Mouse eALA	UC,GUU	c	GUCCU	cagugc	AGGGC	·	AACaGG

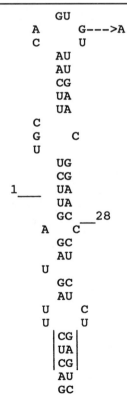

Table 6.1 Homology (%) among IRE (iron responsive element) types in different mRNAs. (Phylogenetic conservation for a single IRE type is 96–99%)[a]

Family Members	Ferritin (% Homology)	TfR-b (% Homology)
Ferritin	100	39
eALAS[a]	48	39
TfR-a[a]	32	62
TfR-b[a]	36	100
TfR-c[a]	36	90
TfR-d[a]	32	72
TfR-e[a]	36	76

[a]eALAS = erythroid aminolevulinate synthase; TfR = transferrin receptor. The number of sequences examined in each IRE type is: ferritin = 9 (human, mouse, rat, chick, frog); eALAs = 2 (human, mouse); TfR-a–e = 3 (human, rat, chick).

Different IRE Contexts in Ferritin MRNA and Transferrin Receptor MRNA

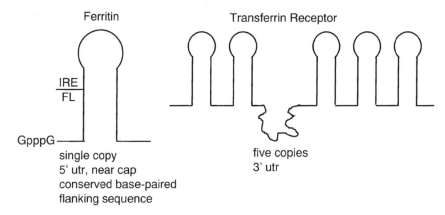

Figure 6.2 Different contexts of IREs. Translation IREs, which control ribosome binding, are near the cap. In the case of the ferritin IRE, base-paired flanking regions (FLs) provide correct position; FLs vary in length in rough proportion to the distance of the IRE from the cap (Dix et al. 1993; Wang et al. 1990). A conserved triplet of base pairs in the FL (Figure 6.1) modulate regulation (Harrell et al. 1991). The 5′-untranslated region in eALAS mRNA is much shorter than ferritin mRNA with the IRE very close to both the cap and the initiator AUG. Stability IREs are present as multiple variants in the 3′ UTR with AU-rich regions that include the canonical AUUUA of other mRNA stability elements. The redundancy of the IRE sequence, which includes a repeated palindrome, permits the prediction of the secondary structure of the five Tfr-IREs or a very long hair pin loop in which the stems of IREs b and c are paired with each other in a different hairpin loop with a very long stem (see Figure 6.6 and Guo et al. 1994).

134

rich double-stranded regions near the cap for translation control or multiple AU-rich sequences in the 3′ UTR for mRNA instability), plus a common hairpin loop to confer the response to cellular iron. The common hairpin loop may reflect convergent evolution of generic translation elements and rapid turnover elements. Coordinate regulation of the mRNAs for iron storage or iron uptake is achieved through interactions of IREs with conserved *trans* factors, the IRPs (IRE-BPs/ FRP/, IRF/, P-90s, iron regulatory proteins) and other proteins. No other example of coordinated genetic regulation of metabolically related proteins using common RNA sequences is known at this time. Either other examples remain yet undiscovered, or the mechanism reflects the unusual importance of regulating iron metabolism.

2. Solution 1: Ferritin mRNA Regulation

Coordinate regulation of iron concentration/storage and iron uptake needs to be in opposite directions, if it is to be effective. As might be predicted, high iron is a positive regulator of ferritin expression for iron storage and a negative regulator of TfR for iron uptake, whereas low iron negatively regulates ferritin expression (no iron is available to store) and positively regulates TfR expression (iron uptake is increased).

2.1 Positive Regulation—High Iron

Ferritin synthesis can be induced as much as 50× when cellular iron levels are high. (Kimata and Theil 1994; Lescure et al. 1991; Proudhon et al. 1996; Zahringer et al. 1976; Schaefer and Theil 1981; Shull and Theil 1982, 1983). There are two components to the induction. First, inhibition of polyribosome binding is reversed (Zahringer et al. 1976). Second, when ferritin mRNA is released for translation, the mRNA is unusually efficient; i.e., it binds a disproportionate share of initiation factors (eIFs) (Schaefer and Theil 1981; Shull and Theil 1982, 1983). Information for both aspects of regulation is encoded in the IRE region (Dix et al. 1992). Encoding both translation enhancement (eIF interactions) and translation repression (IRP binding) in a single RNA (IRE) element is analogous to encoding transcription enhancement and repression in the same DNA element (Diamond et al. 1990).

The effect of excess iron on ribosome binding to ferritin mRNA is illustrated by the observation that a larger fraction of ferritin mRNA is found in polyribosomes of cells from iron-treated animals (Zahringer et al. 1976). That the information for iron regulation of ferritin synthesis resides in mRNA was shown by observing that iron had no effect on the cellular concentration or subcellular distribution of ferritin mRNA (Schaefer and Theil 1981; Shull and Theil 1982,

IREs - Composite mRNA Regulatory Elements

Generic Regulatory Elements

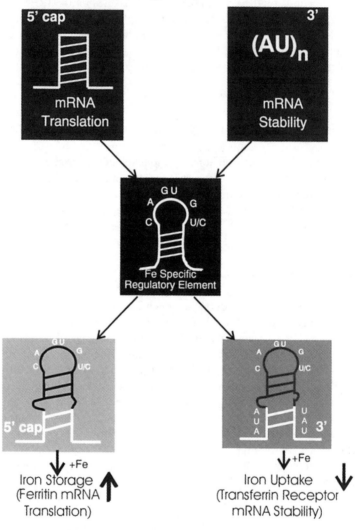

1983; Didsbury et al. 1986). Moreover, using protein-free RNA, ferritin mRNA-specific repression of ferritin synthesis could be established *in vitro* by selecting the type of cell-free extract used to measure translation with or without the endogenous *trans*factor (Dickey et al. 1988). A variety of experiments, mainly fusing the IRE to CAT or hGH DNA, transfecting cells, and examining the effect of excess iron on the expression of the reporter, were used to show that the dominant regulatory information is in the IRE [exemplified by Aziz and Munro (1987); Hentze et al. (1987). However, modulating information appears to exist in the 3' UTR, when studied in ferritin mRNAs with an altered (CAGUAU) canonical sequence (Dickey et al. 1988). Recent observations that polyA tails on mRNA can affect translation (Gallie and Tanquay 1994) suggest an avenue to studying the importance of the 3' UTR in ferritin regulation.

The efficiency of ferritin mRNA translation was first observed in red cells of iron-treated animals, when the effect of cycloheximide on iron-induced protein synthesis was examined (Schaefer and Theil 1981). In cells with high concentrations of the protein synthesis machinery, positive control is more difficult to observe (Walden and Thach 1986). For most mRNAs, initiation is the rate-limiting step; thus a slight decrease in elongation with cycloheximide will have little effect. On the other hand, if initiation is very efficient and elongation is the rate-limiting step, any change in elongation will have a differentially large effect on translation of the efficient mRNA. Such was the observation when small amounts of cycloheximide were added to red cell suspensions of iron-treated animals (Table 6.2). Ferritin synthesis was decreased 52% by 0.2 μM cycloheximide, which had essentially no effect on total protein synthesis; a 20-fold increase in cycloheximide concentration was required to produce an equivalent inhibition of total protein synthesis.

The high efficiency of ferritin mRNA, from both red cells and liver, was also demonstrated using nonregulating cell-free extracts (wheat germ) and increasing

---

◄ Figure 6.3 IREs as composite regulatory elements. Generic translation elements include stable RNA A-helices (Watson-Crick base pairs) which form near the 5' cap, as in the stems of translation IREs. Generic mRNA stability elements include multiple AU-rich sequences in the 3' UTR, with the canonical sequence AUUUA. Examination of the IRE stems of the translation (ferritin/eALAS) and stability elements (TfR-a–e) fulfill the criteria for generic elements regulating translation or stability. Coordination of iron storage and heme synthesis (translation of ferritin and eALAS mRNAs) with iron uptake (TfR mRNA degradation), when cellular iron levels are high, appears to have been achieved by adding to the generic elements an iron-regulatory-specific hairpin loop with a short stem; note that the internal loop or bulge occurs at the putative junction of the generic and iron-specific RNA regulatory elements. (Reprinted from Theil 1994 with permission.)

Table 6.2 Differential effect of cycloheximide on ferritin and total protein (globin) synthesis in red cells

| | ³H-Leucine (dpm) Incorporation in Red Cell Suspensions | |
	Ferritin	Total Protein (Mostly Globin)
Control	6100±122	74,000±7400
Cycloheximide	2900±32	74,000±7400

Data taken from Schaefer and Theil (1981). To obtain inhibitions of total protein synthesis equivalent to ferritin at 0.2 µM cycloheximide, 4 µM was required.

concentrations of polyA⁺ RNA. As the concentration of mRNA increased, and competition among mRNAs for initiation factors (eIFs) and ribosomes increased, ferritin mRNA was translated disproportionately well (Shull and Theil 1982, 1983). Thus when eIFs and ribosomes were in excess, ferritin was only 3% of the total protein synthesized, but at mRNA excess, ferritin synthesis was 8% of the total (Shull and Theil 1982).

Relief from competition for eIFs between ferritin mRNA and other mRNAs can be obtained by adding crude eIF-4B to wheat germ extracts with polyA⁺ RNA in excess (Theil 1987) (Figure 6.4A). Moving the IRE region 73 nucleotides downstream or deletion of the IRE greatly decreased the translational efficiency of the ferritin mRNA (Dix et al. 1992) (Figure 6.4B). In contrast, the CAGU<u>A</u>U mutation had no effect on translational efficiency (Dix et al. 1992) but blocked

▶

Figure 6.4 Positive control of ferritin mRNA translation. Positive control has been observed in reticulocytes (low eIFs and ribosomes) and cell-free extracts (Schaefer and Theil 1981; Shull and Theil 1983). When polyA⁺ mRNA from embryonic erythroid cells is translated at limiting concentrations of initiation factors [(mRNA) >> (ribosomes and factors)], ferritin synthesis was disproportionately high in wheat germs or rabbit reticulocyte extracts (Dix et al. 1992; Schaefer and Theil 1981; Shull and Theil 1983) and in tadpole reticulocytes (Zahringer et al. 1976). A. Relief of competition between globin and mRNA, in polyA⁺ mRNA. Partially purified initiation factors (4B + 4F from wheat germ) was added to a homologous extract; an increase in globin synthesis occurred with no effect on ferritin synthesis. (Initiation factors were kindly provided by J. Ravel and K. Browning.) B. The effect of the IRE on translation rate. Full-length ferritin mRNA transcripts were translated faster than globin mRNA (individual reaction mixtures) except at 27 nM; ferritin transcripts saturated at 12 nM. In contrast, ferritin IRE ferritin transcripts were translated poorly at all concentrations [data taken from Dix et al. (1992) with permission].

(a)

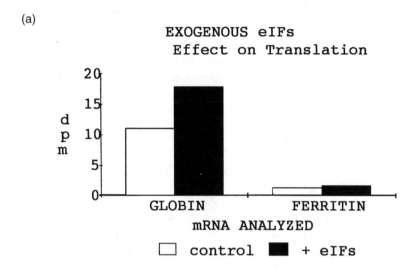

(b)

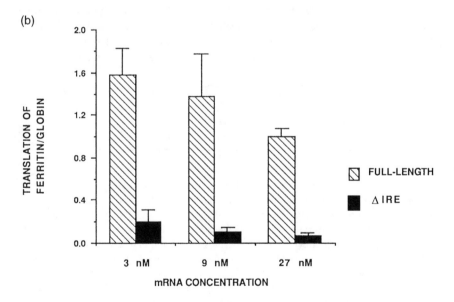

IRP binding (Harrell et al. 1991). Thus, the IRE contains information for both positive control (eIF binding) and negative control (IRP) binding.

mRNA competition is important *in vivo* in cells with relatively low rates of protein synthesis (low concentrations of ribosomes and eIFs), such as reticulocytes. Competition is less important in cells with high rates of protein synthesis, e.g., normal hepatocytes or fibroblasts (Shull and Theil 1983; Walden and Thach 1986). However, competition appears to be important in liver ferritin synthesis of starved rats where the concentration of eIFs and ribosomes (rates of protein synthesis) is lower (Theil 1987).

2.2 Negative Regulation—Low Iron

When iron is low, ferritin synthesis is repressed. The magnitude of the repression is greater for ferritin IREs than for, e.g., the IRE eALAS mRNA (Bhasker et al. 1993; Melefors et al. 1993). A *trans* factor, first observed in tadpole red cells (Shull and Theil 1982, 1983) and in rabbit reticulocyte lysates (Dickey et al. 1988; Walden et al. 1989), was shown to block ribosome binding (Dickey et al. 1988) (Figure 6.5A,B). Subsequent studies have shown the factors to be present in a wide variety of animal cells (Rothenberger et al. 1990). The protein responsible for repressing ferritin synthesis was isolated from rabbit liver, using a functional assay (inhibition of ferritin synthesis in wheat germ extracts (Walden et al. 1989) or RNA affinity chromatography (Rouault et al. 1989). Variously called P-90, FRP, IRE-BP or IRF, most workers have recently agreed to use IRP (iron regulatory protein).

The IRP "footprint" covers the entire IRE (Figure 6.6); the "toeprint" was

---

▶

Figure 6.5 Negative control of ferritin mRNA translation (initiation). A. Translation of endogenous mRNA in whole red cells (upper panel) or isolated polyA$^+$ in wheat germ extracts. Open circles represent data from reticulocytes or polyA$^+$ RNA of control animals; closed circles are data from iron-treated animals (Shull and Theil 1983). Note the high level of functional ferritin mRNA, which is repressed when iron levels are relatively low. B. Distribution of ferritin and globin mRNA during translation of red cell polyA$^+$ RNA in wheat germ or rabbit reticulocyte lysates. Translation mixtures, under the direction of reticulocyte polyA$^+$ RNA, were sedimented through sucrose gradients, the RNA extracted, and the location of globin or ferritin mRNA determined by hybridization to specific cDNAs (Diamond et al. 1990; Brown and Theil 1977). The data show the presence of a specific ferritin mRNA *trans* regulatory factor in red cell extracts, now known to be the IRP. In addition, the data show that the step in ferritin synthesis regulated by the IRP is initiation.

(a) [³H]Leucine Incorporation into Protein of Reticulocytes of Embryos (Tadpoles)

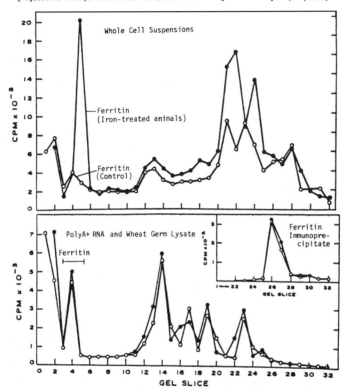

(b)

mRNA DISTRIBUTION
Supernatent-->Polyribosomes

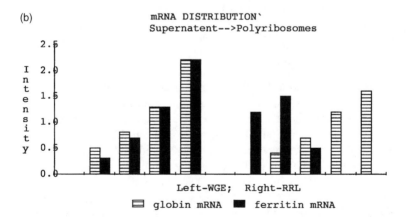

Left-WGE; Right-RRL

⊟ globin mRNA   ■ ferritin mRNA

141

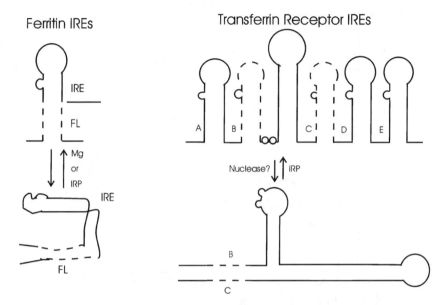

Figure 6.6 Alternate conformations of IREs. The ferritin IRE has a structure similar to the predicted secondary structure when [Mg] is high (> 5 mM) or when the hairpin sequence is changed from CAGUGU to CAGUAU[55]; the native structure is folded and apparently stabilized by intra-IRE tertiary interactions (Rouault et al. 1989; Harrell et al. 1991; Dix et al. 1993; Wang et al. 1990; Weiss et al. 1993). Computer prediction of the secondary structure of the transferrin receptor regulatory region shows that either a long hairpin loop can be obtained by the default parameters (Guo et al. 1994), or five small IRE hairpin loops and one large stem/loop can be predicted by fixing the CAGUGU/C as single-stranded regions (Zheng et al. 1992).

observed at the 5′ end of the IRE (Harrell et al. 1991). When the IRP binds to the IRE of ferritin mRNA, changes occur in the conformation of the base-paired flanking region (Harrell et al. 1991). Ferritin is unique among the IREs in having a base-paired flanking region (FL) (Figure 6.1B). Embedded in the FL, 11 nucleotides from the IRE, is a phylogenetically conserved triplet of base pairs (TBP), CUC/GAG, GAG/CUC, or CCC/GGG at the site of the conformational change associated with IRP binding (Dix et al. 1993). Disruption of the base pairs (GAG/GAG) leads to a 60% decrease in repression by the IRP, which was reversed by a derivative mutation with restored base pairs (GAG/CUC) (Dix et al. 1993). While the IRP binds to the TBP mutant, the conformation of is different: a new Fe-bleomycin site appears in the IRE 20 nucleotides from the mutation site (Dix et al. 1993). Clearly, the FL modulates IRP repression and this may explain why ferritin mRNA is more completely repressed by the IRP than the eALAS mRNA

(Bhasker et al. 1993; Melefors et al. 1993). As the IRE+FL can exist in several different conformations (Wang et al. 1990, 1991), the role of the TBP may be to "nucleate" the conformation that best binds the IRP.

Cloning and sequencing the cDNA for the IRP of human, rabbit, and rat revealed high (97% sequence identity) phylogenetic conservation (Rouault et al. 1990; Patino and Walden 1992; Yu et al. 1992). In addition, the tissue distribution was found to vary (Patino and Walden 1992), and the expression was observed to be iron independent (Yu et al. 1992). At least two IRPs are known (Rouault et al. 1990), IRP1 and IRP2, which have overlapping but distinct binding specificities (Henderson et al. 1994; Butts et al. 1996). Recently, phosphorylation of IRP1 by protein kinase C has been demonstrated (Eisenstein et al. 1993; Schalinske and Eisenstein 1996), which changes IRE binding. In addition, IRP1 can form an FeS cluster, which is associated with an increase in aconitase activity of crude cell extracts (Kaptain et al. 1991; Haile et al. 1992). Peptides accounting for 10% of the sequence of cytoplasmic aconitase have a 97–100% sequence identity (Kennedy et al. 1992) with IRP1. Although nothing is known of the other 90% or whether the sequence in the remaining 90% would alter the relationship to both IRP1 and IRP2, the identity of IRP1 and c-aconitase is widely accepted. The apo forms of c-aconitase, IRP1 and IRP2, bind all natural IREs tested, but IRP2 is more selective than IRP1 in binding variant-sequence IREs (Henderson et al. 1994; Butts et al. 1996). IRP2 does not acquire aconitase activity when mixed with Fe-S and is regulated differently than IRP1 (Guo et al. 1994; Samaniego et al. 1994; Iwai et al. 1995). However, the aconitase mRNA has an IRE that binds the IRP (Zheng et al. 1992). The apo form of mitochondrial aconitase, which also has sequence homology to c-aconitase (Zheng et al. 1992), does not appear to bind the IRE. Whether the aconitase family has more than three members (m-aconitase, c-aconitase/FeSIRP-1, IRP2) remains to be sorted out in the future. A more complete discussion of the IRPs, Fe-S proteins, and aconitases is provided in Chapters 8 and 9.

2.3 Iron-IRP Interactions

Apoaconitases such as the IRP can form either an Fe_3S_4 or an Fe_4S_4 cluster (Kaptain et al. 1991). An attractive hypothesis (Rouault et al. 1991), the conversion of the 3Fe to the 4Fe cluster when cellular iron levels are high, has not been supported by the data (Kaptain et al. 1991; Kennedy et al. 1992; Zheng et al. 1992) under the conditions so far used. Direct interactions of iron with the RNA (IRE) do not appear to occur, based on the fact that ferritin mRNA 5′ UTR sequences in cells with iron-induced or constitutive ferritin mRNA synthesis are the same (Didsbury et al. 1986).

2.4 Ferritin mRNAs Without IREs

Several types of ferritin mRNAs do not contain IREs: yolk (von Darl et al. 1994), soybean, and maize (Proudhon et al. 1996). Moreover, downstream sequences interfere with IRE function (Dix et al. 1992; Kimata and Theil 1994); a chimeric mRNA containing the animal IRE near the cap and fused to soybean ferritin mRNA had only a small effect on regulation *in vitro* (Dix et al. 1992; Kimata and Theil 1994). There is also no evidence for translational regulation of soybean ferritin *in vivo,* in spite of the fact that high mRNA concentrations can be associated with developmentally regulated decreases in ferritin protein (Theil 1994). (A more complete discussion of regulated ferritin expression in plants is given in Chapter 17.) The ferritin mRNA for snail egg yolk (von Darl et al. 1994) shares with plant ferritin mRNAs a signal (von Darl et al. 1994) or transit peptide (Ragland and Theil 1993; Ragland et al. 1990), which targets the protein for secretion from the liver (snail egg yolk) or the plastid (soybean or maize). It is possible that the targeting sequence interferes with IRE activity. However, deletion of the transit peptide RNA sequence in the chimera of the animal (frog red cell) IRE with plant (soybean) ferritin mRNA did not restore IRE regulatory activity (Danger and Theil, unpublished observations). Clues to the RNA sequences that interfere with IRE function may be hidden among those encoding *nonconserved* amino acids.

What is clear at the moment is that translational regulation is not shared by all ferritin mRNAs, although iron and developmental regulation are. In animals, NO also appears to regulate ferritin mRNA through effects on the IRP (Pantopoulos and Hentze 1995; Weiss et al. 1993; Henze and Kuhn 1996). Possibly the IRE is an ancient sequence that was lost during evolution after insertion of secretion or plastid targeting sequences; there would be no selective advantage to retaining the ferritin IRE if, as in plants, it has no functional effect. One alternative explanation is to imagine that only those genes that encoded intracellular, animal ferritin acquired the IRE sequence, even though iron signals appear to regulate ferritin expression with or without the IRE in other organisms. A second alternative explanation is that iron can regulate ferritin mRNAs with transit/target sequences by interacting with the targeting pathway itself, independently of IREs.

3. Solution II: Transferrin (TfR) Receptor Regulation

Iron-dependent regulation of transferrin receptors depends on changes in mRNA turnover. TfR mRNA concentrations decrease when cellular iron increases (Hentze and Kuhn 1996; Mullner and Kuhn 1988; Casey et al. 1989; Owen and Kuhn 1987), whereas ferritin mRNA concentrations do not change in iron excess

(Shull and Theil 1982, 1983; Didsbury et al. 1986). The decrease in TfR mRNA with iron excess was originally thought to relate to repression of transcription, but was shown, in fact, to occur because excess iron caused TfR mRNA destabilization; deleted portions of the 3′ UTR of the TfR mRNA stabilized the mRNA even when iron was in excess (Hentze and Kuhn 1996; Mullner and Kuhn 1988; Casey et al. 1989). Thus the conditions that induce ferritin synthesis decrease TfR synthesis because the TfR template concentration decreases.

Regulation of gene expression by controlled turnover of mRNA occurs in both animals and plants [reviewed in Beelman and Parker (1995)]. Changes in mRNA stability share with changes in mRNA translation rate, rapid response to regulatory signals. Iron is a regulatory signal that apparently requires a rapid response (Shull and Theil 1983), illustrated by the use of *both* iron-dependent instability (TfR mRNA) and iron-dependent translation activation (ferritin mRNA). Genes for growth factors such as c-myc (Green 1993) or light-regulated genes such as insecticidal crystal protein (Herrick and Ross 1994; Murray et al. 1991), ferredoxins, phytochromes, and ribulose-1,5-bisphosphate carboxylase small subunits also contain mRNA instability elements.

A common, but not exclusive mRNA rapid turnover motif, is the multiple $(AU)_n$ and AUUUA sequences found among the five TfR-IREs a–e (Figures 6.1A and 6.7). Such motifs can confer rapid turnover on chimeric mRNAs (Hentze and Kuhn 1996; Mullner and Kuhn 1988; Casey et al. 1989; Owen and Kuhn 1989). In general, regulated mRNA degradation appears to involve digestion at both the 3′ and 5′ ends as well as endonucleolytic degradation, based on the few RNA degradation products that could be characterized for unstable mRNAs (Herrick and Ross 1994; Beeling and Park (1995). One mechanism proposed for specific mRNA degradation is the RNA binding of specific *trans* factors that target or protect the rapid turnover sequence from cytoplasmic RNAses. An alternative mechanism involves translation itself because degradation is enhanced in mRNAs associated with polyribosomes, as exemplified by c-myc (Green 1993). A single site for degradation has been observed in TfR mRNA near a TfR-IREc in which the canonical CAGUGx was CAGUAx (Binder et al. 1994).

The conserved sequences of the TfR mRNA rapid turnover element are found in ~700 nucleotides in the 3′ UTR that can be predicted to form a long hairpin loop. The AU-rich regions and the CAGUGU/C sequences are also in the stem of the predicted structure (Mullner and Kuhn 1988; Casey et al. 1989). When the ferritin IRE sequence was identified, the similarity to sequences in the TfR regulatory region led to a reexamination of the predicted folding; an alternate structure of similar predicted stability in the TfR 3′ UTR was the five IREs (Figure 6.6). Two of the IREs, b and c, which are absolutely required for regulation (Mullner and Kuhn 1988; Casey et al. 1988, 1989; Owen and Kuhn 1989), contain identical palindromic sequences; identical base pairs can thus form either *within* an IRE or *between* IREs b and c (Figure 6.1, Figure 6.6) (Theil 1994). TfR-IREs

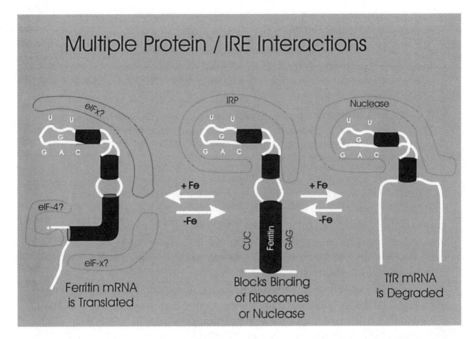

Figure 6.7 Multiple proteins recognized by IRE regions. The IRE structures recognize not only the IRP, which blocks ribosome binding (5′ IREs) or degradation (3′ IREs), but must also recognize (or be recognized by) initiation factors that enhance translation or nucleolytic activity, which enhances degradation.

b and c have the highest sequence identity (Table 6.1). Others of the TfR-IREs (IREa, IREe) can be deleted in transfected constructs that display regulation which leave unexplained the sequence conservation of IREa and IREe.

Each individual TfR-IRE can bind the IRP *in vitro,* and all five can bind multiple IRPs *in vitro* (Casey et al. 1988), indicating that the IRE form of the TfR regulatory region likely can occur *in vivo.* However, protein binding varies among the IREs when the TfR-IRE-c is compared to the ferritin IRE; a smaller fraction is complexed with IRP (Theil, McKenzie, and Walden, to be published). TfR-IRE-c can form intra-IRE base pairs or inter-IRE base pairs with TfR-IREb using the conserved palindrome. It is easy to imagine that mobile equilibrium exists between the two predicted secondary structures (Figure 6.7) and that IRP binding to one IRE leads to cooperative IRP binding by the other IRPs and shifts the equilibrium to a nuclease-inaccessible RNA conformation. Transfection experiments with chimeric mRNAs containing altered TfR IREs showed that cycloheximide or puromycin stabilized the mRNA containing the TfR element (Murray et al. 1991). Such results suggest that translation does not enhance degradation or that the nuclease required for mRNA degradation is short lived (Herrick and Ross 1994).

Structural studies of the sort used with ferritin IREs (Harrell et al. 1991; Dix et al. 1993; Wang et al. 1990, 1991; Rouault et al. 1990; Thorp et al. 1996) are difficult for the TfR regulatory element because of the large size. Difficulties in measuring regulation of mRNA stability *in vitro* (Green 1993; Herrick and Ross 1994) also hamper molecular analysis of the TfR-IREs. However, progress in studies of the ferritin IRE will form a strong platform on which to build future studies of the TfR regulatory region.

4. Solution III: Erythroid Aminolevulinate Synthase (eALAS) Regulation

The first step in heme biosynthesis is catalyzed by aminolevulinate synthase. Two genes encode the enzyme, one expressed in erythroid cells and one in other cell types. Cell specificity of expression occurs for some other red cell proteins, including ferritin; the red cell-specific ferritin is the major source of embryonic storage iron in vertebrates (Theil 1973, 1976, 1981; Theil and Tosky 1979; Brown and Theil 1977; Theil and Brenner 1981). A number of general metabolic features, such as protein synthesis, heme biosynthesis, receptor (transferrin)-mediated endocytosis, receptor regulation, and mRNA-specific *trans* factors which block translation initiation, were first observed in red cells. The control of heme biosynthesis by iron at the rate-limiting step of eALAS synthesis was suggested more than 30 years ago, based on interrelated changes in both iron storage and heme synthesis in certain anemias (Vogel et al. 1960).

cDNA cloning of eALAS revealed an IRE sequence (Cox et al. 1991; Dandekar et al. 1991) that is seven nucleotides from the 5′ cap. The eALAs IREs differ from ferritin IREs in lacking the long base-paired flanking region with the conserved triplet of base pairs and the four-nucleotide internal loop; eALAS IREs have a single nucleotide internal bulge, as do the TfR-IREs, which contrasts with the four-nucleotide internal loop in ferritin IREs (Figure 6.1A). The eALAS IRE showed sequence-specific interactions with red cell proteins (MEL and K562 (Bhasker et al. 1993) and fibroblast proteins (Melefors et al. 1993), although the fraction of the RNA complexed by the IRP appeared to be lower than for the ferritin IRE (Bhasker et al. 1993; Melefors et al. 1993).

Translational regulation of eALAs mRNA was shown in several ways. First, iron treatment of MEL cells increased the fraction of eALAS mRNA in the polyribosomes and increased the amount of immunoprecipitable eALAS. Either Fe-transferrin (ferric ammonium citrate) or hemin could increase synthesis of eALAS, but whether the magnitude of the induction was similar to that for ferritin was not analyzed (Melefors et al. 1993). However, in wheat germ extracts to which purified (human placental) IRP was added, translation of a ferritin IRE-CAT chimeric mRNA was consistently repressed more than the chimera with the eALAS-IRE-CAT, whereas in a rabbit reticulocyte extract (endogenous IRP and erythroid eIFs), a chimeric mRNA of the eALAS IRE and hGH was repressed 95% (Bhasker et al. 1993; Melefors et al. 1993). Mutation of the canonical CAGUGU hairpin loop sequences to CCUAG decreased translational regulation because the IRP could not bind (Bhasker et al. 1993). In analogy to the ferritin IRE, moving the eALAS IRE downstream diminished the effect on translation (Bhasker et al. 1993), emphasizing the role of the IRE in control of translation initiation.

In red cells at least, ferritin and heme synthesis can be coordinately regulated during translation initiation by iron signals through IRE/IRP interactions. The isoforms of the IREs (ferritin and eALAS) (shown in Table 6.1) have different affinities for the IRP and quantitatively different translational effects in chimeric transcripts (Melfors et al. 1993). However, in cell extracts with excess endogenous IRP, the eALAS IRE is very effective in blocking ribosome binding. Questions for the future include: Does the eALAS encode positive control, as does the ferritin IRE? What is the functional significance of the absence of base-paired flanking regions, conserved triplet of base pairs, and the internal loop in the eALAS-IRE? Do IRP1, IRP2 (IRPX?) make functionally significant distinctions between the iso-IREs?

5. The Iron Signal

Identifying the iron signal that regulates ferritin and eALAS mRNA translation or TfR mRNA degradation will not be simple. Intracellular forms of iron, between the endosome containing the transferrin/TfR complex and proteins such as ferritin and hemoglobin are essentially uncharacterized. Opportunities for mischievous free radical (Fenton) chemistry between dioxygen and ferrous ions, coupled with insolubility of the ferric ions, necessitates a carrier of some kind. However, the molecular characterization of the intracellular "iron carrier" or "transit pool" has made little progress over many decades of investigation.

Two types of Fe complexes have been suggested as regulating the IRP: the Fe-S cluster (Kaptain et al. 1991; Haile et al. 1992; Kennedy et al. 1992; Rouault et al. 1991) and heme (Lin et al. 1991; Goessling et al. 1992). Little is known about regulated turnover of Fe-S clusters. Fe-S turnover could be fast enough to fulfill the regulatory requirements, but the question is difficult to fully analyze at this time. However, it is known that Fe-S clusters, once formed, can be structurally stable and can be released from the protein intact (Beinert and Kennedy 1989). In addition, conditions required for release of Fe-S from the IRP are harsh enough (Kaptain et al. 1991) to change the protein. Thus, *if complete disruption of the IRP-Fe-S interaction is required for regulation,* it is very likely that only under extreme conditions of iron excess will the Fe-S cluster be a metabolic iron signal. Heme, a second putative regulatory iron complex, leads to degradation of the IRP (Lin et al. 1991; Goessling et al. 1992). As a result, excess iron in the form of heme will induce ferritin synthesis and decrease TfR synthesis. Such a regulatory mechanism is relatively insensitive as reversal requires the synthesis of new IRP. On the other hand, IRP2, which does not appear to form (i.e., S cluster) is regulated by a degradative pathway (Guo et al. 1994; Samaniego et al. 1994; Iwai et al. 1995). It is likely that heme contributes to the regulation of IRP stability as a "fail-safe" mechanism, used when iron (heme) levels are very, very high; regulation of the IRP by heme or Fe-S at moderate iron levels obviates the sensitivity of regulating ferritin and TfR mRNAs rather than transcription. FeS and heme may, then, be part of an iron toxicity regulatory response, rather than regulation of iron uptake and storage for more ordinary metabolic needs. More extensive discussions of Fe-S and heme as metabolic regulators can be found in Chapters 7 and 8.

Because the IRP can be phosphorylated by protein kinase C *in vitro* and in cell culture (Eisenstein et al. 1993; Schalinske and Eisenstein 1996), a phenomenon that influences the IRE/IRP interaction, it is possible that under ordinary metabolic conditions, iron enters a signal transduction pathway several steps removed from the IRE/IRP interaction. Such a possibility is particularly attractive in explaining the dual pathways of iron on ferritin expression with either DNA or RNA as the target in plants and animals. If iron enters a signal transduction pathway several steps away from the protein/nucleic acid interactions, a common signaling pathway could exist for iron which would only need to diverge at the end. Moreover, such a mechanism would have the salubrious effect of keeping iron away from informational macromolecules. That iron damages nucleic acids is actually exploited therapeutically with the use of Fe-bleomycin to target tumor DNA or as a structure probe for RNA (Theil 1994; Carter et al. 1990; Hecht 1994).

6. Perspective

Iron uptake and iron storage in animals are linked through *trans* factor (IRP) recognition by a noncoding element, the IRE, for specific regulation of translation initiation or mRNA stability. In red cells, heme synthesis is also linked to iron storage and uptake by control of translation initiation of the aminolevulinate synthase mRNA, which encodes the catalyst for the rate-limiting step in heme biosynthesis. The RNA *cis* elements, the IREs, that recognize the IRP are a family of sequences 28–30 nucleotides long which combine a common hairpin loop with stems that reflect generic features of regulated translation or RNA turnover. Evidence is accumulating that the IRP negative regulatory RNA binding protein is also a family (of at least two members). The up or down regulation of ferritin or TfR synthesis by *cis* (IRE)/*trans* (IRP) interactions, coupled to interactions with other proteins (eIFs, nucleases), strengthens the analogies between regulation of mRNA and DNA. IREs recognize several classes of proteins (Figure 6.7), of which the specific IRP regulator is only one. Because the three-dimensional structure of RNAs has wider variations in shape than DNA, the number of proteins necessary to achieve specificity could be lower than for DNA. Understanding of the molecular events in (metal-dependent) mRNA regulation can be extended to increase understanding of (metal-dependent) DNA regulation. If the RNA world is the more ancient, than transcriptional regulation is in fact likely a modification of RNA regulation.

Questions to be answered in the future for the full understanding of iron-regulated mRNA function include: What is the mechanism for IRE/IRP control of RNA turnover? Are the isoIRPs (IRP1, IRP2) and isoIREs (ferritin, TfR, eALAS) matched *in vivo?* What is the functional consequence of IRE variation? What is the significance of an IRE that is not phylogenetically conserved in all examples of the mRNA (e.g., succinic dehydrogenase?) (Kohler et al. 1995; 1996). What is the metabolic consequence of FeS-IRP aconitase activity? How many IRPs occur? How many IRPs are aconitases? Are all the metabolic iron signals encompassed by heme or FeS clusters on the IRP? What is the metabolic significance of the phosphorylation of the IRP? Does IRP phosphorylation mean that iron also enters the signaling pathway at the membrane or transferrin receptor/transferrin interaction? Finally, is coordinated control of the synthesis of proteins of iron metabolism at the site of mRNA translation/stability a paradigm for groups of other metabolically related proteins?

Acknowledgments

The author is grateful for research support by both the NIH Hematology Extramural Program (DK-20251) and the North Carolina Agricultural Research Service, and for the opportunity to work with so many talented students, postdoctoral associates, and research assistants.

References

Aziz, N. and H. N. Munro. 1987. Iron regulates ferritin mRNA translation through a segment of its 5′-untranslated region. *Proc. Natl. Acad. Sci. U.S.A.* 84:8478–8482.

Balla, G., H. S. Jacob, J. Balla, M. Rosenberg, K. Nath, F. Apple, J. W. Eaton, and G. M. Vercellotti. 1992. Ferritin: A cytoprotective antioxidant stratagem of endothelium. *J. Biol. Chem.* 267:18148–18153.

Beelman, C. A., and R. Parker. 1995. Degradation of RNA in eukaryotes. Cell 81:179–183.

Beinert, H., and M. C. Kennedy. 1989. Engineering of protein bound iron-sulfur clusters. *Eur. J. Biochem.* 186:5–15.

Bhasker, C. R., G. Burgil, B. Neupert, A. Emery-Goodman, L. C. Kuhn, and B. K. May. 1993. The putative iron-responsive element in the human erythroid 5-aminolevulinate synthase mRNA mediates translational control. *J. Biol. Chem.* 268:12669–12705.

Binder, R., J. A. Horowitz, J. B. Basilion, D. M. Koeller, R. D. Klausner, and J. B. Harford. 1994. Evidence that the pathway of transferrin receptor mRNA degradation involves an endonucleolytic cleavage within the 3' UTR and does not involve polyA tail shortening. *EMBO J.* 18:1969–1980.

Brown, J. E., and E. C. Theil. 1977. Red cells, ferritin and iron storage during amphibian development. *J. Biol. Chem.* 253:2673–2678.

Butts, J., H.-Y. Kim, J. P. Basilion, S. Cihenm, K. Iwai, C. C. Philpott, S. Altschul, R. D. Klausner, T. A. Rousault. 1996. Differences in the RNA binding sites of iron regulatory proteins and potential target diversity. *Proc. Natl. Acad. Sci.* 93:4345–4349.

Carter, B. J., E. deVroom, E. C. Long, G. A. van der Marl, J. H. van Boom, and S. M. Hecht. 1990. Site-specific cleavage of RNA by Fe(II) bleomycin. *Proc. Natl. Acad. Sci. U.S.A.* 87:9373–9377.

Casey, J. L., M. W. Hentze, D. M. Koeller, S. W. Caughman, T. A. Rouault, R. D. Klausner, and J. B. Harford. 1988. Iron responsive elements: Regulatory RNA sequences that control mRNA levels and translation. *Science* 240:924–928.

Casey, J. L., D. M. Koeller, V. C. Ramin, R. D. Klausner, and J. B. Harford. 1989. Iron regulation of transferrin receptor levels requires iron-responsive elements and a rapid turnover in the 3'-untranslated region of the mRNA. *EMBO J.* 8:3693–3699.

Cox, T. C., M. J. Bawden, A. Martin, and B. K. May. 1991. Human erythroid 5-aminolevulnate synthase: Promoter analysis and identification of an iron-responsive element in the mRNA. *EMBO J.* 10:1891–1902.

Dandekar, T., R. Stripecke, N. K. Gray, B. Goossen, A. Constable, H. E. Johansson, and M. W. Hentze. 1991. Identification of a novel iron-responsive element in murine and human erythroid delta aminolevulinic acid synthease mRNA. *EMBO J.* 10:1903–1909.

Diamond, M. I., J. N. Miner, S. K. Yoshinagu, and K. R. Yamamato. 1990. Transcription factor interactions: Selectors of positive or negative regulation from a single DNA element. *Science* 249:1266–1272.

Dickey, L. F., Y.-H. Wang, G. E. Shull, I. A. Wortmann III, and E. C. Theil. 1988. The importance of the 3'-untranslated region in the translational control of ferritin mRNA. *J. Biol. Chem.* 263:3071–3074.

Didsbury, J. R., E. C. Theil, R. E. Kaufman, and L. F. Dickey. 1986. Multiple red cell ferritin mRNAs, which code for an abundant protein in the embryonic cell type, analyzed by cDNA sequence and by primer extension of the 5'-untranslated regions. *J. Biol. Chem.* 261:949–955.

Dix, D. J., P. N. Lin, Y. Kimata, and E. C. Theil. 1992. The iron regulatory region (IRE) of ferritin mRNA is also a positive control element for iron-independent translation. *Biochem.* 31:2818–2822.

Dix, D. J., P. N. Lin, A. R. McKenzie, W. E. Walden, and E. C. Theil. 1993. The influence of the base-paired flanking region (FL) on structure and function of the ferritin mRNA iron regulatory element (IRE). *J. Mol. Biol.* 231:230–240.

Eisenstein, R. S., P. T. Tuazon, K. L. Schalinske, S. A. Anderson, and J. A. Traugh. 1993. Iron-responsive element-binding protein. *J. Biol. Chem.* 268:27363–27370.

Gallie, D. R., and R. Tanquay. 1994. Poly(a) binds to initiation factors and increases cap-dependent translation in vitro. *J. Biol. Chem.* 269:17166–17173.

Goessling, L. S., S. Daniels-McQueen, M. Chattaacharayya-Pakrasi, J.-J. Lin, and R. E. Thach. 1992. Enhanced degradation of the ferritin repressor protein during induction of ferritin repressor protein during induction of ferritin messenger RNA translation. *Science* 256:670–673.

Green, P. J. 1993. Control of mRNA stability in higher plants. *Plant Physiol.* 102:1065–1070.

Guo, B., Y. Yu, and E. A. Leibold. 1994. Iron regulates cytoplasmic levels of a novel iron-responsive element-binding protein without aconitase activity. *J. Biol. Chem.* 268:24252–24260.

Haile, D. J., T. A. Rouault, J. B. Harford, M. C. Kennedy, G. A. Blondin, H. Beinert, and R. D. Klausner. 1992. Cellular recognition of the iron-responsive element binding protein: Dissembly of the cubane iron-sulfur cluster results in high affinity RNA binding. *Proc. Natl. Acad. Sci.* 89:11735–11739.

Harrell, C. M., A. R. McKenzie, M. M. Patino, W. E. Walden, and E. C. Theil. 1991. Ferritin mRNA: Interactions of iron regulatory element (IRE) with the translational regulatory protein (P-90) and the effect on base-paired flanking regions. *Proc. Natl. Acad. Sci. U.S.A.* 88:4166–4170.

Harrison, P. M., and T. H. Lilley. 1990. Ferritin. In *Iron Carriers and Iron Proteins,* ed. T. M. Loller. pp. 353–452. VCH Weinheim, New York.

Hecht, S. M. 1994. RNA degradation by bleomycin, a naturally occurring bioconjugate. *Bioconjugate Chemistry* 5:513–526.

Henderson, B. R., E. Menotti, C. Bonnard, and L. C. Kuhn. 1994. Optimal sequence and structure of iron responsive elements. *J. Biol. Chem.* 269:17481–17489.

Hentze, M. W., S. W. Caughman, T. A. Rouault, J. G. Barriocanal, A. Dancis, J. B. Harford, and R. D. Klausner. 1987. Identification of the iron-responsive element for the translational regulation of human ferritin mRNA. *Science* 238:1570–1573.

Hentze, M. W., and L. C. Kuhn. 1996. Molecular control of vertebrate iron metabolism: mRNA-based regulatory circuits operated by iron, nitric oxide and oxidative stress. *Proc. Natl. Acad. Sci.* 93:8175–8182.

Herrick, D. J., and J. Ross. 1994. The half-life of c-myc mRNA in growing and serum-stimulated cells: Influence of the coding and 3′ untranslated regions and role of ribosome translocation. *Mol. Cell. Biol.* 14:2119–2128.

Iwai, K., R. D. Klausner, and T. A. Rouault. 1995. Requirements for iron-regulated degradation of the RNA binding protein, iron regulatory protein-2. *EMBO J.* 14:5350–5367.

Kaptain, S., W. E. Downey, C. Tang, C. Philpott, D. Haile, D. Orloff, J. B. Harford, T. A. Rouault, and R. D. Klausner. 1991. A regulated RNA binding protein also possesses aconitase activity. *Proc. Natl. Acad. Sci. U.S.A.* 88:10109–10113.

Kennedy, M. C., L. Mend-Mueller, G. A. Blondin, and H. Beinert. 1992. Purification and characterization of cytosolic aconitase from beef liver and its relationship to the iron-responsive element binding protein. *Proc. Natl. Acad. Sci.* 89:11730–11734.

Kimata, Y., and E. C. Theil. 1994. Posttranslational regulation of ferritin during nodule development in soybean. *Plant Physiol.* 104:263–270.

Klausner, R. D., T. A. Rouault, and J. B. Harford. 1993. Regulating the rate of mRNA: The control of cellular iron metabolism. *Cell* 72:19–29.

Koeller, D. M., J. A. Horowitz, J. L. Casey, R. D. Klausner, and J. B. Harford. 1991. Translation and the stability of mRNAs encoding the transferrin receptor and c-fos. *Proc. Natl. Acad. Sci. U.S.A.* 88:7778–7782.

Kohler, S. A., B. R. Henderson, and L. C. Kuhn. 1995. Succinic dehydrogenase b mRNA of *Drosophila melanogaster* has a functional iron-responsive element of the 5′ untranslated region. *J. Biol. Chem.* 270:30781–30786.

Kohler, S. A., B. R. Henderson, and L. C. Kuhn. 1996. Translational regulation of mammalilan and *Drosophila* citric acid cycle enzymes via iron-responsive elements. *Proc. Natl. Acad. Sci. U.S.A.*

Lescure, A. M., D. Proudhon, H. Pesey, M. Ragland, E. C. Theil, and J. F. Briat. 1991. Ferritin gene transcription is regulated by iron in soybean cell cultures. *Proc. Natl. Acad. Sci. U.S.A.* 88:8222–8226.

Lin, J.-J., M. M. Patino, L. Gaffield, W. E. Walden, A. Smith, and R. E. Thach. 1991. Crosslinking of hemin to a specific site on the 90-kDa ferritin repressor protein. *Proc. Natl. Acad. Sci. U.S.A.* 88:6068–6071.

Mascotti, D. P., Rup, D., and Thach, R. E. Regulation of Iron Metabolism: Translational Effects Mediated by Iron, Heme, and Cytokines. Annu. Rev. Nutr., 1995. 15. p. 239–261.

Melefors, O., B. Goossen, H. E. Johansson, R. Stripecke, N. K. Gray, and M. W. Hentze. 1993. Translational control of 5-aminolevulinate synthase mRNA by iron-responsive elements in erythroid cells. *J. Biol. Chem.* 268:5974–5978.

Mullner, E. W., and L. C. Kuhn. 1988. A region in the 3′ untranslated region mediates iron dependent regulation of transferrin receptor mRNA stability in the cytoplasm. *Cell* 53:815–825.

Munro, H. N. 1990. Iron regulation of ferritin gene expression. *J. Cell. Biochem.* 44:107–115.

Murray, E. E., T. Rocheleau, M. Eberle, C. Stock, V. Sekar, and M. Adang. 1991. Analysis of unstable RNA transcripts of insecticidal crystal protein genes of *Bacillus thuringiensis* in transgenic plants and electroporated protoplasts. *Plant Mol. Biol.* 16:1035–1050.

Owen, D., and L. C. Kuhn. 1987. Noncoding 3′ sequences of the transferrin receptor gene are required for mRNA regulation by iron. *EMBO J.* 6:1287–1293.

Pantopoulos, K., and M. W. Hentze. 1995. Nitric oxide signaling to iron-regulatory protein:

Direct control of ferritin mRNA translation and transferrin receptor mRNA stability in transfected fibroblasts. *Proc. Natl. Acad. Sci. U.S.A.* 92:1267–1271.

Patino, M. M., and W. E. Walden. 1992. Cloning of a functional cDNA for the rabbit ferritin mRNA repressor proteins: Demonstration of a tissue-specific pattern of expression. *J. Biol. Chem.* 267:19011–19016.

Proudhon, D., J. Wei, J.-F. Briat, and E. C. Theil. 1996. Ferritin gene organization: Differences between plants and animals suggest possible kingdom-specific selective constraints. *J. Mol. Evol.* 42:325–336.

Ragland, M., J.-F. Briat, J. Gagnon, J.-P. Laulhere, O. Massenet, and E. C. Theil. 1990. Evidence for conservation of ferritin sequences among plants and animals and for a transit peptide in soybean. *J. Biol. Chem.* 263:18339–18344.

Ragland, M. and E. C. Theil. 1993. Ferritin and iron are developmentally regulated in nodules. *Plant Mol. Biol.* 21:555–560.

Rothenberger, S., E. W. Mullner, and L. C. Kuhn. 1990. The mRNA-binding protein which controls ferritin and transferritin receptor expression is conserved during evolution. *Nucl. Acids Res.* 18:1175–1179.

Rouault, T. A., M. A. Hentze, D. J. Haile, J. B. Harford, and R. D. Klausner. 1989. The iron-responsive element binding protein: A method for affinity purification of a regulatory RNA-binding protein. *Proc. Natl. Acad. Sci.* 86:5718–5722.

Rouault, T. A., Klausner, R. D., Post-transcriptional regulation of genes of iron metabolism in mammalian cells. J. Biol. Inorg. Chem., 1996, 1. p. 494–499.

Rouault, T. A., C. D. Stout, S. Kaptain, J. B. Harford, R. D. Klausner. 1991. Structural relationship between iron-regulated RNA-binding protein (IRE-BP) and aconitase: Functional implications. *Cell* 64:881–883.

Rouault, T. A., C. K. Tang, S. Kaptain, W. H. Burgeess, D. J. Haile, F. Samniego, O. W. McBride, J. B. Harford, and R. D. Klausner. 1990. Cloning of the cDNA encoding an RNA regulatory proteins—The human iron-responsive element binding protein. *Proc. Natl. Acad. Sci. U.S.A.* 87:7958–7962.

Samaniego, F., J. Chin, K. Iwai, T. A. Rouault, and R. D. Klausner. 1994. Molecular characterization of a second iron responsive element binding protein (IRP2): Structure, function and posttranslational regulation. *J. Biol. Chem.* 269:30904–30910.

Schaefer, F. V., and E. C. Theil. 1981. The effect of iron on the synthesis and amount of ferritin in red blood cells during ontogeny. *J. Biol. Chem.* 256:1711–1715.

Schalinske, K. L., and R. S. Eisenstein. 1996. Phosphorylation and activation of both iron regulatory proteins 1 and 2 in HL-60 cells. *J. Biol. Chem.* 271:7168–7176.

Shull, G. E., and E. C. Theil. 1982. Translational control of ferritin synthesis by iron in embryonic reticulocytes of the bullfrog. *J. Biol. Chem.* 257:14187–14191.

Shull, G. E., and E. C. Theil. 1983. Regulation of ferritin mRNA: A possible gene-sparing phenomenon. *J. Biol. Chem.* 258:7921–7923.

Sierzputowska-Gracz, H., R. A. McKenzie, and E. C. Theil. 1995. The importance of a single G in the hairpin loop of the iron responsive element (IRE) in ferritin mRNA for structure: An NMR spectroscopy study. *Nucl. Acids Res.* 23:145–152.

Theil, E. C. 1973. Red cell ferritin content during the hemoglobin transition of amphibian metamorphosis. *Dev. Biol.* 34:282–288.

Theil, E. C. 1976. The abundance of ferritin in yolk-sac derived red blood cells of the embryonic mouse. *Brit. J. Haem.* 33:437–442.

Theil, E. C. 1981. Red cell ferritin and iron storage during the early hemoglobin switch. In *Hemoglobins in Development and Differentiation,* eds. A. Nienhuis and G. Stamato-yannopoulis. pp. 423–431. Alan R. Liss, New York.

Theil, E. C. 1987. Storage and translation of ferritin messenger RNA. In *Translational Regulation of Gene Expression,* ed. J. Ilan. pp. 141–163. Plenum Press, New York.

Theil, E. C. 1987. Ferritin: Structure, gene regulation, and cellular function in animals, plants, and microorganisms. *Ann. Rev. Biochem.* 56:289–315.

Theil, E. C. 1990. The ferritin family of iron storage proteins. *Adv. Enzymol.* 63:421–449.

Theil, E. C. 1990. Regulation of ferritin and transferrin receptor (TR) mRNAs. *J. Biol. Chem.* 265:4771–4774.

Theil, E. C. 1993. The IRE (iron regulatory element) family: Structures which regulate mRNA translation and stability. *Biofactors* 4:87–93.

Theil, E. C. 1994. IREs, a family of composite mRNA regulatory elements. *Biochem. J.* 304:1–11.

Theil, E. C. 1994. Transition metal coordination complexes as probes of mRNA structure: The IRE (iron regulatory element) of ferritin mRNA as a case study. *New J. Chemistry* 18:435–441.

Theil, E. C. 1998. The Iron Responsive Element (IRE) family of mRNA Regulators in Sigel, E. and Sigel, H., eds. Metal Ions in Biological Systems, Volume 35: Iron Transport and Storage in Microorganisms, Plants and Animals. Marcel Dekker, New York, (in press).

Theil, E. C., and W. E. Brenner. 1981. The ferritin content of human red blood cells during the replacement of embryonic cells by fetal cells. *Dev. Biol.* 84:481–484.

Theil, E. C., and G. M. Tosky. 1979. Red cell ferritin and iron storage during chick embryonic development. *Dev. Biol.* 69:666–672.

Thorp, H. H., R. A. McKenzie, P.-N. Lin, W. E. Walden, and E. C. Theil. 1996. Cleavage of functionally relevant sites in ferritin mRNA by oxidizing metal complexes. *Inorg. Chem.* 35:2773–2779.

Vogel, W., D. A. Richert, B. Q. Pixley, and M. P. Schulman. 1960. Heme synthesis in iron-deficient duck blood. *J. Biol. Chem.* 235:1769–1775.

von Darl, M., P. M. Harrison, W. Bottke. 1994. cDNA cloning and deduced amino acid sequence of two ferritins: soma ferritin and yolk ferritin, from the snail *Lymnaea stagnalis* L. *Eur. J. Biochem.* 222:353–366.

Walden, W. E., M. H. Patino, and L. Gaffield. 1989. Purification of a specific repressor of ferritin mRNA translation from rabbit liver. *J. Biol. Chem.* 264:13765–13769.

Walden, W. E., and R. E. Thach. 1986. Translational control of gene expression in a normal fibroblast characterization of a subclass of mRNAs with unusual kinetic properties. *Biochem.* 25:2003–2041.

Waldo, G. S., and E. C. Theil. 1996. Ferritin and iron biomineralization. In *Comprehensive Supramolecular Chemistry,* Vol. 5, *Bioinorganic Systems,* vol. ed. K. S. Suslick. pp. 65–89. Pergamon Press, Oxford, U.K.

Wang, Y.-H., P.-N. Lin, S. R. Sczekan, R. A. McKenzie, and E. C. Theil. 1991. Ferritin mRNA probed, near the iron regulatory region (IRE), with protein and chemical (1,10-penanathroline-Cu) nucleases: A possible role for base-paired flanking regions. *Biol. Metals* 4:56–61.

Wang, Y.-H., S. R. Sczekan, and E. C. Theil. 1990. Structure of the 5′ untranslated regulatory region of ferritin mRNA studies in solution. *Nucl. Acids Res.* 18:4463–4468.

Weiss, G., B. Goossen, W. Dopple, D. Fuchs, K. Pantopoulos, G. Werner-Felmayer, H. Wachter, and M. W. Hentze. 1993. Translational regulation via iron-responsive elements by the nitric oxide/NO-synthase pathway. *EMBO J.* 12:3651–3657.

Yu, Y., E. Radisky, and E. A. Leibold. 1992. The iron-responsive element binding protein: Purification, cloning and regulation in rat. *J. Biol. Chem.* 267:19005–19010.

Zahringer, J., B. S. Baliga, and H. N. Munro. 1976. Novel mechanisms for translational control in regulation of ferritin synthesis by iron. *Proc. Natl. Acad. Sci.* 73:857–861.

Zheng, L., M. C. Kennedy, G. A. Blondin, H. Beinert, and H. Zalkin. 1992. *Arch. Biochem. Biophys.* 299:356–360.

7

The Iron Responsive Element (IRE), the Iron Regulatory Protein (IRP), and Cytosolic Aconitase
Posttranscriptional Regulation of Mammalian Iron Metabolism

Richard S. Eisenstein, M. Claire Kennedy, and Helmut Beinert

1. Introduction

The fundamental role of iron in the maintenance of human health has been apparent for many years. The requirement for iron in growth and development of organisms from bacteria to humans arises because it is an essential component of proteins that perform redox and nonredox roles in a number of cellular functions. Given that iron is one of the most abundant, chemically versatile, and reactive elements in our environment, it is not surprising that nature has made extensive use of its properties. However, there are two significant problems regarding the use of iron in biological systems: its low solubility, particularly as Fe(III), and its toxicity when present in excess because of its ability to induce formation of damaging free radicals. Organisms have developed a variety of mechanisms to acquire and make use of iron for a large number of necessary functions while simultaneously reducing the incidence of inappropriate effects of this micronutrient on cell viability. Recent investigations of the regulation of iron homeostasis in mammals have identified two unique proteins: the iron regulatory proteins or IRPs,[1] which act as central regulators of iron utilization. IRPs appear to represent the only members of the aconitase family of proteins that function in gene regulation (Frishman and Hentze 1996; Rouault et al. 1992). IRPs are cytosolic RNA binding proteins that modulate synthesis of proteins that

[1]For purposes of this review we have focused on IRP1, which is the iron-free form of cytoplasmic aconitase and was the first IRE binding protein purified by a number of laboratories (Rouault et al. 1989; Neupert et al. 1990; Walden et al. 1989; Yu et al. 1992). When we refer to IRP or IRPs we make reference to both IRP1 and IRP2.

function in the uptake, storage, and utilization of iron by binding to their mRNAs, thereby affecting their translation or stability. Posttranslational regulation of IRP function by iron and phosphorylation, with subsequent effects on iron metabolism, are topics of current inquiry to those interested in posttranscriptional gene regulation, iron-sulfur protein structure and function, and regulation of iron homeostasis.

2. Iron: An Essential but Potentially Toxic Nutrient

2.1 Roles for Iron in Cellular Function

Iron is one of the most prevalent elements in our environment, and we require relatively low levels in our diet. However, in humans, iron absorption is not very efficient, and for some populations iron-containing foods are not abundant; as a result, dietary iron intake has a significant impact on human growth and development and on the maintenance of health during aging (Baynes and Bothwell 1990; Bothwell 1995; Dallman 1986; INACG 1990). The necessity for adequate iron intake in order to achieve optimal health is illustrated by the fact that iron deficiency is one of the most common human nutritional diseases. More than 600 million individuals suffer from anemia as a result of iron deficiency with more than 1.5 billion people believed to have inadequate iron intake (INACG 1990). In mammals, deficiency of iron, which can be caused by inadequate dietary intake of iron or excessive loss of blood, is associated with impairment of multiple physiological processes including development and function of erythroid cells, skeletal and cardiac muscle function, cell and humoral mechanisms of immunity, and possibly cognitive function (Baynes and Bothwell 1990; Dallman 1986; Kretchmer et al. 1996; Pollitt 1993). Impairment of some physiological or developmental processes due to iron deficiency, such as cognitive function, can occur before the development of anemia, suggesting that a moderate level of iron deficiency can be detrimental to health at least during the years of early development (Baynes and Bothwell 1990; Dallman 1986). All in all such data indicate that iron deficiency is an important public health issue throughout the world.

Iron has both beneficial and potentially deleterious roles in human health. Although it is clear that iron is an important component of a healthy diet, when present at levels that exceed the capacity of organisms to safely utilize it, it can be toxic because of its ability to induce oxidation of lipids, proteins, and other cellular components (Aisen et al. 1990; Dougherty et al. 1981; Halliwell and Gutteridge 1986; Imlay et al. 1988; Reilly et al. 1991; Stadtman and Oliver 1991). High levels of iron within the body have been associated with increased incidence of certain cancers and with dysfunction of organs such as the heart, pancreas, or liver, and may be associated with the development of some neurodegenerative disorders (Bacon and Britton 1990; Halliwell 1989; Stevens et al.

1988). Furthermore, in contrast to the expected result, it has become apparent that providing iron to individuals deficient in the mineral does not always promote optimal health. There has been concern that iron supplementation might actually increase the risk of infection because such individuals frequently have multiple health problems and can be less able to resist an invading organism (Harvey et al. 1985, 1989; Weinberg 1993). In such situations, the effect of iron supplementation may even favor the foreign organism. Thus, in the face of variations in the availability of iron, humans are presented with a dilemma between the levels of iron needed for optimal health as opposed to the amounts of this nutrient that may increase the incidence of pathological insults.

The requirement for iron in order to promote the growth and development of various organisms occurs because of the multiplicity of enzymatic or nonenzymatic functions that iron-containing proteins perform (Crichton 1991; Howard and Rees 1991; Williams 1990). The number of species of iron centers in proteins is quite large, as are the types of chemical reactions in which iron centers participate. Many aspects of cellular function including DNA synthesis, oxygen transport, respiration, synthesis and some neurotransmitters and hormones, and some aspects of host defense are performed by iron-containing proteins. In order to promote the conservation of iron and thereby facilitate synthesis of various essential iron-containing proteins, the human body utilizes iron in a tightly controlled system where it is efficiently reutilized (Bothwell et al. 1979; Fairbanks 1994). Levels of iron contained within the body are largely modulated by changes in intestinal absorption, and there is no regulated mechanism for iron excretion (Fairbanks 1994). The requirement for iron in order to attain optimal health of organisms from bacteria to humans has led to the development of mechanisms to promote adequate uptake, transport, metabolic utilization, and storage of this essential nutrient. A role for IRPs in the regulation of intestinal absorption of iron has not been identified, but these regulatory RNA binding proteins serve as important modulators of the interorgan metabolism of iron.

2.2 Problems Regarding the Utilization of Iron in Biological Systems

The use of iron by organisms requires that several potential problems with the presence of this mineral in aqueous aerobic environments be overcome (Crichton 1991; Williams 1990). First, although iron is one of the most abundant elements on earth, its bioavailability is low. One of the major factors affecting the biological utilization of iron is its low solubility. Several factors affect the solubility of iron, including its oxidation state [iron(II) versus iron(III)] as well as the pH and redox potential of the medium in which it is present. As the free metal, both iron(II) and iron(III) are soluble only at levels several orders of magnitude below the micromolar range required biologically. In solution, under physiological conditions, iron(II) is oxidized to iron(III), and if present as the uncomplexed ion, iron(III) forms insoluble iron hydroxide precipitates, thereby rendering it largely unusable. Thus, chelation of iron, particularly iron(III), by proteins and other biological molecules serves to facilitate its solubilization under physiological conditions (Williams 1990).

Second, iron toxicity can result from generation of toxic oxygen metabolites, some of which are free radicals, causing oxidation of unsaturated lipids, proteins, nucleic acids, and other cell constituents (Aisen 1991; Bacon and Britton 1990; Dougherty et al. 1981; Halliwell and Gutteridge 1986; Imlay et al. 1988; Reilly et al. 1991; Stadtman and Oliver 1991). Although it is clear that some physiological processes, such as phagocytic killing of bacteria (Babior 1984), utilize free radicals for beneficial reasons, excessive and/or ectopic production of these toxicants can be detrimental to cellular function. Free radicals such as the superoxide anion (O_2^-) are produced as side products of normal cellular metabolism, and cells contain mechanisms for elimination of such potentially toxic agents. However, interaction of iron with O_2^- and its metabolites produces more highly reactive radicals that can cause extensive damage. Evidence from animal models of iron overload and cases of genetically induced iron overload in humans has demonstrated the potentially detrimental effects of iron on health (Aisen 1991; Bacon and Britton 1990; Dougherty et al. 1981; Halliwell 1989; Stevens et al. 1988). Furthermore, some evidence suggests that tissue damage as a result of ischemic events (i.e., stroke or heart attacks), apparently including metal-catalyzed oxidation of some proteins, is caused, in part, by generation of free radicals during tissue reperfusion; antioxidants and/or iron chelators can ameliorate some of these effects (Hedlund and Hallaway 1993; Oliver et al. 1990; Reilly et al. 1991). The mechanism(s) of free radical generation and the species of radical produced remain to be fully defined, but generation of hydroxyl radical as a result of the "iron-catalyzed Haber-Weiss reaction," otherwise known as Fenton chemistry, contributes to the damage of cellular macromolecules (Aisen 1991; Halliwell and Gutteridge 1986; Reilly et al. 1991). It is these potentially deleterious effects of iron that are likely to have been a significant driving force for the evolution of systems to transport, store, and utilize iron in a safe and available form.

3. Modes for Regulating Iron Homeostasis in Vertebrates

3.1 Systems Evolved to Provide Metabolic Balance and Diversity of Function

The biological mechanisms that have been developed to maintain iron homeostasis have similar purposes in numerous organisms. First, some systems are aimed at scavenging low amounts of iron from the environment in order to promote viability of the organism. Second, other mechanisms serve to increase the solubility of iron and, in higher eukaryotes, facilitate its intercellular and/or interorgan transport. Third, most, if not all, eukaryotes contain the means to store iron in a safe and available manner. Fourth, antioxidant systems have evolved in order to reduce damage from free radicals produced from a variety of sources including those related to iron. The roles of antioxidants and antioxidant systems has been amply covered by others (DiMascio et al. 1991; Sies 1991; Sohal and Orr 1992). We will focus on the first three functions, uptake, transport, and storage of iron, and the role of proteins that directly (ferritin, transferrin, IRP1) or indirectly (transferrin receptor) bind iron in the maintenance of iron homeostasis in mammals.

3.2 Overview of Cellular Iron Metabolism

In mammals, interorgan transport and uptake of iron is performed largely by transferrin (Tf) and transferrin receptor (TfR) (Figure 7.1A). Generally speaking, once iron is delivered to the cytoplasm there are three possible fates for it: (1) short- or long-term storage, (2) metabolic utilization, and (3) export from the cell. Iron is stored by ferritin, a multisubunit protein that retains iron in a safe but available manner (Figure 7.1C). Metabolic utilization of iron refers to its incorporation, as nonheme and/or heme iron centers, into proteins that require the mineral as a cofactor (Figure 7.1B). Synthesis of heme-containing proteins requires mitochondrial incorporation of iron into protoporphyrin IX to form heme. Lastly, certain cell types such as hepatocytes and Kupffer cells release iron in the form of transferrin (Tf) or ferritin, for transport to other areas of the body (Aisen 1991) (Figure 7.1B). Iron homeostasis is established through coordinate changes in the abundance and/or activity of proteins that control the uptake, utilization, and storage of this essential nutrient. IRPs influence the metabolic fate of iron by acting as modulators of the rate of iron uptake from the blood and the flux of iron through the storage and metabolic utilization pathways. Agents that affect IRP function in a cell-type specific manner can promote selective regulation of iron metabolism in various tissues of the body.

Multiple processes contribute to the maintenance of iron homeostasis. First, the rate of iron uptake can be regulated (Figure 7.1A). On the cell surface TfR binds diferric Tf (Fe_2Tf), the serum iron transport protein. The distribution of TfR between cell surface and internal pools and the rate at which it cycles between these pools is affected by specific growth factors and other agents (D'Souza-Schorey et al. 1995; Dargemont et al. 1993; Davis et al. 1987; Iacopetta et al. 1986; Klausner et al. 1984; Wiley and Kaplan 1984). In most cells the TfR synthesis rate is directly related to cellular iron requirement. In response to an increased need for iron, as occurs during the onset of cell proliferation or as a consequence of differentiation of cell types like erythroid cells, TfR synthesis

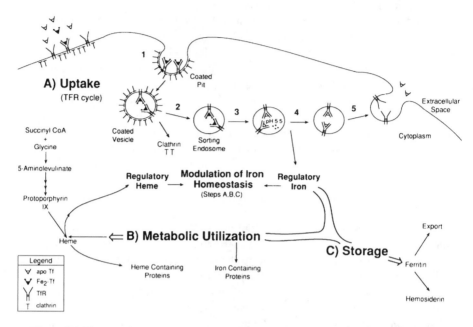

Figure 7.1 The transferrin receptor cycle and the metabolic fate of iron. The pathway for iron uptake from serum diferric transferrin is shown. This figure represents a highly idealized view of the process of endocytotic uptake, sorting, and transfer back to the cell membrane of the Tf-TfR complex (Bomford and Munro 1985; Huebers and Finch 1987; Trowbridge et al. 1993). A. TfR on the cell surface preferentially binds Fe₂-Tf, and together they are internalized into the cell first in coated pits that bud off into the cytoplasm, forming coated vesicles (Step 1). Coated pit formation is facilitated by the protein clathrin, which forms a basket-like network around the coated vesicle. After internalization the clathrin coat is shed, and the coated vesicle is now termed an endosome (Step 2). Endosomes appear to be composed of vesicular and tubular structures (Trowbridge et al. 1993). The Tf-TfR complex is sorted away from other internalized receptors that are destined for lysosomal degradation, such as the ∝−2-macroglobulin receptor (Step 2) (Trowbridge et al. 1993). Once the vesicles become acidified, iron is released from Tf in a manner that is facilitated by TfR (Step 3) (Bali et al. 1991). Iron is released to the cytoplasm (Step 4), and the Tf-TfR eventually re-fuses with the cell membrane, where at physiological pH apo-Tf dissociates from TfR (Step 5). Once iron is delivered into the cytoplasm there are several metabolic fates for it: metabolic utilization (B), and storage (C) or export. The rate of uptake, metabolic utilization, and storage of iron is influenced by the level of nonheme and/or heme iron in the cell.

and/or display is increased (Andreesen et al. 1984; Casey et al. 1988a,b; Cox et al. 1990; Davis et al. 1987; Hu et al. 1977; Iacopetta et al. 1986; Larrick and Cresswell et al. 1979; Núñez et al. 1977; Owen and Kühn 1987; Pelosi-Testa et al. 1988; Seiser et al. 1993; Teixeira and Kühn 1991; Testa et al. 1989, 1991; Trowbridge and Omary 1981; Wiley and Kaplan 1984). The effect of iron and other agents on TfR synthesis is regulated largely through altered stability of its mRNA (Casey et al. 1988a,b; Owen and Kühn 1987). However, alterations in TfR gene transcription do contribute to the observed changes in expression of the receptor (Rao et al. 1985; Seiser et al. 1993; Taetle et al. 1987). Synthesis of the serum iron binding protein transferrin (Tf) is also inversely related to cellular iron levels largely through transcriptional mechanisms, at least in liver, but altered translation of Tf mRNA may also be important (Cox and Adrian 1993; Cox et al. 1995; McKnight et al. 1980).

Second, the metabolic fate of iron is modulated according to the needs of the cell. The extent to which iron delivered to the cytoplasm is then directly incorporated into ferritin (Figure 7.1C) as opposed to being shunted toward metabolic utilization (Figure 7.1B) pathways varies in a cell-type specific manner. Many cell types initially deposit much of the newly taken-up iron into ferritin, whereas in reticulocytes little iron is deposited for storage and most appears to be directly used for heme formation (Adams et al. 1989; Mattia et al. 1986). In liver and especially erythroid cells a large portion of iron delivered to the cytoplasm is ultimately used for heme formation. The rate-limiting enzyme for heme formation in most, if not all cells, is 5-aminolevulinate synthase (ALAS) (Kappas et al. 1995). Control of heme formation occurs largely through cell-type specific regulation of ALAS synthesis by heme and iron (Bhasker et al. 1993; Cox et al. 1991; Dandekar et al. 1991; Lim et al. 1994; Melefors et al. 1993). Synthesis of the erythroid-specific isoform of ALAS (eALAS) is translationally regulated by iron (Bhasker et al. 1993; Melefors et al. 1993). In addition to modulating the flow of iron towards metabolic utilization, an important component of the means for establishing iron homeostasis is through alterations in iron storage. Regulation of iron storage occurs, in part, as a result of changes in ferritin synthesis. Production of ferritin is directly coupled to iron status through changes in translation of a preexisting pool of *H*- and L-ferritin mRNAs, thereby allowing cells to respond rapidly to changes in iron status, particularly to iron overload (Aziz and Munro 1986, 1987; Drysdale and Munro 1966; Rogers and Munro 1987; Zähringer et al. 1976). Alterations in the uptake, utilization, and storage of iron are largely modulated by IRPs because of their function as iron-regulated RNA binding proteins that modulate TfR, eALAS, and ferritin synthesis.

3.3 Roles for IRPs in Whole Body Iron Metabolism

In studies of whole body iron kinetics and in cell and molecular biological investigations of the proteins that function to transport, store, or utilize iron, several points illustrate the central role of IRPs in the homeostatic regulation of iron metabolism within the body. First, more than 80% of iron flux within the body involves the formation and turnover of erythrocytes (Fairbanks 1994). Thus, regulation of protoporphyrin IX formation in erythrocyte precursors through IRP-mediated changes in eALAS synthesis is likely to have a significant impact on whole body iron utilization (Figure 7.2). Second, it is apparent that during certain physiological or pathological situations the synthesis of TfR is affected by iron and other factors (Casey et al. 1988a,b; Müller-Eberhard et al. 1988; Owen and Kühn 1987; Pelosi-Testa et al. 1988; Rao et al. 1985; Seiser et al. 1993; Taetle et al. 1987; Teixeira and Kühn 1991; Testa et al. 1989, 1991), and that the number of TfR molecules displayed on the cell surface is directly related to the rate of iron uptake by cells (Morgan and Baker 1986; Testa et al. 1984). IRPs

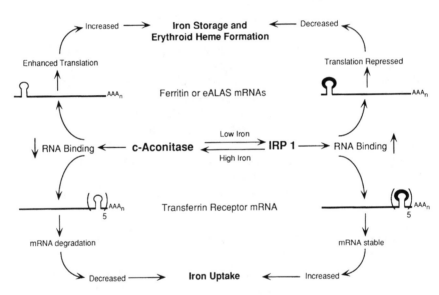

Figure 7.2 Iron regulatory protein 1: Central regulator of iron utilization. Intracellular iron levels regulate the interconversion between IRP1 and c-aconitase. When iron levels are low, the protein is present as IRP1. It binds to mRNAs encoding ferritin or erythroid 5-aminolevulinate synthetase (eALAS) and represses their translation and hence decreases levels of newly synthesized ferritin and eALAS proteins. This reduces the capacity of the cell to store and, in the case of erythroid cells, utilize iron. Simultaneously, IRP1 is believed to bind TfR mRNA, leading to stabilization of the message and increased synthesis of TfR protein. This increases the capacity of the cell to transport Fe_2-Tf into the cytoplasm. Conversely, when iron levels are high, IRP1 is converted to c-aconitase and its high-affinity RNA binding activity is lost. Synthesis of ferritin and eALAS is derepressed, and iron storage and erythroid heme formation are increased. TfR mRNA is degraded after dissociation of IRP1, and levels of newly synthesized TfR decline.

are important effectors of TfR gene expression and therefore of iron uptake by cells as a result of their function in modulating the half-life of TfR mRNA (Figure 7.2). Despite the important role of IRPs in modulating TfR mRNA stability, it is important to also consider that TfR protein may, under some conditions, have a longer half-life than TfR mRNA, implying that changes in TfR mRNA concentration may not immediately be reflected in alterations in TfR protein level. Third, alterations in ferritin abundance promote its function as a depot for short- and long-term storage of iron and as a protectant from iron-induced oxidative damage (Balla et al. 1992, 1993; Bomford et al. 1981; Kohgo et al. 1980). Interaction of IRP with ferritin mRNAs affects their translation and, as a consequence, the levels of newly synthesized ferritin shells used to store increased levels of iron (Figure 7.2). Fourth, in healthy adults, the amount of iron absorbed and excreted is in balance and represents less than 10% of the iron flux in the body on a daily basis (Fairbanks 1994). Absorption of heme and especially nonheme iron by the intestine represents the only point at which changes in body iron levels can be achieved in a regulated manner. Thus, the body promotes the active conservation and reutilization of iron, and IRPs serve as key sensors of iron status of various tissues and as a modulator of interorgan cooperation in the utilization of the mineral by the body. Fifth, natural or induced mutation of the IRP regulatory system leads to alterations in cellular metabolism of iron (Beaumont et al. 1995; DeRusso et al. 1995; Girelli 1996). Many observations have established that IRP-mediated changes in abundance of TfR and ferritin provide the means to alter the uptake and utilization of iron in response to both general and specific needs of a variety of cell types. Synthesis of TfR and ferritin is regulated by intracellular (e.g., iron) and extracellular agents (e.g., hormones and growth factors) as well as alterations during proliferation or differentiation of specific cell types. Because they act to modulate TfR and ferritin synthesis, IRPs serve as focal points for integrating the uptake and utilization of iron by tissues of the body in response to a variety of regulatory signals.

3.4 Receptor-Ligand Interaction and the Regulation of Iron Uptake

IRON UPTAKE SYSTEMS INDEPENDENT OF TRANSFERRIN

Mammalian and other eukaryotic cells possess multiple mechanisms for transporting nonheme or heme iron into the cytoplasm from the extracellular medium (De Silva 1996). Uptake of nonheme iron into cells occurs through transferrin-dependent and independent pathways.[2] Transferrin-independent pathways for uptake of iron are likely to be important when cells are exposed to non-transferrin-bound iron and when this occurs under both physiological and pathological

[2]For purposes of this section iron refers to nonheme iron and heme refers to heme iron.

situations (Gutteridge 1994, 1996; Berger 1995). There are multiple Tf-independent pathways for iron uptake. We shall briefly discuss two such pathways that have important roles in whole body iron homeostasis. First, cells lining the intestinal lumen, cerebrospinal cavity and, in some cases, other organs are exposed to significant levels of non-transferrin-bound iron, and at least the luminal surface of mucosal cells lining the duodenum and other areas of the small intestine have been shown to be capable of specifically transporting non-Tf-bound iron (Conrad et al. 1990, 1993; Griffiths 1987; Nichols et al. 1992; Teichmann and Stremmel 1990; Wolf 1994; Wright et al. 1986). As noted above, the regulation of intestinal iron uptake represents a key point at which whole body iron homeostasis is modulated. Recent work has identified a novel intestinal iron binding protein as well as a putative transporter of non-Tf-bound iron, but much remains to be learned concerning how intestinal uptake of iron occurs and how it is regulated (Conrad et al. 1990; Teichmann and Stremmel 1990; Wolf 1994). In addition to intestine, tissues such as liver as well as certain cultured cell lines are capable of transporting non-Tf-bound iron (Randell et al. 1994; Sturrock et al. 1990). It appears that liver uptake of non-Tf-bound iron is an important contributor to the large accumulation of iron in this organ in the iron storage disease, hemochromatosis, when serum transferrin can become fully saturated and non-protein-bound iron accumulates in the serum (Bonkovsky 1991; Grootveld et al. 1989). However, in healthy individuals, non-Tf-bound iron is not present in the serum at appreciable levels. Therefore, iron uptake by Tf-independent pathways would appear to not be important for most interorgan transport of iron in healthy individuals. The second pathway is the phagocytic uptake of senescent red cells by macrophages of the reticuloendothelial system, which makes a large contribution to whole body iron flux (Fairbanks 1994; Young and Aisen 1988). These pathways are important contributors to the daily flux of iron within the body. It remains to be determined if IRPs have a role in modulating the activity of these pathways.

TRANSFERRIN-MEDIATED PATHWAYS FOR IRON UPTAKE

Another contributor to interorgan transport of iron in the body utilizes serum transferrin (Bomford and Munro 1985; Huebers and Finch 1987; Thorstensen and Romslo 1990). The evidence suggests the existence of at least two means for uptake of transferrin-bound iron. One may involve reductive release of iron at the cell surface such that transferrin is not internalized in order for iron to be released and to enter the cell (Randell et al. 1994; Thorstensen and Romslo 1990). However, the more extensively characterized transferrin-dependent system for iron uptake involves receptor-mediated endocytotic uptake of transferrin via the transferrin receptor cycle (Figure 7.1). There are several pieces of evidence that argue in favor of the transferrin receptor cycle as the major means of interorgan iron transport. First, interaction of Fe_2-Tf with its receptor represents the major mechanism by which many cell types in mammals take up non-heme iron from the extracellular environment. This is reflected in the observations that iron-labeled Tf rapidly loses its iron after intravenous injection into rats and that Tf is an essential physiological donor of iron to cells in culture. Second, on the basis of independent measurements of the turnover of the protein and mineral components of Fe_2-Tf, it is apparent that Tf itself is reutilized many times on a daily basis in the interorgan transport of iron. Third, as noted above, under physiological conditions Tf represents the major transport form of nonheme iron in the body. Fourth, the rate of iron accumulation by reticulocytes and other cell types is proportional to the surface display of TfR.

THE TRANSFERRIN RECEPTOR CYCLE

Delivery of iron into the cytoplasm by the endocytotic uptake of serum Tf represents one of the best-characterized examples of receptor-mediated endocytosis. Initial investigations on the fractionation and characterization of human plasma identified transferrin as the major iron-containing protein and its function as transporter of iron from the site of absorption (i.e., intestine) or storage (i.e., liver) to the sites of utilization, primarily the bone marrow (Jandl et al. 1959). It was observed that Tf bound iron with high affinity ($k_A = 10^{29}$), but *in vivo* kinetic measurements demonstrated that this iron turned over much more rapidly than did the iron binding protein itself (Davis et al. 1962; Katz 1961). Thus, it was apparent that the Tf molecule could be reutilized many times to deliver and release iron to cells before being degraded. Further evidence regarding the dynamic nature of the interaction of Tf with iron came from the observation that, in a cell-dependent manner, the interaction between protein and metal ion could be labilized, thereby permitting iron utilization and Tf recycling (Jandl and Katz 1963). Additional key observations included the demonstration that reticulocytes (a) concentrated iron several-fold, on a molar basis, over their ability to bind Tf; (b) contained receptors for Tf that were destroyed by trypsin; (c) possessed large numbers of receptors, whereas erythrocytes, which no longer synthesize heme, lacked receptors; (d) could be used to show that Fe_2-Tf bound to the receptor more tightly than did apo-Tf. Taken together, these observations established that Tf was capable of efficiently donating iron to cells in a manner that was dependent on its interaction with specific cell surface Tf binding protein(s) (Jandl and Katz 1963; Morgan and Appleton 1969).

A key series of observations includes those detailing the isolation and characterization of the membrane binding protein for Tf, the transferrin receptor (TfR), and additional aspects of its internalization by cells (Enns et al. 1983; Hamilton et al. 1979; Hemmaplardh and Morgan 1977; Iacopetta and Morgan 1983; Karin and Mintz 1981; Klausner et al. 1983b; McClelland et al. 1984; Seligman et al.

1979). TfR is a disulfide-linked homodimer of an M_r = 95,000 glycoprotein that is comprised of 760 amino acids with 671 amino acids in the extracellular domain, 28 amino acids in the intramembrane domain, and 61 amino acids in the cytoplasm (McClelland et al. 1984). Because the receptor binds Fe_2-Tf with nanomolar affinity and the concentration of Fe_2-Tf in blood is micromolar, then the TfR should always be saturated with ligand (Dautry-Varsat et al. 1983; Klausner et al. 1983a; Morgan and Baker 1986; Testa et al. 1984; Thorstensen and Romslo 1990). This result helps explain the observations that TfR display is directly related to the rate of iron uptake. Active internalization of the receptor is constitutive and independent of ligand. Rapid internalization is dependent on a specific amino acid sequence contained within residues 19–28 of the cytoplasmic tail of the receptor (Collawn et al. 1990, 1993; Gironés et al. 1991; Jing et al. 1990; Trowbridge 1991; Trowbridge et al. 1993). TfR can be phosphorylated within the cytoplasmic domain, but the effect of this on its function remains to be elucidated (Cox et al. 1985; Dargemont et al. 1993; Davis and Meisner 1987b; Rothenberger et al. 1987).

Internalization of TfR is initiated by protein-dependent formation of invaginations within the cell membrane called coated pits (Figure 7.1, Step 1) (Hopkins et al. 1994; Miller et al. 1991a; Trowbridge et al. 1993). The pits are internalized forming clathrin coated vesicles with the Tf/TfR complex inside the vesicle and the amino-terminal tail still exposed to the cytoplasm (Figure 7.1A, Step 1). Once internalized, coated vesicles lose their clathrin coat and form membrane-bound vesicles termed endosomes (Figure 7.1A, Step 2) (Trowbridge et al. 1993). Before its return to the cell surface Tf releases its iron in part as a result of acidification (pH $\approx$ 5.5) of the endosomal compartment and reduction of iron(III) to iron(II) (Figure 7.1A, Step 3) (Bomford et al. 1985; Ciechanover et al. 1983; Núñez et al. 1990; Watkins et al. 1992). A membrane-bound proton ATPase acidifies the endosomal compartment, and some evidence suggests that this proton translocase may itself transfer iron into the cytoplasm (Figure 7.1A, Step 4) (Li et al. 1994; Watkins et al. 1992). Apparently, interaction of Tf with TfR induces structural changes in Tf such that the rate of iron release at low pH is greatly enhanced and sufficient to account for the rate of cellular iron uptake (Bali et al. 1991). Changes in pH have another important role in the TfR cycle. At low pH, apo-Tf remains associated with TfR as tightly as Fe_2-Tf, but at physiological pH Fe_2-Tf binds approximately 20-fold more tightly to TfR than does apo-Tf (Thorstensen 1990). Therefore, when endosomes re-fuse with the cell membrane and the Tf-TfR complex is exposed to the higher pH of the extracellular medium, apo-Tf dissociates from TfR (Figure 7.1A, Step 5). Considered as a whole, the TfR cycle is very efficient. More than 99% of internalized receptor returns to the cell surface, and the entire endocytic cycle for TfR takes as little as 4–6 min with nearly 100% of the iron being unloaded from Tf in the process (Ciechanover et al. 1983; Iacopetta and Morgan 1993; Jing et al. 1990; Klausner et al. 1983a,b).

Finally, it is important to note that the distribution of receptor between cell surface and intracellular compartments and the rate at which it cycles between these compartments appears to be regulated in a complex manner (D'Souza-Schorey et al. 1995; Dargemont et al. 1993; Davis et al. 1987a; Iacopetta et al. 1986; Klausner et al. 1984; Wiley and Kaplan 1984).

Many questions remain to be answered regarding the trafficking of TfR between the cell surface and cytoplasm. Some questions relate to the mechanism by which TfR is concentrated in coated pits and internalized into the cell, whereas others focus on regulation of the TfR cycle in response to various extracellular effectors. Of particular relevance to this review are questions concerning the mechanism of iron release from the endosome and the routing of iron to various metabolic fates in the cell. How is iron directed through the cytoplasm so that it ends up in specific locations such as the mitochondrion? Does the cytoplasmic tail of TfR play a role in the unloading of iron from Tf or perhaps in directing proteins or other factors involved in iron trafficking to the endosomal compartments that contain iron? How and when does IRP come into contact with newly delivered iron? Studies addressing these and related issues should shed further light on the roles of IRP in modulating iron utilization.

3.5 Ferritin: Ubiquitous Iron-Storage Molecule

ASPECTS OF FERRITIN STRUCTURE AND FUNCTION

Ferritin or ferritin-like molecules have been found in primitive organisms including fungi and also in bacteria such as *E. coli*. In most higher eukaryotes, ferritin is a large macromolecule composed of 24 subunits of two types, termed H- and L, that forms a hollow sphere. The cavity within ferritin can contain up to 4500 iron atoms in the form of ferric oxyhydroxide, but *in vivo* the average number of iron atoms is lower (Harrison et al. 1987; Munro and Linder 1978; Munro et al. 1988). The relative proportions of H- and L-subunits in the assembled ferritin sphere can vary. Ferritin macromolecules of differing subunit composition are called *isoferritins*. H-rich isoferritins appear to take up and release iron rapidly, perhaps reflecting a role as short-term depot for iron storage (Bomford et al. 1981; Kohgo et al. 1980; Levi et al. 1994; Wagstaff et al. 1978). In contrast, isoferritins rich in the L-subunit are more prevalent in iron storage tissues, and iron within these species of ferritin may turn over slowly (Bomford et al. 1981; Kohgo et al. 1980; Levi et al. 1994). The synthesis rates of the individual ferritin subunits can be selectively regulated and have been observed to change in response to hormonal treatment and during cell proliferation or differentiation (Beaumont et al. 1994; Chazenbalk et al. 1990; Chou et al. 1986; Coccia et al. 1992; Miller et al. 1991b; White and Munro 1988).

Each ferritin subunit has a molecular weight between 19,000 and 21,000 with

the H-subunit being the larger of the two proteins. Thus, the ferritin macromolecule has a molecular weight of about 500,000. There are a few exceptions, as
heart ferritin appears to contain 36 subunits (Linder et al. 1981) and in the
bullfrog, there is a third subunit, denoted M, that is quite similar to the H subunit
(Dickey et al. 1987). The assembled macromolecular sphere has two types of
channels, hydrophilic and hydrophobic, penetrating the wall that function in iron
entry and perhaps exit from the cavity (Harrison et al. 1987). Iron can apparently
be released from ferritin either by degradation of the ferritin shell, presumably
leaving hemosiderin (Andrews et al. 1988; Drysdale and Munro 1970; Roberts
and Bomford 1988), or perhaps by removal of iron without loss of the protein
shell (Jones et al. 1978). The importance of ferritin as a storage site for iron
becomes apparent when we realize that, depending on age and sex, as much as
25 to 30% of whole body iron is present as ferritin (Fairbanks 1994). Thus, next
to hemoglobin ($\approx$65–70% of total body iron) the largest amount of iron in the
body can be found in ferritin.

3.6 Regulation of Iron Metabolism at Specific Levels of Gene Regulation

Numerous studies illustrate that the steady-state levels of ferritin and TfR
protein are modulated in order to promote iron homeostasis in individual cells
and in the whole organism. Iron uptake is required in probably all proliferating
cells, and there are large changes in TfR expression in response to growth
activation (Kühn et al. 1990; Larrick and Cresswell 1979; Owen and Kühn 1987;
Pelosi-Testa et al. 1988; Seiser et al. 1993; Teixeira and Kühn 1991; Testa et
al. 1991; Trowbridge and Omary 1981). Conversely, iron uptake decreases during
differentiation except in cell types that exhibit specialized needs for the nutrient
such as some hematopoietic cells (Andreesen et al. 1984; Hu et al. 1977; Núñez
et al. 1977). Storage of iron is also modulated as a function of proliferation or
differentiation of cells. Quiescent cells stimulated to proliferate show decreased
levels of ferritin protein, implying that iron is more heavily used for synthesis
of essential iron proteins (Testa et al. 1991). Specific cell types contain different
levels of ferritin and they exhibit relative differences in abundance of ferritin
subunits depending on their specialized requirements for iron (Beaumont et al.
1994; Bomford et al. 1981; Chou et al. 1986; Coccia et al. 1992; Kohgo et al.
1980; Munro and Linder 1978; Theil 1987). Cell-specific differences in iron
metabolism can be achieved through programmed changes in ferritin or TfR abundance.

Iron status, hormones, growth factors, and other agents affecting cell function
can influence the rate of uptake and metabolic fate of iron by acting to alter
ferritin and TfR gene expression transcriptionally and/or posttranscriptionally.
What physiological advantages are provided to an organism in promoting the

regulation of expression of ferritin or TfR at multiple levels? First, the potential advantages are centered on the extent to which the uptake, metabolic utilization, or storage of iron are coordinately regulated and the ability to selectively modulate the metabolic fate of iron in a cell-type specific manner. Second, it provides multiple means by which intracellular (i.e., iron) or extracellular (i.e., growth factors) effectors can impinge on the homeostatic regulators of iron metabolism. Transcriptional regulation of the ferritin or TfR genes provides a means for selective regulation of their expression. Regulation at the level of transcription provides a mechanism by which iron and other factors can discriminate between the H- and L-ferritin genes and consequently promote changes in relative abundance of the isoferritin species present in different tissues (Beaumont et al. 1994; Chazenbalk et al. 1990; Chou et al. 1986; Coccia et al. 1992; Miller et al. 1991b; White and Munro 1988). These effects are likely to be important in the general prevention of iron toxicity and also in interorgan cooperation in the storage of iron. TfR gene expression can also be modulated by changes in gene transcription, and this contributes to alterations in receptor expression during cell proliferation or differentiation (Rao et al. 1985; Seiser et al. 1993; Taetle et al. 1987). Clearly, a major role of posttranscriptional regulation through IRPs is to coordinately alter the rate of uptake, storage, and, in erythroid cells, metabolic utilization of iron. Most investigations on posttranscriptional regulation of ferritin or TfR synthesis have centered on the effects of iron and the extent to which IRPs serve as a frontline defense to resist changes in intracellular iron status. However, other studies have begun to describe how IRPs act as focal points through which agents other than iron can alter uptake, utilization, and storage of iron. Taken together, these investigations illustrate the central role of IRPs in modulation of iron metabolism by a variety of agents or circumstances.

4. The Iron Responsive Element (IRE) Influences mRNA Utilization

4.1 Translational Control of Ferritin Synthesis Predicts the Presence of a Cytoplasmic Sensor of Iron Status

The basis for investigations of the roles of posttranscriptional control in the regulation of cellular iron metabolism originates from the early observations on the effects of iron status on the abundance or synthesis of ferritin in tissues of the rat. Studies by Granick, more than 50 years ago, demonstrated that oral administration of iron led to increased abundance of ferritin in intestinal epithelium (Granick 1946). A number of investigators subsequently demonstrated that iron stimulated de novo synthesis of ferritin. Of particular relevance were the observations that iron induction of ferritin synthesis was not dependent on new

RNA synthesis, thereby suggesting that iron acted cytoplasmically to stimulate accumulation of additional ferritin protein (Chu and Fineberg 1969; Drysdale and Munro 1965; Drysdale et al. 1968; Saddi and Decken 1965). Convincing evidence indicating that iron acted at the translational level was provided by Munro and colleagues (Ariz and Munro 1986, 1987; Drysdale and Munro 1966; Rogers and Munro 1987; Zähringer et al. 1976), who demonstrated that ferritin mRNA was found on polyribosomes in the liver of rats injected with iron but was present in the translationally inactive ribonucleoprotein particle (RNP) fraction in control rats (Aziz and Munro 1986; Zähringer et al. 1976). As iron was subsequently shown to induce translational activation of H- and L-ferritin mRNAs, it appeared that a common mechanism regulated the iron-dependent synthesis of both ferritin subunit proteins (Aziz and Munro 1986). With the advent of molecular cloning procedures, sequences for ferritin cDNAs and genes were obtained. These investigations provided evidence on the evolutionary relatedness of the H- and L-subunits and, more strikingly, have demonstrated the presence of a highly conserved 28-nucleotide sequence in the 5′ untranslated region (UTR) of both H- and L-ferritin mRNAs in species from molluscs to humans[3] (Eisenstein et al. 1990; Leibold and Guo 1992; Theil 1994).

4.2 Ferritin and TfR mRNAs Contain a Common Regulatory Element

The presence of this highly conserved motif in the 5′UTR of ferritin mRNAs suggested that it represented a common iron-sensitive control point for the regulation of ferritin synthesis by iron and perhaps other effectors. As will be discussed in the following paragraphs, this RNA sequence element, called the iron responsive element or IRE, is a binding site for IRP, which acts as a repressor of ferritin synthesis. The IRE sequence is about 90% identical between both H- and/or L-ferritin mRNAs from molluscs to humans (Eisenstein et al. 1990; Leibold and Guo 1992; Theil 1994). This level of nucleotide sequence identity between the IRE region was observed to be significantly higher than was the level of nucleotide sequence identity between the coding regions of the two ferritin subunit mRNAs (about 55% identical), strongly implying that there was a role for the IRE in the regulation of ferritin mRNA translation (Leibold and Munro 1987). Further support for the central role of the IRE in iron metabolism came from the exciting observation that mRNAs encoding other proteins that serve to establish iron homeostasis also contained IREs. After IREs were discovered in ferritin mRNAs it was found that TfR mRNA contains multiple IREs. However, in contrast to

---

[3]Recent evidence indicating that, in humans, an mRNA from an expressed L-ferritin pseudogene does not contain an IRE represents the only example, to date, of a "natural" ferritin mRNA that appears to not be translationally regulated by iron (Renaudie et al. 1992).

ferritin, TfR mRNA contains IRE elements in its 3' UTR, where they regulate its stability in response to iron (Casey et al. 1988b). In addition to ferritin and TfR mRNAs, two other mammalian mRNAs, mitochondrial aconitase (m-aconitase) (Dandekar et al. 1991) and, as noted previously, erythroid 5-aminolevulinate synthase (eALAS) (Cox et al. 1991; Dandekar et al. 1991; Lim et al. 1994) contain an IRE in their 5' UTR. Similarly, the mRNA encoding the iron-protein subunit of *Drosophila melanogaster* succinate dehydrogenase (SDH) has an IRE in its 5'UTR, and synthesis of this protein is translationally regulated by iron (Kohler et al. 1995; Gray et al. 1996). ALAS is the rate-limiting enzyme in heme formation in most cells. In the case of eALAS, the IRE serves to regulate eALAS synthesis in erythroid cells (Bhasker et al. 1993; Melefors et al. 1993a). m-Aconitase and SDH are Krebs cycle enzymes that convert citrate to isocitrate and succinate to fumarate, respectively. Regarding SDH and m-aconitase, recent evidence indicates that the synthesis and/or abundance of these proteins are sensitive to changes in cellular or whole body iron status (Chen et al. 1997; Kim et al. 1996; Kohler et al. 1995; Gray 1996). Thus, the IRE is a critical determinant of the iron-dependent synthesis of TfR, ferritin, eALAS, the iron protein subunit of SDH, and m-aconitase.

Analyses of IRE structure and function have confirmed its role as an iron-dependent regulator of mRNA translation or stability. Computer predictions, nuclease mapping, and mutagenesis results show that the IRE forms a region of secondary structure comprised of a double-stranded stem interrupted at several locations by bulged nucleotides and "topped" by a six-nucleotide loop (Barton et al. 1990; Bettany et al. 1992; Caughman et al. 1988; Dix et al. 1993; Harrell et al. 1991; Henderson et al. 1994; Hentze et al. 1987, 1988; Jaffrey et al. 1993; Leibold et al. 1990; Rouault et al. 1988; Sierzputowska-Gracz et al. 1995; Thorp et al. 1996; Wang et al. 1990). Additional results obtained by selecting high affinity ligands for IRP1 indicate that it prefers RNAs in which the first and fifth nucleotides of the loop have the capacity to base-pair (Butt et al. 1996; Henderson et al. 1996). Thus, the structure that IRP1 binds may in fact be a smaller 3 nt loop with a single nucleotide bulge on the 3' side (Henderson et al. 1996). IREs from various mRNAs have similar secondary structure with the apparent exception of the size of a bulged nucleotide region that starts four base pairs 5' of the loop. Ferritin IREs are predicted to have a three-nucleotide bulge UGC at this location, whereas IREs in TfR, eALAS, m-aconitase, and SDH iron-protein mRNAs have a single-bulged C residue (Bettany et al. 1992; Theil 1994). It is presently unclear if the size of this bulged nucleotide region confers functional differences on these IREs, although recent evidence indicates that IRPs appear to bind less well to the IRE in m-aconitase or eALAS IREs as compared to the ferritin IREs, and that iron intake differentially influences the abundance of m-aconitase and ferritin protein in rat liver (Chen et al. 1997; Cox et al. 1991; Kim et al. 1996). Studies have shown that other aspects of IRE secondary or primary structure influence its interaction with IRP. Mutations that disrupt the upper stem region of the IRE stem-loop ablate high-affinity interaction of the IRE with the IRP (Bettany et al. 1992; Leibold et al. 1990). However, when the stem is allowed to reform with a different primary sequence, IRP binds with high affinity, indicating that the ability to form an appropriate stem structure in this region is more important for IRP binding than is primary sequence of the stem structure (Bettany et al. 1992; Leibold et al. 1990). The loop sequence is CAGUGN in nearly all IREs examined, and deletions or substitutions within this region can lower the affinity of interaction of IRP with the IRE and impair the ability of the binding proteins to modulate ferritin synthesis in cell-free systems (Dix et al. 1992, 1993; Eisenstein et al. 1997; Thorp et al. 1996). No mutations in the wild-type sequence have been found that increase the affinity of the IRE/IRP interaction. However, recent studies have identified sequences significantly differ-ent from the wild-type loop that interact nearly as well with the binding protein. Using PCR-mediated selection of high-affinity RNA ligands for IRP, a signifi-cantly different loop sequence was identified (UAGUAN) that binds to IRP1 as well as the wild-type sequence (CAGUGN) (Butt et al. 1996; Henderson et al. 1994, 1996). This approach has also yielded the loop sequence GGGAGU that binds specifically to IRP2 (Butt et al. 1996; Henderson et al. 1996). The occur-rence and function of these novel loop sequences in a naturally occurring mRNA has not yet been demonstrated.

Analysis of IRE structure and function using deletion or site-directed mutations has revealed that the IRE is a major determinant of the utilization of ferritin and TfR mRNAs (Casey et al. 1988a,b, 1989; Caughman et al. 1988; Hentze et al. 1988; Müllner et al. 1989; Owen and Kühn 1987). The evidence demonstrating that the IRE/IRP complex, composed of IRP1 or IRP2, could repress translation of ferritin mRNA came from studies using *in vitro* translation systems and gene transfection approaches (Brown et al. 1989; Dix et al. 1992; Eisenstein et al. 1997; Guo et al. 1994; Walden et al. 1988). Deletion of the IRE from ferritin mRNA or disruption of its secondary structure results in a high level of ferritin synthesis in intact cells, irrespective of the presence of excess iron or iron chelator (Caughman et al. 1988; Kikinis et al. 1995). Insertion of an IRE into the 5' UTR of a heterologous mRNA renders translation of that mRNA responsive to cellular iron status (Aziz and Munro 1987; Hentze et al. 1987; Rouault et al. 1988). Gene transfection experiments have also demonstrated that the IRE must be close to the 5' end of an mRNA in order to confer iron regulation of translation (Goossen et al. 1990; Oliveira et al. 1993). Thus, when bound to the IRE at the 5' end of an mRNA, IRP1 appeared to interfere with an early step in the initiation of mRNA translation (Dickey et al. 1988; Gray and Hentze 1994). Investigations using *in vitro* translation of natural and synthetic ferritin mRNAs demonstrated that in the repressed state ferritin mRNA is found in fractions apparently lacking the 40 S ribosomal subunit and more specifically that IRP blocks binding of the

43 S preinitiation complex to the 5' end of the mRNA (Gray and Hentze 1994). When the IRE is more distant from the 5' end the small ribosomal subunit can apparently scan through the RNA/protein complex, presumably displacing the binding protein in the process.

For TfR, initial experiments pointed out the importance of the 3'UTR, and of posttranscriptional processes, in the iron-dependent accumulation of this mRNA (Owen and Kühn 1987). Subsequent studies identified IREs in the TfR 3'UTR that were essential for iron-regulated mRNA stability and showed that they could act as regulators of translation when placed in the 5'UTR of a heterologous mRNA (Binder et al. 1994; Casey et al. 1988a,b, 1989; Caughman et al. 1988; Hentze et al. 1988; Müllner et al. 1989). Deletion analysis of the TfR 3'UTR produced two different phenotypes of mRNA. Removal of IREs produced a constitutively unstable mRNA, whereas deletion of a sequence 5' of the IREs produced a constitutively stable species of messenger (Casey et al. 1989). These results identified a "rapid turnover determinant" in the TfR mRNA 3'UTR that appears to function as a site for nuclease attack. When IRP is bound to TfR mRNA the 3'UTR is resistant to endonucleolytic cleavage and degradation of the mRNA (Binder et al. 1994). Recent studies have identified a specific region within the TfR 3'UTR that appears to be the initial site of nuclease attack for subsequent degradation of TfR mRNA. Unlike in the case of some other mRNAs, degradation of TfR mRNA involves an endonucleolytic attack not preceded by shortening of the poly (A) tail.

4.3 Additional Sequences Function with IREs to Regulate mRNA Utilization

Although it is clear that the IRE is a critical determinant of the posttranscriptional regulation of ferritin and TfR synthesis by iron, other sequences within these mRNAs contribute to the full range of regulation of their utilization. As noted above, the rapid turnover determinant of TfR mRNA functions together with IREs to regulate stability of the message. In the case of ferritin it appears that sequences flanking the IRE sequence contribute to the overall secondary structure of the stem-loop containing the IRE and to binding of IRPs (Dix et al. 1993; Harrell et al. 1991; Sierzputowska-Gracz et al. 1995; Theil 1994; Thorp et al. 1996). Furthermore, when translated in the wheat germ translation system, which lacks an endogenous IRP, synthetic ferritin mRNAs containing an IRE are translated more efficiently than those lacking an IRE (Dix et al. 1992). These results suggest that when iron derepresses ferritin mRNA translation the IRE promotes efficient translation of the message. The end result is that the IRE/IRP system permits a large difference in the rate of ferritin synthesis between situations of iron deficiency and those of iron excess.

5. Iron Regulatory Proteins: Central Regulators of Iron Homeostasis

5.1 IRPs Are Iron-regulated Sequence-specific RNA Binding Proteins

The IRE affects ferritin mRNA translation and TfR mRNA stability because of its interaction in a sequence-specific manner with IRPs (Leibold and Munro 1988; Rouault et al. 1988). IRPs are present in all vertebrate tissues examined, and IRE binding activity has been observed in species from *Drosophila* to humans (Müllner et al. 1992; Rothenberger et al. 1990). IRP1 has been purified from numerous sources (Hu and Connor 1996; Eisenstein et al. 1997; Neupert et al. 1990; Rouault et al. 1989; Walden et al. 1989; Yu et al. 1992), and the rabbit (Walden et al. 1989) and rat (Guo et al. 1994) proteins have been shown to act as translational repressors *in vitro*. When iron supply is low, IRPs are active for RNA binding. Under these conditions the IRP binds ferritin mRNA, thereby inhibiting its translation while simultaneously binding to the 3'UTR of TfR mRNA, protecting it from degradation (Klausner et al. 1993; Kühn and Hentze 1992). When iron is in excess, the RNA binding activity of IRP1 and IRP2 is inactivated, it is released from ferritin and TfR mRNAs, and synthesis of ferritin is initiated while TfR levels decline, presumably because of dissociation of IRP(s) from TfR mRNA (Eisenstein et al. 1990; Klausner et al. 1993; Kühn and Hentze 1992; Theil 1994). By providing a common site for the modulation of ferritin and TfR synthesis, IRPs are central regulators of cellular iron homeostasis in higher eukaryotes.

5.2 Roles for Chelatable Iron and Redox State in Affecting IRP Function

For our purposes we will focus on IRP1, the first IRP identified, and on its regulation by nonheme iron and PKC-dependent phosphorylation. The relationship between intracellular heme concentration and IRP function is the subject of Chapter 8 in this volume by Mascotti et al. Along with its discovery came the demonstration that IRP could be regulated by addition of iron to cells in culture (Haile et al. 1989; Hentze et al. 1989; Leibold and Munro 1988). Hemin is frequently used in cell culture studies of ferritin and transferrin receptor gene expression because it is an efficient donor of iron to cells. The ability of heme to influence ferritin or TfR synthesis can be blocked by the iron chelator Desferal, indicating that the iron released from heme by the microsomal enzyme heme oxygenase (HO) is necessary for the induction process (Eisenstein et al. 1991). Indeed, induction of ferritin synthesis in rat fibroblasts is preceded by a large increase in HO synthesis, suggesting that release of iron from heme is a prerequisite for the derepression of ferritin synthesis (Eisenstein et al. 1991). Furthermore, agents that block HO enzyme activity or promote reformation of heme greatly reduce the ability of heme to induce ferritin synthesis (Bottomley et al. 1985; Eisenstein et al. 1991).

A key step in advancing our understanding of how iron modulates IRP RNA binding activity was the observation that the RNA binding activity of IRP1 was dependent on the oxidation state of one or more cysteines in the binding protein (Haile et al. 1989; Hentze et al. 1989). These studies first illustrated the importance of posttranslational event(s) in the modulation of RNA binding by IRP1. Addition of hemin to cells in culture converts IRP1 to a form that requires high levels ($\approx$300 mM) of thiol (2-mercaptoethanol; 2-ME) in order to bind RNA with high affinity in the gel shift assay. In extracts from cells treated with the permeable iron chelator Desferal, IRP1 binds RNA with high affinity in the presence of low ($\approx$10 mM) concentrations of thiol. Several studies have illustrated that the reduced form of IRP1 (see Section 6.6) binds RNA with a K_D = 20–100 pM, whereas the form found in iron-treated cells binds with much lower affinity (K_D = 1–3 nM) (Haile et al. 1989; Schalinske et al. 1997). It was observed that *in vitro* treatment of low-affinity IRP1 with high concentrations of thiol (300 mM) *reversibly* activated the RNA binding function (Barton et al. 1990; Haile et al. 1989, 1992a; Hentze et al. 1989). In contrast, manipulation of the iron status of cultured cells led to more stable changes in IRP1 function as measured in extracts from these cells. These studies and others established the idea that cellular iron status could regulate IRP1 RNA binding through reversible changes in the oxidation state of one or more critical cysteines in the molecule. It was subsequently shown that the presence or absence of an iron-sulfur cluster in IRP1 imparted stable changes in RNA binding (see Section 6).

5.3 Multiple IRPs May Regulate Iron Utilization

There are at least two IRPs, and there are differences in the mechanisms by which iron and other factors modulate their function (Guo et al. 1994; Henderson et al. 1993; Rouault et al. 1990, 1992; Samaniego et al. 1994). First, we will discuss common characteristics of the two known IRPs, IRP1 and IRP2. Second, we will briefly describe what is known concerning iron regulation of IRP2. Third, in Section 6, we will provide a more extensive description of the regulation of the function of IRP1 and the relationship between this RNA binding protein and c-aconitase.

Studies by Leibold and Munro that first identified IRE binding proteins found two species of IRP in the rat (Leibold and Munro 1988). IRP1 appears to be much more abundant, at least in some tissues, as the first four reports of purified binding protein all identified this species. Klausner and associates demonstrated that a separate mRNA and gene encoded a second IRP, which was subsequently named IRP2 (Rouault et al. 1990, 1992). Rat, human, and mouse IRP2 have been purified and/or characterized (Guo and Leibold 1994, 1995a; Henderson et al. 1993; Rouault et al. 1992; Samaniego et al. 1994). IRP2 is 57% identical in amino acid sequence to IRP1 (Samaniego et al. 1994). As compared to IRP1, IRP2 is a slightly larger protein because of the presence of a 73-amino-acid insertion near the amino-terminus of the protein (Guo et al. 1995a; Hirling et al. 1992; Patino and Walden 1992; Rouault et al. 1990, 1992; Samaniego et al. 1994; Yu et al. 1992). Both IRP1 and IRP2 exhibit similar ability to repress translation of IRE containing mRNAs in *in vitro* protein synthesis systems (Guo et al. 1994; Kim et al. 1995). With the use of a large number of IRE variants, a difference in RNA binding specificity of the two proteins has been observed, but this has not yet been extended to IREs in any natural mRNAs (Henderson et al. 1994; Butt et al. 1996). According to gel shift analysis, IRP2 appears to be upregulated during liver regeneration and after lymphocyte activation, suggesting a role for the protein in regulating iron metabolism during cell proliferation (Cairo and Pietrangelo 1994; Teixeira and Kühn 1991).

Changes in cellular iron status lead to alterations in the RNA binding activity of IRP1 and IRP2 through fundamentally different mechanisms. On the basis of an apparent lack of certain residues found in IRP1 that are required for aconitase activity, IRP2 should not be able to function as an aconitase, but it is not clear if it can assemble an iron-sulfur (Fe-S) cluster (see Section 6) (Guo and Leibold 1994; Henderson et al. 1993; Samaniego et al. 1994). One of the first indications that cellular iron status might regulate the RNA binding activity of the two IRPs in different ways came from examinations of the ability of 2-ME to restore IRE binding activity in iron-treated cells. As noted above, iron inactivates the high-affinity RNA binding form of IRP1 by converting it to c-aconitase, but *in vitro* addition of high levels of 2-ME reactivates the inactive pool of binding protein (Hentze et al. 1989). In contrast, in iron-treated cells, the loss in IRP2 RNA binding activity cannot be recovered by addition of 2-ME (Guo et al. 1995a; Iwai et al. 1995). Furthermore, immunoblotting of IRPs in control and iron-treated cells confirmed that iron reduces the steady-state level of IRP2 protein but fails to significantly affect IRP1 (Guo and Leibold 1994; Henderson et al. 1993). The iron-dependent variation in steady-state level of IRP2 was found to be due to significant changes in the rate of degradation of this binding protein with no change in its relative rate of synthesis (Guo et al. 1995b; Samaniego et al. 1994). The proteasome is intimately involved in degradation of IRP2, and the 73-amino-acid insertion found in IRP2 but not IRP1 contains the necessary and sufficient information for iron regulation of its rate of degradation (Guo et al. 1995b; Iwai et al. 1995).

It is apparent that IRP2 can be redox regulated in response to extracellular factors such as phorbol ester or Desferal (Henderson and Kühn 1995; Schalinske and Eisenstein 1996) and its redox state varies between different cell types (Guo and Leibold 1994; Schalinske and Eisenstein 1996). Titration of cell extracts with diamide or *N*-ethylmaleimide leads to inactivation of IRP2, indicating that one or more Cys thiols are required for the binding protein to interact with RNA

(Henderson and Kühn 1995; Schalinske and Eisenstein 1996). Stimulation of IRP2 phosphorylation in HL60 cells was associated with an increase in the amount of stably reduced protein, suggesting that cell-type specific differences in phosphorylation state contribute to the spontaneous RNA activity of IRP2 (Schalinske and Eisenstein 1996) (see Section 7). Considered indicate that IRP2 exhibits several functional similarities to IRP1 but differs significantly in the means by which iron and other factors modulate its function. Further elucidation of the mechanisms by which various agents modulate the function of these two regulatory RNA binding proteins will permit a better understanding of their overlapping and possibly specific roles in the maintenance of cellular iron homeostasis.

6. Relationship Between IRP1 and Cytoplasmic Aconitase

6.1 IRP1 is Apo-Cytoplasmic Aconitase

In March 1991 notes appeared in the journal *Cell* and shortly thereafter in *Nucleic Acids Research* reporting that the amino acid sequence of IRP1 is 30% identical to that of the Fe-S protein, mitochondrial (m-) aconitase from pig heart (Figure 7.3) (Hentze and Argos 1991; Rouault et al. 1991). Furthermore,

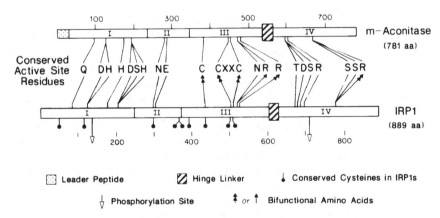

Figure 7.3 Amino acid sequence identity between m-aconitase and IRP1. Linear demonstration of amino acid sequence identity between IRP1 and m-aconitase. The four-domain structure of m-aconitase is shown; a similar four-domain structure is predicted for IRP1 (Beinert and Kennedy 1993; Lauble et al. 1994; Rouault et al. 1991). Only the conserved active site residues between the two proteins are shown in this representation. Location of putative phosphorylation sites, conserved cysteines in IRP1s, and amino acids functioning in the RNA binding and c-aconitase forms of IRP (bifunctional amino acids) are shown. Double-headed arrow indicates amino acid residues with stronger bifunctional roles than those with single arrowhead.

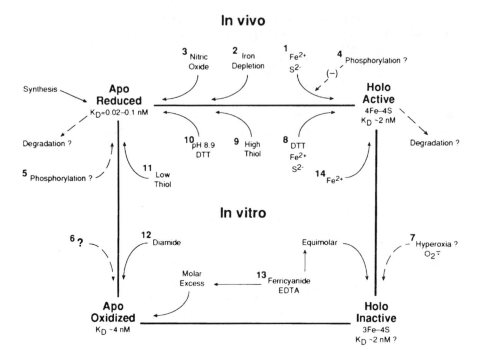

Figure 7.4 Forms of iron regulatory protein 1 (IRP1) observed *in vitro* and *in vivo*. IRP1 exists in four forms; the apo-reduced and holo-active forms are known to exist *in vivo*. Interconversion of IRP1 can occur *in vivo* (Steps 1–5 and 7) and *in vitro* (Steps 8–14). *Apo-reduced* binds RNA with high affinity (K_D = 0.02–0.1 nM) and is the form of IRP1 that regulates ferritin or eALAS synthesis or TfR mRNA stability. *Holo-active* is c-aconitase and contains the 4Fe-4S cluster and binds RNA with low affinity (K_D = 2–5 nM). *Holo-inactive* contains cluster in the 3Fe-4S form and is considered inactive with respect to RNA binding and aconitase activity. It is not known to exist *in vivo*. A K_D for this form has not been directly determined but appears to be similar to that exhibited by the holo-active form. *Apo-oxidized* is produced *in vitro* from holo-protein by oxidation (Step 12). It binds RNA with low affinity. It appears that several forms of oxidized IRP1 can exist depending on which thiols in the protein are affected. *In vivo processes*. Step 1: Addition of iron and sulfide can convert apo-reduced into active 4Fe-4S or c-aconitase form. Step 2: Low levels of intracellular iron, including that induced by chelators of intracellular iron, may lead to loss of cluster from holo-active and its conversion to apo-reduced. Step 3: Intracellular generation of nitric oxide leads to loss of enzyme activity and increases in IRE RNA binding activity. *In vitro* nitric oxide also causes inactivation of enzyme and destruction of cluster, but the final species of IRP1 formed remains to be defined. Step 4: Phosphorylation of IRP1 by PKC prefers the apo-form of IRP1 and may block assembly of cluster. Step 5: Phosphorylation of apo-oxidized may reduce concentration of reductant required for conversion to apo-reduced and thereby facilitate accumulation of high-affinity RNA binding form of IRP1. Step 6: Oxidants may perturb

molecular modeling of IRP1 to the m-aconitase structure indicated that all of the active site residues identified in m-aconitase were conserved in IRP1. This aroused great interest in the IRP as well as the Fe-S community. However, IRP1 does not contain iron (Basilion et al. 1994a), and m-aconitase has been shown not to substitute for IRPs in RNA binding (Zheng et al. 1992b). Also, IRPs are not identical to apo-m-aconitase according to denaturing gel electrophoresis (Kennedy et al. 1992). Yet, because IRP1 is a cytoplasmic protein and there is a cytoplasmic aconitase with properties very similar (though distinct in detail) to those of the mitochondrial enzyme, the obvious question was raised: Is cytoplasmic apo-aconitase IRP1? This was indeed subsequently shown to be the case (Haile et al. 1992b; Kennedy et al. 1992). We will briefly mention here the critical experiments and then, immediately following, discuss the properties of aconitase that are relevant to the theme of this chapter. A detailed summary of what is known about the interrelationships of the various forms of IRP1/c-aconitase is provided in Figure 7.4 at the end of Section 6.1.

Soon after the sequence homology between IRP1 and aconitase was noted, there were several reports showing that in cell lysates from Desferal (DF)-treated cells or in purified preparations of IRP1, after incubation with an iron salt and

---

◀ Figure 7.4 *continued*
the equilibrium between apo-reduced and apo-oxidized and promote accumulation of apo-oxidized. Step 7: Hyperoxia inactivates aconitase through action of superoxide anion (Gardner et al. 1994; Liochev and Fridovich 1993). *In vitro processes.* Step 8: Addition of dithiothreitol (DTT), iron, and sulfide leads to generation of 4Fe-4S cluster and activation of aconitase function. Step 9: High levels of 2-mercaptoethanol (2-ME) (about 300 mM) can *transiently* convert holo-active into form that mimics RNA binding characteristics of apo-reduced but may still contain cluster. Aconitase substrates effectively antagonize this process. Step 10: High pH with high DTT (100 mM) can lead to *stable* conversion of holo-active into apo-reduced with loss of cluster. Step 11: Apo-oxidized can be converted to apo-reduced with loss of S^0 by relatively low (10 to 50 mM) levels of 2-ME. Step 12: Diamide-induced formation of disulfides, particularly those including Cys 437, leads to inactivation of high-affinity RNA binding. Additional disulfides produced include intermolecular species. Thus, apo-oxidized "form" of IRP1 appears to contain different molecular forms of the protein. Step 13: Addition of ferricyanide with EDTA can convert holo-active to holo-inactive if ratio of ferricyanide to protein is 1:1; when ratio of ferricyanide to protein is increased (20:1) cluster is destroyed, iron is lost, and apo-oxidized is produced. Step 14: Further addition of iron can promote completion of cluster and formation of holo-active from holo-inactive. See references (Barton et al. 1990; Gardner et al. 1994; Haile et al. 1992b; Hentze et al. 1989; Kennedy and Beinert 1988; Liochev and Fridovich 1993; Rouault et al. 1991) for further details. Equilibrium binding affinities for oxidized apoprotein and the holoprotein were determined by gel shift analysis with varying concentrations of rat L-ferritin IRE (Eisenstein 1993; Schalinske 1996).

thiol, aconitase activity increased several-fold over that observed with iron-free controls (Emergy-Goodman et al. 1993; Gray et al. 1993; Haile et al. 1992a; Kaptain et al. 1991). Preparations from cells exposed to hemin and then treated with DTT at pH 8.9 (Figure 7.4, Step 10) showed minimal aconitase activity, unless citrate was included in the alkaline incubation mixture; high pH and DTT destroy aconitase activity and citrate is known to protect aconitase (see below). Preparations that had aconitase activity did not exhibit appreciable binding to RNAs containing an IRE, whereas those that bound IREs with high affinity had low aconitase activity. On the other hand, *in vitro* studies using purified c-aconitase in three distinct states, [4Fe-4S] (active enzyme), [3Fe-4S] (inactive enzyme), and apo-enzyme, showed that only the apoform had the ability to bind RNA in the presence of thiol concentrations similar to those found in the cytoplasm of intact cells (Figure 7.4) (Haile et al. 1992b). At high thiol concentrations such as 2% ($\approx$300 mM) 2-mercaptoethanol (2-ME), the aconitase form of the protein was transiently converted into the RNA binding form (Figure 7.4, Step 9), and this conversion was prevented by aconitase substrates, agents known to stabilize the cluster (see below) (Haile et al. 1992b). Thus, the two activities of the protein appeared to be mutually exclusive; i.e., enzymatic activity and high-affinity RNA binding cannot occur simultaneously in an individual molecule.

Direct evidence for the identity between c-aconitase and IRP1 came from several sources. First, amino acid sequence analysis of seven random peptides obtained on cleavage of c-aconitase from bovine liver (12% of the total sequence) showed that the peptide sequence from bovine c-aconitase differed by only 3% in sequence as compared to human IRP1 from liver or placenta, a difference commonly found between IRP1s from mammalian species (Kennedy et al. 1992). Second, recombinant IRP1 could be converted to active aconitase by incorporating an Fe-S cluster *in vitro* by addition of iron, sulfide, and thiol (Figure 7.4, Step 8). After conversion of the active aconitase to its inactive 3Fe form with ferricyanide (Figure 7.4, Step 13), the EPR signal typical of this form of aconitase was observed (Basilion et al. 1994a). Thus the conclusion was that the difference between c-aconitase and IRP1 is the presence of an Fe-S cluster in aconitase. In other words, IRP1 is apo-aconitase (see below for qualification of this statement). Because these and other results (Basilion et al. 1994b; Hirling et al. 1994; Philpott et al. 1994) indicated that substrate and RNA appear to bind in the same region of the protein, we will discuss what is known about the substrate binding site from studies of m-aconitase.

6.2 *Cytoplasmic vs. Mitochondrial Aconitase*

It had long been known and was definitively established in 1971 that there are two distinct species of aconitase: a mitochondrial and a cytoplasmic (c-) one

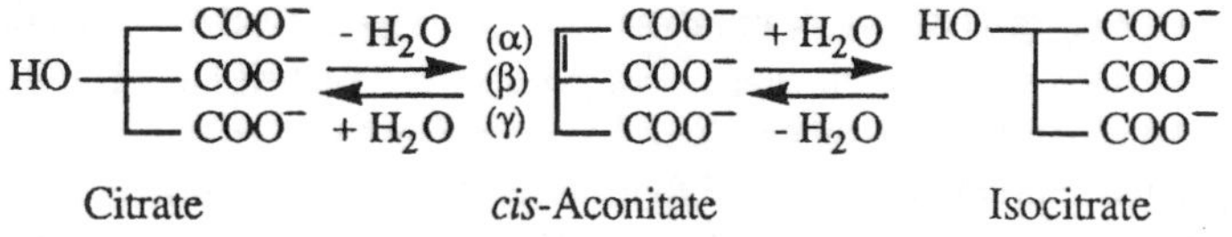

$$HO - \begin{array}{c} COO^- \\ COO^- \\ COO^- \end{array} \quad \begin{array}{c} -H_2O \ (\alpha) \\ \xrightleftharpoons{\ \ \ (\beta)\ \ \ } \\ +H_2O \ (\gamma) \end{array} \quad \begin{array}{c} COO^- \\ COO^- \\ COO^- \end{array} \quad \begin{array}{c} +H_2O \\ \xrightleftharpoons{\quad\quad} \\ -H_2O \end{array} \quad HO - \begin{array}{c} COO^- \\ COO^- \\ COO^- \end{array}$$

| Citrate | *cis*-Aconitate | Isocitrate |

Scheme 7.1

(Eanes and Kun 1974). In scheme 7.1 we show the reactions catalyzed by aconitase, and in Table 7.1 we compare some properties of the (m)- and (c)-enzymes. Both are Fe-S proteins, containing a $[4Fe-4S]^{2+}$ cluster in their active forms and a $[3Fe-4S]^+$ cluster in their reversibly inactivated forms (Scheme 7.2). This inactive form can be reconverted to the active 4Fe form with Fe under reducing conditions (Kennedy et al. 1983). The specific activities of these enzymes with all substrates are very similar (Kennedy et al. 1992). However, c-aconitase is more stable than its mitochondrial counterpart, as on purification under aerobic

Table 7.1 Comparison of properties of m- and c-aconitase

Property	m-Aconitase	c-Aconitase
M_r	82,754	98,400
Number of residues	754	889
Number of Cys (3 conserved)	12	9-12 (human-rat)
Vicinal Cys	0	1 pair
Form tightly binding substrate	[4Fe-4S]	[4Fe-4S] and [3Fe-4S]
Relative stability of [4Fe-4S]	−	+
Extra iron binding sites	0	1

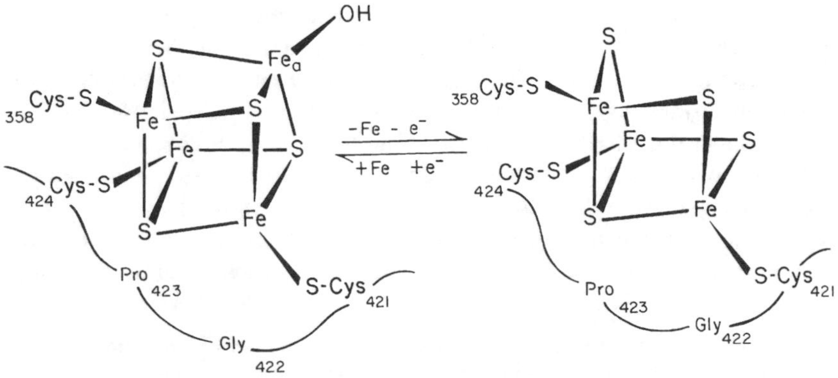

Scheme 7.2

conditions the c-enzyme is obtained up to 90% in its active form (Kennedy et al. 1992), whereas the m-enzyme is obtained largely as the inactive 3Fe form (Kennedy et al. 1983). It is of particular interest that the c-enzyme contains one extra, noncluster iron in both the 3Fe and 4Fe forms.[4] The function of this Fe as well as its binding site are not known. The m-enzyme has been much more thoroughly investigated (Beinert and Kennedy 1989, 1993; Beinert et al. 1996), because it is more readily purified. It has been crystallized in different states with bound substrates, analogs, and inhibitors, and high-resolution structures are available from X-ray diffraction studies (Lauble et al. 1992, 1994; Robbins and Stout 1989). It has also been extensively investigated using UV-visible (Emptage et al. 1983a), EPR (Emptage et al. 1983a; Surerus et al. 1989), ENDOR (Kennedy et al. 1987; Werst et al. 1990a,b), Mössbauer (Emptage et al. 1983b; Kent et al. 1982, 1985; Surerus et al. 1989), resonanceRaman (Kilpatrick et al. 1994), MCD (Johnson et al. 1984), and EXAFS (Beinert et al. 1983) spectroscopies.

6.3 Aspects of the Structure and Reaction Mechanism of m-Aconitase

The stereochemistry of the reaction was established 30 years ago (England 1960; Gawron et al. 1961). Most kinetic studies were also carried out before it became known that the enzyme is an Fe-S protein (Kennedy et al. 1972) or before implications of this finding were appreciated. A synopsis of all this work, which spanned several decades, results in the following picture for the structure and the reaction mechanism of aconitase.

The aconitase structure has been interpreted as consisting of four domains, all of which contribute residues to the active site (Figures 7.3 and 7.5). This site is located at the base of a cleft between domains 1–3 and the largest domain, 4. This domain is linked to domain 3 by a 25-residue peptide loop (the "hinge linker") and might therefore allow motion of domain 4 with respect to the other domains. Such motion could open or close the cleft between domains, where the Fe-S cluster is located. This feature might be of importance in the release of the Fe-S cluster and a docking of RNA onto the c-aconitase apoprotein. Indeed, the recent demonstration of significant structural changes in the putative cleft and hinge linker region of IRP1 as a function of the presence or absence of the Fe-S cluster supports these suggestions (Schalinske et al. 1997).

The cluster is ligated by three Cys residues, all from domain 3; two Cys residues are present in the -Cys(421)-xx-Cys(424)-pattern, typical of the majority of Fe-S proteins. The third Cys, (358), is furnished by another part of domain 3.

---

[4]When we speak of the 3Fe, 4Fe, or apoenzyme, we exclusively consider cluster iron, ignoring the extra iron that, although consistently observed only in c-aconitase, is not the subject of any useful discussion yet at this point.

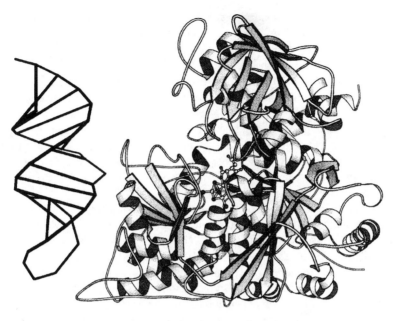

Figure 7.5 Comparison of m-Aconitase Structure with RNA of Size Similar to IRE: "Ribbon" structure of pig heart mitochondrial aconitase (right), showing the polypeptide chain, [4Fe-4S] cluster, and bound isocitrate. Water and three cysteine sulfurs also bonded to the cluster are shown. The protein secondary structure is represented by coils for helices and ribbons for strands within the β sheets. The N terminus is at the lower right and the C terminus at the upper right. In this view the fourth domain is at the top and the three N terminal domains are arranged from right to left in the lower portion of the molecule. For comparison there is also shown (left) a schematic view of an RNA molecule of the size of an IRE, drawn to scale, so that the size relationships between the protein and the nucleotide structure can be assessed. This is not to suggest any mode of binding. Note that the Mr of m-aconitase is smaller than that of c-aconitase (Table 7.1). (This image was calculated by G. Gippert at Scripps Research Institute using the program Molscript written by P. Kraulis (Kraulis, 1991) and was kindly supplied by Dr. C.D. Stout.). A color version of the protein structure is on the cover of the book and the legend for the cover figure is on p. i.

In c-aconitase the corresponding residues are Cys 503, 506, and 437, respectively (Figure 7.3). Individual replacement of any of these three Cys with a Ser eliminates enzyme activity but does not affect the RNA binding function in the c-enzyme (Philpott et al. 1994). For the c-enzyme, Cys(437) has attracted special attention, because it was found that binding of bulky groups, e.g., *N*-ethylmaleimide (NEM), to the sulfhydryl group of this residue prevents RNA binding (Hirling et al. 1994; Philpott et al. 1994). However, replacement of Cys 437 with a serine does not block RNA binding activity but makes it impossible for NEM to inactivate RNA binding (Hirling et al. 1994; Philpott et al. 1994). Further evidence for a role of Cys 437 in influencing RNA binding comes from studies indicating that formation of a disulfide bond between Cys 437 and either Cys 503 or 506 prevents RNA binding in the absence of thiol (Hirling et al. 1994).

In the active enzyme one of the four cluster irons has no Cys, but a solvent derived ligand: OH when free of substrate (Scheme 7.2), and water when substrate or an analog is also bound (Figure 7.6). The Fe that lacks the Cys ligand (called Fe_a) is the one that is readily lost on oxidation (Kennedy et al. 1983). Mössbauer spectroscopy has unambiguously shown that this is also the iron that binds the substrate (Emptage et al. 1983b) by reorganizing its coordination sphere from tetrahedral (4-coordinate) to octahedral (6-coordinate) (Figures 7.6 and 7.7). With the substrates, citrate or isocitrate, a five-membered chelate structure is formed, involving the iron, one of the carboxyls (β for citrate and α for isocitrate) and

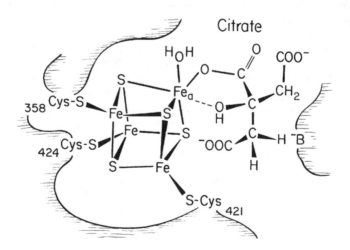

Figure 7.6 Representation of information derived from ENDOR about the enzyme-citrate complex. When isocitrate is bound, the carboxyl group now on the bottom is bound to Fe_a and the $-CH_2-COO^-$ group, now pointing upward (right), would point downward (180° flip).

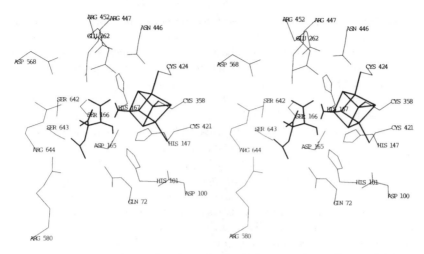

Figure 7.7 Stereo view of the active site region in pig heart aconitase. The α carbon atoms and side chains of 19 adjacent residues are shown. The substrate isocitrate is shown in heavy lines along with the [4Fe-4S] cluster and bound H_2O with hydrogens included. (Reproduced with permission from Lauble et al. 1992. Copyright 1992, American Chemical Society).

the hydroxyl, with the carboxyl and hydroxyl becoming ligands for Fe_a (Kennedy et al. 1987; Lauble et al. 1992; Lauble and Stout 1994; Werst et al. 1990a). Fe_a now has three sulfide and three oxygen ligands, from the carboxyl and the hydroxyl of the substrate and a water molecule. The remaining two carboxyl groups of the substrate are bound by Arg residues, of which Arg 580, which binds the γ-carboxyl, seems to be the critical anchoring point.

After the substrate, say, citrate, is bound, the proton at the α-carbon and the OH at the β-carbon have to be abstracted. According to stereochemical studies these groups are located *trans* with respect to the double bond that is formed in the dehydration reaction, which leads to *cis-* aconitate (Gawron et al. 1961). The base that removes the proton during formation of cis-aconitate has been found to be Ser 642. The proton shifts required for the abstraction of OH and H are obviously facilitated by two His-Asp pairs that are strategically positioned in the active site (Figure 7.7) (Cleland and Kreevoy 1994; Lauble et al. 1992, 1994). The four active site Arg residues involved in hydrogen bonding of other residues and in the binding of the anionic substrate may well prove to be important in the binding of RNA to IRP1 (see below). Inasmuch as the active site residues of m- and c-aconitase are identical, there is little doubt that the enzymatic mechanisms [reviewed in Beinert et al. (1996), Lauble et al. (1992), and Zheng et al. (1992a)] of both enzymes are identical.

6.4 Binding of Substrate Protects the Cluster from Degradation

The enzymatic activities of aconitase are not of primary importance in the present context; however, we have, in the preceding paragraph, given some detail, which may help to convey an impression of the features of the active site, because there is strong evidence that the RNA binding region of IRP1 must include at least parts of the active site. The Fe-S cluster, the center of the active site, has to be removed if RNA binding is to occur; and UV-cross-linking experiments (Basilion et al. 1994b) as well as mutational studies have identified residues in this site as being required for interaction with RNA (see below). As we have pointed out, there are Args involved in the binding of substrate to aconitase. After removal of the cluster and the bound substrate, these Arg residues become more accessible and may be used in RNA binding as some RNA-protein interactions involve Arg residues (Puglisi et al. 1992). However, substrate is very tightly bound, more so in c-aconitase than in m-aconitase. It is very difficult to obtain completely substrate-free enzyme, and several steps are necessary to accomplish this. The substrate protects the Fe-S cluster from degradation and, therefore, enters into considerations of the conversion of aconitase into its apo-enzyme, which as pointed out previously, is IRP1 in the case of the c-enzyme (Table 7.2). Exactly what role citrate (substrate) plays in this conversion is not clear. Because the enzyme binds citrate and isocitrate in a similar manner at Fe_a, the orientation of the $-\gamma\ CH_2COO^-$ must be different for each. This necessitates that the active site must be flexible to accommodate both structures. It is also apparent from the three-dimensional structures that, when substrate is bound, domain 4 moves in toward the other domains and tightens up the structure somewhat (Lauble et al. 1992; Lauble and Stout 1994).

Table 7.2 Comparison of the properties of purified cytosolic aconitase with cellular form of IRPI

Protein	Aconitase Activity	RNA Binding in the Presence of 2% 2-ME	
		−Substrate	+Substrate
Pure aconitase			
Apoprotein	—	+[a]	+
[3Fe-4S]	—	+	—
[4Fe-4S]	+	+	—
Cytosol			
Df-treated cells	—	+[b]	+
Hemin-treated cells	+	+	—

[a]Only 0.02% 2-ME is required for activation of RNA binding by the apoprotein.
[b]No 2-ME is required.

6.5 Interaction with Substrate or RNA Involves Overlapping Regions of c-Aconitase

The observations that RNA binding and aconitase functions of the c-enzyme are mutually exclusive (Haile et al. 1992a,b; Kaptain et al. 1991; Kennedy et al. 1992) suggested that similar regions of the protein were involved in these activities. According to the X-ray structure of the m-enzyme and the evidence suggesting the similarity in tertiary structure of the m- and c-enzymes, the Fe-S cluster is in the solvent-exposed cleft in both proteins. As noted above, it is clear that the presence of the cluster blocks high-affinity RNA binding and that mutations that prevent cluster assembly produce a stably active RNA binding protein. However, several pieces of evidence have demonstrated that the IRE specifically interacts with regions of the c-enzyme containing and including those amino acid residues essential for aconitase function. UV-cross-linking coupled with proteolytic digestion of the cross-linked complex, as well as RNA and protein sequencing, have revealed areas of overlap of the RNA binding and active site of the c-enzyme (Basilion et al. 1994b; Swenson and Walden 1994). Limited chymotryptic digests of IRP1 produced a fragment spanning residues 480 to 623, thus including Cys 503 and 506, that binds IRE RNA specifically but with low affinity (Swenson and Walden 1994). Other investigators have observed a specific interaction between IRE RNA and residues 121–130 of IRP1; residues 125 and 126 are believed to be essential for binding of substrate by c-aconitase (Basilion et al. 1994b). More recently, a scanning mutagenesis approach of IRP1 identified residues 116–151 as being essential for high-affinity interaction of the binding protein with its RNA ligand (Neupert et al. 1995). In addition, it is of interest that residue 132 displays cluster-dependent changes in cleavage by chymotrypsin supporting the contention that cellular iron status influences accessibility of the RNA binding site(s) of IRP1 to RNA (Schalinske et al. 1997). Some RNA binding proteins contain Arg residues that function in the specific interaction of protein and RNA (Puglisi et al. 1992). It is of interest that aconitase contains four Arg that are necessary for its enzymatic activity and that in c-aconitase three of these residues, Arg 541 and 780 and to a lesser extent Arg 536, appear to be essential for high-affinity RNA binding (Figure 7.3) (Basilion et al. 1994b; Hirling et al. 1994). Taken together, these studies begin to define a molecular basis for the mutually exclusive metabolic and gene-regulatory functions of c-aconitase.

6.6 Fe-S Cluster Chemistry and State of the Apoenzyme

Most important in the context of this chapter is, of course, the Fe-S cluster chemistry that may occur in the active site, as it is the presence or absence of the cluster that determines the function of the protein. Little is known about the *in vivo* process of the assembly/disassembly of Fe-S clusters. *In vitro* the formation of clusters is spontaneous, and it is probably reasonable to assume that the course of the assembly/disassembly reactions depends largely on the environment that an individual protein provides and on the distribution pattern of binding groups, primarily sulfhydryls. This has been confirmed by mutational studies (Hirling et al. 1994; Philpott et al. 1994; Zheng et al. 1992a) and is also apparent from the demonstrated propensity of SH groups in proteins to enter into cluster or intra- or intermolecular disulfide formation (Kennedy et al. 1984; Martin et al. 1990).

As pointed out above, 3Fe c-aconitase does not bind RNA more efficiently than 4Fe c-aconitase. The only stable RNA binding form that has been observed is the apoenzyme without any cluster iron. Experiments on the conversion of holo-c-aconitase to IRP1 with 2% 2-ME have, however, indicated that there probably can exist an intermediate form of the Fe-S protein, in which 2-ME has replaced some of the cluster ligands (Barton et al. 1990; Hentze et al. 1989) such that, on removal of 2-ME by desalting, RNA binding is again lost, indicating that the RNA binding activity of the 2-ME inducible form of IRP1 (i.e., the holoprotein) is reversibly activated by the excess thiol.

We have just emphasized that the only stable form of the enzyme that binds RNA is the apoenzyme. However, we must consider that "apoenzyme" is not a sufficient definition. The term 'apo-' implies only that the iron sulfur cluster has been removed. However, the three Cys ligands of the cluster furnished by the protein are still present and their state is undefined. They could be present as CysSH, Cys-S-S-Cys or Cys-S-S_x-S-Cys (x=1-4), if the sulfides have become trapped on oxidative degradation (see legend to Figure 7.4). That this is not irrelevant is shown by the necessity to "reduce" the chemically prepared apoenzyme (Haile et al. 1992b; Kennedy et al. 1992) to observe RNA binding. The standard chemical method of preparation involves oxidation by ferricyanide (Figure 7.4, Step 13) (Kennedy and Beinert 1988). There are three SH groups in the site and six to nine elsewhere in the molecule, depending on the source of IRP1. There are thus several possibilities of interaction between sulfhydryls, primarily of course between those in the active site. We have mentioned that Cys 437 has to be unencumbered for RNA binding to be possible, whereas the other two Cys residues involved in cluster formation, 503 and 506 may be blocked by NEM with no interference arising in RNA binding ability. When the apoenzyme is prepared by oxidation of the cluster with ferricyanide, there are di- or tri- and tetrasulfides formed from the Cys residues and sulfides present (Kennedy and Beinert 1988; Petering et al. 1971). One may therefore conclude from the unique sensitivity toward blockage of Cys 437 that, if this residue is involved in a di- (or higher) sulfide bridge, RNA binding is not possible. This is as much as can be said at this stage about the involvement of the Fe-S cluster in RNA binding in *in vitro* experiments. What happens *in vivo* is likely to be a problem of much greater complexity.

6.7 Cluster Disassembly and Reassembly in vivo

In vitro experiments on c-aconitase have shown that the Fe-S cluster of this enzyme is very stable under anaerobic conditions. It is also quite stable under aerobic conditions, and this stability is enhanced when substrate is present (Kennedy et al. 1992). One may expect that substrate is always present in the cytoplasm, except under very unusual circumstances, so it would be likely that the cluster is stable *in vivo*. The question, therefore, arose of whether RNA binding activity of IRE could be generated at a sufficient rate—other than by *de novo* synthesis— if the need for IRP1 should arise. Spontaneous decay of the cluster appears to be too inefficient. The possibility of enzymatic catalysis promoting the conversion was considered, as well as the intervention of a reactant of low molecular weight. In 1993 two reports demonstrated that the induction of nitric oxide synthase (NOS), producing NO, can bring about the conversion of the aconitase form of the protein to the RNA binding form in macrophages activated with interferon-γ and lipopolysaccharide (Figure 7.4, Step 3) (Drapier et al. 1993; Weiss et al. 1993). These observations have found support from additional cellular experiments (Cairo and Pietrangelo 1995; Jaffery et al. 1994; Oria et al. 1995; Pantopoulos and Hentze 1995a; and Phillips et al. 1996; Richardson et al. 1995) and also from the demonstration that pure c-aconitase exposed to NO *in vitro* loses its activity and Fe-S cluster at a rate that makes it appear feasible that NO is indeed one of the *in vivo* agents that can bring about the interconversion (Kennedy et al. 1994). The efficiency of this interconversion and of its reversal is yet to be ascertained. There have also been suggestions that peroxynitrite, which is formed from superoxide anion and NO, might be the active agent (Beckman et al. 1990). It was found in *in vitro* studies that the rate of inactivation of aconitase by peroxynitrite is higher than that by NO.[5] More recent cellular studies (macrophage-like RAW 264.7 cells) (Bouton et al. 1996) have shown that although superoxide, peroxynitrite, and hydrogen peroxide inactivate aconitase activity, IRE binding by IRP1 does not increase. In this same study, binding activity was increased by NO donors, spermine NONOate, and nitrosothiols. There is now evidence that in addition to NO and iron depletion, oxidative stress to cells, in some but not all circumstances, results in the activation of RNA binding by IRP1 (Cairo et al. 1995; Martins et al. 1995; Pantopoulos and Hentze 1995b). This cellular activation is relatively rapid. However, when cell extracts were exposed to H_2O_2, IRP1 activation did not occur. This led the authors (Martins et al. 1995)

---

[5]Recently published *in vitro* experiments (Castro et al. 1994; Hausladen and Fridovich 1994), in which no significant effect of NO on m-aconitase was found, were carried out with preparations containing less than 1% aconitase and large amounts of contaminating heme and nonheme iron. They can, therefore, not be accepted as valid evidence for a lack of effect of NO on aconitase.

to conclude that the effects of H_2O_2 are indirect. Similar conclusions were reached in another study where it was suggested that this rapid response may be due to a phosphorylation-dependent reaction (Pantopoulos et al. 1996). Lastly, it has been shown (Weiss et al. 1994) that, in a mouse macrophage cell line (J774), increased intracellular iron levels lead to a decrease in NOS activity, whereas depletion of intracellular iron strongly enhances NOS activity after interferon/ lipopolysaccharide stimulation; and these differences were due to NO synthase mRNA levels. The authors suggest that there may be a regulatory loop between iron metabolism and the NO/NO-synthase pathway. How general such a regulatory mechanism might be is yet to be explored.

6.8 Role of Citrate and c-Aconitase in the Cytoplasm

We have repeatedly mentioned the protection that bound substrate provides for the Fe-S cluster of aconitase. This, together with the fact that citrate is also a fairly efficient chelator of iron, points to citrate as yet another potentially important player *in vivo*. This topic has been the subject of interesting proposals in the recent literature (Melefors and Hentze 1993; O'Halloran 1993). Along similar lines one also may raise the question about the function or necessity for an aconitase in the cytoplasm, while the tricarboxylic acid cycle, where the classical function of aconitase is required, is exclusively part of the mitochondrial metabolic machinery. Citrate is required in the cytoplasm to provide acetyl CoA for fatty acid synthesis via the citrate cleavage enzyme (Srere 1992). The required citrate is produced in mitochondria via the tricarboxylic acid cycle and exported to the cytoplasm. One may ask: Where does aconitase enter here? However, in view of the enormous complexity of metabolism under *in vivo* conditions (Srere 1994), particularly in a tissue such as the liver, speculation on this topic appears to be of very limited value.

6.9 In vivo *Abundance of the Two Forms of c-Aconitase*

In this context it may be germane to mention observations on the *in vivo* distribution of the two forms of IRP1, RNA binding and active enzyme. In various organs of the mouse there was on average 20% in the RNA binding form and 80% as latent IRP1, which can be converted to active IRP1 by 2-ME and can reasonably be assumed to be aconitase (Müllner et al. 1992). In the authors' laboratory much the same numbers have been found as average of more than 20 c-aconitase preparations from bovine liver, namely about 80% holoenzyme (4Fe and 3Fe) and 20% apoenzyme, corresponding to IRP1 (Kennedy et al. 1992). There were definite fluctuations in these values between individual cows, ranging from 10 to 50% IRP1. These values were obtained on the purified enzyme by determination of enzyme activity, after activation to the 4Fe form, together with analyses for iron, sulfide, and protein.

7. IRP1 and IRP2 as Sites of Action for Modulation of Iron Homeostasis by Multiple Effectors

Because they can coordinately regulate ferritin, TfR and eALAS synthesis IRPs are a likely focal point for the modulation of iron metabolism by extracellular agents acting to program the uptake and metabolic fate of iron. Considered from a broad perspective, regulation of the function of IRPs through the action of specific protein kinases or through other signal transduction mechanisms would significantly expand the agents and circumstances under which iron metabolism could be modulated. First, changes in the phosphorylation state of key regulatory proteins, induced by extracellular factors, is frequently used as a means for overriding the effects of intracellular allosteric effectors (Krebs 1986). In the case of IRP1 and/or IRP2, iron and perhaps citrate could be considered allosteric regulators of their function, and phosphorylation of IRPs might impinge on the ability of these low-molecular-weight effectors to influence the RNA binding function. Phosphorylation might provide a means for altering the set-point at which cytoplasmic iron concentration affects the function of IRPs (Eisenstein et al. 1993). Second, phosphoregulation of IRPs by one or more kinases provides the opportunity for altering iron metabolism through the use of specific signal transduction cascades. Thus, hormones, cytokines, or growth factors could alter iron uptake or utilization by activating signal cascades that affect the phosphorylation state of IRPs (Eisenstein et al. 1993). It has also been suggested that the response of IRP1 to oxidative stress may involve alterations in phosphorylation state of this RNA binding protein or other proteins that influence the function of IRP1 (Pantopoulos et al. 1996). Third, new opportunities could be exploited for the cell-type specific regulation of iron metabolism, as illustrated in the effects of thyroid hormone on IRP function (Leedman et al. 1996). Coupled with the effects of various hormones or cytokines on ferritin or TfR gene transcription, the action of such agents on the phosphorylation state of IRPs could promote a greater selectivity in the cell-type specific modulation of the metabolic fate of iron. Fourth, although it is clear that iron mediates the function of IRP1 and IRP2 under the nonphysiological conditions used in cell culture, it is not obvious that the daily changes in iron status of normal individuals is sufficient to affect function of IRPs. Recent studies have begun to more fully address the physiological regulation of the IRPs by dietary iron (Chen et al. 1997). The directed action of other extracellular agents on the phosphorylation state of IRP1 and IRP2 may prove to be of equal or greater physiological importance to the regulation of iron homeostasis.

7.1 Evidence for Iron-independent Modulation of IRPs

Circumstantial support for the posttranslational regulation of IRP function by factors other than iron comes from studies of the regulation of ferritin and TfR synthesis during changes in the state of proliferation or differentiation of cells in culture. First, in splenic lymphocytes stimulated to proliferate, TfR mRNA and protein levels increase significantly, and this is accompanied by increases in the RNA binding activity of IRP1 and IRP2 (Seiser and Kühn 1993; Teixeira and Kühn 1991; Testa et al. 1991). This is associated with the increased requirement for iron as the lymphocytes progress from a quiescent to a proliferating state. It is of interest that phorbol 12-myristate 13-acetate (PMA) significantly accelerates the increase in TfR mRNA in lymphocytes as compared to the effects of the mitogen phytohemagglutin, and the effects of PMA are not blocked by cycloheximide (Kumagai et al. 1987, 1988). PMA is a potent activator of several isoforms of protein kinase C. In our hands PMA plus ionomycin leads to a rapid and sustained increase in IRP1 phosphorylation and IRE binding in splenic lymphocytes.[6] Furthermore, in an IL-2 dependent lymphocyte cell line, IL-2 leads to large posttranscriptional increases in TfR mRNA, and it was suggested that this occurred partly as a result of posttranslational regulation of IRP1 by factors other than iron (Seiser and Kühn 1993; Teixeira and Kühn 1991). The IL-2 receptor activates signal transduction cascades that include the MAP kinases and possibly some isoforms of PKC (Sato and Miyajima 1994). It is clear that multiple factors contribute to the enhanced accumulation of TfR mRNA after lymphocyte activation, but the studies noted above suggest that phosphoregulation of IRPs could provide an iron-independent signal for enhancement of iron transport.

Second, during the differentiation of monocytes into macrophages, synthesis of both ferritin and TfR is increased (Andreesen et al. 1984; Testa et al. 1989, 1991). This appears to reflect a cell-specific requirement to simultaneously enhance the uptake and storage of iron (Andreesen et al. 1984; Testa et al. 1989, 1991). Macrophages appear to accomplish this by transcriptionally and posttranscriptionally increasing TfR and ferritin gene expression (Testa et al. 1989, 1991). IRE RNA binding activity increases significantly and in a manner that seems to reduce the ability of iron to regulate the RNA binding function, perhaps reflecting a need to increase iron uptake without inactivating IRPs in order to maintain TfR mRNA concentration (Testa et al. 1991). This increase in IRP function likely contributes to accumulation of TfR mRNA and protein. To simultaneously increase ferritin synthesis, macrophages greatly increase ferritin mRNA accumulation, but the increase in ferritin synthesis appears to be less than the increase in ferritin mRNA levels. A constitutively active IRP might therefore enhance accumulation of TfR mRNA and the rate of iron uptake; the increased accumula-

---

[6]K. P. Blemings, K. L. Ross, and Eisenstein, in preparation.

tion of ferritin mRNA will by mass action increase the amount of ferritin message in the polysomal pool even in the presence of active IRPs. Phosphorylation of IRPs during monocytic differentiation might provide a mechanism for maintaining the binding protein in an active RNA binding form even in the presence of increased rates of iron uptake.

7.2 Phosphorylation of IRPs is Associated with Alterations in Their Function

Both IRP1 and IRP2 are phosphoproteins in intact cells, and alterations in their phosphorylation state are associated with changes in IRE RNA binding activity (Eisenstein et al. 1993; Schalinske and Eisenstein 1996). Highly purified IRP1 is an efficient substrate for phosphorylation by PKC. PKC phosphorylates IRP1 up to a level of 1.3 mole phosphate per mole protein, and the kinase has a k_m of 0.5 μM for the binding protein (Eisenstein et al. 1993). Phosphopeptide mapping, phosphoamino acid analysis, and the use of synthetic peptides as substrates for PKC have identified two major sites of phosphorylation of purified IRP1 by this kinase (Eisenstein et al. 1993). The results of the experiments with synthetic peptides indicate that Ser 138 and especially Ser 711 may be PKC phosphorylation sites (Eisenstein et al. 1993). Ser 138 is of particular interest because it lies within a region of the protein shown to be essential for high-affinity interaction with RNA and is close to Asp 125 and His 126, which are believed to be involved in catalysis. Using limited proteolysis as a probe of structure, we found that a chymotryptic cleavage site near Ser 138 (Phe 133) is accessible in IRP1 but not in c-aconitase (Schalinske et al. 1997). This suggests that Ser 138 would be able to be phosphorylated in IRP1 but not in c-aconitase. Interestingly, IRP1 is four to five times more efficiently phosphorylated by PKC *in vitro* then is c-aconitase (Schalinske et al. 1997). It is also of interest that both Ser 138 and Ser 711 are predicted to be near the entrance to the cleft in IRP1 (Figure 7.5), suggesting that phosphorylation of these residues may influence RNA binding and/or its regulation by iron. Further support for the localization of the phosphorylation sites near the entrance to the cleft comes from the recent demonstration that IRE containing RNA specifically blocks phosphorylation of IRP1 by PKC (Schalinske et al., 1997). Taken together, these results support the hypothesis that phosphorylation by PKC represents a means to partition IRP1 between its metabolic and genetic functions.

HL60 cells, a human promyelocytic cell line that can be induced to differentiate into macrophages with PMA, have been used to investigate the relationship between the phosphorylation state of IRPs and alterations in expression of proteins involved in iron metabolism (Eisenstein et al. 1993; Schalinske and Eisenstein 1996). PMA stimulates phosphorylation of IRP1 within 30 min and the protein is not phosphorylated in untreated cells. Phosphorylation of IRP2 is also rapidly stimulated in HL60 cells by PMA treatment, but in contrast to IRP1, IRP2 is highly phosphorylated in untreated HL60 cells. In PMA-treated HL60 cells IRP1 is phosphorylated at a single site, whereas IRP2 appears to have two sites of phosphorylation.[7] The stimulation of IRP phosphorylation is associated with a two-fold increase in IRE RNA binding activity that is independent of protein synthesis. PMA affects the maximal capacity (B_{max}) for RNA binding but not the k_D and acts to recruit a latent pool of IRE binding activity into a form that binds RNA with high affinity. In HL60 cells about 70% of the spontaneous IRE binding activity is attributable to IRP2. After PMA treatment IRP2 IRE binding activity increases 2.2-fold, whereas IRP1 increases by 60% and, in the case of IRP2 phosphorylation, appears to regulate its function through alterations in the oxidation state of one or more Cys in the binding protein (Schalinske and Eisenstein 1996). These increases in RNA binding by IRP1 and IRP2 occurred in the presence of a significant stimulation in phosphorylation of both proteins and a 50% increase in TfR mRNA abundance. Thus changes in IRP phosphorylation state appear to represent a mechanism by which agents other than iron can influence the activity of these regulatory RNA binding proteins.

A number of issues need be addressed regarding phosphoregulation of IRPs. How might phosphorylation affect the function of IRP1 or IRP2? What kinase(s) other than PKC phosphorylate IRP in intact cells? To what extent does iron-regulation and phosphoregulation of IRP overlap? Are critical Cys residues, such as Cys 437, more accessible in phosphorylated vs nonphosphorylated IRP1? It is of interest that, according to sequence homology, Ser 711 appears to be near the entrance to the active site and RNA binding cleft (Figure 7.5) as well as being close to Arg 699, which is essential for substrate binding. Furthermore, Ser 138 is near or within the region recently identified to be important for RNA binding (Basilion et al. 1994b; Neupert et al. 1995) and also appears, based on sequence homology, to be near the entrance to the cleft (Figure 7.5). The fact that IRP1 is a better substrate for PKC as opposed to c-aconitase raises the question as to whether or not phosphorylation affects cluster formation. The proximity of Ser 711 and Arg 699 suggests that, should a cluster be assembled, decreased binding of substrate as a result of phosphorylation might make the cluster more labile. For IRP2 it remains to be determined where it is phosphorylated, but the presence of putative PKC and MAP kinase sites in the 73-amino-acid region that is responsible for iron regulation of IRP2 stability raises the intriguing possibility that, as for IRP1, phosphoregulation and iron regulation may overlap in their actions on this regulatory RNA binding protein. These and other issues addressing the effects of phosphorylation on the function of IRPs

---

[7]In PMA-treated rat 2 fibroblasts, IRP1 appears to be phosphorylated at three sites (K. L. Schalinske, P. T. Tuazon, J. A. Traugh, and R. S. Eisenstein, unpublished observations).

will expand our understanding of how iron metabolism can be regulated by extracellular agents.

Many of the questions, of a chemical or biological nature, that we had to leave unanswered, will undoubtedly be subjects for ongoing or future investigations.

Acknowledgments

The recent experimental work (1985–present) of M. Claire Kennedy and Helmut Beinert was supported by the National Institutes of General Medical Sciences (Helmut Beinert: GM-34812 and M. Claire Kennedy: GM51831) and by the Medical College of Wisconsin. We acknowledge with appreciation the collaborators and colleagues whose names appear on our joint publications. Richard S. Eisenstein acknowledges support from the University of Wisconsin College of Agricultural and Life Sciences and Graduate School, NIH R29-DK47219, and USDA # 94-37200-0361.

References

Adams, M. L., I. Ostapiuk, and J. A. Grasso. 1989. The effects of inhibition of heme synthesis on the intracellular localization of iron in rat reticulocytes. *Biochim. Biophys. Acta* 1012:243–253.

Aisen, P. 1991. Ferritin receptors and the role of ferritin in iron transport. *Targeted Diagnosis and Therapy* 4:339–354.

Aisen, P., G. Cohen, and J. O. Kang. 1990. Iron toxicosis. *Int. Rev. Exp. Pathol.* 31:1–46.

Andreesen, R., J. Osterholz, H. Bodermann, J. Bross, U. Costabel, and G. W. Löhr. 1984. Expression of transferrin receptors and intracellular ferritin during terminal differentiation of human monocytes. *Blut* 49:195–202.

Andrews, S. C., A. Treffry, and P. M. Harrison. 1988. Siderosomal ferritin: The missing link between ferritin and haemosiderin. *Biochem. J.* 245:439–446.

Aziz, N., and H. N. Munro. 1986. Both subunits of rat liver ferritin are regulated at a translational level by iron induction. *Nucl. Acids. Res.* 14:915–927.

Aziz, N., and H. N. Munro. 1987. Iron regulates ferritin mRNA translation through a segment of its 5' untranslated region. *Proc. Natl. Acad. Sci. U.S.A.* 84:8478–8482.

Babior, B. M. 1984. The respiratory burst of phagocytes. *J. Clin. Invest.* 73:599–601.

Bacon, B. R., and R. S. Britton. 1990. The pathology of iron overload: A free radical-mediated process? *Hepatology* 11:127–136.

Bali, P. K., O. Zak, and P. Aisen. 1991. A new role for the transferrin receptor in the release of iron from transferrin. *Biochemistry* 30:324–328.

Balla, G., H. S. Jacob, J. Balla, M. Rosenberg, K. Nath, K. Apple, J. W. Eaton, and G. M. Vercellotti. 1992. Ferritin: A cytoprotective antioxidant stratagem of endothelium. *J. Biol. Chem.* 267:18148–18153.

Balla, J., H. S. Jacob, G. Balla, K. Nath, J. W. Eaton, and G. M. Vercellotti. 1993. Endothelial cell heme uptake from heme proteins: Induction of sensitization and desensitization to oxidant damage. *Proc. Natl. Acad. Sci. U.S.A.* 90:9285–9289.

Barton, H. A., R. S. Eisenstein, A. B. Bomford, and H. N. Munro. 1990. Determinants of the interaction of the iron regulatory element binding protein with its binding site in rat L-ferritin mRNA. *J. Biol. Chem.* 265:7000–7008.

Basilion, J. P., M. C. Kennedy, H. Beinert, C. M. Massinople, R. D. Klausner, and T. A. Rouault. 1994a. Overexpression of iron-responsive element binding protein and its analytical characterization as the RNA-binding form, devoid of an iron-sulfur cluster. *Arch. Biochem. Biophys.* 311:517–522.

Basilion, J. P., T. A. Rouault, C. M. Massinople, R. D. Klausner, and W. H. Burgess. 1994b. The iron-responsive element-binding protein: Localization of the RNA-binding site to the aconitase active-site cleft. *Proc. Natl. Acad. Sci. U.S.A.* 91:574–578.

Baynes, R. D., and T. H. Bothwell. 1990. Iron deficiency. *Ann. Rev. Nutr.* 10:133–148.

Beaumont, C., P. Leneuve, I. Devaux, J-Y. Scoazec, M. Berthier, M-N. Loiseau, B. Grandchamp, and D. Bonneau. 1995. Mutation in the iron responsive element of the L ferritin mRNA in a family with dominant hyperferritinaemia and cataract. *Nature Genetics* 11:444–446.

Beaumont, C., A. Seyhan, A. K. Yachou, B. Grandchamp, and R. Jones. 1994. Mouse ferritin H subunit gene. Functional analysis of the promoter and identification of an upstream regulatory element active in erythroid cells. *J. Biol. Chem.* 269:20281–20288.

Beckman, J. S., T. W. Beckman, J. Chan, P. A. Marshall, and B. A. Freeman. 1990. Apparent hydroxyl radical formation by peroxynitrite: Implications for endothelial injury from nitric oxide and super-oxide. *Proc. Natl. Acad. Sci. U.S.A.* s87:1620–1624.

Beinert, H., M. H. Emptage, J.-L. Dreyer, R. A. Scott, J. E. Hahn, K. O. Hodgson, and A. J. Thomson. 1983. Iron-sulfur stoichiometry and structure of iron-sulfur clusters in three-iron proteins: Evidence for [3Fe-4S] clusters. *Proc. Natl. Acad. Sci. U.S.A.* 80:393–396.

Beinert, H., and M. C. Kennedy. 1989. Engineering of protein bound iron-sulfur clusters: A tool for the study of protein and cluster chemistry and mechanism of iron-sulfur enzymes. *Eur. J. Biochem.* 186:5–15.

Beinert, H., and M. C. Kennedy. 1993. Aconitase, a two-faced protein: Enzyme and iron regulatory factor. *FASEB J.* 7:1442–1449.

Beinert, H., M. C. Kennedy, and C. D. Stout. 1996. Aconitase as iron-sulfur protein, enzyme, and iron-regulatory protein. *Chem. Reviews.* 96:2335–2373.

Berger, H. M., S. Mumby, and J. M. Gutteridge. 1995. Ferrous ions detected in iron-overloaded cord blood plasma from preterm and term babies: implications for oxidative stress. *Free Radical Research.* 22:555–559.

Bettany, A. J. E., R. S. Eisenstein, and H. N. Munro. 1992. Mutagenesis of the iron regulatory element binding protein further defines a role for RNA secondary structure in the regulation of ferritin and TfR expression. *J. Biol. Chem.* 267:16531–16537.

Bhasker, C. R., G. Burgiel, B. Neupert, A. Emery-Goodman, L. C. Kühn, and B. K. May.

1993. The putative iron-responsive element in the human erythroid 5-aminolevulinate synthase mRNA mediates translational control. *J. Biol. Chem.* 263:12699–12705.

Binder, R., J. A. Horowitz, J. P. Basilion, D. M. Koeller, R. D. Klausner, and J. B. Harford. 1994. Evidence that the pathway of transferrin receptor mRNA degradation involves an endonucleolytic cleavage within the 3′ UTR and does not involve poly(A) tail shortening. *EMBO J.* 13:1969–1980.

Bomford, A., C. Conlon-Hollingshead, and H. N. Munro. 1981. Adaptive response of rat tissue isoferritins to iron administration. *J. Biol. Chem.* 256:948–955.

Bomford, A. B., and H. N. Munro. 1985. Transferrin and its receptor: Their roles in cell function. *Hepatology* 5:870–875.

Bomford, A., S. P. Young, and R. Williams. 1985. Release of iron from the two iron-binding sites of transferrin by cultured cells: Modulation by methylamine. *Biochem.* 24:3472–3478.

Bonkovsky, H. L. 1991. Iron and the liver. *Amer. J. Med. Sci.* 301:32–43.

Bothwell, T. H. 1995. Overview and mechanisms of iron regualtion. *Nutr. Rev.* 53:237–245.

Bothwell, T. H., R. W. Charlton, J. D. Cook, and C. A. Finch. 1979. Iron Metabolism in Man. Blackwell Scientific Publications, St. Louis.

Bottomley, S. S., L. C. Wolfe, and K. R. Bridges. 1985. Iron metabolism in K562 erythroleukemia cells. *J. Biol. Chem.* 260:6811–6815.

Bouton, C., M. Raveau, and J.-C. Drapier. 1996. Modulation of iron regulatory protein functions. *J. Biol. Chem.* 271:2300–2306.

Brown, P. H., S. Daniels-McQueen, W. E. Walden, M. M. Patino, L. Gaffield, D. Bielser, and R. E. Thach. 1989. Requirements for the translational repression of ferritin transcripts in wheat germ extracts by a 90-kDa protein from rabbit liver. *J. Biol. Chem.* 264:13383–13386.

Butt, J., H-Y. Kim, J. P. Basilion, S. Cohen, K. Iwai, C. C. Philpott, S. Altschul, R. D. Klausner, and T. A. Rouault. 1996. Differences in the RNA binding sites of iron regulatory proteins and potential target diversity. *Proc. Natl. Acad. Sci. U.S.A.* 93:4345–4349.

Cairo, G., and A. Pietrangelo. 1994. Transferrin receptor gene expression during liver regeneration. *J. Biol. Chem.* 269:6405–6409.

Cairo, G., and A. Pietrangelo. 1995. Nitric-oxide-mediated activation of iron regulatory protein controls hepatic iron metabolism during acute inflammation. *Eur. J. Biochem.* 232:358–363.

Cairo, G., L. Tacchini, G. Pogliaghi, E. Anzon, A. Tomasi, and A. Bernelli-Zazzera. 1995. Induction of ferritin synthesis by oxidative stress. *J. Biol. Chem.* 270:700–703.

Casey, J. L., B. Di Jeso, K. Rao, R. D. Klausner, and J. B. Harford. 1988a. Two genetic loci participate in the regulation by iron of the gene for the human transferrin receptor. *Proc. Natl. Acad. Sci. U.S.A.* 85:1787–1791.

Casey, J. L., M. W. Hentze, D. M. Koeller, S. W. Caughman, T. A. Rouault, R. D. Klausner, and J. B. Harford. 1988b. Iron-responsive elements: Regulatory RNA sequences that control mRNA levels and translation. *Science* 240:924–928.

Casey, J. L., D. M. Koeller, V. C. Ramin, R. D. Klausner, and J. B. Harford. 1989. Iron regulation of transferrin receptor mRNA levels requires iron-responsive elements and a rapid turnover determinant in the 3′-untranslated region of the mRNA. *EMBO J.* 8:3693–3699.

Castro, L., M. Rodriguez, and R. Radi. 1994. Aconitase is readily inactivated by peroxynitrite, but not by its precursor, nitric oxide. *J. Biol. Chem.* 269:29409–29415.

Caughman, S. W., M. W. Hentze, T. A. Rouault, J. B. Harford, and R. D. Klausner. 1988. The iron-responsive element is the single element responsible for iron-dependent translational regulation of ferritin biosynthesis. *J. Biol. Chem.* 263:19048–19052.

Chazenbalk, G. D., H. L. Wadsworth, D. Foti, and B. Rapoport. 1990. Thyrotropin and adenosine 3′5′-monophosphate stimulate the activity of the ferritin H-promoter. *Molec. Endo.* 4:1117–1124.

Chen, O. S., K. L. Schalinske, and R. S. Eisenstein. 1997. Dietary Iron Intake Modulates the Activity of Iron Regulatory Proteins (IRPs) and the Abundance of Ferritin and Mitochondrial Aconitase in Rat Liver. *J. Nutrition* 127:238–248.

Chou, C.-C., R. A. Gatti, M. L. Fuller, P. Concannon, A. Wong, S. Chada, R. C. Davis, and W. A. Salser. 1986. Structure and expression of ferritin genes in a human promyelocytic cell line that differentiates in vitro. *Mol. Cell. Biol.* 6:566–573.

Chu, L. L. H., and R. A. Fineberg. 1969. On the mechanism of iron-induced synthesis of apoferritin in HeLa cells. *J. Biol. Chem.* 244:3847–3854.

Ciechanover, A., A. L. Schwartz, A. Dautry-Varsta, and H. Lodish. 1983. Kinetics of internalization and recycling of transferrin and the transferrin receptor in a human hepatoma cell line. *J. Biol. Chem.* 258:9681–9689.

Cleland, W. W., and M. M. Kreevoy. 1994. Low-barrier hydrogen bonds and enzymic catalysis. *Science* 264:1887–1890.

Coccia, E. M., V. Profita, G. Fiorucci, G. Romeo, E. Affabris, U. Testa, M. W. Hentze, and A. Battistini. 1992. Modulation of ferritin H-chain expression in friend erythroleukemia cells: Transcriptional and translational regulation by hemin. *Mol. Cell. Biol.* 12:3015–3022.

Collawn, J. F., A. Lai, D. Domingo, M. Fitch, S. Hatton, and I. S. Trowbridge. 1993. YTRF is the conserved internalization signal of the transferrin receptor, and a second YTRF signal at position 31–34 enhances endocytosis. *J. Biol. Chem.* 268:21686–21692.

Collawn, J. F., M. Stangel, L. A. Kühn, V. Esekogwu, S. Jing, I. S. Trowbridge, and J. A. Tainer. 1990. Transferrin receptor internalization sequence YXRF implicates a tight turn as the structural recognition motif for endocytosis. *Cell* 63:1061–1072.

Conrad, M. E., J. N. Umbreit, and E. G. Moore. 1993. Regulation of iron transport: Proteins involved in duodenal mucosal uptake and transport. *J. Am. Coll. Nutr.* 12:720–728.

Conrad, M. E., J. N. Umbreit, E. G. Moore, R. D. A. Peterson, and M. B. Jones. 1990. A newly identified iron-binding protein in duodenal mucosa of rats. *J. Biol. Chem.* 265:5273–5279.

Cox, L. A., and G. S. Adrian. 1993. Posttranscriptional regulation of chimeric transferrin genes by iron. *Biochemistry* 32:4738–4745.

Cox, L. A., M. C. Kennedy, and G. S. Adrian. 1995. The 5′-untranslated region of human transferrin mRNA, which contains a putative iron-regulatory element, is bound by purified iron-regulatory protein in a sequence-specific manner. *Biochem. Biophys. Res. Comm.* 212:925–932.

Cox, T. C., M. J. Bawden, A. Martin, and B. K. May. 1991. Human erythroid 5-aminolevulinate synthase: Promoter analysis and identification of an IRE in the mRNA. *EMBO J.* 10:1891–1902.

Cox, T. C., M. W. O'Donnell, P. Aisen, and I. London. 1985. Hemin inhibits internalization of transferrin by reticulocytes and promotes phosphorylation of the membrane transferrin receptor. *Proc. Natl. Acad. Sci. U.S.A.* 82:5170–5174.

Cox, T. M., P. Ponka, and H. M. Schulman. 1990. Erythroid cell iron metabolism and heme synthesis. In *Iron Transport and Storage,* eds. P. Ponka, H. M. Schulman, and R. C. Woodworth, CRC Press, Boca Raton, FL, pp. 271–288.

Crichton, R. R. 1991. *Inorganic Biochemistry of Iron Metabolism.* Ellis Horwood, New York.

D'Souza-Schorey, D., G. Li, M. I. Colombo, and P. D. Stahl. 1995. A regulatory role for ARF6 in receptor-mediated endocytosis. *Science* 267:1175–1177.

Dallman, P. R. 1986. Biochemical basis for the manifestations of iron deficiency. *Ann. Rev. Nutr.* 6:13–40.

Dandekar, T., R. Stripecke, N. K. Gray, B. Goossen, A. Constable, H. E. Johansson, and M. W. Hentze. 1991. Identification of a novel IRE in murine and human eALAS mRNA. *EMBO J.* 10:1903–1909.

Dargemont, C., A. LeBivic, S. Rothenberger, B. Iacopetta, and L. C. Kühn. 1993. The internalization signal and the phosphorylation site of transferrin receptor are distinct from the main basolateral sorting information. *EMBO J.* 12:1713–1721.

Dautry-Varsat, A., A. Ciechanover, and H. F. Lodish. 1983. pH and the recycling of transferrin during receptor-mediated endocytosis. *Proc. Natl. Acad. Sci. U.S.A.* 80:2258–2262.

Davis, B., P. Saltman, and S. Benson. 1962. The stability constants of iron-transferrin complex. *Biochem. Biophys. Res. Comm.* 8:56–60.

Davis, R. J., M. Faucher, L. K. Racaniello, A. Carruthers, and M. P. Czech. 1987. Insulin-like growth factor I and epidermal growth factor regulate the expression of TfRs at the cell surface by distinct mechanisms. *J. Biol. Chem.* 262:13126–13134.

Davis, R. J., and H. Meisner. 1987b. Regulation of transferrin receptor recycling by protein kinase C is independent of receptor phosphorylation at serine 24 in Swiss 3T3 fibroblasts. *J. Biol. Chem.* 262:16041–16047.

DeRusso, P. A., C. C. Philpott, K. Iwai, H. S. Mostowski, R. D. Klausner, and T. A. Rouault. 1995. Expression of a constitutive mutant of iron regulatory protein 1 abolishes iron homeostasis in mammalian cells. *J. Biol. Chem.* 270:15451–15454.

De Silva, D. M., C. C. Askwith, and J. Kaplan. 1996. Molecular mechanisms of iron uptake in eukaryotes. *Physiol. Rev.* 76:31–47.

Dickey, L. F., S. Sreedharan, E. C. Theil, J. R. Didsbury, Y. H. Wang, and R. E. Kaufmann. 1987. Differences in the regulation of messenger RNA for housekeeping and specialized-cell ferritin. *J. Biol. Chem.* 262:7901–7907.

Dickey, L. F., Y.-H. Wang, G. E. Shull, I. A. Wortmann III, and E. C. Theil. 1988. The importance of the 3′-untranslated region in the translational control of ferritin mRNA. *J. Biol. Chem.* 263:3071–3074.

DiMascio, P., M. E. Murphy, and H. Sies. 1991. Antioxidant defense systems: The role of carotenoids, tocopherols, and thiols. *Amer. J. Clin. Nutr.* 53:194S–200S.

Dix, D. J., P.-N. Lin, Y. Kimata, and E. C. Theil. 1992. The iron regulatory region of ferritin mRNA is also a positive control element for iron-independent translation. *Biochemistry* 31:2818–2822.

Dougherty, J. J., W. A. Croft, and W. G. Hoekstra. 1981. Effects of ferrous chloride and iron-dextran on lipid peroxidation in vivo in vitamin E and selenium adequate and deficient rats. *J. Nutr.* 111:1784–1796.

Drapier, J.-C., H. Hirling, J. Wietzerbin, P. Kaldy, and L. C. Kühn. 1993. Biosynthesis of nitric oxide activates iron regulatory factor in macrophages. *EMBO J.* 12:3643–3649.

Drysdale, J. W., and H. N. Munro. 1965. Failure of actinomycin D to prevent induction of liver apoferritin after iron administration. *Biochim. Biophys. Acta.* 103:185–188.

Drysdale, J. W., and H. N. Munro. 1966. Regulation of ferritin synthesis and turnover in rat liver. *J. Biol. Chem.* 241:3630–3637.

Drysdale, J. W., and H. N. Munro. 1970. Role of iron in the regulation of ferritin metabolism. *Fed. Proc.* 29:1469–1473.

Drysdale, J. W., E. Olafsdottir, and H. N. Munro. 1968. Effect of RNA depletion on ferritin induction in rat liver. *J. Biol. Chem.* 243:552–555.

Eanes, R. Z., and E. Kun. 1974. Inhibition of liver aconitase isozymes by (−)-erythrofluorocitrate. *Mol. Pharmacol.* 10:130–139.

Eisenstein, R. S., H. A. Barton, W. P. Pettingell, and A. B. Bomford. 1997. Isolation, characterization, and functional studies of rat liver iron regulatory protein 1. *Arch. Biochem. Biophys.* 343:81–91.

Eisenstein, R. S., A. J. E. Bettany, and H. N. Munro. 1990. Regulation of ferritin gene expression. In Metal Ion Induced Regulation of Gene Expression, vol. 8. In *Advances in Inorganic Biochemistry,* eds. G. L. Eichorn and L. G. Marzilli, pp. 91–138. Elsevier, New York.

Eisenstein, R. S., D. G. Garcia-Mayol, W. P. Pettingell, and H. N. Munro. 1991. Regulation of ferritin and heme oxygenase synthesis by different forms of iron. *Proc. Natl. Acad. Sci. U.S.A.* 88:688–692.

Eisenstein, R. S., P. T. Tuazon, K. L. Schalinske, S. A. Anderson, and J. A. Traugh. 1993. Iron-responsive element binding protein: Phosphorylation by protein kinase C. *J. Biol. Chem.* 268:27363–27370.

Emery-Goodman, A., H. Hirling, L. Scarpellino, B. Henderson, and L. C. Kühn. 1993. Iron regulatory factor expressed from recombinant baculovirus: Conversion between the RNA-binding apoprotein and Fe-S cluster containing aconitase. *Nucl. Acids. Res.* 21:1457–1461.

Emptage, M. H., J.-L. Dreyer, M. C. Kennedy, and H. Beinert. 1983a. Optical and EPR characterization of different species of active and inactive aconitase. *J. Biol. Chem.* 258:11106–11111.

Emptage, M. H., T. A. Kent, M. C. Kennedy, H. Beinert, and E. Münck. 1983b. Mössbauer and EPR studies of activated aconitase: Development of a localized valence state at a subsite of the [4Fe-4S] cluster on binding of citrate. *Proc. Natl. Acad. Sci. U.S.A.* 80:4674–4678.

Englard, S. 1960. Configurational considerations in relation to the mechanisms of the stereospecific enzymatic hydrations of fumarate and *cis*-aconitate. *J. Biol. Chem.* 235:1510–1516.

Enns, C. A., J. W. Larrick, H. Suomalainen, J. Schroder, and H. H. Sussman. 1983. Co-migration and internalization of transferrin and its receptor on K562 cells. *J. Cell. Biol.* 97:579–585.

Fairbanks, V. F. 1994. Iron in medicine and nutrition. In *Modern Nutrition in Health and Disease,* 8th edition, eds. M. E. Shils, J. A. Olson, and M. Shike. Lea and Febiger, Philadelphia, pp. 185–213.

Frishman, D., and M. W. Hentze. 1996. Conservation of aconitase residues revealed by multiple sequence analysis: Implications for structure/function relationships. *Eur. J. Biochem.* 239:197–200.

Gardner, P. R., D.-D. Nguyen H., and C. W. White. 1994. Aconitase is a sensitive and critical target of oxygen poisoning in cultured mammalian cells and in rat lungs. *Proc. Natl. Acad. Sci. U.S.A.* 91:12248–12252.

Gawron, O., A. J. Glaid III, and T. P. Fondy. 1961. Stereochemistry of Krebs' cycle hydrations and related reactions. *J. Am. Chem. Soc.* 83:3634–3640.

Girelli, D., O. Olivieri, P. Gasparini, and R. Corrocher. 1996. Molecular basis for the hereditary hyperferritinemia-cataract *Blood.* 87:4912–4913.

Gironés, N., E. Alvarez, A. Seth, I.-M. Lin, D. A. Latour, and R. J. Davis. 1991. Mutational analysis of the cytoplasmic tail of the human transferrin receptor. *J. Biol. Chem.* 266:19006–19012.

Goossen, B., S. W. Caughman, J. B. Harford, R. D. Klausner, and M. W. Hentze. 1990. Translational repression by a complex between the iron-responsive element of ferritin mRNA and its specific cytoplasmic binding protein is position dependent in vivo. *EMBO J.* 9:127–133.

Granick, S. 1946. Protein apoferritin and ferritin in iron feeding and absorption. *Science* 103:107.

Gray, N. K., K. Pantopoulos, T. Dandekar, B. Ackrell, and M. W. Hentze. 1996. Translational regulation of mammalian and drosophila citric acid cycle enzymes via iron-responsive elements. *Proc. Natl. Acad. Sci. USA.* 93:4925–4930.

Gray, N. K., and M. W. Hentze. 1994. Iron regulatory protein prevents binding of the 43S translation pre-initiation complex to ferritin and eALAS mRNAs. *EMBO J.* 13:3882–3891.

Gray, N. K., S. Quick, B. Goossen, A. Constable, H. Hirling, L. C. Kühn, and M. W. Hentze. 1993. Recombinant iron-regulatory factor functions as an iron-responsive-element-binding protein, a translational repressor and an aconitase: A functional assay for translational repression and direct demonstration of the iron switch. *Eur. J. Biochem.* 218:657–667.

Griffiths, E. 1987. Iron in biological systems. In *Iron and Infection,* eds. J. J. Bullen and E. Griffiths. John Wiley and Sons, New York.

Grootveld, M., J. D. Bell, B. Halliwell, O. I. Aruoma, A. Bomford, and P. J. Sadler. 1989. Non-transferrin-bound iron in plasma or serum from patients with idiopathic hemochromatosis. *J. Biol. Chem.* 264:4417–4422.

Guo, B., F. M. Brown, J. D. Phillips, Y. Yu, and E. A. Leibold. 1995a. Characterization and expression of iron regulatory protein 2 (IRP2). *J. Biol. Chem.* 270:16529–16535.

Guo, B., J. D. Phillips, Y. Yu, and E. A. Leibold. 1995b. Iron regulates the intracellular degradation of iron regulatory protein 2 by the proteasome. *J. Biol. Chem.* 270:21645–21651.

Guo, B., Y. Yu, and E. A. Leibold. 1994. Iron regulates cytoplasmic levels of a novel iron-responsive element-binding protein without aconitase activity. *J. Biol. Chem.* 269:24252–24260.

Gutteridge, J. M., G. J. Quinlan, and T. W. Evans. 1994. Transient iron overload with bleomycin detectable iron in the plasma of patients with adult respiratory distress syndrome. *Thorax.* 49:707–710.

Gutteridge, J. M., S. Mumby, M. Koizumi, and N. Taniguchi. 1996. "Free" iron in neonatal plasma activates aconitase: evidence for biologically reactive iron. *Biochem. Biophys. Res. Comm.* 229:806–809.

Haile, D. J., M. W. Hentze, T. A. Rouault, J. B. Harford, and R. D. Klausner. 1989. Regulation of interaction of the IRE binding protein with iron-responsive RNA elements. *Mol. Cell. Biol.* 9:5055–5061.

Haile, D. J., T. A. Rouault, C. K. Tang, J. Chin, J. B. Harford, and R. D. Klausner. 1992a. Reciprocal control of RNA-binding and aconitase activity in the regulation of the iron-responsive element binding protein: Role of the iron-sulfur cluster. *Proc. Natl. Acad. Sci. U.S.A.* 89:7536–7540.

Haile, D. J., T. A. Rouault, J. B. Harford, M. C. Kennedy, G. A. Blondin, H. Beinert, and R. D. Klausner. 1992b. Cellular regulation of the iron-responsive element binding protein: Disassembly of the cubane iron-sulfur cluster results in high-affinity RNA binding. *Proc. Natl. Acad. Sci. U.S.A.* 89:11735–11739.

Halliwell, B. 1989. Oxidants and the CNS: Some fundamental questions. *Acta. Neurol. Scand.* 126:23–33.

Halliwell, B., and J. M. C. Gutteridge. 1986. Oxygen free radicals and iron in relation to biology and medicine: Some problems and concepts. *Arch. Biochem. Biophys.* 246:501–514.

Hamilton, T. A., T. G. Wada, and H. H. Sussman. 1979. Identification of transferrin receptors on the surface of human cultured cells. *Proc. Natl. Acad. Sci. U.S.A.* 76:6406–6410.

Harrell, C. M., A. R. McKenzie, M. M. Patino, W. E. Walden, and E. C. Theil. 1991. Ferritin mRNA: Interactions of iron regulatory element with translational regulator protein P-90 and the effect on base-paired flanking regions. *Proc. Natl. Acad. Sci. U.S.A.* 88:4166–4170.

Harrison, P. M., G. C. Ford, D. W. Rice, J. M. A. Smith, A. Treffry, and J. L. White. 1987. Structural and functional studies on ferritins. *Biochem. Soc. Trans.* 15:744–748.

Harvey, P. W., R. G. Bell, and M. C. Nesheim. 1985. Iron deficiency protects inbred mice against infection with *Plasmodium chabaudi*. *Infect. Imm.* 50:932–934.

Harvey, P. W., P. F. Heywood, M. C. Nesheim, K. Galme, M. Zegans, J. P. Habicht, L. S. Stephenson, K. L. Radimer, B. Brabin, and K. Forsyth. 1989. The effect of iron therapy on malarial infection in Papua New Guinean school children. *Amer. J. Trop. Med. Hygiene* 40:12–18.

Hausladen, A., and I. Fridovich. 1994. Superoxide and peroxynitrite inactivate aconitases, but nitric oxide does not. *J. Biol. Chem.* 269:29405–29408.

Hedlund, B., and P. E. Hallaway. 1993. High-dose systemic iron chelation attenuates reperfusion injury. *Biochem. Soc. Trans.* 21:340–353.

Hemmaplardh, D., and E. H. Morgan. 1977. The role of endocytosis in transferrin uptake by reticulocytes and bone marrow cells. *Brit. J. Haem.* 36:85–96.

Henderson, B. R., and L. C. Kühn. 1995. Differential modulation of the RNA-binding proteins IRP-1 and IRP-2 in response to iron. *J. Biol. Chem.* 270:20509–20515.

Henderson, B. R., E. Menotti, C. Bonnard, and L. C. Kühn. 1994. Optimal sequence and structure of iron responsive elements. *J. Biol. Chem.* 269:17481–17489.

Henderson, B. R., E. Menotti, and L. C. Kühn. 1996. Iron regulatory proteins 1 and 2 bind distinct sets of RNA target sequences. *J. Biol. Chem.* 271:4900–4908.

Henderson, B. R., C. Seiser, and L. C. Kühn. 1993. Characterization of a second RNA-binding protein in rodents with specificity for iron responsive elements. *J. Biol. Chem.* 268:27327–27334.

Hentze, M. W., and P. Argos. 1991. Homology between IRE-BP, a regulatory RNA-binding protein, aconitase and isopropylmalate isomerase. *Nucl. Acids Res.* 19:1739–1740.

Hentze, M. W., S. W. Caughman, J. L. Casey, D. M. Koeller, T. A. Rouault, J. B. Harford, and R. D. Klausner. 1988. A model for the structure and function of iron responsive elements. *Gene* 72:201–208.

Hentze, M. W., T. A. Rouault, S. W. Caughman, A. Dancis, J. B. Harford, and R. D. Klausner. 1987. A *cis*-acting element is necessary and sufficient for translational regulation of human ferritin expression in response to iron. *Proc. Natl. Acad. Sci. U.S.A.* 84:6730–6734.

Hentze, M. W., T. A. Rouault, J. B. Harford, and R. D. Klausner. 1989. Oxidation-reduction and the molecular mechanism of a regulatory RNA-protein interaction. *Science* 244:357–359.

Hirling, H., A. Emery-Goodman, N. Thompson, B. Neupert, C. Seiser, and L. C. Kühn. 1992. Expression of active iron regulatory factor from a full-length human cDNA by *in vitro* transcription/translation. *Nucl. Acids Res.* 20:33–39.

Hirling, H., B. R. Henderson, and L. C. Kühn. 1994. Mutational analysis of the [4Fe-4S]-cluster converting iron regulatory factor from its RNA-binding form to cytoplasmic aconitase. *EMBO J.* 13:453–461.

Hopkins, C. R., A. Gibson, M. Shipman, D. K. Strickland, and I. S. Trowbridge. 1994. In migrating fibroblasts, recycling receptors are concentrated in narrow tubules in the pericentriolar area, and then routed to the plasma membrane of the leading lamella. *J. Cell. Biol.* 125:1265–1274.

Howard, J. B., and D. C. Rees. 1991. Perspectives on non-heme iron protein chemistry. *Adv. Protein Chem.* 42:199–280.

Hu, H-Y, Y., J. Gardner, P. Aisen, and A. I. Skoultchi. 1977. Inducibility of transferrin receptors on friend erythroleukemia cells. *Science* 197:559–561.

Hu, J., and J. R. Connor. 1996. Demonstration and characterization of the iron regulatory protein in human brain. *J. Neurochem.* 67:838–844.

Huebers, H. A., and C. A. Finch. 1987. The physiology of transferrin and transferrin receptors. *Physiol. Rev.* 67:520–582.

Iacopetta, B., J. L. Carpentier, T. Pozzan, D. P. Lew, P. Gorden, and L. Orci. 1986. Role of intracellular calcium and protein kinase C in the endocytosis of transferrin and insulin by HL60 cells. *J. Cell Biol.* 103:851–856.

Iacopetta, B. J., and E. H. Morgan. 1983. The kinetics of transferrin endocytosis and iron uptake in rabbit reticulocytes. *J. Biol. Chem.* 258:9108–9115.

Imlay, J. A., S. M. Chin, and S. Linn. 1988. Toxic DNA damage by hydrogen peroxide through the Fenton reaction in vivo and in vitro. *Science* 240:640–642.

International Nutritional Anemia Consultive Group (INACG). 1990. Combating iron deficiency anemia through food fortification technology. *INACG report.* XII INACG Meeting, Washington, D.C.

Iwai, K., R. D. Klausner, and T. A. Rouault. 1995. Requirements for iron-regulated degradation of the RNA binding protein, iron regulatory protein 2. *EMBO J.* 21:5350–5357.

Jaffrey, S. R., N. A. Cohen, T. A. Rouault, R. D. Klausner, and S. H. Snyder. 1994. The iron-responsive element binding protein: A target for synaptic actions of nitric oxide. *Proc. Natl. Acad. Sci. U.S.A.* 91:12994–12998.

Jaffrey, S. D., D. J. Haile, R. D. Klausner, and J. B. Harford. 1993. The interaction between the iron-responsive element binding protein and its cognate RNA is highly dependent upon both sequence and structure. *Nucl. Acids. Res.* 21:4627–4631.

Jandl, J. H., J. K. Inman, R. L. Simmons, and D. W. Allen. 1959. Transfer of iron from serum iron-binding protein to human reticulocytes. *J. Clin. Invest.* 38:161–185.

Jandl, J. H., and J. H. Katz. 1963. The plasma to cell cycle of transferrin. *J. Clin. Invest.* 42:314–326.

Jing, S., T. Spencer, K. Miller, C. Hopkins, and I. S. Trowbridge. 1990. Role of the transferrin receptor cytoplasmic domain in endocytosis: Localization of a specific signal for internalization. *J. Cell Biol.* 110:283–294.

Johnson, M. K., A. J. Thomson, A. J. M. Richards, J. Peterson, A. E. Robinson, R. R. Ramsay, and T. P. Singer. 1984. Characterization of the Fe-S cluster in aconitase using low temperature magnetic circular dichroism spectroscopy. *J. Biol. Chem.* 259:2274–2282.

Jones, T., R. Spencer, and C. Walsh. 1978. Mechanism and kinetics of iron release from ferritin by dihydroflavins and dihydroflavin analogues. *Biochemistry* 17:4011–4017.

Kappas, A., S. Sassa, R. A. Galbraith, and Y. Nordman. 1995. The porphyrias. In *The Metabolic Basis of Inherited Disease,* eds. C. R. Scriver, A. L. Beaudet, W. S. Sly, and D. Valle. Vol. 2, Ch. 66, pp. 2103–2161. McGraw-Hill, New York.

Kaptain, S., W. E. Downey, C. Tang, C. Philpott, D. Haile, D. G. Orloff, J. B. Harford, T. A. Rouault, and R. D. Klausner. 1991. A regulated RNA binding protein also possesses aconitase activity. *Proc. Natl. Acad. Sci. U.S.A.* 88:10109–10113.

Karin, M., and B. Mintz. 1981. Receptor-mediated endocytosis of transferrin in developmentally totipotent mouse teratocarcinoma stem cells. *J. Biol. Chem.* 256:3245–3252.

Katz, J. H. 1961. Iron and protein kinetics studied by means of doubly labeled human crystalline transferrin. *J. Clin. Invest.* 40:2143–2152.

Kennedy, C., R. Rauner, and O. Gawron. 1972. On pig heart aconitase. *Biochem. Biophys. Res. Commun.* 47:740–745.

Kennedy, M. C., W. E. Antholine, and H. Beinert. 1997. An EPR investigation of the products of the reaction of cytosolic and mitochondrial aconitases with nitric oxide. *J. Biol. Chem.* 272:20340–20347.

Kennedy, M. C., and H. Beinert. 1988. The state of cluster SH and S^{2-} of aconitase during cluster interconversions and removal: A convenient preparation of apoenzyme. *J. Biol. Chem.* 263:8194–8198.

Kennedy, M. C., M. H. Emptage, J.-L. Dreyer, and H. Beinert. 1983. The role of iron in the activation-inactivation of aconitase. *J. Biol. Chem.* 258:11098–11105.

Kennedy, M. C., T. A. Kent, M. Emptage, H. Merkle, H. Beinert, and E. Münck. 1984. Evidence for the formation of a linear [3Fe-4S] cluster in partially unfolded aconitase. *J. Biol. Chem.* 259:14463–14471.

Kennedy, M. C., L. Mende-Mueller, G. A. Blondin, and H. Beinert. 1992. Purification and characterization of cytosolic aconitase from beef liver and its relationship to the iron-responsive element binding protein. *Proc. Natl. Acad. Sci. U.S.A.* 89:11730–11734.

Kennedy, M. C., M. Werst, J. Telser, M. H. Emptage, H. Beinert, and B. M. Hoffman. 1987. Mode of substrate carboxyl binding to the [4Fe-4S]$^+$ cluster of reduced aconitase as studied by ^{17}O and ^{13}C electron-nuclear double resonance spectroscopy. *Proc. Natl. Acad. Sci. U.S.A.* 84:8854–8858.

Kent, T. A., J.-L. Dreyer, M. C. Kennedy, B. H. Huynh, M. H. Emptage, H. Beinert, and E. Münck. 1982. Mössbauer studies of beef heart aconitase: Evidence for facile interconversions of iron-sulfur clusters. *Proc. Natl. Acad. Sci. U.S.A.* 79:1096–1100.

Kent, T. A., M. H. Emptage, H. Merkle, M. C. Kennedy, H. Beinert, and E. Münck. 1985. Mössbauer studies of aconitase: Substrate and inhibitor binding, reaction intermediates, and hyperfine interactions of reduced 3Fe and 4Fe clusters. *J. Biol. Chem.* 260:6871–6881.

Kikinis, Z., R. S. Eisenstein, A.J.E. Bettany, and H. N. Munro. 1995. Role of RNA secondary structure of the iron-responsive element in translational regulation of ferritin synthesis. *Nucl. Acids Res.* 4190–4195.

Kilpatrick, L. K., M. C. Kennedy, H. Beinert, R. S. Czernuszewicz, D. Qiu, and T. G. Spiro. 1994. Cluster structure and H-bonding in native, substrate-bound, and 3Fe forms of aconitase as determined by resonance Raman spectroscopy. *J. Am. Chem. Soc.* 116:4053–4061.

Kim, H-Y., R. D. Klausner, and T. A. Rouault. 1995. Translational repressor activity is equivalent and is quantitatively predicted by *in vitro* RNA binding for two iron-responsive element-binding proteins, IRP1 and IRP2. *J. Biol. Chem.* 270:4983–4986.

Kim, H-Y., T. LaVaute, K. Iwai, R. D. Klausner, and T. A. Rouault. 1996. Identification of a conserved and functional iron-responsive element in the 5′-untranslated region of mammalian mitochondrial aconitase. *J. Biol. Chem.* 271:24226–24230.

Klausner, R. D., G. Ashwell, J. van Renswoude, J. B. Harford, and K. R. Bridges. 1983a. Binding of apotransferrin to K562 cells: Explanation of the transferrin receptor cycle. *Proc. Natl. Acad. Sci. U.S.A.* 80:2263–2266.

Klausner, R. D., J. van Renswoude, G. Ashwell, G. Kempf, A. N. Schechter, A. Dean, and K. R. Bridges. 1983b. Receptor-mediated endocytosis of transferrin in K562 Cells. *J. Biol. Chem.* 258:4715–4724.

Klausner, R. D., J. B. Harford, and J. van Renswoude. 1984. Rapid internalization of the transferrin receptor in K562 cells is triggered by ligand binding or treatment with a phorbol ester. *Proc. Natl. Acad. Sci. U.S.A.* 81:3005–3009.

Klausner, R. D., T. A. Rouault, and J. B. Harford. 1993. Regulating the fate of mRNA: The control of cellular iron metabolism. *Cell* 72:19–28.

Kohgo, Y., M. Yokota, and J. W. Drysdale. 1980. Differential turnover of rat liver isoferritins. *J. Biol. Chem.* 255:5195–5200.

Köhler, S. A., B. R. Henderson, and L. C. Kuhn. 1995. Succinate dehydrogenase B mRNA of *Drosophilia melanogaster* has a functional iron-responsive element in its 5′-untranslated region. *J. Biol. Chem.* 270:30781–30786.

Kraulis, P. J. 1991. Molscript: A program to produce detailed and schematic plots of protein structures. *J. Appl. Cryst.* 24:946–950.

Krebs, E. G. 1986. In *The Enzymes: Control by Phosphorylation,* eds. P. D. Boyer and E. G. Krebs. Vol. XVII, Part A, pp. 3–20. Academic Press, New York.

Kretchmer, N., J. L. Beard, and S. Carlson. 1996. The role of nutrition in the development of normal cognition. *Amer. J. Clin. Nutr.* 63:997S–1001S.

Kühn, L. C., and M. W. Hentze. 1992. Coordination of cellular iron metabolism by post-transcriptional gene regulation. *J. Inorg. Biochem.* 47:183–192.

Kühn, L. C., H. M. Schulman, and P. Ponka. 1990. Iron transferrin requirements and TfR expression in proliferating cells. In *Iron Storage and Transport,* eds. P. Ponka, H. M. Schulman, and R. C. Woodworth. pp. 155–198. CRC Press, Boca Raton, FL.

Kumagai, N., S. H. Benedict, G. B. Mills, and E. W. Gelfand. 1987. Requirements for the simultaneous presence of phorbol esters and calcium ionophores in the expression of human T-lymphocyte proliferation-related genes. *J. Immunol.* 139:1393–1399.

Kumagai, N., S. H. Benedict, G. B. Mills, and E. W. Gelfand. 1988. Comparison of phorbol ester and phytohemagglutinin-induced signaling in human T-lymphocytes. *J. Immunol.* 140:37–43.

Larrick, J. W., and P. Cresswell. 1979. Modulation of cell surface iron TfRs by cellular density and state of activation. *J. Supramol. Struct.* 11:579–586.

Lauble, H., M. C. Kennedy, H. Beinert, and C. D. Stout. 1992. Crystal structures of aconitase with isocitrate and nitroisocitrate bound. *Biochemistry.* 31:2735–2748.

Lauble, H., M. C. Kennedy, H. Beinert, and C. D. Stout. 1994. Crystal structures of aconitase with *trans*-aconitate and nitocitrate bound. *J. Mol. Biol.* 237:437–451.

Lauble, H., and C. D. Stout. 1994. Steric and conformational features of the aconitase mechanism. *Proteins: Structure, Function, Genetics* 22:1–11.

Leedman, P. J., A. R. Stein, W. R. Chin, and J. T. Rogers. 1996. Thyroid hormone modulates the interaction between iron regulatory proteins and the ferritin mRNA iron-responsive element. *J. Biol. Chem.* 271:12017–12023.

Leibold, E. A., and B. Guo. 1992. Iron-dependent regulation of ferritin and transferrin receptor expression by the iron-responsive element binding protein. *Ann. Rev. Nutr.* 12:345–368.

Leibold, E. A., A. Laudano, and Y. Yu. 1990. Structural requirements of iron-responsive elements for binding of the protein involved in both TfR and ferritin mRNA post-transcriptional regulation. *Nucl. Acids Res.* 18:1819–1824.

Leibold, E. A., and H. N. Munro. 1987. Characterization and evolution of the expressed rat ferritin light subunit gene and its pseudogene family. *J. Biol. Chem.* 262:7335–7341.

Leibold, E. A., and H. N. Munro. 1988. Cytoplasmic protein binds *in vitro* to a conserved sequence in the 5' untranslated region of ferritin H- and L-chain mRNAs. *Proc. Natl. Acad. Sci. U.S.A.* 85:2171–2175.

Levi, S., P. Santabrogio, A. Cozzi, E. Rovida, B. Corsi, E. Tamborini, S. Spada, A. Albertini, and P. Arosio. 1994. The role of the 1-chain in ferritin iron incorporation. Studies of homo and heteropolymers. *J. Mol. Biol.* 238:649–654.

Li, C-Y., A. Watkins, and J. Glass. 1994. The H^+-ATPase from reticulocyte endosomes reconstituted into liposomes acts as an iron transporter. *J. Biol. Chem.* 269:10242–10246.

Lim, K-C., H. Ishihara, R. D. Riddle, Z. Yang, N. Andrews, M. Yamamoto, and J. D. Engel.

1994. Structure and regulation of the chicken erythroid δ-aminolevulinate synthase gene. *Nucl. Acids. Res.* 22:1226–1233.

Linder, M. C., G. M. Nagel, M. Roboz, and D. M. Hungerford, Jr. 1981. The size and shape of heart and muscle ferritins analyzed by sedimentation, gel filtration, and electrophoresis. *J. Biol. Chem.* 256:9104–9110.

Liochev, S. I., and I. Fridovich. 1993. The role of O_2^- in the production of HO^-: In vitro and in vivo. *Free Rad. Biol. Med.* 16:29–33.

Martin, A., B. K. Burgess, C. D. Stout, V. Cash, D. R. Dean, G. M. Jensen, and P. J. Stephens. 1990. Site-directed mutagenesis of *Azotobacter vinelandii* ferredoxin I: [Fe-S] cluster driven protein rearrangement. *Proc. Natl. Acad. Sci. U.S.A.* 87:598–602.

Martins, E. A. L., R. L. Robalino, and R. Meneghini. 1995. Oxidative stress induces activation of a cytosolic protein responsible for control of iron uptake. *Arch. Biochem. Biophys.* 316:128–134.

Mascotti, D. P., L. S. Goessling, D. Rup, and R. E. Thach. 1997. Mechanisms for induction and re-repression of ferritin synthesis. In *Metal Ions in Gene Regulation,* eds. S. Silver and W. Walden. Ch. 8, pp. 217–230, Chapman & Hall, New York.

Mattia, E., D. Josic, G. Ashwell, R. Klausner, and J. van Renswoude. 1986. Regulation of intracellular iron distribution in K562 erythroleukemia cells. *J. Biol. Chem.* 261:4587–4593.

McClelland, A., L. C. Kühn, and F. H. Ruddle. 1984. The human transferrin receptor gene: Genomic organization, and the complete primary structure of the receptor deduced from a cDNA sequence. *Cell* 39:267–274.

McKnight, G. S., D. C. Lee, D. Hemmaplardh, C. A. Finch, and R. D. Palmiter. 1980. Transferrin gene expression: Effects of nutritional iron deficiency. *J. Biol. Chem.* 255:144–147.

McRee, D. E. (1992). A visual protein crystallographic software system for XII/Xview. *J. Mol. Graphics* 10:44–47.

Melefors, Ö., B. Goossen, H. E. Jonansson, R. Stripecke, N. K. Gray, and M. W. Hentze. 1993. Translational control of 5-aminolevulinate synthase mRNA by iron-responsive elements in erythroid cells. *J. Biol. Chem.* 268:5974–5978.

Miller, K., K. Shipman, I. S. Trowbridge, and C. R. Hopkins. 1991a. Transferrin receptors promote the formation of clathrin lattices. *Cell* 65:621–632.

Miller, L. L., S. C. Miller, S. V. Torti, Y. Tsuji, and F. M. Torti. 1991b. Iron-independent induction of ferritin H chain by tumor necrosis factor. *Proc. Natl. Acad. Sci. U.S.A.* 88:4946–4950.

Morgan, E. H., and T. C. Appleton. 1969. Autoradiographic localization of 125-I-labeled transferrin in rabbit reticulocytes. *Nature* 223:1371–1372.

Morgan, E. H., and E. Baker. 1986. Iron uptake and metabolism by hepatocytes. *Fed. Proc.* 45:2810–2816.

Müller-Eberhard, U., H. H. Liem, J. A. Grasso, S. Giffhorn-Katz, M. G. DeFalco, and

N. R. Katz. 1988 Increase in surface expression of transferrin receptors on cultured hepatocytes of adult rats in response to iron deficiency. *J. Biol. Chem.* 263:14753–14756.

Müllner, E. W., B. Neupert, and L. C. Kühn. 1989. A specific mRNA binding factor regulates the iron-dependent stability of cytoplasmic transferrin receptor mRNA. *Cell* 58:373–382.

Müllner, E. W., S. Rothenberger, A. M. Müller, and L. C. Kühn. 1992. In vivo and in vitro modulation of the mRNA-binding activity of iron regulatory factor. *Eur. J. Biochem.* 208:597–605.

Munro, H. N., N. Aziz, E. A. Leibold, M. Murray, J. Rogers, J. K. Vass, and K. White. 1988. The ferritin genes: Structure, expression and regulation. *Ann. N.Y. Acad. Sci.* 526:113–123.

Munro, H. N., and M. C. Linder. 1978. Ferritin: Structure, biosynthesis, and role in iron metabolism. *Physiol. Rev.* 58:317–396.

Neupert, B., E. Menotti, and L. C. Kühn. 1995. A novel method to identify nucleic acid binding sites in proteins by scanning mutagenesis: Application to iron regulatory protein. *Nucl. Acids Res.* 14:2579–2583.

Neupert, B., N. A. Thompson, C. Meyer, and L. C. Kühn. 1990. A high yield affinity purification method for specific RNA-binding proteins: Isolation of the iron regulatory factor from human placenta. *Nucl. Acids Res.* 18:51–55.

Nichols, G. M., A. R. Pearce, X. Alverez, N. K. Bibb, K. Y. Nichols, C. B. Alfred, and J. Glass. 1992. The mechanisms of nonheme iron uptake determined in IEC-6 rat intestinal cells. *J. Nutr.* 122:945–952.

Núñez, M-T., V. Gaete, J. A. Watkins, and J. Glass. 1990. Mobilization of iron from endocytic vesicles. *J. Biol. Chem.* 265:6688–6692.

Núñez, M-T., J. Glass, S. Fischer, L. M. Lavidor, E. M. Lenk, and S. H. Robinson. 1977. Transferrin receptors in developing erythroid cells. *Brit. J. Haem.* 36:519–526.

O'Halloran, T. V. 1993. Transition metals in control of gene expression. *Science* 261:715–725.

Oliveira, C. C., B. Goossen, N. I. Zanchin, J. E. McCarthy, M. W. Hentze, and R. Stripecke. 1993. Translational repression by human iron-regulatory factor (IRF) in *Saccharomyces cerevisiae. Nucl. Acids. Res.* 21:5316–5322.

Oliver, C. N., P. E. Starke-Reed, E. R. Stadtman, G. J. Liu, G. M. Carney, and R. A. Floyd. 1990. Oxidative damage to brain proteins, loss of glutamine synthetase activity and production of free radicals during ischemia/reperfusion-induced injury to gerbil brain. *Proc. Natl. Acad. Sci. U.S.A.* 87:5144–5147.

Oria, R., L. Sanchez, T. Houston, M. W. Hentze, F. Y. Liew, and J. H. Brock. 1995. Effect of nitric acid on expression of transferrin receptor and ferritin and on cellular iron metabolism in K562 human erythroleukemia cells. *Blood* 85:2962–1271.

Owen, D., and L. C. Kühn. 1987. Noncoding 3′ sequences of the transferrin receptor gene are required for mRNA regulation by iron. *EMBO J.* 6:1287–1293.

Pantopoulos, K., and M. W. Hentze. 1995a. Nitric oxide signaling to iron-regulatory protein: Direct control of ferritin mRNA translation and transferrin receptor RNA stability in transfected fibroblasts. *Proc. Natl. Acad. Sci. U.S.A.* 92:1267–1271.

Pantopoulos, K., and M. W. Hentze. 1995b. Rapid responses to oxidative stress mediated by iron regulatory protein. *EMBO J.* 14:2917–2924.

Pantopoulos, K., G. Weiss, and M. W. Hentze. 1996. Nitric oxide and oxidative stress (H_2O_2) control mammalian iron metabolism by different pathways. *Mol. Cell. Biol.* 16:3781–3788.

Patino, M. M., and W. E. Walden. 1992. Cloning of a functional cDNA for the rabbit ferritin mRNA repressor protein. *J. Biol. Chem.* 267:19011–19016.

Pelosi-Testa, E., P. Samoggia, G. Giannella, E. Montesoro, T. Caravita, G. Salvo, A. Camagna, G. Isacchi, and U. Testa. 1988. Mechanisms underlying T-lymphocyte activation: Mitogen initiates and IL-2 amplifies the expression of transferrin receptors via intracellular iron levels. *Immunology* 64:273–279.

Petering, D., J. A. Fee, and G. Palmer. 1971. The oxygen sensitivity of spinach ferredoxin and other iron-sulfur proteins: The formation of protein-bound sulfur-zero. *J. Biol. Chem.* 246:643–653.

Phillips, J. D., D. V. Kinikini, Y. Yu, B. Guo, and E. A. Leibold. 1996. Differential regulation of IRP1 and IRP2 by nitric oxide in rat hepatoma cells. *Blood* 87:2983–2992.

Philpott, C. C., R. D. Klausner, and T. A. Rouault. 1994. The bifunctional iron responsive element-binding protein/cytosolic aconitase (IRP1): The role of active site residues in ligand binding and regulation. *Proc. Natl. Acad. Sci. U.S.A.* 91:7321–7325.

Pollitt, E. 1993. Iron deficiency and cognitive function. *Ann. Rev. Nutr.* 13:521–537.

Puglisi, J. D., R. Tan, B. J. Calnan, A. D. Frankel, and J. R. Williamson. 1992. Conformation of the TAR RNA-arginine complex by NMR spectroscopy. *Science* 257:76–80.

Randell, E. W., J. G. Parkes, N. F. Oliver, and D. M. Templeton. 1994. Uptake of non-transferrin-bound iron by both reductive and nonreductive processes is modulated by intracellular iron. *J. Biol. Chem.* 269:16046–16053.

Rao, K. K., D. Shapiro, E. Mattia, K. Bridges, and R. Klausner. 1985. Effects of alterations in cellular iron on biosynthesis of the transferrin receptor in K562 cells. *Mol. Cell. Biol.* 5:595–600.

Reilly, P. M., H. J. Schiller, and G. B. Bulkley. 1991. Reactive oxygen metabolites in shock. In *Trauma IV*. Ch. 8, pp. 1–30. Scientific American Books, New York.

Renaudie, F., A. K. Yachov, B. Grandchamp, R. Jones, and C. Beaumont. 1992. A second ferritin L subunit is encoded by an intronless gene in the mouse. *Mamm. Genome* 2:143–149.

Richardson, D. R., V. Neumannova, E. Nagy, and P. Ponka. 1995. The effect of redox-related species of nitrogen monoxide on transferrin and iron uptake and cellular proliferation of erythroleukemia (K562) cells. *Blood* 86:3211–3219.

Robbins, A. H., and C. D. Stout. 1989. The structure of aconitase. *Proteins: Structure, Function, Genetics* 5:289–312.

Roberts, S., and A. Bomford. 1988. Ferritin iron kinetics and protein turnover in K562 cells. *J. Biol. Chem.* 263:19181–19187.

Rogers, J. T., and H. N. Munro. 1987. Translation of ferritin light and heavy subunit mRNAs is regulated by intracellular iron levels in rat hepatoma cells. *Proc. Natl. Acad. Sci. U.S.A.* 84:2277–2281.

Rothenberger, S., B. J. Iacopetta, and L. C. Kühn. 1987. Endocytosis of the transferrin receptor requires the cytoplasmic domain but not its phosphorylation site. *Cell* 49:423–431.

Rothenberger, S., E. W. Müllner, and L. C. Kühn. 1990. The mRNA-binding protein that controls ferritin and TfR expression is conserved during evolution. *Nucl. Acids Res.* 18:1175–1179.

Rouault, T. A., D. J. Haile, W. E. Downey, C. C. Philpott, C. Tang, F. Samaniego, J. Chin, I. Paul, D. Orloff, J. B. Harford, and M. W. Hentze. 1992. An iron-sulfur cluster plays an unusual regulatory role in the iron regulatory element binding protein. *Biometals* 5:131–140.

Rouault, T. A., M. W. Hentze, S. W. Caughman, J. B. Harford, and R. D. Klausner. 1988. Binding of a cytosolic protein to the iron-responsive element of human ferritin messenger RNA. *Science* 241:1207–1210.

Rouault, T. A., M. W. Hentze, D. J. Haile, J. B. Harford, and R. D. Klausner. 1989. The iron-responsive element binding protein: A method for the affinity purification of a regulatory RNA-binding protein. *Proc. Natl. Acad. Sci. U.S.A.* 86:5768–5772.

Rouault, T. A., C. D. Stout, S. Kaptain, J. B. Harford, and R. D. Klausner. 1991. Structural relationship between an iron-regulated RNA-binding protein (IRE-BP) and aconitase: Functional implications. *Cell* 64:881–883.

Rouault, T. A., C. K. Tang, S. Kaptain, W. H. Burgess, D. J. Haile, F. Samaniego, O. W. McBride, J. B. Harford, and R. D. Klausner. 1990. Cloning of the cDNA encoding an RNA regulatory protein—The human iron-responsive element-binding protein. *Proc. Natl. Acad. Sci. U.S.A.* 87:7958–7962.

Saddi, R., and A. von der Decken. 1965. The effect of iron administration on the incorporation of [^{3}H]leucine into ferritin by rat liver systems. *Biochim. Biophys. Acta.* 111:124–133.

Samaniego, F., J. Chin, K. Iwai, T. A. Rouault, and R. D. Klausner. 1994. Molecular characterization of a second iron responsive element binding protein, iron regulatory protein 2. *J. Biol. Chem.* 269:30904–30910.

Sato, N., and A. Miyajima. 1994. Multimeric cytokine receptors: Common versus specific functions. *Current Biol.* 6:174–179.

Schalinske, K. L., S. A. Anderson, P. T. Tuazon, O. S. Chen, M. C. Kennedy, and R. S. Eisenstein. 1997. The iron-sulfur cluster of iron regulatory protein 1 modulates the accessibility of rna binding and phosphorylation sites. *Biochemistry.* 36:3950–3958.

Schalinske, K. L., and R. S. Eisenstein. 1996. Phosphorylation and activation of both iron regulatory protein 1 (IRP1) and IRP2 in HL60 cells. *J. Biol. Chem.* 271:7168–7176.

Seiser, C., S. Teixeira, and L. Kühn. 1993. Interleukin-2-dependent transcriptional and post-transcriptional regulation of transferrin receptor mRNA. *J. Biol. Chem.* 268:13074–13080.

Seligman, P. A., R. B. Schleicher, and R. H. Allen. 1979. Isolation and characterization of the transferrin receptor from human placenta. *J. Biol. Chem.* 254:9943–9946.

Sierzputowska-Gracz, H., R. A. McKenzie, and E. C. Theil. 1995. The importance of a single G in the hairpin loop of the iron responsive element (IRE) in ferritin mRNA for structure: An NMR spectroscopy study. *Nucl. Acids. Res.* 23:146–153.

Sies, H. 1991. Oxidative stress: From basic science to clinical application. *Amer. J. Med.* 91:31S–38S.

Sohal, R. S., and W. C. Orr. 1992. The relationship between antioxidants, prooxidants and the aging process. *Ann. N.Y. Acad. Sci.* 663:74–84.

Srere, P. A. 1992. The molecular physiology of citrate. In *Current Topics in Cellular Regulation*, eds. E. R. Stadtman and P. BoonChock. Vol. 33, pp. 261–275. Academic Press, New York.

Srere, P. A. 1994. Complexities of metabolic regulation. *Trends in Biochem. Sci.* 19:519–520.

Stadtman, E. R., and C. N. Oliver. 1991. Metal-catalyzed oxidation of proteins. *J. Biol. Chem.* 266:2005–2008.

Stevens, R. G., D. Y. Jones, M. S. Micozzi, and P. R. Taylor. 1988. Body iron stores and the risk of cancer. *N. Engl. J. Med.* 319:1047–1052.

Sturrock, A., J. Alexander, J. Lamb, C. Craven, and J. Kaplan. 1990. Characterization of a transferrin-independent uptake system for iron in HeLa cells. *J. Biol. Chem.* 265:3139–3145.

Surerus, K. K., M. C. Kennedy, H. Beinert, and E. Münck. 1989. Mössbauer study of the inactive Fe_3S_4 and Fe_3Se_4 and the active Fe_4Se_4 forms of beef heart aconitase. *Proc. Natl. Acad. Sci. U.S.A.* 86:9846–9850.

Swenson, G. R., and W. E. Walden. 1994. Localization of an RNA binding element of the iron responsive element binding protein within a proteolytic fragment containing iron coordination ligands. *Nucl. Acids Res.* 22:2627–2633.

Taetle, R., S. Ralph, S. Smedsrud, and I. Trowbridge. 1987. Regulation of transferrin receptor expression in myeloid leukemia cells. *Blood* 70:852–859.

Teichmann, R., and W. Stremmel. 1990. Iron uptake by human upper small intestine microvillous membrane vesicles. *J. Clin. Invest.* 86:2145–2153.

Teixeira, S., and L. C. Kühn. 1991. Post-transcriptional regulation of the TfR and 4F2 antigen heavy chain mRNA during growth activation of spleen cells. *Eur. J. Biochem.* 202:819–826.

Testa, U., L. Kühn, M. Petrini, M. T. Quaranta, E. Pelosi, and C. Peschle. 1991. Differential regulation of IRE-BP(s) in extracts of activated lymphocytes versus macrophages. *J. Biol. Chem.* 266:13925–13930.

Testa, U., M. Petrini, M. T. Quaranta, E. Pelosi-Testa, G. Mastroberardino, A. Camagna, G. Boccoli, M. Sargiacomo, G. Isacchi, A. Cozzi, P. Arosio, and C. Peschle. 1989.

Iron up-modulates the expression of TfRs during monocyte-macrophage maturation. *J. Biol. Chem.* 264:13181–13197.

Testa, U., M. Titieux, F. Louache, P. Thomopoulos, and H. Rochant. 1984. Effect of phorbol esters on iron uptake in human hematopoietic cell lines. *Cancer Res.* 44:4981–4986.

Theil, E. C. 1987. Ferritin: Structure, gene regulation and cellular function in animals, plants, and microorganisms. *Ann. Rev. Biochem.* 56:289–315.

Theil, E. C. 1994. Iron regulatory elements (IREs): A family of mRNA non-coding sequences. *Biochem. J.* 304:1–11.

Thorp, H. H., R. A. Mckenzie, P. N. Lin, W. E. Walden, and E. C. Theil. 1996. Cleavage of functionally relevant sites in ferritin mRNA by oxidizing metal complexes. *Inorg. Chem.* 35:2773–2779.

Thorstensen, K., and I. Romslo. 1990. The role of transferrin in the mechanism of iron uptake. *Biochem. J.* 271:1–10.

Trowbridge, I. S. 1991. Endocytosis and signals for internalization. *Current Biol.* 3:634–641.

Trowbridge, I. S., J. F. Collawn, and C. R. Hopkins. 1993. Signal-dependent membrane protein trafficking in the endocytic pathway. *Ann. Rev. Cell Biol.* 9:129–161.

Trowbridge, I. S., and M. B. Omary. 1981. Human cell surface glycoprotein related to cell proliferation is the receptor for transferrin. *Proc. Natl. Acad. Sci. U.S.A.* 78:3039–3043.

Wagstaff, M., M. Worwood, and A. Jacobs. 1978. Properties of human tissue isoferritins. *Biochem. J.* 173:969–977.

Walden, W. E., S. Daniels-McQueen, P. H. Brown, L. Gaffield, D. A. Russell, D. Bielser, L. C. Bailey, and R. E. Thach. 1988. Translational repression in eucaryotes: Partial purification and characterization of a repressor of ferritin mRNA translation. *Proc. Natl. Acad. Sci. U.S.A.* 85:9503–9507.

Walden, W. E., M. M. Patino, and L. Gaffield. 1989. Purification of a specific repressor of ferritin mRNA translation from rabbit liver. *J. Biol. Chem.* 264:13765–13769.

Wang, Y.-H., S. R. Sczekan, and E. C. Theil. 1990. Structure of the 5' untranslated regulatory region of ferritin mRNA studied in solution. *Nucl. Acids Res.* 18:4463–4468.

Watkins, J. A., J. D. Altazan, P. Elder, C.-Y. Li, M-T. Núñez, X.-X. Cui, and J. Glass. 1992. Kinetic characterization of reductant dependent processes for iron mobilization from endocytic vesicles. *Biochemistry* 31:5820–5830.

Weinberg, E. D. 1993. The development of awareness of iron-withholding defense. *Perspec. Bio. Med.* 36:215–221.

Weiss, G., B. Goossen, W. Doppler, D. Fuchs, K. Pantopoulos, G. Werner-Felmayer, H. Wachter, and M. W. Hentze. 1993. Translational regulation via iron-responsive elements by the nitric oxide/NO-synthase pathway. *EMBO J.* 12:3651–3657.

Weiss, G., G. Werner-Felmayer, E. R. Werner, K. Grünewald, H. Wachter, and M. W. Hentze. 1994. Iron regulates nitric oxide synthase activity by controlling nuclear transcription. *J. Exp. Med.* 180:969–976.

Werst, M. M., M. C. Kennedy, H. Beinert, and B. M. Hoffman. 1990a. ^{17}O, ^{1}H, and ^{2}H

electron nuclear double resonance characterization of solvent substrate and inhibitor binding to the [4Fe-4S]$^+$ cluster of aconitase. *Biochemistry* 29:10526–10532.

Werst, M. M., M. C. Kennedy, A. L. P. Houseman, H. Beinert, and B. M. Hoffman. 1990b. Characterization of the [4Fe-4S]$^+$ cluster at the active site of aconitase by ^{57}Fe, ^{33}S, and ^{14}N electron nuclear double resonance spectroscopy. *Biochemistry* 29:10533–10540.

White, K., and H. N. Munro. 1988. Induction of ferritin subunit synthesis by iron is regulated at both the transcriptional and translational levels. *J. Biol. Chem.* 263:8938–8942.

Wiley, H. S., and J. Kaplan. 1984. Epidermal growth factor rapidly induces a redistribution of transferrin receptor pools in human fibroblasts. *Proc. Natl. Acad. Sci. U.S.A.* 81:7456–7460.

Williams, R. J. P. 1990. An introduction to the nature of iron transport and storage. In *Iron Transport and Storage,* eds. P. Ponka, H. M. Schulman, and R. C. Woodworth. pp. 1–16, CRC Press, Boca Raton, FL.

Wolf, G. 1994. An integrin-mobilferrin iron transport pathway in intestine and hematopoietic cells. *Nutr. Rev.* 52:387–389.

Wright, T. L., P. Brissot, W. L. Ma, and R. A. Weisiger. 1986. Characterization of non-transferrin-bound iron clearance by rat liver. *J. Biol. Chem.* 261:10909–10914.

Young, S. P., and P. Aisen. 1988. The liver and iron. In *The Liver: Biology and Pathophysiology,* 2nd edition, eds. I. W. Arias, W. B. Jakoby, H. Popper, D. Schacter, and D. A. Shafritz. pp. 535–550. Raven Press, New York.

Yu, Y., E. Radisky, and E. A. Leibold. 1992. The IRE binding protein. *J. Biol. Chem.* 267:19005–19010.

Zähringer, J., B. S. Baliga, and H. N. Munro. 1976. Novel mechanism for translational control in regulation of ferritin synthesis by iron. *Proc. Natl. Acad. Sci. U.S.A.* 73:857–861.

Zheng, L., M. C. Kennedy, G. A. Blondin, H. Beinert, and H. Zalkin. 1992a. Binding of cytosolic aconitase to the iron responsive element of porcine mitochondrial aconitase mRNA. *Arch. Biochem. Biophys.* 299:356–360.

Zheng, L., M. C. Kennedy, H. Beinert, and H. Zalkin. 1992b. Mutational analysis of active site residues in pig heart aconitase. *J. Biol. Chem.* 267:7895–7903.

8

Mechanisms for Induction and Rerepression of Ferritin Synthesis

David P. Mascotti, Lisa S. Goessling, Diane Rup, and Robert E. Thach

1. Introduction

An iron responsive element (IRE) located within the 5′ untranslated region (5′ UTR) of certain mRNAs (e.g., ferritin) serves as a binding site for a class of specific binding proteins [referred to as the iron regulatory proteins (IRPs)] which, when bound to an IRE, repress translation of those mRNAs [reviewed in Leibold and Guo (1992), Melefors and Hentze (1993), Klausner et al. (1993), Munro (1993), Theil (1993), Mascotti et al. (1995)]. Binding of an IRP to an IRE located proximally to the 5′ end of an mRNA is believed to prevent access of eIF-4F (the cap binding protein) to the 5′ cap structure, resulting in a blockage of initiation (Goossen et al. 1990; Bhasker et al. 1993; Gray and Hentze 1994). When chelatable iron levels are increased in the cell, an IRP is induced to dissociate from the IRE and allow initiation of translation of ferritin [reviewed in Leibold and Guo (1992), Melefors and Hentze (1993), Klausner et al. (1993), Munro (1993), Theil (1993)]. Messages that contain functional IREs in their 5′ UTRs code for ferritin (Leibold and Guo 1992; Melefors and Hentze 1993; Klausner et al. 1993; Munro 1993; Theil 1993), erythroid δ-aminolevulinic acid synthase (δ-ALAS) (Cox et al. 1991; Bhasker et al. 1993), mitochondrial aconitase (m-acon) (Dandekar et al. 1991), *Drosophila melanogaster* succinate dehydrogenase (Kohler et al. 1995; Gray et al. 1996), and possibly transferrin (Tf) (Cox and Adrian 1993; Cox et al. 1995).

The transferrin receptor mRNA (TfR) contains several functional IREs; however, these IREs are located in the 3′ untranslated region (3′ UTR) (Casey et al.

1988; Mullner and Kuhn 1988). IRPs also bind to these IREs in the absence of iron and dissociate in the presence of added iron. While bound to these IREs, the IRP's confer protection to the TfR mRNA from degradation by nucleases (Casey et al. 1989; Mullner et al. 1989; Koeller et al. 1989). Dissociation of IRP by iron results in rapid degradation of TfR mRNA, which is correlated with an activation of ferritin translation in the same cells. Thus, a coordinately regulated cycle for iron uptake and storage is created [reviewed in Klausner et al. (1993)].

The question of how IRPs sense iron availability has been addressed by several laboratories. One IRP (IRP1) has been established to be a cytosolic aconitase (c-acon) (Kaptain et al. 1991; Kennedy et al. 1992; Haile et al. 1992b; Constable et al. 1992). The mitochondrial version of c-acon (referred to here as m-acon) is a Krebs cycle ferroenzyme that catalyzes the conversion of citrate to isocitrate via a *cis*-aconitate intermediate. Mitochondrial aconitase contains a 4Fe-4S iron-sulfur cluster that is required for aconitase activity. These facts have led to a widely accepted model regarding the iron-sensing mechanism of IRP1, which involves iron in a direct manner. The mechanism assumes that IRP1 contains either zero, three, or four iron atoms per polypeptide monomer. It is the 4Fe-4S form that has been shown to have aconitase activity. When IRP1 contains a fully loaded 4Fe-4S cluster, it is incapable of binding to IREs, presumably because of occlusion of the RNA binding domain of IRP1 resulting from a conformational change (Klausner et al. 1993; Basilion et al. 1994a). The form of IRP1 that binds IREs *in vivo* with high affinity is most likely the apo-IRP1 (Kennedy et al. 1992; Haile et al. 1992a; Klausner et al. 1993; Hirling et al. 1994). It has been shown that the RNA binding site on the IRP1 lies in close proximity to the aconitase active site (Basilion et al. 1994a; W. Walden, personal communication). Several laboratories have elaborated upon the mechanism for assembly and disassembly of the iron-sulfur center, which is central to the function of IRP1 as either an RNA binding protein or an aconitase (Klausner and Rouault 1993; Emery-Goodman et al. 1993; Philpott et al. 1993; Beinert and Kennedy 1993; Drapier et al. 1993; Weiss et al. 1993; Hirling et al. 1994).

A second IRP, IRP2, shares considerable homology with IRP1 except for the presence of a 75-amino-acid peptide region within IRP2 (Guo et al. 1994). The similarity of the amino acid sequences is reflected by similar functions. IRP2 binds to the same wild-type IREs as does IRP1, although its specificity varies slightly (Henderson et al. 1994). The most notable difference discovered thus far is how the two IRPs respond to the presence of iron (Cairo and Pietrangelo 1994; Cairo et al. 1995; Guo et al. 1994; Henderson et al. 1993; Samaniego et al. 1994). Although iron administration down-regulates the IRE binding activity of IRP2, this process occurs by way of a specific decrease in the IRP2 protein level. Recent evidence indicates that, unlike IRP1, IRP2 cannot acquire aconitase activity *in vivo* or *in vitro* (Guo et al. 1994).

2. Alternative Mechanisms of Induction

An alternative mechanism has been proposed to explain the inactivation of IRP1 in the presence of iron. This mechanism is not related to the aconitase homology discussed above; rather, it postulates that heme is a form of iron that can bind to IRP1 and induce dissociation from IREs (Lin et al. 1990a,b, 1991; Battistini et al. 1991; Goessling et al. 1994). An early basis for this mechanism is that optimal induction of ferritin synthesis in response to iron salts is not as great as optimal induction by heme (Hoffman et al. 1980; Eisenstein et al. 1991; Lin et al. 1990a,b, 1991; Battistini et al. 1991). Heme has been subsequently shown to inactivate IRP1 *in vitro* (Lin et al. 1990a,b; Mullner et al. 1992; Haile et al. 1990). Inactivation is probably due to a spontaneous binding of heme to the IRP1, which forms, within a few minutes, a complex that is resistant to boiling in SDS (Lin et al. 1991). Of the variety of metalloporphyrins tested, only heme and Co^{3+}-protoporphyrin IX had significant activity (Lin et al. 1990b; Haile et al. 1990). Under appropriate conditions, the action of heme is highly specific to IRP1 (Lin et al. 1990/91, 1991). Inactivation of IRP by heme or iron salts *in vivo* is essentially irreversible under standard growth conditions (Goessling et al. 1994). This salient observation is important in that it applies to the iron-dependent (aconitase-based) mechanism as well. Therefore, the evidence for irreversibility should be considered in more detail.

Evidence for irreversibility comes from several laboratories. The first suggestion of irreversibility was the finding that inhibitors of RNA or protein synthesis, such as actinomycin D, cordycepin, and cycloheximide cause a slow induction of ferritin synthesis (Rogers and Munro 1987; Daniels-McQueen et al. 1992; Goessling et al. 1992; Kuhn, personal communication). That so many disparate agents induce the same effect argues for a general mechanism. A likely explanation for all these observations is that continuous protein synthesis is necessary to maintain high levels of active IRP1. A corollary would be that the presence of iron enhances the turnover of IRP1. Direct evidence for the degradation of IRP1 in the presence of iron was obtained by immune precipitation studies using antibodies against IRP1 (Goessling et al. 1992). This observation has been more recently confirmed both by Western analysis and by RNA binding activity measurement (Goessling et al. 1994). A very recent reinvestigation of this iron/heme-dependent degradation phenomenon has revealed that it is more readily observed in: a) cells that have not been carried in culture for more than a year, and b) cells that have been pulse-labeled with ^{35}S-Met in the absence of serum, followed by immune precipitation analysis of labeled protein.

Attempts to identify intermediates in the degratory pathway of IRP1 have been only partially successful. The earliest intermediates detectable are high-molecular-weight species (HMS) of immune precipitable proteins which migrate in SDS-

PAGE at between 200–400 kDa. The inability to remove these complexes from SDS-PAGE gel matrix, as is required for western analysis, suggests that the HMS are multiply branched or cross-linked protein aggregates rather than simple end-to-end dimers or trimers. Although HMS have also been detected by gel exclusion chromatography (Hu et al., unpublished), further analysis has been difficult. It may be significant that heme readily induces IRP1 oligomerization *in vitro* (Lin et al. 1991). Heme is a stable component of these structures *in vitro* (Lin et al. 1991). Although the interaction of heme with IRP1 is enhanced by moderately reducing conditions, formation of these oligomers is enhanced by the presence of oxidizing agents (Lin et al. 1991). The conditions favoring heme-IRP1 binding correlate well with the observation that the heme-induced derepression of ferritin synthesis *in vitro* is enhanced by reducing agents (Lin et al. 1990a,b). By contrast, although tight heme binding to monomeric IRP1 also occurs *in vivo*, its presence is greatly diminished in HMS (Figure 8.1). The reason for the difference in *in vitro* and *in vivo* results is not clear.

The conversion of monomeric IRP1 to HMS *in vivo* is rapid, being detectable after only 15 minutes of exposure to heme. By one hour of exposure, this process can be virtually complete (at 50–100-μM heme concentrations). Interestingly, little, if any, ferritin synthesis is detectable during this time, raising the question of whether the HMS may still be active as a repressor of ferritin translation; however, until the molar ratio of IRP1 to mRNA is known more precisely, such a correlation cannot be drawn. (It may be that IRP1 bound to mRNA reacts with heme more slowly than unbound IRP1.)

Surprisingly, HMS formation is readily reversible. Simply washing heme out of cell culture media causes HMS to revert to the monomeric form (Goessling et al. 1994). This occurs within one hour, and the resulting monomer is stable and active. The ready reversibility of this step stands in marked contrast to the irreversibility of the induction process as a whole. It suggests that some subsequent reaction involving HMS is irreversible. Inasmuch as HMS slowly disappears in the prolonged presence of heme, it was concluded that this irreversible step was probably the proteolytic degradation of HMS. However, efforts to detect stable intermediates have not been entirely satisfactory.

We have considered the possibility that the formation and subsequent degradation of HMS may be ubiquitin-related phenomena. To this end, studies using antibodies specific for ubiquitin *per se*, or for polyubiquitinated proteins, were conducted. It has been shown that proteosome-dependent degradation of other proteins can occur independently of ubiquitination (Murakami et al. 1992). However, in spite of great effort, no conclusive results could be drawn due to technical difficulties (unpublished observations).

It is important to note that the formation and subsequent disappearance of HMS is also inducible by iron salts (Goessling et al. 1994). This effect is enhanced by δ-aminolevulinic acid (ALA), a precursor in heme synthesis. This effect of

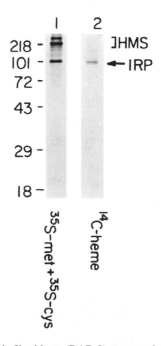

Figure 8.1 For lane 1, rabbit fibroblasts (RAB-9) were prelabeled with 200 µCi/ml ^{35}S-methionine and ^{35}S-cysteine (Trans-^{35}S; ICN) in MEM minus Met and Cys for 2 hours. This labeling medium was then replaced with Earle's MEM + 2% FCS containing hemin (42.6 µM) for 15 minutes. For lane 2, RAB-9 cells were labeled with 2 µCi/ml ^{14}C-hemin (42.6 µM; University of Leeds, U.K.) in Earle's MEM supplemented with 2% FCS for 15 minutes. All cells were then washed twice with phosphate buffered saline (PBS) and lysed by the addition of immune buffer (PBS with 1% deoxycholate, 1% triton X-100, 0.1% SDS, and 1mM EDTA). Radiolabeled IRP1 was detected by immunoprecipitation of the lysates [pretreated with 1.1% SDS at 100°C for 5 minutes then diluted 5-fold with immune buffer that contained bovine serum albumin (13 mg/ml) and no SDS] with rat antirabbit IRP1 serum, SDS-PAGE, and fluorography. The relative migration of IRP1 (labeled as IRP in the figure) and high-molecular-weight species (HMS) are indicated.

ALA is inhibited by succinyl acetone (SA), which blocks heme synthesis specifically. Figure 8.2 shows an example of these effects. These observations suggest that iron is converted to heme *in vivo* in order to stimulate these two reactions (Goessling et al. 1992). Although no experiments have been performed to detect binding of heme to IRP2, recent experiments show that incubation of cells with 7.5 mM SA inhibits the iron-mediated degradation of IRP2 when iron salts are used as the iron source (manuscript in preparation). The degradation of IRP2 is known to occur via the ubiquitin-proteosome pathway (Guo et al. 1995), so

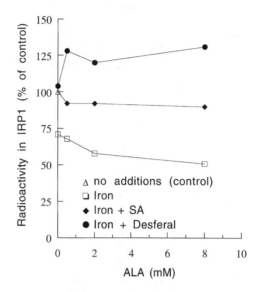

Figure 8.2 Effect of ALA on IRP-1 degradation. RAB-9 cells were radiolabeled with 200 μCi/ml ³⁵S-methionine and ³⁵S-cysteine (Tran-³⁵S; ICN) in Earle's Minimal Essential Medium (MEM) minus Met and Cys for 2 hours. Cells were then chased with excess (1 mM) unlabeled methionine for 6 hours in the presence of Earle's MEM + 2% fetal calf serum (FCS), Δ; or 100 μM ferric ammonium citrate + 0.2 mg/ml transferrin (iron), ☐; or iron + 7.5 mM succinyl acetone (SA), ◆; or iron + 200 μM Desferal, ●; all with varying concentrations of δ-aminolevulinic acid (ALA). Labeled IRP1 from lysed cells was immune precipitated with rat antirabbit IRP1 antibody and analyzed by SDS-PAGE and fluorography. Radioactivity remaining in IRP1 was determined by densitometry, and is expressed as a percent of the control (no additions) on the ordinate.

this offers circumstantial support for the mechanism of heme-induced IRP1 degradation postulated above.

3. Coexistence of Two Major Pathways?

The preceding results raise major questions regarding the interrelationship of the two induction mechanisms proposed. In theory, the aconitase and heme models of IRP inactivation could be complementary (Mullner et al. 1992). Insights into this problem were recently gained when it was observed that in quiescent cells, the heme-dependent pathway was greatly diminished relative to that seen in growing cells (Goessling et al. 1994). For example, not only was induction of ferritin synthesis by heme reduced, but the stability of IRP1 in the presence of heme was enhanced. By contrast, induction of ferritin synthesis by iron salts in quiescent cells was enhanced relative to that in rapidly growing cells. These results strongly argue that whereas in growing cells the heme-dependent mechanism is a significant pathway, in quiescent cells the primary response to chelatable iron is via the aconitase pathway. As disparate as these two pathways seem, they have interesting features in common. These similarities are evident in Figure 8.3. It may be of relevance to note that under conditions of rapid growth, IRP1, but not IRP2, is phosphorylated and increases RNA binding activity during that time (Eisenstein et al. 1993). It is tempting to speculate that the phosphorylated form of IRP1 might respond differently to iron or heme, although further experiments are needed to answer this question.

Pathway A in Figure 8.3 describes the iron-dependent formation of c-acon that has been elucidated in impressive detail by others (Klausner et al. 1993; Klausner and Rouault 1993; Hirling et al. 1994; Basilion et al. 1994b). This pathway is apparently reversible only when special circumstances prevail, such as in the presence of nitric oxide (NO) (Drapier et al. 1993; Weiss et al. 1993) or iron chelating agents (Tang et al. 1992; Goessling et al. 1994). Little is known about the mechanism whereby c-acon is turned over. Indeed, the necessity of a c-acon turnover step is inferred by the irreversibility of the induction process rather than by its having been directly demonstrated. The formation of c-acon from IRP1 may be slow inasmuch as it is thought to be concurrent with the induction of ferritin synthesis. [The latter process begins several hours after iron or heme addition (Goessling et al. 1994)].

Pathway B in Figure 8.3 summarizes the heme-dependent events discussed above. Although this pathway has formal similarity to pathway A, there are striking differences that suggest that the two are not directly related. The most important differences are the rapidity with which heme-dependent alterations in IRP1 can be observed (within 15 minutes), the reversibility of those alterations at early times, and the lack of a correlation in time between these events and the induction of ferritin synthesis. We cannot rule out the possibility that the heme dependence of the IRP1 degradation step in pathway B may be identical to that in pathway A (i.e., they both might be mediated by heme).

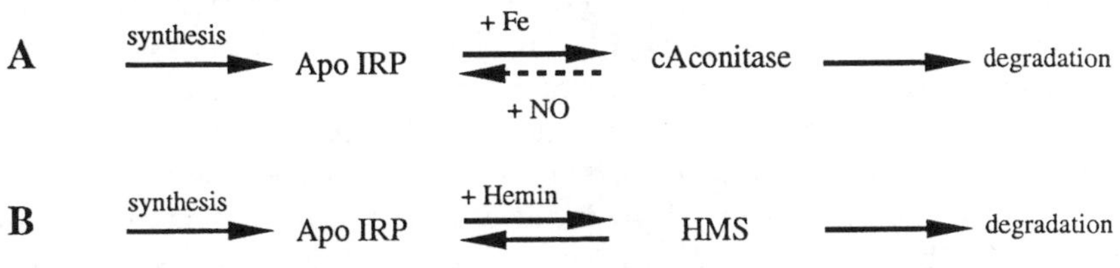

Figure 8.3 Two proposed mechanisms for the induction of the iron regulatory protein. Details are given in the text.

The simultaneous existence of two major induction modes in growing cells can explain many discrepancies found in the literature. Most notably, the variously reported effects of metabolites and inhibitors that influence porphyrin biosynthesis can be rationalized in terms of a dual pathway model. A notable example is the fact that the porphyrin precursor ALA can either inhibit (Eisenstein et al. 1991) or stimulate (Goessling et al. 1992) the iron induction of ferritin synthesis. Which of these two effects is observed may depend on a complex set of conditions. Similarly, it has been reported that the heme synthesis inhibitor SA can either slightly enhance (Eisenstein et al. 1991) or strongly inhibit (Goessling et al. 1992) the iron-dependent induction. (SA had no effect on heme-dependent induction.) This bewildering diversity of effects is entirely consistent with the simultaneous existence of two different pathways, one of which is affected directly by iron, the other by heme.

4. Biological Significance of the Heme-Dependent Pathway

Why two induction pathways should coexist in the same cell is not immediately obvious. One advantage for having a heme-induced mechanism is that this would shorten the overall processing time of excessive amounts of intracellular heme by ~50%. This is due to the fact that the induction of ferritin synthesis by iron is slow, requiring 4–6 hours to attain a maximum rate. If heme can also act as an inducer, then the 3–4 hours required for generation of iron from heme (through action of induced heme oxygenase, HO) can be eliminated. Another advantage to a heme-induction mechanism is that cellular uptake of iron via the transferrin pathway could be rapidly shut down soon after the influx of heme, and before elaboration of the HO pathway. The advantages gained by such shortened response times may not be trivial, especially when the presence of oxygen would enhance the production of free radicals through the Fenton reaction (Rice-Evans and Burdon 1994). Indeed, in view of the high toxicity of iron and heme, an excess of these substances might well be expected to elicit a stress response. The fact that HO is one of the most common stress proteins is consistent with this view. Of course, the implication that iron and iron compounds may be viewed both as essential metabolites and as stressors does not absolutely require that more than one inductive mechanism should exist. Nevertheless, the duality of roles for iron makes it plausible that multiple levels of control may have evolved.

An interesting comparison can be made with zinc and cadmium metabolism. The primary mechanism of storage and/or detoxification of these metals is through chelation by metallothionein. However, a secondary detoxification mechanism has recently been identified (Jungmann et al. 1993). This mechanism involves one or more protein degradative steps, as inferred by the utilization of a special ubiquitin-conjugating enzyme in the detoxification process. Thus the two pathways of cadmium metabolism share striking features with the two iron-dependent pathways shown in Figure 8.3. However, one important difference is that whereas the IRPs are clearly regulatory proteins, metallothionein is not. Nevertheless, as the target(s) of the cadmium-induced degradation is not known, it is still a possibility that one or more regulatory proteins may be preferentially degraded in the presence of cadmium. This notion of a stress response being mediated by the preferential degradation of a regulatory protein (which normally functions to keep the stress response pathway in the "off" mode) is inherently satisfying. An especially pleasing aspect is the ease with which specific stress-response mechanisms could evolve, especially under conditions of repeated challenge by high levels of the stressing agent.

These considerations help explain why both heme and iron have been observed to act as primary inducers *in vivo*. For those cases where iron is present at a low level, its toxicity is minimal, and so it could easily be processed by a slow mechanism suitable for the handling of metabolites. By contrast, in cases where heme (which is much more active than iron in Fenton chemistry) is present at high concentration, the rapid mechanism, which is more closely identified with stress responses, would be expected to predominate (Balla et al. 1992, 1993). According to this rationale, then, it is not at all surprising to see iron taken up and stored by liver tissue without concomitant degradation of IRP1 (Yu et al. 1992). By contrast, it is not surprising to see evidence of IRP1 degradation in brain tissue that has been exposed to lysed erythrocytes that generate free heme (Koeppen et al. 1992, 1993). The proposal by Eisenstein and Munro (1990/91) that macrophages and other reticuloendothelial cells might also employ the heme-dependent degradative pathway is consistent with this hypothesis.

References

Balla, G., H. S. Jacob, J. Balla, M. Rosenberg, K. Nath, F. Apple, J. W. Eaton, and G. M. Vercellotti. 1992. Ferritin: A cytoprotective antioxidant strategem of endothelium. *J. Biol. Chem.* 267:18148–18153.

Balla, J., H. Jacob, G. Balla, K. Nath, J. W. Eaton, and G. M. Vercellotti. 1993. Endothelial-cell heme uptake from heme proteins: Induction of sensitization and desensitization to oxidant damage. *Proc. Natl. Acad. Sci.* 90:9285–9289.

Basilion, J. P., T. A. Rouault, C. M. Massinople, R. D. Klausner, and W. H. Burgess. 1994a. The iron-responsive element-binding protein: Localization of the RNA-binding site to the aconitase active-site cleft. *Proc. Natl. Acad. Sci.* 91:574–578.

Basilion, J. P., M. C. Kennedy, H. Beinert, C. M. Massinople, R. D. Klausner, and T. A. Rouault. 1994b. Overexpression of iron-responsive element-binding protein and its analytical characterization as the RNA-binding form, devoid of an iron-sulfur cluster. *Arch. Biochem. Biophys.* 311:517–522.

Battistini, A. B., E-M. Coccia, G. Marziali, D. Bulgarini, S. Scalzo, G. Fiorucci, G.

Romeo, E. Affabris, U. Testa, G. B. Rossi, and C. Peschle. 1991. Intracellular heme coordinately modulates globin chain synthesis, transferrin receptor number, and ferritin content in differentiating friend erythroleukemia cells. *Blood* 78:2098–2103.

Beinert, H., and M. C. Kennedy. 1993. Aconitase, a two-faced protein: Enzyme and iron regulatory factor. *FASEB J.* 7:1442–1449.

Bhasker, C. R., G. Burgiel, B. Neupert, A. Emery-Goodman, L. C. Kuhn, and B. K. May. 1993. The putative iron-responsive element in the human erythroid 5-aminolevulinate synthase mRNA mediates translational control. *J. Biol. Chem.* 268:12699–12705.

Cairo, G. and A. Pietrangelo. 1994. Transferrin receptor gene expression during rat liver regeneration: Evidence for post-translational regulation by iron regulatory factor$_B$, a second iron-responsive element-binding protein. *J. Biol. Chem.* 269:6405–6409.

Cairo, G., L. Tacchini, G. Pogliaghi, E. Anzon, A. Tomasi and A. Bernelli-Zazzera. 1995. Induction of ferritin synthesis by oxidative stress: Transcriptional and post-transcriptional regulation by expansion of the "free" iron pool. *J. Biol. Chem.* 270:700–703.

Casey, J. L., M. W. Hentze, D. M. Koeller, S. W. Caughman, T. A. Rouault, R. D. Klausner, and J. B. Harford. 1988. Iron responsive elements: Regulatory RNA sequences that control mRNA levels and translation. *Science* 240:924–928.

Casey, J. L., D. M. Koeller, V. C. Ramin, R. D. Klausner, and J. B. Harford. 1989. Iron regulation of transferrin receptor mRNA levels requires iron-responsive elements and a rapid turnover determinant in the 3′ untranslated region of the mRNA. *EMBO J.* 8:3693–3699.

Constable, A., S. Quick, N. K. Gray, and M. W. Hentze. 1992. Modulation of the RNA-binding activity of a regulatory protein by iron in vitro: Switching between enzymatic and genetic function? *Proc. Natl. Acad. Sci.* 89:4554–4558.

Cox, L. A., and G. S. Adrian. 1993. Posttranscriptional regulation of chimeric human transferrin genes by iron. *Biochem.* 32:4738–4745.

Cox, L. A., M. C. Kennedy, and G. S. Adrian. 1995. The 5′-untranslated region of human transferrin mRNA, which contains a putative iron-regulatory element, is bound by purified iron-regulatory protein in a sequence-specific manner. *Biochem. Biophys. Res. Comm.* 212:925–932.

Cox, T. C., M. J. Bawden, A. Martin, and B. K. May. 1991. Human erythroid 5-aminolevulinate synthase: Promoter analysis and identification of an iron-responsive element in the mRNA. *EMBO J.* 10:1891–1902.

Dandekar, T., R. Stripecke, N. K. Gray, B. Goossen, A. Constable, H. E. Johansson, and M. W. Hentze. 1991. Identification of a novel iron-responsive element in murine and human erythroid δ-aminolevulinic acid synthase mRNA. *EMBO J.* 10:1903–1909.

Daniels-McQueen, S., L. S. Goessling, and R. E. Thach. 1992. Inducible expression bovine papilloma virus shuttle vectors containing ferritin translational regulatory elements. *Gene* 122:271–279.

Drapier, J-C., H. Hirling, J. Wietzerbin, P. Kaldy, and L. C. Kuhn. 1993. Biosynthesis of nitric oxide activates iron regulatory factor in macrophages. *EMBO J.* 12:3643–3649.

Eisenstein, R. S., D. Garcia-Mayol, W. Pettingell and H. N. Munro. 1991. Regulation of ferritin and heme oxygenase synthesis in rat fibroblasts by different forms of iron. *Proc. Natl. Acad. Sci.* 88:688–692.

Eisenstein, R. S., and H. N. Munro. 1990/91. Translational regulation of ferritin synthesis by iron. *Enzyme* 44:42–58.

Eisenstein, R. S., P. T. Tuazon, K. L. Schalinske, S. A. Anderson, and J. A. Traugh. 1993. Iron-responsive element-binding protein: Phosphorylation by protein kinase C. *J. Biol. Chem.* 268:27363–27370.

Emery-Goodman, A., H. Hirling, L. Scarpellino, B. Henderson, and L. C. Kuhn. 1993. Iron regulatory factor expressed from recombinant baculovirus: Conversion between the RNA-binding apoprotein and Fe-S cluster containing aconitase. *Nucl. Acid Res.* 21:1457–1461.

Goessling, L. S., S. Daniels-McQueen, M. Bhattacharyya-Pakrasi, J.-J. Lin, and R. E. Thach. 1992. Enhanced degradation of the ferritin repressor protein during induction of ferritin messenger RNA translation. *Science* 256:670–673.

Goessling, L. S., D. P. Mascotti, M. Bhattacharyya-Pakrasi, H. Gang, and R. E. Thach. 1994. Irreversible steps in the ferritin synthesis induction pathway. *J. Biol. Chem.* 269:4343–4348.

Goossen, B., S. W. Caughman, J. B. Harford, R. D. Klausner and M. W. Hentze. 1990. Translational repression by a complex between the iron-responsive element of ferritin mRNA and its specific cytoplasmic binding protein is position-dependent *in vivo*. *EMBO J.* 9:4127–4133.

Gray, N. K., and M. W. Hentze. 1994. Iron regulatory protein prevents binding of the 43S translation pre-initiation complex to ferritin and eALAS mRNAs. *EMBO J.* 13:3882–3891.

Gray, N. K., K. Pantopoulos, T. Dandekar, B. A. Ackrell, and M. W. Hentze. 1996. Translational regulation of mammalian and *Drosophila* citric acid cycle enzymes via iron-responsive elements. *Proc. Natl. Acad. Sci.* 93:4925–4930.

Guo, B., J. D. Phillips, Y. Yu, and E. A. Leibold. 1995. Iron regulates the intracellular degradation of iron regulatory protein 2 by the proteasome. *J. Biol. Chem.* 270:21645–21651.

Guo, B., Y. Yu, and E. A. Leibold. 1994. Iron regulates cytoplasmic levels of a novel iron-responsive element-binding protein without aconitase activity. *J. Biol. Chem.* 269:24252–24260.

Haile, D. J., T. A. Rouault, J. B. Harford, and R. D. Klausner. 1990. The inhibition of the iron responsive element RNA-protein interaction by heme does not mimic in vivo iron regulation. *J. Biol. Chem.* 265:12786–12789.

Haile, D. J., T. A. Rouault, J. B. Harford, M. C. Kennedy, G. A. Blondin, H. Beinert, and R. D. Klausner. 1992a. Cellular regulation of the iron-responsive element binding protein: Disassembly of the cubane iron-sulfur cluster results in high-affinity RNA binding. *Proc. Natl. Acad. Sci.* 89:11735–11739.

Haile, D. J., T. A. Rouault, C. K. Tang, J. Chin, J. B. Harford, and R. D. Klausner. 1992b.

Reciprocal control of RNA binding and aconitase activity in the regulation of the iron responsive element binding protein: Role of the iron-sulfur cluster. *Proc. Natl. Acad. Sci.* 89:7536–7540.

Henderson, B. R., C. Seiser, and L. C. Kuhn. 1993. Characterization of a second RNA-binding protein in rodents with specificity for iron-responsive elements. *J. Biol. Chem.* 268:27327–27334.

Henderson, B. R., E. Menotti, C. Bonnard, and L. C. Kuhn. 1994. Optimal sequence and structure of iron-responsive elements: Selection of RNA stem loops with high affinity for iron regulatory factor. *J. Biol. Chem.* 269:17481–17489.

Hirling, H., B. R. Henderson, and L. C. Kuhn. 1994. Mutational analysis of the [4Fe-4S]-cluster converting iron regulatory factor from its RNA-binding form to cytoplasmic aconitase. *EMBO J.* 13:453–461.

Hoffman, R., N. Ibrahim, M. J. Murnane, A. Diamond, B. G. Forget, and R. D. Levere. 1980. Hemin control of heme biosynthesis and catabolism in a human leukemia cell line. *Blood* 56:567–570.

Jungmann, J., H. A. Reins, C. Schobert, and S. Jentsch. 1993. Resistance to cadmium mediated by ubiquitin-dependent proteolysis. *Nature* 361:369–371.

Kaptain, S., W. E. Downey, C. Tang, C. Philpott, D. Haile, D. G. Orloff, J. B. Harford, T. A. Rouault, and R. D. Klausner. 1991. A regulated RNA binding protein also possesses aconitase activity. *Proc. Natl. Acad. Sci.* 88:10109–10113.

Kennedy, M. C., L. Mende-Mueller, G. A. Blondin, and H. Beinert. 1992. Purification and characterization of cytosolic aconitase from beef liver and its relationship to the iron-responsive element binding protein. *Proc. Natl. Acad. Sci.* 89:11730–11734.

Klausner, R. D., and T. A. Rouault. 1993. A double life: Cytosolic aconitase as a regulatory RNA binding protein. *Molec. Biol.* 4:1–5.

Klausner, R. D., T. A. Rouault, and J. B. Harford. 1993. Regulating the fate of mRNA: The control of cellular iron metabolism. *Cell* 72:19–28.

Koeller, D. M., J. L. Casey, M. W. Hentze, E. M. Gerhaardt, L-N. L. Chan, R. D. Klausner, and J. B. Harford. 1989. A cytosolic protein binds to structural elements within the iron regulatory region of the transferrin receptor mRNA. *Proc. Natl. Acad. Sci.* 86:3574–3578.

Koeppen, A. H., A. C. Dickson, R. C. Chu, and R. E. Thach. 1993. The pathogenesis of superficial siderosis of the central nervous system. *Ann. Neurol.* 34:646–653.

Koeppen, A. H., C. G. Hurwitz, R. E. Dearborn, A. C. Dickson, R. C. Borke, and R. C. Chu. 1992. Experimental superficial siderosis of the central nervous system: Biochemical correlates. *J. Neurol. Sci.* 112:38–45.

Kohler, S. A., B. R. Henderson, and L. C. Kuhn. 1995. Succinate dehydrogenase b mRNA of *Drosophila melanogaster* has a functional iron-responsive element in its 5'-untranslated region. *J. Biol. Chem.* 270:30781–30786.

Leibold, E. A., and B. Guo. 1992. Iron-dependent regulation of ferritin and transferrin receptor expression by the iron-responsive element binding protein. *Ann. Rev. Nutr.* 12:345–368.

Lin, J-J., S. Daniels-McQueen, L. Gaffield, M. M. Patino, W. E. Walden, and R. E. Thach. 1990b. Specificity of the induction of ferritin synthesis by hemin. *Biochim. Biophys. Acta* 1050:146–150.

Lin, J-J., S. Daniels-McQueen, M. M. Patino, L. Gaffield, W. E. Walden, and R. E. Thach. 1990a. Derepression of ferritin messenger RNA translation by hemin *in vitro.* Science 247:74–77.

Lin, J-J., M. M. Patino, L. Gaffield, W. E. Walden, A. Smith, and R. E. Thach. 1991. Crosslinking of hemin to a specific site on the 90-kDa ferritin repressor protein. *Proc. Natl. Acad. Sci.* 88:6068–6071.

Lin, J-J., W. E. Walden, and R. E. Thach. 1990/91. Induction of ferritin synthesis in vitro: Problems of specificity inherent in the use of iron compounds. *Enzyme* 44:59–67.

Mascotti, D. P., D. Rup, and R. E. Thach. 1995. Regulation of iron metabolism: Translational effects mediated by iron, heme and cytokines. *Ann. Rev. Nutr.* 15:239–261.

Melefors, O., and M. W. Hentze. 1993. Translational regulation by mRNA/protein interactions in eukaryotic cells: Ferritin and beyond. *Bioessays* 15:85–90.

Mullner, E. W., and L. C. Kuhn. 1988. A stem-loop in the 3′ untranslated region mediates iron-dependent regulation of transferrin receptor mRNA stability in the cytoplasm. *Cell* 53:815–825.

Mullner, E. W., B. Neupert, and L. C. Kuhn. 1989. A specific mRNA binding factor regulates the iron-dependent stability of cytoplasmic transferrin receptor mRNA. *Cell* 58:373–382.

Mullner, E. W., S. Rothenberger, A. M. Muller, and L. C. Kuhn. 1992. In vivo and in vitro modulation of the mRNA-binding activity of iron-regulatory factor: Tissue distribution and effects of cell proliferation, iron levels and redox state. *Eur. J. Biochem.* 208:597–605.

Munro, H. 1993. The ferritin genes: Their response to iron status. *Nutr. Rev.* 51:65–73.

Murakami, Y., S. Matsufuji, T. Kameji, S.-I. Hayashi, K. Igarashi, T. Tamura, K. Tanaka, and A. Ichihara. 1992. Ornithine decarboxylase is degraded by the 26S proteasome without ubiquitation. *Nature* 360:597–599.

Philpott, C. C., D. Haile, T. A. Rouault, and R. D. Klausner. 1993. Modification of a free Fe-S cluster cysteine residue in the active iron-responsive element-binding protein prevents RNA binding. *J. Biol. Chem.* 268:17655–17658.

Rice-Evans, C. A., and R. H. Burdon. 1994. In *Free Radical Damage and its Control.* Elsevier, New York.

Rogers, J. T., and H. N. Munro. 1987. Translation of ferritin light and heavy subunit mRNAs is regulated by intracellular chelatable iron levels in rat hepatoma cells. *Proc. Natl. Acad. Sci.* 84:2277–2281.

Samaniego, F., J. Chin, K. Iwai, T. A. Rouault, and R. D. Klausner. 1994. Molecular characterization of a second iron-responsive element binding protein, iron regulatory protein 2: Structure, function, and post-translational regulation. *J. Biol. Chem.* 269:30904–30910.

Tang, C. K., J. Chin, J. B. Harford, R. D. Klausner, and T. A. Rouault. 1992. Iron regulates the activity of the iron-responsive element binding protein without changing its rate of synthesis or degradation. *J. Biol. Chem.* 267:24466–24470.

Theil, E. C. 1993. The IRE (iron regulatory element) family: Structures which regulate mRNA translation or stability. *Biofactors* 4:87–93.

Weiss, G., B. Goossen, W. Doppler, D. Fuchs, K. Pantopoulos, G. Werner-Felmayer, H. Wachter, and M. W. Hentze. 1993. Translational regulation via iron-responsive elements by the nitric oxide/NO-synthase pathway. *EMBO J.* 12:3651–3657.

Yu, Y., E. Radisky, and E. A. Leibold. 1992. The iron-responsive element binding protein. *J. Biol. Chem.* 267:19005–19010.

9

Metallothionein Gene Regulation in Mouse Cells

Simon Labbé, Carl Simard, and Carl Séguin

1. Abstract

Heavy metals (Cd, Cu, Zn, etc.) can affect the expression of many genes. The best-known proteins that bind these metal ions are the metallothioneins (MTs). The genes encoding MTs are inducible at the transcriptional level by the same metal ions that the MTs bind. Metal activation of *MT* gene transcription is dependent on the presence of *cis*-acting DNA elements termed Metal Response Elements (MREs), and involves *trans*-acting protein (factor(s) interacting with the MREs, present in six nonidentical copies (MREa through MREf) in the 5′ flanking region of the mouse *MT-I* gene. Different MREs have different transcriptional efficiencies, MREd being the strongest. *In vitro,* footprinting analyses have revealed that one or more nuclear factors can bind to the different MRE elements of the mouse *MT-1* gene. Moreover, the MREd binding activity is inactivated by EDTA and can be restored by addition of Zn^{2+}. Using a Southwestern procedure, we found that a nuclear protein of 108 kDa, termed MEP-1, specifically binds to the different MRE elements of the mouse *MT-I* gene promoter. MEP-1 has been purified, and footprinting studies demonstrated that purified MEP-1 specifically binds to MRE sequences. MEP-1 binding activity is also inhibited by EDTA and can be restored by Zn^{2+}.

Although we have shown that purified MEP-1 is sufficient to generate a footprint over the MREs of the mouse *MT-I* promoter, it is as yet unclear how many classes of nuclear proteins interact with the MRE elements. With the aim of identifying other MRE binding proteins, we developed a purification strategy involving a combination of low-pressure heparin-Sepharose with salt gradient elution, and MRE-DNA affinity chromatography. Using this procedure, we detected, in addition to MEP-1, another MRE-binding protein, named MEP-2, specifically interacting with the MREc element. The two proteins can be distinguished on the basis of their elution profiles, and their respective affinities towards different MRE competitor oligonucleotides, as assayed by DNaseI footprinting. These results show that at least two different nuclear proteins interact with the MRE elements of the mouse *MT-I* gene. The role of these two proteins in the regulation of *MT* gene transcription and the relationship between them or with the other MRE binding proteins reported in the literature remain to be determined.

2. Introduction

Ions of heavy metals such as Cd^{2+}, Cu^{2+}, and Zn^{2+} can affect the intracellular homeostasis. The best-known proteins that bind these metal ions are the metallothioneins (MTs), a class of small cysteine-rich proteins present in many different tissues and cell types. MTs are thought to function in heavy metal detoxification and metabolism. Elimination of *MT-I* and *MT-II* genes by targeted gene disruption renders null mutant mice highly sensitive to cadmium intoxication (Michalska and Choo 1993; Masters et al. 1994). The transcription of a variety of genes responds dramatically to changes in the concentrations of specific metals. For instance, metal ions rapidly stimulate transcription of the genes encoding heat shock proteins, superoxide dismutase, heme oxygenase, several acute phase proteins, and also some *proto*-oncogene products such as c-*jun*, c-*myc*, and c-*fos* [reviewed in Thiele (1992)]. However, the *MT* genes have been the most intensively studied and best-understood examples of metal-regulated genes. Thus, they provide a useful model system for understanding how a eukaryotic gene modulates its expression in response to metal ions, and for characterizing the signal transduction pathway involved in this activation.

All vertebrates examined contain two or more distinct MT isoforms, which are grouped into four classes, MT-I through MT-IV [reviewed in Andrews (1990); Suzuki et al. (1993)]. In many cases, each class consists of several different isoproteins (designated MT-IA, MT-IB, etc.). In mammals, MTs are polypeptides of 61 or 62 amino acids including 20 cysteine (Cys) residues arranged in typical Cys-x-Cys, Cys-x-x-Cys (x: any amino acid), and Cys-Cys motifs that form thiolate complexes with metal ions. MTs have been identified in a wide range of species and are present in various tissues and cell types. MTs are inducible at the transcriptional level by a wide variety of agents such as Cd^{2+}, Zn^{2+}, and Cu^{2+}, hormones, cytokines, alcohols, herbicides, and by a number of stress agents, including UV irradiation, heat and cold exposure, oxidative stress, or tissue injury resulting from exposure to turpentine, carbon tetrachloride, or bacterial endotoxin. Metals are the most general and potent of these inducers.

The ability of vertebrate *MT* genes to be induced by metals is controlled in *cis* by a short DNA sequence (metal regulatory element or MRE) present in multiple imperfect copies in the 5' flanking region of the *MT* genes [reviewed in Hamer (1986)]. Figure 9.1 shows the arrangement of the six MRE elements on the mouse *MT-I* gene, the G-rich sequence that interacts with the major late transcription factor MLTF (Carthew et al. 1987), the antioxidant response element (ARE) (Dalton et al. 1994), and the two Spl (Mueller et al. 1988) sites. Different MREs have different transcriptional efficiencies. MREd is the strongest, MREa and MREc are 50 to 80% weaker, MREb is very weak, and MREe and MREf are apparently nonfunctional (Stuart et al. 1985; Searle 1990). Detailed point mutation analysis of mouse *MT-I* MREa and MREd shows that the highly conserved core sequence, 5'-TGCRCNC-3' (R, purine; N, any nucleotide), is crucial for induction by metals [reviewed in Imbert et al. (1990)]. At least two MREs are required for efficient metal induction, and these elements can be present in different orientation.

It has been suggested that the ability of MREs to modulate transcription in response to metals depends on the ability of a metalloregulatory transcription factor(s) (MRTF) to bind to the sequence in the presence of metals and induce transcription [reviewed in Thiele (1992); Heuchel et al. (1995)]. A critical step in the characterization of the signal transduction pathway by which metal ions regulate gene expression is to isolate a true MRTF regulating metal responsive gene expression. The molecular cloning and characterization of a nuclear regulatory factor that can act as a metal-responsive switch able to sense and translate inorganic signals into changes in metabolism is central to defining the signal transduction pathway mediating metal-regulated transcription.

3. *In Vitro* Footprinting Studies

Footprinting experiments have shown that nuclear proteins can bind to MREs and support the model in which the elevation of metal ion concentrations triggers

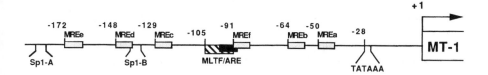

Figure 9.1 Schematic representation of the mouse *MT-I* gene promoter. Arrangement of the six metal regulatory elements (MRE; arrows), the binding sites for the transcription factor Spl, the G-rich sequence interacting with the transcription factor MLTF, the antioxidant response element (ARE), and the TATA box are shown.

the binding of a metal-dependent factor to MREs. For example, in rats, *in vivo* footprinting data suggested that MRTF binds to MRE sites in the presence of Cd^{2+} ions, and that this factor dissociates from the gene upon removal of metal ions from the cells (Andersen et al. 1987). In mouse cells, increased dimethylsulfate protection is dependent on Zn^{2+} (or Cd^{2+}) induction at all MREs except MREd, where metal treatment enhances a preexisting protection (Mueller et al. 1998). *In vitro,* DNA binding assays indicate that mouse nuclear factors can bind to MREs. Using nuclear extracts prepared from heavy-metal-resistant L cells (L50 cells), we found that the region of the *MT-I* gene promoter between positions −173 [relative to the transcriptional start site (tsp)] and +68 contains many binding sites for nuclear proteins, as assayed by DNaseI footprinting (Figure 9.2).

Footprints of different intensities were present on all five MRE elements; the stronger protections were observed over MREd, MREc, and MREa, whereas those on MREb and MREe were weaker or partial. Protections were also present over the G/C-rich MLTF binding site, the TATA box, and the tsp. In addition, using an exonucleaseIII (ExoIII) footprinting assay, we have shown that one or more nuclear factors, present in extracts from L or L50 cells bind to the mouse MREd (Figure 9.3, lane U, and Figure 9.4) (Séguin and Hamer 1987; Séguin 1991). The addition of exogenous $CdCl_2$ to the buffers of noninduced-cell extracts slightly increased the signal (Figure 9.4, compare lanes 7–9 with 10–12), but the chelating agents EDTA (Figure 9.3, lane 0, and Figure 9.4, lane 2) and 1,10-phenanthroline (Séguin 1991) selectively inhibited the binding of this protein to the MRE. Binding activity could be restored by Zn^{2+}, but none of the other cations tested (Cd^{2+}, Cu^{2+}, Mn^{2+}, Mg^{2+}, Ca^{2+}) restored binding activity to the treated extracts (Figure 9.3). This shows that Zn^{2+} ions are required for specific *in vitro* DNA binding of the MREd binding protein and suggests that the MREd binding component is a metalloprotein that requires bound Zn^{2+} for the integrity of its DNA binding domain. The nucleotide sequence recognized by this factor is the same as the one required for *in vivo* transcriptional activity of MREd (Culotta and Hamer 1989), because individual nucleotide substitutions introduced in the core region (5'TgCAcTC3') at either residues G or C (in lowercase) completely abolished binding activity as assayed by competition experiments (Séguin 1991).

Interestingly, extracts prepared with heavy-metal-resistant L50 cells contained approximately four times more MREd binding activity than the extracts prepared from L cells (Figure 9.4, compare lanes 1 and 6, and data not shown) thus showing that an activation or increase of the synthesis of the MRE binding factor had occurred in the L50 cells. Southern-blot hybridization, using L50- and L-cell genomic DNAs and a mouse *MT-I* cDNA, did not reveal any amplification of the *MT* genes in L50 cells (unpublished data). Genetically selected and naturally occurring lines of both cultured mammalian cells and yeast exhibit substantial variations in MT synthesis and heavy metal resistance (Enger et al. 1984). Experiments utilizing cloned *MT* gene hybridization probes have demonstrated that

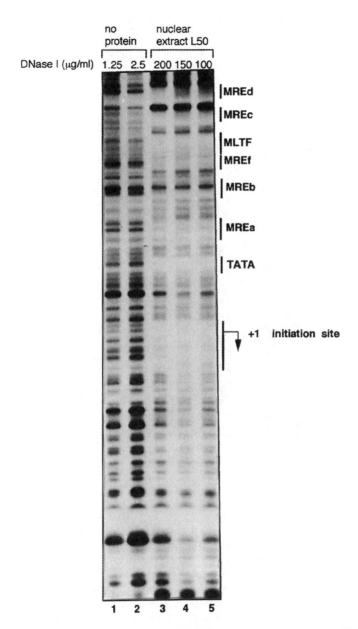

Figure 9.2 DNase I footprinting analysis using L50-cell crude nuclear extracts. The probe was a mouse *MT-I* gene promoter DNA fragment extending from positions −173 to +68. The amount of DNaseI used in each reaction is indicated over the lanes. Lanes: 1 and 2, no protein; 3–5, 15 μl of crude nuclear extract. L50-cell nuclear extracts were used because they contain approximately four times more MREd binding activity than the extracts prepared from induced L cells (see Figure 9.4). The positions of the MRE elements, the MLTF binding site, the TATA box, and the tsp are indicated on the right as determined by Maxam-Gilbert sequencing.

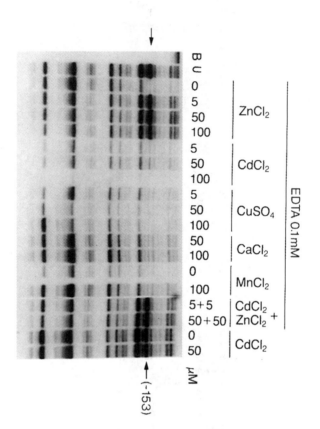

Figure 9.3 Reconstitution experiments. Ability of different cations to restore extracts binding activity at the MREd region following EDTA chelation as assayed by ExoIII footprinting analysis. Samples of extracts prepared from L50 cells were dialyzed overnight against a buffer containing 0.1 mM EDTA as indicated and then incubated in the presence of increasing concentrations of $ZnCl_2$, $CdCl_2$, $CuSO_4$, $CaCl_2$, $MnCl_2$, or $CdCl_2$ plus $ZnCl_2$. After treatment, the binding activity was assayed in footprinting reactions. The DNA probe used spans the mouse *MT-1* promoter sequences from positions −200 to +64, and was ^{32}P-labeled at position +64. Lanes: B, no extract; U, untreated extract. The arrow indicates the protected band at position −153 in the MREd region. (From Séguin 1991).

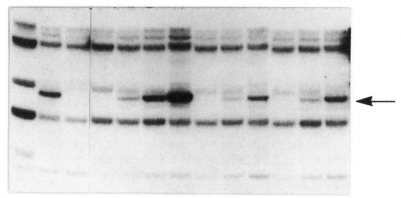

Figure 9.4 Extracts prepared from the heavy-metal-resistant L-cell line L50 show increased MREd-binding activity. ExoIII footprinting experiments were performed with 10 μg (lane 3), 25 μg (lanes 4, 7, 10), 50 μg (lanes 5, 8, 11) and 100 μg (lanes, 1, 2, 6, 9, 12) of extracts from mouse L cells (lanes 1, 2, 7–12) or L50 cells (lanes 3–6). Extracts were prepared from noninduced cells with a final dialysis buffer containing no added EDTA (lanes 7–12) or 0.1 mM EDTA (lane 2) or from Cd^{2+}-induced cells (lanes 1, 3–6). In lanes 10–12, the binding reaction was performed in the presence of 40 μM $CdCl_2$. O, no extract. Probe as in Figure 9.3 The arrowhead to the right indicates the protected band at position −153 in the MREd region.

such differences can be caused by changes in the copy number or methylation status of *MT* genes. Although enhanced MT synthesis was attributed to increased copy number of *MT* genes in some resistant cells (Beach and Palmiter 1981), increased Cd^{2+} resistance does not require increased copy number of *MT* genes (Enger et al. 1984; Chopra et al. 1990) and may involve an increase in the synthesis of other factors such as transcriptional regulatory proteins (Wan et al. 1995). The four-fold increase in the MREd-binding activity detected by the footprinting assay in L50 cells may be the consequence of an increase of the intracellular concentration of the protein(s) binding to this element or, alternatively, it may be caused by an increase in the affinity toward the binding site as a consequence of a mutation. The cloning of the cDNA encoding this MRE binding protein in L and L50 cells will be required to answer this question.

4. Southwestern Analysis

Using a protein-blotting procedure (Southwestern) and synthetic oligonucleotides (oligos), we have shown that a nuclear protein of 108 kDa, termed MEP-

1 2 3 4 5 6 7 8 9 10 11 12 13 14 15 16 17 18 19 20 21 22

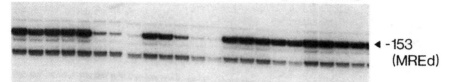

◀ -153
(MREd)

Figure 9.5 Competition experiments in an ExoIII footprinting assay performed with induced L50-cell nuclear extracts and double-stranded unlabeled oligos. The arrowhead indicates the ExoIII stop at the −153 boundary. Competitors used: lanes 1–4, Spl oligo (0, 20, 40, 100 ng); lanes 5–8, mouse MREd (0, 20, 40, 100 ng); lanes 9–13, the MT promoter DNA fragment (−200 to +68) (0, 1, 10, 50, 100 ng); lanes 14–18, the SV40 DNA (0, 3.5, 35, 175, 350 ng); lanes 19–22, a 309 bp-pBR322-*Msp*l fragment (0, 17, 34, 51 ng). Probe as in Figure 9.3 (From Séguin and Prévost 1988).

1, binds specifically to MREd (Séguin and Prévost 1988). In addition to the conserved MRE core sequence TGCRCNC, the MREd region contains a putative binding site (5′-YCCGCCC-3′) (Y = T or C) for the transcription factor Spl. We showed that both the MRE and Spl consensus sequences TGCAC and CCGCC are required for efficient binding of MREd to MEP-1; Mueller et al. (1988) have also shown that, *in vitro*, purified Spl weakly binds to this region. Thus, the possibility that MEP-1 was in fact Spl was tested by performing footprint competition experiments. We showed that neither a DNA fragment of SV40 which contains six Spl binding sites, nor an Spl oligo, competed for the protein binding to the MREd region, as assayed by ExoIII (Figure 9.5) or DNaseI (Figure 9.9A, lanes 4–6) footprinting analyses. Thus, despite the presence of an Spl binding site overlapping the MRE site, the Spl transcription factor does not interact with the MREd probe and therefore is different from MEP-1.

5. Purification of MEP-1

MEP-1 has been purified from L50 cells using footprinting, Southwestern, and UV cross-linking techniques to assay its binding activity (Labbé et al. 1993). The purification scheme, starting from crude nuclear extracts, involved a combination of standard heparin-Sepharose and MRE-DNA affinity chromatography. The enrichment of MEP-1 activity was estimated at about 7000-fold. The purified protein preparation appeared as a single polypeptide band of 108 kDa on polyacrylamide gel electrophoresis. Moreover, MEP-1 does not appear to be glycosylated as it eluted with the flow-through on a wheat germ-Sepharose column. Purified MEP-1 binds specifically to MRE sequences and is sufficient to produce a footprint on the mouse MREd element. As for the MREd binding activity present in crude extracts, MEP-1 binding is sequence specific and zinc dependent (Labbé et al. 1993).

Purified MEP-1 was digested with trypsin, and MEP-1 internal tryptic peptides whose amino acid sequences did not show any significant matches with known protein sequences were used to deduce degenerate oligos that served as primers in RT-PCR experiments aimed at amplifying a specific MEP-1 cDNA fragment. However, so far, all the attempts to amplify an MEP-1 specific fragment have failed.

6. Identification of MEP-2

Each mouse *MT-I* gene MRE has the capacity to confer metal inducibility on a heterologous promoter (Stuart et al. 1985). In addition, mouse MREd has the capacity to respond to the same spectrum of metal ions (Cd^{2+}, Zn^{2+}, and Cu^{2+}) as does the complete *MT* gene promoter (Culotta and Hamer 1989). This suggests that all MRE elements are responsive to the different metals and act to facilitate a strong induction response. However, it is not yet known whether metal induction of *MT* gene transcription involves different factors binding to different MRE sequences or a single factor binding to different MREs with different affinities. It is also unclear how many classes of nuclear proteins interact with the MRE elements.

To determine if different MRE elements can bind common nuclear factors, we performed competition experiments in an ExoIII footprinting assay, using synthetic oligos corresponding to the metal regulatory elements MREa, MREb, MREc, MREd, and MREe of the mouse *MT-I* gene. All of the MREs tested could compete with the protein(s) binding to the mouse MREd region, indicating that the same cellular factor can bind to all of these MRE elements (Figures 9.6 and 9.9A, lanes 7–12). Mouse MREd, MREa, MREb, and MREc, trout MREa, and human MRE4 showed similar competition strength; mouse MREe, a weak element *in vivo*, was approximately 50% weaker (Labbé et al. 1991). Thus, the protein(s) binding to the mouse MREd region can bind the other MREs present in the mouse gene as well as MREs from *MT* genes of other species.

With the aim of identifying other MRE binding proteins, we developed a purification strategy involving a combination of low-pressure heparin-Sepharose chromatography with salt gradient elution, and two steps of affinity chromatography. An important characteristic of MEP-1 binding to MREd is the apparition of a hypersensitive site at position −153 (Figure 9.7 and 9.9A, arrow). Elution of the heparin-Sepharose column led to the gradual disappearance of this hypersensitive site between 350 mM and 480 mM salt, whereas the footprint on MREc remained present in fractions containing up to 650 mM salt (Figure 9.7), thus

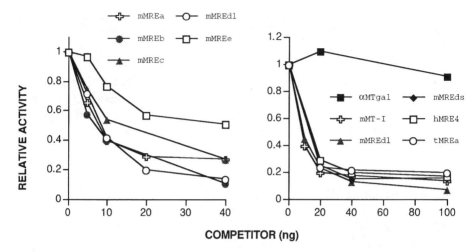

Figure 9.6 ExoIII footprinting competition experiments performed with induced L50-cell nuclear extracts and double-stranded unlabeled oligos corresponding to the various mouse (mMRE), human (hMRE), and trout (tMRE) MRE elements, as indicated, or with the same MT DNA fragment (mMT-I) used as a probe in the ExoIII assay. The probe (as in Figure 9.3) was used at a concentration of approximately 1 ng per assay. The DNA competitor αMTgal is a plasmid in which most of the *MT* promoter sequences have been replaced by a fragment of the human α-*globin* gene (Séguin et al. 1984); no MRE element is present in this DNA. The intensity of the band at −153 (arrow in Figures 9.3, 9.4, and 9.5) relative to that with no competitor DNA (relative activity) is shown as a function of the concentration of competitor DNA. (From Labbé et al. 1991.)

suggesting the existence of two distinct proteins interacting with these two MREs. Fractions 3–13 (Figure 9.7) of the heparin-Sepharose chromatography were pooled and loaded on an MREa-affinity column (Figure 9.8); proteins were eluted with steps of salt, and two distinct MRE-binding activities were detected. In addition to MEP-1, found in the 650-mM salt fractions, another MRE-binding protein, named metal element protein-2 (MEP-2), specifically interacting with the MREc element, was present in the 250-mM salt fractions (Figure 9.8). It is as yet unclear whether the 650-mM salt fractions contain a mixture of both MEP-1 and MEP-2, or whether MEP-1 has a broader binding spectrum and is able to bind to both MREd and MREc. To assess the binding specificity of MEP-2 to MREc, a heparin-Sepharose fraction (Figure 9.7, lane 4, fraction 3) containing both MEP-1 and MEP-2, and a 250-mM MRE-affinity fraction (Figure 9.8, fraction 4) containing only MEP-2 were used in competition experiments. Oligos corresponding to MREd or MREe competed for the protein binding to the MREd element (Figure 9.9A), but none of the oligos, not even MREc, could compete

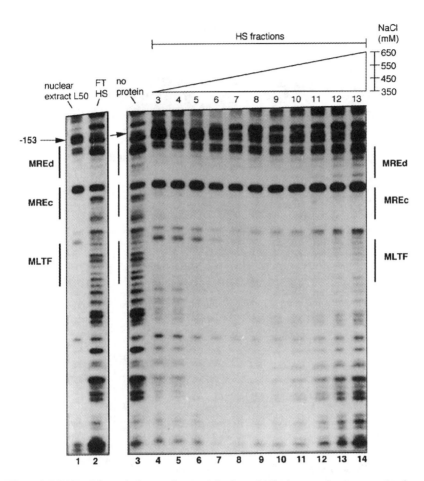

Figure 9.7 DNaseI footprinting analyses of the heparin-Sepharose chromatography fractions eluted with a NaCl gradient. The salt concentration in each fraction is schematically represented over the lanes. Numbers over the lanes correspond to the different chromatography fractions; numbers at the bottom identify each lane. The arrow shows the hypersensitive site at position −153 generated by the binding of MEP-1. The activity generating the footprint on MREc is present in all the fractions shown, whereas most of the MREd binding activity (MEP-1) eluted in fractions 3–5 (compare lanes 4 and 14). Lanes: 1, L50-cell nuclear extract; 2, flow-through of the heparin-Sepharose column (FT HS); 3, no extract. The probe was a mouse *MT-I* gene promoter *Apa*L1-*Hin*f1 DNA fragment extending from positions −173 to −40, and ^{32}P-labeled at the *Hin*f1 site. The positions of the MRE elements and of the USF/MLTF binding site are indicated on each side of the figure as determined by Maxam-Gilbert sequencing.

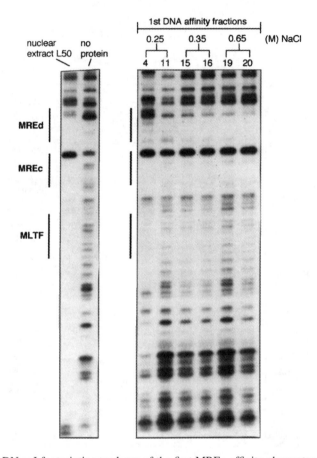

Figure 9.8 DNaseI footprinting analyses of the first MREa-affinity chromatography fractions eluted with steps of salt as indicated. Numbers over the lanes correspond to the different chromatography fractions. Fraction 4 of the 250-mM salt step contains the specific MREc-binding activity (MEP-2); fraction 20 corresponds to the MEP-1 pic activity. Probe as in Figure 9.7 The positions of the MRE elements and of the MLTF binding site are indicated on the left as determined by Maxam-Gilbert sequencing.

Figure 9.9 Competition experiments in DNaseI footprinting assays using (A) fraction 3 of the heparin-Sepharose (HS) column (lanes 3–15) or (B) fraction 4 of the first affinity column (lanes 4–19). Competition was performed with double-stranded unlabeled oligos corresponding to the Spl-A site (5′-gatcCCAAAGGGGCGGTCCCGCTa-3′, positions −192 to −174; the *MT* sequence is in uppercase and flanked with nucleotides, in lowercase, added to generate restriction sites of the mouse *MT-I* gene promoter, wild-type mouse *MT-I* MREa, MREc, MREd (MREds), and MREe (Labbé et al. 1991), and a nonfunctional

A)

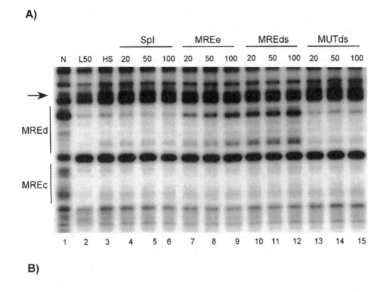

B)

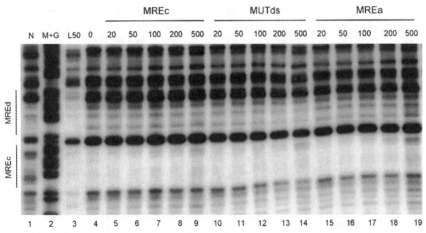

MREd mutant, MUTds (Séguin and Prévost 1988), in which both the Sp1 and the MRE core sequences have been mutated. The MREd oligo and, to a lesser extent, the MREe oligo, competed for the binding of MEP-1 to the MREd element, but neither they nor MREa and MREc competed for MEP-2 binding to the MREc element. The amount (in ng) of competitor used in each reaction is shown over the lanes. The probe concentration was approximately 1 ng/reaction. Lanes: (A) 1, no extract (N); 2, L50-cell nuclear extract. (B) 1, no extract (N); 2, M + G Maxam and Gilbert sequencing reactions; 3, L50-cell nuclear extract. Probe as in Figure 9.7 Only portion of the autoradiogram is shown. The positions of the MRE elements is indicated on the left as determined by Maxam-Gilbert sequencing.

the protein interacting with the MREc element (Figure 9.9, A and B). The MREc oligo competed for MEP-1 binding to the MREd element (Figure 9.6, and data not shown). This shows that the MREc binding protein MEP-2 is distinct from MEP-1. Further studies will be required to define the DNA sequence element required for the binding of MEP-2.

6. Discussion and Conclusion

In addition to MEP-1 and MEP-2, different MRE binding proteins have been detected *in vitro* by different DNA binding assays (Table 9.1). These are ZAP (Searle 1990) and p33 (Datta and Jacob 1993), found in liver extracts; MREBP (Koizumi et al. 1992a), ZRF (Koizumi et al. 1992b), and MREBP-34 (Hahn and Gahl 1993), present in HeLa cells; MBF-1, MafY, M96/ZiRF1, present in mouse L cells (Imbert et al. 1989; Xu, 1993; Inouye et al. 1994); p39, present in rat FAO hepatoma cells (Anderson et al. 1990); and MRE-BF1 and 2, detected in different human cell lines (Czupryn et al. 1992). Finally, MTF-1 is present in HeLa cells and binds, in a Zn^{2+}-dependent manner, to the mouse MREd element (Westin and Schaffner 1988). Amino acid sequences obtained from ZRF tryptic peptides revealed that ZRF corresponds to MTF-1 (Otsuka et al. 1994). The cDNA of MTF-1 was cloned and shown to correspond to a zinc-finger protein with an $M_r = 72,500$ (Radtke et al. 1993; Brugnera et al. 1994). Ectopic expression of MTF-1 strongly enhances transcription of a reporter gene that is driven by four consensus MREd sites. *MTF-1* gene knock-out by homologous recombination in ES cells showed that MTF-1 is a key factor for basal and metal-induced transcription (Heuchel et al. 1994).

It has also been proposed that the interaction of MTF-1 with MRE elements is normally prevented by interaction with a non-DNA-binding inhibitor called MTI (metallothionein transcription inhibitor) (Palmiter 1994). In this model, the metal sensor (MRTF) is the inhibitor protein. In the alternative hypothesis, it is assumed that metal ions cause the conversion of MRTF from an inactive (non-DNA-binding) state to an active, DNA binding configuration, in a way similar to that of the yeast *MT* gene transcription activator ACE1 (Thiele 1992). Further work is required to establish which model is correct.

The precise relationship between the different MRE binding proteins listed in Table 9.1 and MEP-1 and MEP-2 is unclear. Although the data shown in Table 9.1 suggest that there might be a family of transcription factors interacting with the MREs, it has been proposed that MTF-1 is the only factor that binds MREs and the only transcription factor that mediates responsiveness to different metals (Palmiter 1994). However, we have identified at least two MRE binding proteins: MEP1 and MEP2. We also have evidence showing that MEP1 is distinct from MTF-1, as assayed by Western analysis, using specific anti-MTF-1 antibodies

Table 9.1 MRE binding proteins

Proteins	Description
MTF-1	First detected by electrophoresis mobility shift assay (EMSA) in HeLa cells using mouse *MT-I* MREd as a probe; DNA binding is zinc inducible; cDNA cloned; zinc finger protein; MW = 72.5 kDa; *MTA-1* gene knock-out by homologous recombination in ES cells showed that MTF-1 is a key factor for basal and metal-induced transcription; has been proposed to interact with a non-DNA-binding inhibitor (MTI) acting as a metal sensor (Westin and Schaffner 1988; Radtke et al. 1993; Brugnera et al. 1994; Palmiter 1994).
ZRF	Detected in HeLa cells by EMSA using human *MTIIA* MREa as a probe; zinc inducible; purified and partial amino acid sequence was obtained from tryptic peptides; corresponds to MTF-1 (Koizumi et al 1992b; Otsuka et al. 1994).
MBF-1	Prresent in mouse L cells; it has been purified; MW = 74 dKa; cDNA cloned and showed to correspond to RP-A (Imbert et al. 1989; Suzuki et al. 1993).
MafY	Present in mouse cells; cDNA cloned in yeast by genetic complementation; approximately 10 kDa; includes a single zinc finger sequence; constitutively expressed in L cells; induces transcription in a metal-independent fashion (Xu 1993).
M96/ZiRF1	Present in mouse cells; cDNA cloned in yeast by the one-hybrid system; MW = 42 kDa; shows similarity to the trithorax proteins; requires zinc ions for functional DNA binding activity (Inouye et al. 1994; Remondelli and Leone 1997).
p33	Detected in liver nuclear extracts; EMSA using mouse *MT-I* MREc′ as a probe; MW = 33 kDa, as assessed by Southwestern; involved in basal transcription (Datta and Jacob 1993).
ZAP	Detected in nuclei from rat liver cells by EMSA using mouse *MT-I* MREa and MREd oligos as probes; DNA binding is zinc inducible (Searle 1990).
p39	Detected in rat FAO cells by EMSA; binds to several MREs for DNA binding, which is cadmium inducible; MW = 39 kDa, as assessed by Southwestern analysis (Andersen et al. 1990).
MREBP	Detected in HeLa cells by EMSA; binds to several MREs of the human *MTIIA* gene promoter; partially purified, MW = 112 kDa (Koizumi et al. 1992a).
MRE-BF1/2	Present in different human cell lines and interacts with MREa of the human *MT-I* genes; MW = 86 and 28 kDa as assessed by Southwestern analysis (Czupryn et al. 1992).
MREBP-34	Present in cultured human fibroblasts and HeLa cells; detected by EMSA using human *MTIIA* MREa as a probe; DNA binding enhanced by Cu; MW = 34 kDa, as assessed by Southwestern analysis (Hahn and Gahl 1993).

(generously provided to us by W. Schaffner, Zurich) and gel shift assay (data not shown). This brings the minimum number of MRE binding proteins to at least three: MEP-1, MEP-2, and MTF-1. The cloning of the respective cDNAs will be required to establish the precise relationship between these proteins.

Our results show that MEP-1 can bind to multiple MREs in mouse, trout, and human *MT* genes, whereas MEP-2 appears to be specific for the MREc element, and suggest that the MEP proteins are major proteins interacting with MRE sequences. The assessment of MEP-1 and MEP-2 as metal regulatory proteins and the possible interactions with other MRE binding proteins remain challenging problems in the field of *MT* gene transcription.

Acknowledgments

This project is supported by grants 1 RO1 CA61261-01 from the National Cancer Institute, NIH, and MT-12468 from the Medical Research Council.

References

Andersen, R. D., S. J. Taplitz, S. Wong, G. Bristol, B. Larkin, and H. R. Herschman. 1987. Metal-dependent binding of a factor *in vivo* to the metal-responsive elements of the metallothionein 1 gene promoter. *Mol. Cell. Biol.* 7:3574–3581.

Andersen, R. D., S. J. Taplitz, A. M. Oberbauer, K. L. Calame, and H. R. Herschman. 1990. Metal-dependent binding of a nuclear factor to the rat metallothionein-I promoter. *Nucl. Acids Res.* 18:6049–6055.

Andrews, G. K. 1990. Regulation of metallothionein gene expression. *Prog. Food Nutr. Sci.* 14:193–258.

Beach, L. R., and R. D. Palmiter. 1981. Amplification of the metallothionein-I gene in cadmium-resistant mouse cells. *Proc. Natl. Acad. Sci. U.S.A.* 78:2110–2114.

Brugnera, E., O. Georgiev, F. Radtke, R. Heuchel, E. Baker, G. R. Sutherland, and W. Schaffner. 1994. Cloning, chromosomal mapping and characterization of the human metal-regulatory transcription factor MTF-1. *Nucl. Acids Res.* 22:3167–3173.

Carthew, R. W., L. A. Chodosh, and P. A. Sharp. 1987. The major late transcription factor binds to and activates the mouse metallothionein I promoter. *Genes Dev.* 1:973–980.

Chopra, A., J. Thibodeau, Y. C. Tam, C. Marengo, M. Mbikay, and J. P. Thirion. 1990. New mouse somatic cell mutant resistant to cadmium affected in the expression of their metallothionein genes. *J. Cell Physiol.* 142:316–324.

Culotta, V. C., and D. H. Hamer. 1989. Fine mapping of a mouse metallothionein gene metal response element. *Mol. Cell. Biol.* 9:1376–1380.

Czupryn, M., W. E. Brown, and B. L. Vallee. 1992. Zinc rapidly induces a metal response element-binding factor. *Proc. Natl. Acad. Sci. U.S.A.* 89:10395–10399.

Dalton, T., R. D. Palmiter, and G. K. Andrews. 1994. Transcriptional induction of the mouse metallothionein-I gene in hydrogen peroxide-treated Hepa cells involves a composite major late transcription factor antioxidant response element and metal response promoter elements. *Nucl. Acids Res.* 22:5016–5023.

Datta, P. K., and S. T. Jacob. 1993. Identification of a sequence within the mouse metallothionein-I gene promoter mediating its basal transcription and of a protein interacting with this element. *Cell. Mol. Biol. Res.* 39:439–449.

Enger, M. D., C. E. Hildebrand, J. K. Griffith, and R. A. Walters. 1984. Molecular and somatic cell genetic analysis of metal metabolism in cultured cells, p. 7–24. In *Metabolism of Trace Metals in Man.* eds. O. M. Rennart and M. Y. Chan. Vol. 2. CRC Press, Boca Raton, FL.

Hahn, S. H., and W. A. Gahl. 1993. Copper effects on metal regulatory factors of cultured human fibroblasts. *Biochem. Med. Metab. Biol.* 50:346–357.

Hamer, D. H. 1986. *Metallothionein. Ann. Rev. Biochem.* 55:913–951.

Heuchel, R., F. Radtke, O. Georgiev, G. Stark, M. Aguet, and W. Schaffner. 1994. The transcription factor MTF-1 is essential for basal and heavy metal-induced metallothionein gene expression. *EMBO J.* 13:2870–2875.

Heuchel, R., F. Radtke, and W. Schaffner. 1995. Transcriptional regulation by heavy metals, exemplified at the metallothionein genes, p. 206–240. *In Inducible Gene Expression,* ed. P. A. Baeuerly. Vol. 1. Birkhäuser, Boston.

Imbert, J., V. C. Culotta, P. Fürst, L. Gedamu, and D. H. Hamer. 1990. Regulation of metallothionein gene transcription by metals. *Adv. Inorg. Biochem.* 8:139–164.

Imbert, J., M. Zafarullah, V. C. Culotta, L. Gedamu, and D. H. Hamer. 1989. Transcription factor MBF-I interacts with metal regulatory elements of higher eucaryotic metallothionein genes. *Mol. Cell. Biol.* 9:5315–5323.

Inouye, C., P. Remondelli, M. Karin, and S. Elledge. 1994. Isolation of a cDNA encoding a metal response element binding protein using a novel expression cloning procedure: The one hybrid system. *DNA Cell Biol.* 13:731–742.

Koizumi, S., K. Suzuki, and F. Otsuka. 1992a. A nuclear factor that recognizes the metal-responsive element of human metallothionein IIA gene. *J. Biol. Chem.* 267:18659–18664.

Koizumi, S., H. Yamada, K. Suzuki, and F. Otsuka. 1992b. Zinc-specific activation of a HeLa cell nuclear protein which interacts with a metal responsive element of the human metallothionein-IIA gene. *Eur. J. Biochem.* 210:555–560.

Labbé, S., L. Larouche, D. Mailhot, and C. Séguin. 1993. Purification of mouse MEP-1, a nuclear protein that binds to the metal regulatory elements of genes encoding metallothionein. *Nucl. Acids Res.* 21:1549–1554.

Labbé, S., J. Prévost, P. Remondelli, A. Leone, and C. Séguin. 1991. A nuclear factor binds to the metal regulatory elements of the mouse gene encoding metallothionein-I. *Nucl. Acids Res.* 19:4225–4231.

Masters, B. A., E. J. Kelly, C. J. Quaife, R. L. Brinster, and R. D. Palmiter. 1994. Targeted disruption of metallothionein I and II genes increases sensitivity to cadmium. *Proc. Natl. Acad. Sci. U.S.A.* 91:584–588.

Michalska, A. E., and K. H. A. Choo. 1993. Targeting and germ-like transmission of a null mutation at the metallothionein I and II loci in mouse. *Proc. Natl. Acad. Sci. U.S.A.* 90:8088–8092.

Mueller, P. R., S. J. Salser, and B. Wold. 1988. Constitutive and metal-inducible protein: DNA interactions at the mouse metallothionein I promoter examined by *in vivo* and *in vitro* footprinting. *Genes Dev.* 2:412–427.

Otsuka, F., A. Iwamatsu, K. Suzuki, M. Ohsawa, D. H. Hamer, and S. Koizumi. 1994. Purification and characterization of a protein that binds to metal responsive elements of the human metallothionein II(A) gene. *J. Biol. Chem.* 269:23700–23707.

Palmiter, R. D. 1994. Regulation of metallothionein genes by heavy metals appears to be mediated by a zinc-sensitive inhibitor that interacts with a constitutively active transcription factor, MTF-1. *Proc. Natl. Acad. Sci. U.S.A.* 91:1219–1223.

Radtke, F., R. Heuchel, O. Georgiev, M. Hergersberg, M. Gariglio, Z. Dembic, and W. Schaffner. 1993. Cloned transcription factor MTF-1 activates the mouse metallothionein-I promoter. *EMBO J.* 12:1355–1362.

Remondelli, P., and A. Leone. 1997. Interactions of the Zinc Regulated Factor (ZiRF1) with the mouse metallothionein I$_a$ promoter. *Biochem. J.* 323:79–85.

Searle, P. F. 1990. Zinc dependent binding of a liver nuclear factor to metal response element MRE-a of the mouse metallothionein-I gene and variant sequences. *Nucl. Acids Res.* 18:4683–4690.

Séguin, C. 1991. A nuclear factor requires Zn^{2+} to bind a regulatory MRE element of the mouse gene encoding metallothionein-I. *Gene* 97:295–300.

Séguin, C., B. K. Felber, A. D. Carter, and D. H. Hamer. 1984. Competition for cellular factors that activate metallothionein gene transcription. *Nature* 312:781–785.

Séguin, C., and D. H. Hamer. 1987. Regulation *in vitro* of metallothionein gene binding factors. *Science* 235:1383–1387.

Séguin, C., and J. Prévost. 1988. Detection of a nuclear protein that interacts with a metal regulatory element of the mouse metallothionein I gene. *Nucl. Acids Res.* 168:10547–10560.

Stuart, G. W., P. F. Searle, and R. D. Palmiter. 1985. Identification of multiple metal regulatory elements in mouse metallothionein-I promoter by assaying synthetic sequences. *Nature* 317:828–831.

Suzuki, K. T., N. Imura, and M. Kimura (ed.) 1993. *Metallothionein III: Biological Roles and Medical Implications.* Birkhäuser, Basel.

Thiele, D. J. 1992. Metal-regulated transcription in eukaryotes. *Nucl. Acids Res.* 20:1183–1191.

Wan, M., R. Heuchel, F. Radtke, P. E. Hunziker, and J. H. R. Kagi. 1995. Regulation of metallothionein gene expression in Cd-or Zn-adapted RK-13 cells. *Experientia* 51:606–611.

Westin, G., and W. Schaffner. 1988. A zinc-responsive factor interacts with a metal-regulated enhancer element (MRE) of the mouse metallothionein-I gene. *EMBO J.* 7:3763–3770.

Xu, C. 1993, cDNA cloning of a mouse factor that activates transcription from a metal response element of the mouse metallothionein-I gene in yeast. *DNA Cell Biol.* 12:517–525.

10

Menkes' and Wilson's Diseases: Genetic Disorders of Copper Transport

Julian F. B. Mercer and James Camakaris

1. Introduction

Copper is an essential element required by a number of important enzymes, including lysyl oxidase, cytochrome c oxidase, superoxide dismutase, and dopamine β-hydroxylase. Copper deficiency during development can prove lethal to developing mammals, and the multiple-organ-system effects can be explained by the reduced activity of these important enzymes. The severity of copper deficiency is illustrated by the lethal X-linked genetic disorder of copper transport, Menkes' disease (MD). The molecular basis of this disease will be discussed in this chapter. The same properties that make copper a useful element for the redox reactions carried out by enzymes, render it dangerous in a free ionic state. Free copper has the potential to catalyze the formation of the highly reactive hydroxyl radicals, which damage many cell components, including membranes, proteins, and nucleic acids (Kumar et al. 1978). All organisms must have developed mechanisms for supplying copper to essential enzymes without damaging cellular constituents. It is most probable that this delivery is achieved by maintaining the copper in a complex at all times. Thus the Cu ion must be transferred from one complex to another as it moves within cells or between one compartment of the body to another. This movement of copper can be likened to a pathway; the various molecules involved in complexing and transferring copper form the steps of this pathway. The challenge for current research in copper transport is to identify these molecules and understand how the regulation of their concentrations and activity maintains copper supplies within acceptable ranges in the face of widely varying dietary intakes.

2. Outline of the Copper Transport Pathway

Two of the key components of the copper transport pathway have been clarified recently by the positional cloning of the genes affected in the inherited human defects of copper transport, Menkes' disease (MD) and Wilson's disease (WD) (Bull et al. 1993; Chelly et al. 1993; Mercer et al. 1993; Tanzi et al. 1993; Vulpe et al. 1993b; Yamaguchi et al. 1993). These genes encode very similar Cu-transporting P-type ATPases that have been termed *MNK* or ATP7A, for the product of *MNK,* the Menkes gene; and *WND* or ATP7B, the product of *WND,* the Wilson gene. For clarity, we shall use *MNK* and *WND* in this chapter, which will be principally concerned with the structure, regulation, and role of these proteins and the consequences of mutations in the corresponding genes.

Figure 10.1 shows a model of the overall copper pathway in mammals and the points at which *MNK* and *WND* are thought to function. Each of the steps in the pathway is probably regulated by various factors, including the concentration of copper in the tissue, but the details of this regulation are only beginning to be understood.

Most dietary copper is absorbed in the small intestine. Transport from the mucosal cells into the portal circulation requires metabolic energy (Van Campen 1971) (Figure 10.1, Step 1). This step is likely to be performed by *MNK,* as patients with Menkes' disease accumulate copper in the mucosal cells because copper uptake is normal but efflux into the blood is blocked. Thus copper accumulates in the intestinal cells, but there is an overall deficiency in the body (Danks 1995). *MNK* is also involved in the efflux of copper from the periferal tissues into the circulation for return to the liver (Figure 10.1, Step 7), and in transfer of copper across the blood-brain barrier (not shown).

Copper is carried to the liver via albumin (Owen 1965), copper histidine complexes (Harris and Sass-Kortsak 1967), or possibly by a protein termed transcuprein (Goode et al. 1989; Weiss and Linder 1985). Copper uptake by hepatocytes (Figure 10.1, Step 2), is a carrier-mediated process, not dependent on metabolic energy, and can occur from the copper-histidine complex (Darwish et al. 1983; McArdle 1992, McArdle et al. 1988). The liver is the key organ of copper homeostasis. The major copper protein in the plasma, ceruloplasmin, is synthesized in the liver and requires *WND* for copper to be incorporated (Figure 10.1, Step 6). Excess copper is removed by excretion in the bile (Figure 10.1, Steps 4 and 5). In Wilson's disease (WD), both the incorporation of copper into ceruloplasmin and biliary excretion are greatly reduced [for review of both MD and WD see Danks (1995)].

Within hepatocytes, copper is found distributed between a number of components, the pattern of distribution depending on the copper status of the liver. Some copper is found associated with the small metal binding proteins called metallothioneins (MTs), and this is the predominant form in livers loaded with

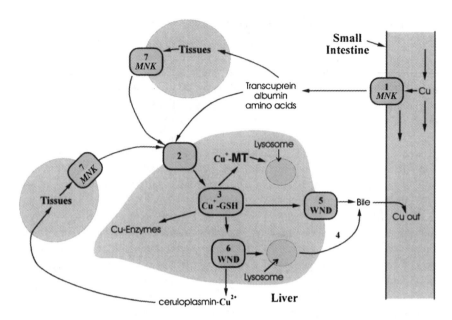

Figure 10.1 Model of the copper transport pathway in mammals. The general flow of copper, indicated by the arrows, is shown from absorption in the small intestine to excretion in the bile. The proposed sites of action of the protein affected in Menkes' disease (*MNK*) and Wilson's disease (*WND*) are indicated. The small shaded figures represent actual or hypothetical components of the pathway, and the numbers refer to the various steps discussed in the text. GSH is glutathione and MT is metallothionein.

copper by dietary supplementation (Evering et al. 1991) or as a result of Wilson's disease (Elmes et al. 1989). In these copper-loaded livers, the copper also accumulates in lysosomes (Evering et al. 1991). Lysosomal exocytosis into the biliary canaliculi is thought to form part of the pathway for biliary excretion of copper (Gross et al. 1989). There is an additional pathway involving direct excretion of cytoplasmic copper from the cytoplasm into the biliary canaliculi (Figure 10.1, Step 5) (Dijkstra et al. 1995; Nederbragt 1989), which we have indicated as possibly being mediated by *WND*, but this has not been demonstrated directly. Cytoplasmic copper is also found associated with glutathione (GSH, Figure 10.1, Step 3), and this form of copper may serve as the major intracellular carrier (Freedman et al. 1989). GSH appears to be involved in the biliary excretion of copper in copper-loaded livers (Dijkstra et al. 1995; Houwen et al. 1990; Nederbragt 1989). There may be other intracellular carriers of copper, and the possibility of the copper pathway involving other vesicular components of the cell has been suggested (Davidson et al. 1994).

3. Features of Menkes' Disease and Occipital Horn Syndrome

There are two closely related X-linked copper deficiency disorders known in man: Menkes' disease (MD) and occipital horn syndrome, also known as X-linked cutis laxa. As we have noted, patients with MD are profoundly copper deficient as a result of reduced intestinal absorption of copper (Figure 10.1, Step 1) compounded by inefficient transport across other epithelial cell layers. For example, *MNK* is required to transport copper across the blood-brain barrier. Thus, the copper deficiency in the brain is very severe because the delivery of the ion is blocked at two steps, absorption from the small intestine and transport into the brain. Paradoxically, some tissues of MD patients have higher than normal amounts of copper despite the overall copper deficiency. This phenomenon is explained if one recognizes that the uptake of copper by cells from Menkes' patients is normal, but efflux is blocked (Herd et al. 1987). As demonstrated in the mouse models of MD, the cells that accumulate copper are those that are normally involved in copper transport, such as the small intestine enterocytes (Yoshimura 1994), cells of the blood-brain barrier (Kodama, 1993), the proximal tubules of the kidney, which are active in resorption of copper from urine (Yoshimura 1994), and the placental cells of the affected fetus (Mann et al. 1980; Xu et al. 1994). The small amount of copper that is absorbed from the diet becomes trapped in these peripheral tissues and gradually accumulates, but is unavailable to the rest of the body.

Most cells from Menkes' patients will accumulate copper if supplies are adequate, and, as noted below, this phenomenon is demonstrated in cultured cells and can be used for prenatal diagnosis [for review see Danks (1995)]. The notable exception to the copper accumulation tendency is the liver, which has been shown not to retain injected copper in mouse models of Menkes (Mann et al. 1979). This observation led Danks to suggest that the liver was not affected by the disorder (Danks 1986). This proposal was subsequently shown to be correct, because *MNK* is only expressed at very low levels in the liver, and, as indicated in Figure 10.1, *WND* substitutes for *MNK* in this organ.

The milder X-linked disorder, occipital horn syndrome copper deficiency is not as profound, and the effects are largely confined to connective tissue. The differences between Menkes' disease and occipital horn syndrome are not fully understood, but molecular analysis shows that both diseases result from mutations of *MNK,* as will be discussed in the following sections. An intermediate phenotype between occipital horn syndrome and MD, termed mild Menkes', has been described (Procopis et al. 1981), and in one case a mutation has been identified.

The clinical features of Menkes' disease were first described in 1962 by John Menkes (Menkes et al. 1962). In 1972 David Danks realized that MD is a genetic copper-deficiency disorder, based on similar features found in copper-deficient sheep and pigs (Danks et al. 1972). A major clinical feature of patients with MD is profound mental retardation, and this most probably is due to the low activity of cytochrome oxidase, a copper-dependent enzyme in the electron transport chain. Low activity of other copper-dependent enzymes in the brain such as superoxide dismutase, peptidylglycine α-amidating monooxygenase, and dopamine β-hydroxylase may also contribute to the brain abnormalities (Kaler 1994). As noted above, the copper deficiency in the brain is even more profound than expected from the reduced uptake of copper from the small intestine, because copper cannot traverse the blood-brain barrier in patients with MD, and the death of MD patients is frequently due to the brain abnormalities. Patients with MD are hypo-pigmented from reduced activity of tyrosinase, a copper-dependent enzyme required for melanin synthesis. Hypothermia is also a feature of MD, and is thought to be due to low cytochrome c oxidase activity. Patients also have connective tissue defects, such as abnormalities of bone and weak vascular walls, resulting from reduced activity of lysyl oxidase. The affected boys usually die by the age of three or four years.

The connective tissue defects found in MD patients become the predominant features of occipital horn syndrome. Lysyl oxidase catalyzes the cross-linking of collagen and elastin, and the low activity of this enzyme is responsible for the aberrant connective tissue. Occipital horn syndrome patients have hyperelastic skin, arterial aneurisms, hernias, bladder diverticulae, and multiple skeletal abnormalities. Bony abnormalities of the occiput are a feature giving rise to the name. Occipital horn syndrome patients may be mildly mentally retarded.

MD patients can be treated by daily injections of copper salts, which can prolong survival and result in clinical improvement in some patients provided treatment is commenced before significant brain damage has occurred (Sherwood et al. 1989; Tumer et al. 1995a). The treatment must be commenced as soon as possible after birth, as the critical period of copper requirements in the brain apparently occurs at this time (Danks 1995). The first Menkes' child in a family will usually not be recognized early enough for successful treatment, but prenatal diagnosis of subsequent pregnancies is used to identify affected fetuses, which may be induced at 35 weeks to allow even earlier treatment. Not all patients respond to therapy, however, and Kaler has suggested that the responding patients are those with some residual *MNK* activity (Kaler 1996; Kaler et al. 1995). However, there have not been enough studies of treatment of Menkes' patients with defined mutations to make a definite conclusion. In responding patients, copper therapy does not correct the connective tissue abnormalities, and patients develop the clinical features of occipital horn syndrome (Tumer et al. 1995a; Danks, unpublished data).

3.1 Mouse Models of MD and Occipital Horn Syndrome: The Mottled Mice

Mutations of the X-linked mottled locus (*Mo*) in mice result in varying degrees of copper deficiency and widely different phenotypic presentations (Lyon and

Searle 1990). Analysis of the mutations has demonstrated that at least some involve the murine homolog of the Menkes' disease gene (see Section 3.6). The brindled mouse (Mo^{br}) is a close homolog to Menkes' disease with most of the typical features of the human disorder; the affected male dies around 15 days after birth from the effects of profound copper deficiency. Figure 10.2 shows a brindled male pup and a normal litter mate. The depigmented coat and the "kinky" whiskers of the mutant are apparent. The blotchy mouse (Mo^{blo}) has a less severe copper deficiency, and the defect primarily involves the connective tissue like occipital horn syndrome (Rowe et al. 1977; Starcher et al. 1978). A number of the mottled mutants are lethal in utero; e.g., the dappled (Mo^{dp}) mutant male dies between day 15 and 17 of gestation. Analysis of the *MNK* mRNA in fetuses shows that there is little if any *MNK* mRNA in the dappled mutants, and this may be due to a small deletion in the *MNK* gene (Mercer et al. 1994). The fetal death of the mutants suggests that the lack of *MNK* has a more severe effect on developing mice than on developing humans, as humans lacking *MNK* develop Menkes' disease, rather than dying in utero.

3.2 The Cellular Copper Phenotype of the Menkes and Mottled Cells

The phenotype of copper hyperaccumulation is observed in a number of cultured cell lines derived from MD patients and from the mottled mouse mutants (Camakaris et al. 1980; Goka et al. 1976; Horn 1976). The excess copper is associated with metallothioneins. Copper hyperaccumulation in the mutant cells is likely to be a consequence of defective copper efflux (Camakaris et al. 1982). Estimation of efflux of copper forms a more reliable method of prenatal diagnosis than copper accumulation (Camakaris et al. 1982). A role of *MNK* in copper efflux is strongly supported by the enhanced copper efflux observed in cultured cell lines that overexpress the Menkes gene (see Section 3.7).

Cultured fibroblasts from patients with occipital horn syndrome also accumulate excess copper (Peltonen et al. 1983) and have defective copper efflux (Camakaris, unpublished data). Surprisingly, it is not possible to distinguish between cells affected by occipital horn syndrome and by MD by these copper measurements, despite the much milder phenotype of occipital horn syndrome. Paradoxically, cultured fibroblasts from MD patients, which possess elevated levels of copper, excrete reduced levels of lysyl oxidase into the culture medium (Royce et al. 1980), suggesting that the copper is not in a cellular compartment that is readily accessible to lysyl oxidase. Lysyl oxidase levels in the blotchy mouse are lower than in the more severely affected brindled mouse. Although this is consistent with the preponderance of connective tissues defects, the reason for the selective reduction of lysyl oxidase in the less severely affected mutant is not clear (Rowe et al. 1977; Royce et al. 1982).

Figure 10.2 Comparison of a brindled mutant with a normal mouse. The brindled male (top) at about 12 days of age has a depigmented coat, and normally straight whiskers seen in the normal litter mate (bottom) are quite twisted. A similar effect on hair structure is seen in human babies with Menkes' disease.

3.3 Isolation of the Gene Affected in Menkes' Disease

The Menkes gene was isolated by a positional cloning procedure based on a chromosomal translocation that had disrupted the Menkes gene and caused MD in a female (Kapur et al. 1987). Affected females are very unusual in an X-linked condition; only six females with Menkes' disease have been reported (Gerdes et al. 1990; Kapur et al. 1987). The Menkes gene had been mapped to between Xq11 and Xq21.3 (Tonneson et al. 1992), and comparative gene mapping suggested that the *MNK* locus was in band Xq13, close to the gene phosphoglycerate kinase-1 (PGK-1). Fine localization was achieved when Verga et al. (1991) showed that the translocation breakpoint lay on a restriction fragment of 300 kilobases containing PGK-1. This suggested the strategy for cloning the gene, which is outlined in Figure 10.3. YAC clones containing the PGK gene were selected and one YAC (Figure 10.3) was found that crossed the translocation breakpoint and therefore hybridized to the translocation chromosome from the patient's cells, suggesting it would include the Menkes gene. This 320-kb YAC was digested into 20-kb pieces and subcloned into lambda phage. One lambda clone, which crossed the breakpoint, was identified by hybridizing to the translocated chromosome. Using this lambda clone to screen cDNA libraries, cDNA clones encoding *MNK* were identified (Mercer et al. 1993). Two other groups also succeeded in obtaining *MNK* cDNA clones by using similar strategies (Chelly et al. 1993; Vulpe et al. 1993b), and one isolated cDNAs that included the complete coding sequence of *MNK* (Vulpe et al. 1993b).

The structure of the deduced 1500-amino-acid polypeptide indicates that *MNK* is a member of the family of cation translocating membrane proteins termed P-type ATPases. Homologies were found to a number of bacterial proteins that confer resistance to heavy metals such as mercury and cadmium. In particular, the cadmium transporter CadA has a very similar predicted sequence to that of the Menkes protein (Silver et al. 1989). More recently, bacterial copper-transporting P-type ATPases, CopA and CopB, have been described (Odermatt et al. 1993; Solioz et al. 1994) as well as copper ATPases in yeast (Fu et al. 1995; Rad et al. 1994). Hydrophobicity analysis of *MNK* predicts six or eight transmembrane domains (Vulpe et al. 1993a,b). A careful analysis of a heavy metal P-type ATPase recently cloned in *Helicobacter pylori* has indicated the presence of eight transmembrane domains (Melchers et al. 1996).

A hypothetical representation of the transmembrane arrangement of *MNK* is shown in Figure 10.4, based on the model proposed for the calcium ATPases (MacLennan et al. 1992). The eight transmembrane domains are thought to form a channel through which the copper traverses the membrane. All P-type ATPases have an ATP binding site and an invariant aspartic acid residue (D in Figure 10.4) that is phosphorylated and dephosphorylated during the reaction cycle of cation translocation. Heavy metal P-type ATPases have the motif CysProCys in

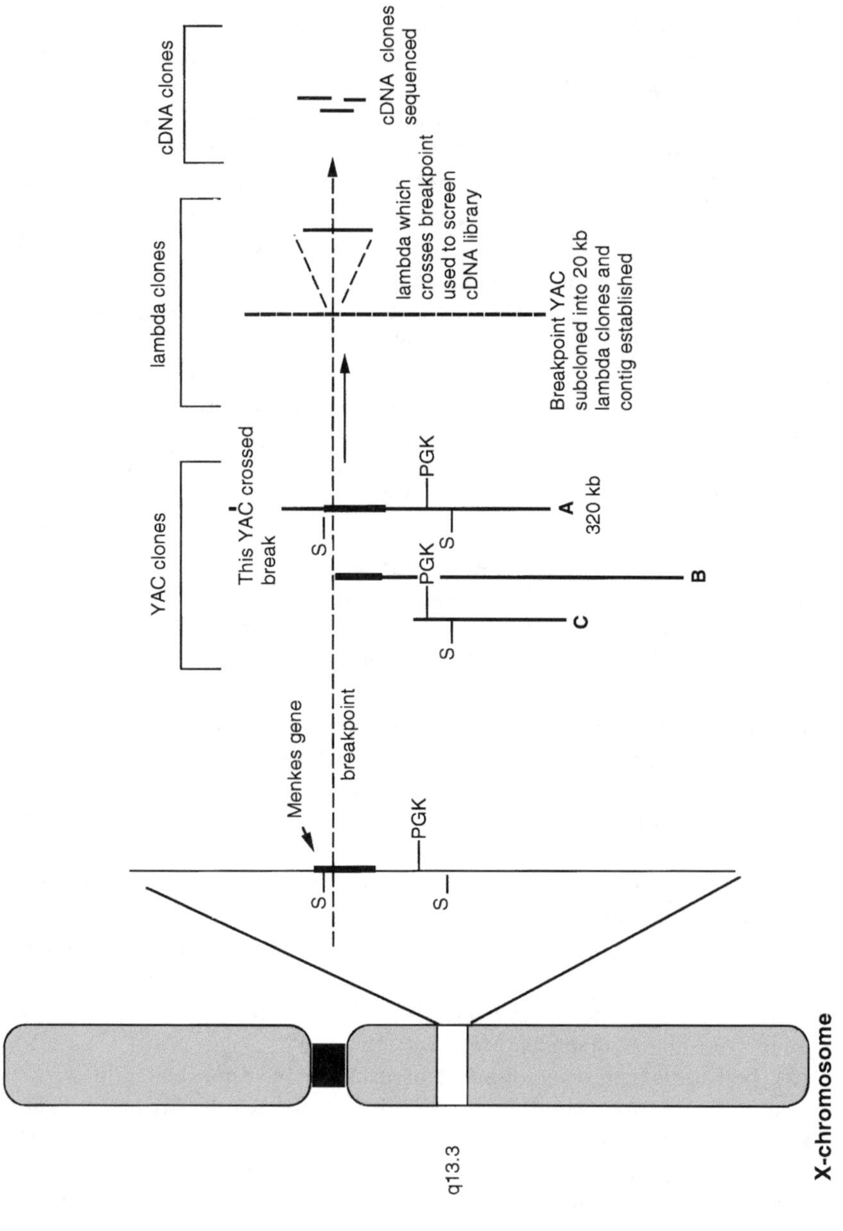
cDNA clones
cDNA clones sequenced
lambda clones
lambda which crosses breakpoint used to screen cDNA library
Breakpoint YAC subcloned into 20 kb lambda clones and contig established
YAC clones
This YAC crossed break
S
PGK
A
320 kb
B
C
Menkes gene
breakpoint
q13.3
X-chromosome

the membrane channel, and the cysteines are likely to bind the metal ion during the transmembrane transfer. Other well-known P-type ATPases, e.g., the Ca^{2+} ATPases have a proline in this position, located either in the membrane or just at the membrane surface, but lack the cysteines. There is evidence that the enzyme alternates between two conformation states termed E1 and E2. In the E1 state (as in Figure 10.4), the cation has access to the high-affinity cation binding sites. In *MNK* these are depicted as the cysteines in the channel, but some of the N-terminal metal binding sites may also participate in the initial copper binding (see next paragraph). In the E2 state (not illustrated), a substantial conformation change occurs that shifts the metal ion from the high-affinity sites to lower-affinity sites, perhaps the methionine and histidines on the outer side of the channel (shown in Figure 10.4). The cation can then no longer access the high-affinity sites (MacLennan et al. 1992). At the N-terminus of the CadA protein there is a dicysteine motif CXXC which is proposed to bind cadmium. Six copies of the motif GMTCXXC (denoted Cu1-6 in Fig. 10.4) are found in the N-terminal portion of *MNK* and are presumed to bind copper. Interestingly, only one or two of these putative copper binding sites are present in the bacterial and yeast copper ATPases (Fu et al. 1995; Odermatt et al. 1993; Rad et al. 1994) and in Cad A. A putative Cu ATPase in *H. pylori* does not possess this motif (Ge et al. 1995), although another *H. pylori* ATPase does have this structure (Melchers et al. 1996). Apparently, the copper transport function of the ATPases does not require more than one N-terminal dicysteine structure, so it is not clear why there are six in the mammalian Cu ATPases. An interesting possibility is that the multiple metal binding sites have a function in addition to that of direct involvement in cation transport. One suggestion is that these sites form a copper-sensing region important for the maintenance of copper homeostasis in mammalian cells.

Figure 10.3 Strategy for positional cloning of the Menkes gene. The Menkes gene had been mapped in the band Xq13.3, and the translocation breakpoint that caused Menkes' disease in the female had been positioned on a 300-kb Sfi *I* fragment (S) containing the gene PGK. YAC clones containing PGK were mapped by fluorescent in situ hybridization to chromosomes from the translocation patient to identify YAC (A), which crossed the breakpoint and therefore was most likely to include the Menkes gene. YAC B contained part of the gene but did not cross the breakpoint and was rejected, as was C, which did not contain the gene or cross the breakpoint. YAC A was digested into pieces of about 20 kb with a restriction enzyme and subcloned into a lambda phage vector. The lambda clones were mapped to the breakpoint, and the one that crossed was used to screen cDNA libraries to identify cDNA clones from the Menkes gene (Mercer et al. 1993).

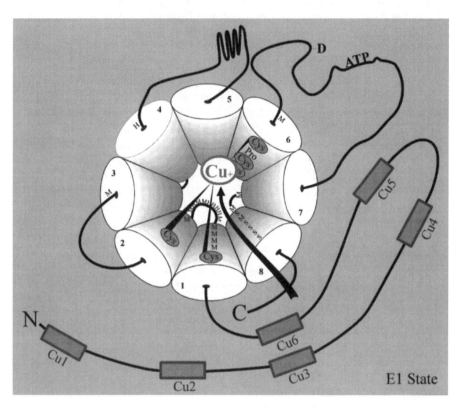

Figure 10.4 Schematic figure of *MNK,* looking from the cytoplasmic side of the membrane through the channel. The molecule is considered to have eight transmembrane (TM) loops (1–8), which form a copper channel. The six N-terminal copper binding sites (Cu1–6) are shown as floating above the membrane channel. The Cu^+ ion is shown bound to the cysteines on the transmembrane domains, presumably forming a transient complex with the copper; the large arrow indicates the possible entry path of copper. The molecule is shown in the E1 conformation. Also indicated are some methionine and histidine residues in the channel and on the loop between TM1 and 2 which may also be involved in copper binding. The aspartic acid that is phosphorylated during the reaction cycle is indicated (D), as is the ATP binding site (ATP).

3.4 Structure and Alternate Splicing of the Menkes Gene

The Menkes gene is large and complex, spanning about 150 kilobases (kb) and consisting of 23 exons (Dierick et al. 1995; Tumer et al. 1995b). The metal binding sites are found on individual exons, except for binding sites 3 and 4 which are both on exon 4. This arrangement suggests that the multiple metal binding sites have evolved by amplification of a single exon (Dierick et al. 1995). An interesting feature is the large intron of about 60 kb separating the first and second exons; this region has only recently been characterized and it contains at least one exon that is involved in alternate splicing to produce an mRNA with an additional sequence in the 5' untranslated region (H. Dierick, L. Ambrosini, J. Mercer, unpublished data). The role of this alternate exon is not known. Other alternately spliced forms are normally formed which affect the coding sequence; for example, exon 10 is sometimes spliced out, producing an mRNA encoding a protein that would lack transmembrane segments 3 and 4 (Dierick et al. 1995). Because the reading frame is not disrupted, a potentially active ATPase could be formed, but it is not clear if such forms have any physiological function or if they represent nonfunctional side products of splicing. Alternate splicing of the Wilson gene has been reported (see Section 4.1), and again the function of these alternate forms is not understood.

3.5 Mutations That Cause Menkes' Disease and Occipital Horn Syndrome

Table 10.1 summarizes the range of mutations found to cause Menkes' disease and occipital horn syndrome as well as the mouse models of these diseases. Partial or complete deletions of *MNK* are found in about 15% of MD patients (Chelly et al. 1993), indicating that the disease is caused by complete loss of *MNK*, the Menkes protein. This conclusion is supported by the analysis of other mutations of *MNK* in MD patients. RNA splicing abnormalities are quite common and usually cause an almost complete loss of normal *MNK* mRNA. Other mutations found include premature stop codons and frame shifts and missense mutations affecting important functional domains of *MNK* (Das et al. 1994) (See Table 10.1). All of these mutations are predicted to result in loss of functional *MNK* (Das et al. 1994). A recent paper identifying mutations in forty-one Menkes' patients supports this general conclusion (Tümer et al., 1997). A minor splice site mutation that still permitted some normal mRNA production was found in a case of mild Menkes' (Kaler et al. 1994), but a splice donor site mutation in the same exon, which resulted in only nonfunctional transcripts, caused severe Menkes' disease (Das et al. 1994). The limited number of occipital horn syndrome cases examined have all been due to mild splice site mutations, which allow some normal mRNA production (Das et al. 1995; Kaler et al. 1994). These cases

Table 10.1 Type of mutations in *MNK* that have been shown to produce Menkes' disease, occipital horn syndrome, and the mottled mice

Disease or Mutant	Type of Mutation[a]	Molecular Consequences
Menkes'	Premature termination	Decreased *MNK* mRNA, no *MNK*[b]
	Frameshift	Decreased *MNK* mRNA, no *MNK*[b]
	Missense	Normal *MNK* mRNA, protein produced but inactive[b]
	Major splice site[c]	Exon skipping, no *MNK* produced[b]
	Gene deletions	No *MNK* mRNA or protein
Mild Menkes'	Minor splice site[c]	Exon skipping, some normal transcript
Occiptal horn	Minor splice site[c]	Exon skipped, some normal transcript
Blotchy mouse	Minor splice site[c]	Some normal MRNA, two larger, some *MNK*[b]
Dappled mouse	Deletion?	No *MNK* mRNA
Brindled mouse[d]	Six-base deletion	Normal mRNA, *MNK* produced, low activity[b]

[a]Mutations described in Chelly et al. (1993); Das et al. (1994); Kaler et al. (1994); Das et al. (1995); Mercer et al. (1994); and Levinson et al. (1994).
[b]Effect of mutations on the protein *MNK* is predicted, no antibody studies have been reported.
[c]Major splice site refers to mutations affecting the major splice signals AG or GT, minor splice site refers to mutations affecting the flanking regions of the major splice site.
[d]Mercer and Grimes, unpublished data.

suggest that retention of even a small amount of *MNK* activity is sufficient to prevent severe Menkes' disease and result in a milder disorder, with predominant connective tissue manifestations. Das et al. (1995) speculate that lysyl oxidase is especially sensitive to the lack of *MNK*, hence explaining the predominance of connective tissue disorders with the less severe mutations.

3.6 Mutations in the Mottled Mice

The coding sequence of the murine Mnk has been obtained by sequencing cDNA clones, and mRNA abnormalities of the expression of Mnk in some of the mottled mutants have been demonstrated (Levinson et al. 1994; Mercer et al. 1994). Comparison of the predicted amino acids sequence of the murine Mnk with human *MNK* show striking areas of conservation, indicating structurally and functionally important regions of the protein. The most conserved region is found in the C-terminal half, particularly around the ATP binding domain. All of the six metal binding sites are conserved, but between these metal binding sites the amino acid conservation is less marked, indicating that some structural flexibility in this region is compatible with activity of the ATPase (Mercer et al. 1994).

As noted in Table 10.1, two larger-molecular-weight forms of Mnk mRNA in addition to mRNA of the normal size were detected in tissues of the blotchy mouse (Levinson et al. 1994; Mercer et al. 1994), and a mutation of the minor splice signals has been found, which is presumably responsible for the aberrant splicing. As noted above similar mild splice site mutations have been found in occipital horn syndrome patients, further supporting the concept that mutations in which a small amount (say 5%) of Mnk is still produced cause pronounced connective tissue abnormalities (Das et al. 1995; Levinson et al. 1994; Mercer et al. 1994).

The mottled mutant with the closest phenotype to Menkes' disease, the brindled mouse, was found to have normal size and quantity of Mnk mRNA (Mercer et al. 1994). We have recently identified a six-base-pair deletion in the Mnk mRNA from this mutant. This mutation causes the deletion of two highly conserved amino acids close to the fourth transmembrane domain and is likely to be the causative mutation (Grimes and Mercer, unpublished data). Males with the dappled (Mo^{dp}) mutation lack Mnk mRNA and die before birth, perhaps due to a genomic rearrangement in the center of the gene (Mercer et al. 1994). Thus absence of Mnk causes fetal death in the mouse, rather than neonatal death as found with MD in humans. The earlier death of the mouse mutant may suggest that mice have a greater requirement for copper during development. Two recent papers have appeared that report mutations in other mottled mutants also showing that severe mutations in *MNK* cause prenatal lethality (Reed and Boyd 1997; Cecchi et al. 1997).

3.7 Properties of Cell Lines Overexpressing the MNK Gene

Cultured cells derived from MD patients are copper sensitive, hyperaccumulate copper, and show evidence of defective copper efflux (Camakaris et al. 1980, 1982). However, copper-resistant variants of cultured Chinese hamster ovary cells (CHO cells) have reduced Cu accumulation at elevated copper concentrations in the medium, and the basis of this is enhanced copper efflux (Camakaris et al. 1995). As is shown in Figure 10.5, the copper-resistant cell lines produce much larger amounts of *MNK* mRNA, up to 100-fold more in the case of the most resistant variant CUR3, and this increased mRNA is due to amplification of the *MNK* gene. Western-blot analysis has shown an equivalently high level of *MNK* protein in the resistant cells (Camakaris et al. 1995). These data provide strong evidence for an efflux role for *MNK*.

The *CopB* gene product in the bacterium *E. hirae* is a P-type ATPase that shows significant amino acid sequence identity to *MNK*. The *CopB* ATPase has been shown to function in copper efflux (Odermatt et al. 1993) and catalyzes

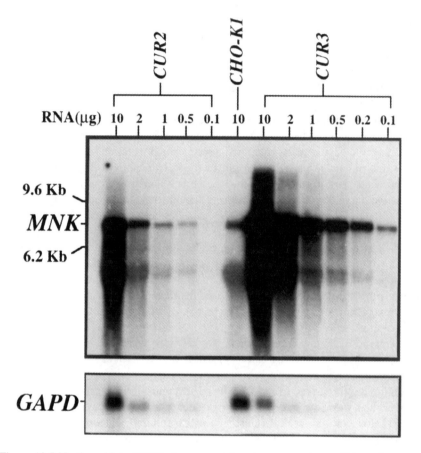

Figure 10.5 Northern blot of RNA from parental and copper-resistant Chinese hamster ovary cells. RNA was extracted from parental (K1), moderately copper-resistant (CUR2), and highly copper-resistant (CUR3) cell lines. Northern blots were prepared and probed with a hamster *MNK* cDNA as described by Camakaris et al. (1995). The amounts of RNA (μg) loaded are indicated above each lane. The blot was also probed with the housekeeping gene GAPD (glyceraldehyde-3-phosphate dehydrogenase).

ATP-driven copper and silver accumulation by inside-out membrane vesicles. Recent studies by Shen and Camakaris (unpublished data) have demonstrated ATP-driven copper accumulation by inside-out vesicles isolated from cultured CHO cell lines. The extent of copper accumulation depended on the degree of expression of *MNK* in the cells.

3.8 Regulation of the Menkes Gene

Analysis of the factors that regulate the Menkes' gene is not extensive. The gene is expressed to varying extents in a wide range of tissues in humans and mice (Paynter et al. 1994; Vulpe et al. 1993b). The liver is an interesting exception; little *MNK* mRNA can be detected in this organ in adult mice or humans (Paynter et al. 1994; Vulpe et al. 1993b). It seems clear that the role of *MNK* in copper efflux from the liver is taken by the closely related ATPase, *WND* (see Figure 10.1), and defects of that gene consequently lead to copper accumulation in the liver, leading to the copper toxicosis disorder Wilson's disease (see Section 4). Interestingly, it appears that the liver in fetal mice expresses quite high levels of *MNK* mRNA (Paynter et al. 1994). It could be that in the immature liver *MNK* substitutes to some extent for *WND*, but the cell types expressing *MNK* and *WND* need to be determined to gain a clearer idea of possible roles of *MNK* in the developing liver. Nothing has been reported of the expression of *MNK* in other tissues of the developing mouse.

Transcription of *MNK* does not appear to be regulated by copper. In the kidney of the brindled mouse, the copper concentration is about five times higher than normal and yet the *MNK* messenger RNA levels are at the same levels as in the normal kidney. Similarly, *MNK* mRNA levels in normal mice are unaffected by extreme levels of injected copper. Normal amounts of *MNK* mRNA are found in the copper-deficient brain of the brindled mouse. These results suggest that copper excess or deficiency does not affect the mRNA transcription (Paynter et al. 1994). Consistent with this conclusion is the observation that levels of *MNK* mRNA and *MNK* protein in CHO cells (normal and copper-resistant) are not altered by marked changes in the media copper concentrations. Efflux of copper, however, is elevated in high copper culture media (Shen, Petris, Camakaris, and Mercer, unpublished data), suggesting some regulation of *MNK* at a posttranslational stage.

Recent studies using cultured CHO cell lines overexpressing the *MNK* protein have provided strong evidence for a novel rapid-response regulatory system based on relocalization of the *MNK* protein. Studies using confocal immunofluorescence microscopy have shown that the *MNK* protein is localized in the *trans*-Golgi network (TGN) in the case of cells growing in basal medium (Petris et al. 1996). Localization of *MNK* to the TGN has been reported in other cell types (Yamaguchi et al. 1996; Dierick et al. 1997). Significantly we found that when cells are transferred to growth media containing elevated copper concentrations, the *MNK* protein is relocalized to the plasma membrane in an ATP-dependent process that does not involve new protein synthesis (Petris et al. 1996). The process is rapid; substantial relocalization occurs within 15 minutes. The trafficking process appears to involve vesicles. When cells are transferred back to basal medium, *MNK* protein recycles to the TGN (Petris et al. 1996). This copper-regulated trafficking of the *MNK* protein represents a novel system where the ligand is regulated the trafficking of its own transporter. Presumably, in the TGN, *MNK* functions to deliver copper to Cu-requiring enzymes that are being processed and are destined for secretion (e.g., lysyl amine oxidase). The rapid trafficking of *MNK* to the plasma membrane allows cells to respond to potentially toxic copper levels by efflux of copper at the plasma membrane (Camakaris et al. 1995).

4. Clinical Features of Wilson's Disease

Wilson's disease (WD) is a genetic copper toxicosis disorder affecting both the liver and central nervous system. The disease is caused by mutations of a gene designated *WND* or *ATP7B* that is closely related in sequence to *MNK*. Mutations of *WND* result in reduction of biliary excretion of copper and low incorporation of copper into ceruloplasmin in the liver (Figure 10.1) (Danks 1995). Because the uptake of copper from the small intestine is normal in Wilson's disease, the reduced copper excretion results in a net positive copper balance in the body, and copper gradually accumulates to high concentrations in the liver (Danks 1995). Despite being bound to metallothionein and sequestered in lysosomes (Elmes et al. 1989), the excess copper eventually causes severe liver damage, liver failure, and death. Copper also deposits in the brain, and neurological abnormalities are a major clinical feature of some patients. Fortunately, WD can be treated by reducing the uptake of copper in the small intestine with large doses of oral zinc (Brewer et al. 1987) or by treating the hepatic damage with the copper chelators, penicillamine (Walsh 1973) or ammonium tetrathiomolybdate (Brewer et al. 1994).

4.1 Identification and Structure of the Gene Affected in Wilson's Disease

Three groups have independently isolated cDNA clones from the Wilson's disease gene using the YAC clones from the chromosomal region where Wilson's disease gene had been mapped, and primers based on the Menkes gene sequence (Bull et al. 1993; Tanzi et al. 1993; Yamaguchi et al. 1993). These workers made the reasonable assumption, which proved to be correct, that the two genes might be related enough to allow PCR amplification or probing of clones from the candidate YACs. There is, however, not enough sequence identity for direct Southern-blot detection of *WND* with *MNK* probes.

The coding sequence reported by Bull et al. (1993) predicts a 1411-amino-acid product (*MNK* is 1500), although subsequent work by Petrukhin et al. (1994) suggests the actual starting methionine is 33 amino acids farther upstream. The original cDNA reported by Tanzi et al. (1993) was actually from the brain and lacks some regions found in both *MNK* and the Bull *WND* sequence from the liver. The missing segments correspond exactly to exons, and it appears that the brain contains an alternately spliced form of the gene, but it is not known if this is functional. The C-terminal region of the Bull sequence differs markedly from the Tanzi sequence, the latter being very similar to *MNK,* suggesting that the C-terminal sequence obtained by Bull et al. (1993) is either from an alternately spliced form or could be a cloning artifact. The predicted protein (*WND*) is 60% identical to the Menkes' disease gene product and has the same predicted functional domains, including the six metal binding sites in the N-terminus region of the protein. The rat *WND* sequence, however, has lost copper binding site 4 and is still functional, demonstrating that there is some redundancy in the N-terminal metal binding sites (Wu et al. 1994).

As might be expected from the similarity in the structure of *WND* and *MNK,* the gene structures are also very similar. *WND* contains 21 exons and the exon/intron boundaries closely match those of *MNK,* underscoring the likely evolution of each from a common ancestor (Dierick et al., 1995; Petrukhin et al. 1994). As noted above, *WND* is alternately spliced, however, and some of the predicted products could not function as P-type ATPases, as they lack key ATPase domains. Petrukhin et al. (1994) consider the possibility that these forms may have some regulatory function in copper transport.

Because WD is primarily a liver disease, it was a surprise that *WND* is expressed quite well in the kidney, and to a lesser extent in the placenta and brain (Bull et al. 1993). Because *MNK* is also expressed in these other tissues, it raises questions about the role of the copper transporters in cells or tissues where both are found. Copper accumulates in the kidneys of patients with Menkes' disease, so it appears that the *WND* in that organ cannot substitute for the missing *MNK.* If the proteins are expressed in the same cells of the kidney, then they are likely to have distinct roles. Alternatively, *WND* and *MNK* may be found in different cell types. We await the results of immunocytochemical localization of each with interest.

A study of Wilson's disease has demonstrated a wide range of mutations, although, in contrast to Menkes' disease, large deletions of *WND* have not been found in patients (Thomas et al. 1995a). The condition is autosomal recessive, and most patients have different mutations at each locus (i.e., they are compound heterozygotes). The different mutations complicate the phenotype/genotype analysis, but a limited number of cases of homozygous mutations suggest that the mutations predicted to cause greatest loss of *WND* activity result in earlier age of onset. Early onset can lead to misdiagnosis, because WD is not expected to occur in early childhood. The age of onset and severity of the disease can also be affected by the amount of copper in the diet, viral illnesses, and possibly genetic variability in the inducibility of metallothionein and the levels of glutathione in the liver. In some cases, the alternate splicing noted above appears to have allowed splicing out of the exon containing a severe mutation, resulting in a milder phenotype (Thomas et al. 1995b).

4.2 Animal Models of Wilson's Disease

A number of animal models of Wilson's disease have the common feature of accumulation of copper in the liver. However, only two of these have been shown to have mutations of *WND*. These animal models are useful for studies of treatment of WD and the role of *WND* in copper transport. The Long-Evans cinnamon (LEC) rat is a mutant originally identified as having a genetic predisposition to hepatitis at four months of age. The surviving animals suffer from hepatocarcinomas at around two years of age (Yoshida et al. 1987). Subsequent research has shown that the mutants accumulate high levels of hepatic copper in the first few months after birth (Li et al. 1991). This excess copper is the likely cause of the early-onset hepatitis. The predisposition for hepatoma development, however, has been recently shown to be genetically independent of the copper toxicosis (Hattori et al. 1995). A deletion of the 3′ portion of *WND* has been found in the LEC rat, thus demonstrating that this is a true model of the disease (Wu et al. 1994).

The mouse mutant toxic milk (*tx*) accumulates hepatic copper in a similar manner to the LEC rat. The name *toxic milk* was derived from the observed death of pups suckling mutant dams, and it was shown that copper-deficient milk was produced by the mutant mother (Rauch 1983). In fact, both placental transfer and delivery of copper to milk are reduced in the mutant dam. We have recently found a mutation affecting the eighth transmembrane domain of the *WND* homolog in the *tx* mouse, showing it to be a true model of WD (Theophilos et al. 1996). Copper-deficient milk has not been reported to be produced by the LEC rat dams or human mothers with Wilson's disease, and it would be of interest to further investigate whether the *tx* mouse is unique in this respect.

4.3 Are Other Genes Involved in Copper Toxicosis?

Two other proposed animal models of WD, the normal sheep and copper toxicosis in the Bedlington terrier dog, have some features which set them apart from the rodent models. In particular, much of the excess hepatic copper is found in lysosomes, whereas the excess copper is bound to MTs in the cytoplasm of the LEC rat and *tx* mouse. This difference in the pattern of copper accumulation may simply represent diversity in the way individual species transport and store copper. A more interesting possibility, however, is that the sheep and the Bedlington terrier may have a variant of a yet undiscovered hepatic copper transporter. The strongest evidence for this has been provided by Yuzbasiyan et al. (1993), who demonstrated that canine copper toxicosis may not be genetically linked to *WND*.

The presence of another genetic copper toxicosis of humans could be involved in cases of copper-associated childhood cirrhosis, which include Indian childhood cirrhosis (ICC) (Pandit and Bhave 1996). ICC is a disease of infancy characterized by massive accumulation of copper in the liver, but in contrast to Wilson's disease, ceruloplasmin levels are normal. Ingestion of milk contaminated with copper is a common factor in ICC. A recent analysis of cases of copper-associated childhood cirrhosis in the Austrian Tyrol with many of the features of ICC has demonstrated a convincing case for autosomal recessive inheritance (Muller et al. 1996). High copper intake in early infancy is apparently required for the disease to develop. In both India and Austria, the disease has virtually disappeared with the replacement of copper-lined vessels used for boiling milk with noncopper containers. It is hoped that in the next few years the genetic basis of the copper toxicity in these disorders will be determined and may identify another component of the copper pathway. Recently a novel human gene (*HAH1*) homologous to a copper binding protein which may deliver copper to superoxide dismutase in yeast has been reported (Klomp et al. 1997) but the role of *HAH1* in mammalian Cu-transport and possible involvement in genetic disorders of copper remains to be established.

5. Conclusions

The study of copper transport has clearly entered a new era with the isolation of the genes affected in Menkes' and Wilson's diseases and study of the protein products encoded by these genes. Analysis of the mutations that cause MD and WD has aiding understanding of the molecular basis of the various disease phenotypes and will be of assistance in prenatal diagnosis. In addition, studies on the function of *MNK* and *WND* are revealing new aspects of copper transport. The finding of a copper-regulated trafficking system for *MNK* is the first defined mechanism for explaining cellular copper homeostasis in mammalian cells. It is possible that copper uptake systems may also function by copper-regulated vesicular trafficking. Specific sequences on the *MNK* protein are likely to be important in the exocytic and endocytic *MNK* trafficking pathways.

The models proposed for the function of these proteins are based on comparison with other P-type ATPases, so it is important that structural features of these copper-transporting ATPases required for activity are experimentally determined. In particular, are all the six N-terminal metal binding sites required for both copper transport and copper-regulated trafficking of *MNK*? Also the functional significance of the alternate spliced forms needs to be investigated: in particular, does the alternate splicing result in the production of different forms of the proteins with distinct roles? The differences between the animal models of WD and the non-Wilson copper toxicosis syndromes in children suggest that there are other copper transport proteins to be discovered. In any event, this is an exciting period in the study of copper transport, and many of the fundamental details are likely to be resolved in the next few years.

Acknowledgment

We are grateful to Michelle Winsor for help in preparation of the figures.

References

Brewer, G. J., R. D. Dick, V. Johnson, Y. Wang, V. Yuzbasiyan-Gurkan, K. Kluin, J. K. Fink, and A. Aisen. 1994. Treatment of Wilson's disease with ammonium tetrathiomolybdate. *Arch. Neurol.* 51:545–554.

Brewer, G. J., V. Yuzbasiyan-Gurkan, and A. B. Young. 1987. Treatment of Wilson's disease. *Sem. in Neurol.* 7:209–220.

Bull, P. C., G. R. Thomas, J. M. Rommens, J. R. Forbes, and D. C. Cox. 1993. The Wilson's disease gene is a putative copper transporting P-type ATPase similar to the Menkes gene. *Nature Genet.* 5:327–337.

Camakaris, J., D. M. Danks, L. Ackland, E. Cartwright, P. Borger, and R. G. H. Cotton. 1980. Altered copper metabolism in cultured cells from human Menkes' syndrome and mottled mouse mutants. *Biochem. Genet.* 18:117–131.

Camakaris, J., D. M. Danks, M. Phillips, S. Herd, and J. R. Mann. 1982. Copper metabolism in Menkes' syndrome and mottled mouse mutants. In *Inflammatory Diseases and Copper*, ed. J. R. J. Sorenson. pp. 85–96. Humana Press, Clifton, New Jersey.

Camakaris, J., M. Petris, L. Bailey, P. Shen, P. Lockhart, T. W. Glover, C. L. Barcroft, J. Patton, and J. F. B. Mercer. 1995. Gene amplification of the Menkes (*MNK*; ATP7A) P-type ATPase gene of CHO cells is associated with copper resistance and enhanced copper efflux. *Hum. Mol. Genet.* 4:2117–2123.

Cecchi, C., M. Biasotto, M. Tosi, P. Auner. 1997. The *mottled* mouse as a model for human Menkes' disease: identification of mutations in the *ATP7a* gene. *Hum. Molec. Genet.* 6:425–433.

Chelly, J., Z. Turmer, T. Tonnerson, A. Petterson, Y. Ishikawa-Brush, N. Tommerup, N. Horn, and A. P. Monaco. 1993. Isolation of a candidate gene for Menkes' disease that encodes a potential heavy metal binding protein. *Nature Genet.* 3:14–19.

Danks, D. M. 1986. Of mice and men, metals and mutations. *J. Med. Genet.* 23:99–106.

Danks, D. M. 1995. Disorders of copper transport, p. 2211–2235. In The Metabolic and Molecular Basis of Inherited Disease, 7th edition, eds. C. R. Scriver, A. L. Beaudet, W. M. Sly, and D. Valle. McGraw-Hill, New York.

Danks, D. M., P. E. Campbell, B. J. Stevens, V. Mayne, and E. Cartwright. 1972. Menkes' kinky hair syndrome: An inherited defect in copper absorption with widespread effects. *Pediatrics* 50:188–201.

Darwish, H. M., J. E. Hoke, and M. J. Ettinger. 1983. Kinetics of Cu(II) transport and accumulation by hepatocytes from copper-deficient mice and the brindled mouse model of Menkes' disease. *J. Biol. Chem.* 258:13621–13626.

Das, S., B. Levinson, C. Vulpe, S. Whitney, J. Gitschier, and S. Packman. 1995. Similar splicing mutations of the Menkes/mottled copper-transporting ATPase gene in occipital horn syndrome and the blotchy mouse. *Amer. J. Hum. Genet.* 56:570–576.

Das, S., B. Levinson, S. Whitney, C. Vulpe, S. Packman, and J. Gitschier. 1994. Diverse mutations in patients with Menkes' disease often lead to exon skipping. *Amer. J. Hum. Genet.* 55:883–889.

Davidson, L. A., S. L. McOrmond, and E. D. Harris. 1994. Characterization of a particulate pathway for copper in K562 cells. *Biochim. Biophys. Acta* 1221:1–6.

Dierick, H. A., L. Ambrosini, J. Spencer, T. W. Glover, and J. F. B. Mercer. 1997. Molecular structure of the Menkes disease gene (ATP7A). *Human Molecular Genetics* 6:409–417.

Dierick, H. A., A. N. Adam, J. F. Escara-Wilke, T. W. Glover. 1997. Immunochemical localization of the Menkes' copper transporter (ATP7a) to the transGolgi compartment of the Golgi complex. *Proc. Natl. Acad. Sci. U.S.A.* (in press).

Dijkstra, M., G. In't Velt, G. J. van den Berg, M. Muller, F. Kuipers, and R. J. Vonk. 1995. Adenosine triphosphate-dependent copper transport in isolated rat liver plasma membranes. *J. Clin. Invest.* 95:412–416.

Elmes, M. E., J. P. Clarkson, N. J. Mathy, and B. Jasani. 1989. Metallothionein and copper in liver disease with copper retention—A histopathological study. *J. Path.* 158:131–137.

Evering, W. E. N. D., S. Haywood, I. Bremner, and J. Trafford. 1991. The protective role of metallothionein in copper-overload: I. Differential distribution of immunoreactive metallothionein in copper-loaded rat liver and kidney. *Chem.-Biol. Interactions* 78:283–295.

Freedman, J. H., M. R. Ciriolo, and J. Peisach. 1989. The role of glutathione in copper metabolism and toxicity. *J. Biol. Chem.* 264:5598–5605.

Fu, D., T. J. Beeler, and T. M. Dunn. 1995. Sequence, mapping and disruption of CCC2, a gene that cross-complements the Ca^{2+} sensitive phenotype of csg1 mutants and encodes a P-type ATPase belonging to the Cu^{2+} ATPase subfamily. *Yeast* 11:283–292.

Ge, Z., K. Hiratsuka, and D. E. Taylor. 1995. Nucleotide sequence and mutational analysis indicate that two *Helicobacter pylori* genes encode a P-type ATPase and a cation-binding protein associated with copper transport. *Mol. Microbiol.* 15:97–106.

Gerdes, A.-M., T. Tonnesen, N. Horn, T. Grisar, W. Marg, A. Muller, R. Reinsch, N. W. Barton, P. Guiraud, A. Joannard, M. J. Richard, and F. Guttier. 1990. Clinical expression of Menkes' syndrome in females. *Clin Genet.* 38:452–459.

Goka, T. J., R. E. Stevenson, P. M. Hefferan, and R. R. Howell. 1976. Menkes' disease: A biochemical abnormality in cultured human fibroblasts. *Proc. Natl. Acad. Sci. U.S.A.* 73:604–606.

Goode, C. A., C. T. Dinh, and M. C. Linder. 1989. Mechanism of copper transport and delivery in mammals: Review and recent findings. *Adv. Exp. Med. Biol.* 258:131–144.

Gross, J. B., Jr., B. M. Myers, L. J. Kost, S. M. Kuntz, and N. F. LaRusso. 1989. Biliary copper excretion by hepatocyte lysosomes in the rat. Major excretory pathway in experimental copper overload. *J. Clin. Invest.* 83:30–39.

Harris, D. I. M., and A. Sass-Kortsak. 1967. The influence of amino acids on copper uptake by rat liver slices. *J. Clin. Invest.* 46:659–677.

Hattori, A., M. Sawaki, K. Enomoto, N. Tsuzuki, H. Isomura, T. Kojima, Y. Kamibayashi, N. Sugawara, T. Sugiyama, and M. Mori. 1995. The high hepatocarcinogen susceptibility of LEC rats is genetically independent of abnormal copper accumulation in the liver. *Carcinogenesis* 16:491–494.

Herd, S. M., J. Camakaris, R. Christofferson, P. Wookey, and D. M. Danks. 1987. Uptake and efflux of copper-64 in Menkes'-disease and normal continuous lymphoid cell lines. *Biochem. J.* 247:341–347.

Horn, N. 1976. Copper incorporation studies on cultured cells for prenatal diagnosis of Menkes' disease. *Lancet* 1:1156–1158.

Houwen, R., M. Dijkstra, F. Kuipers, E. P. Smit, R. Havinga, and R. J. Vonk. 1990. Two pathways for biliary copper excretion in the rat. The role of glutathione. *Biochem. Pharmacol.* 39:1039–1044.

Kaler, S. G. 1994. Menkes' disease. *Advances Pediatr.* 41:262–303.

Kaler, S. G. 1996. Menkes' disease mutations and response to early copper histidine treatment. *Nature Genetics* 13:21–22.

Kaler, S. G., N. R. M. Buist, C. S. Holmes, D. S. Goldstein, R. C. Miller, and W. A. Gahl. 1995. Early copper therapy in classic Menkes' disease patients with a novel splicing mutation. *Ann. Neurol.* 38:921–928.

Kaler, S. G., L. K. Gallo, V. K. Proud, A. K. Percy, Y. Mark, N. A. Segal, D. S. Goldstein, C. S. Holmes, and W. A. Gahl. 1994. Occipital horn syndrome and a mild Menkes phenotype associated with splice site mutations at the *MNK* locus. *Nature Genet.* 8:195–202.

Kapur, S., J. V. Higgins, K. Delp, and B. Rogers. 1987. Menkes' syndrome in a girl with X-autosome translocation. *Amer. J. Med. Genet.* 26:503–510.

Klomp, L. W., S.-J. Lin, D. S. Yuan, R. D. Klausner, V. C. Culotta, and J. D. Gitlin. 1997. Identification and functional expression of *HAH*1, a novel human gene involved in copper homeostasis. *J. Biol. Chem.* 272:9221–9226.

Kodama, H. (1993). Recent developments in Menkes' disease. *J. Inher. Metab. Disease* 16:791–799.

Kumar, S., C. Rowse, and P. Hochstein. 1978. Copper-induced generation of superoxide in human red cell membrane. *Biochem. Biophys. Res. Comm.* 83:587–593.

Levinson, B., C. Vulpe, B. Elder, C. Martin, F. Verley, S. Packman, and J. Gitschier. 1994. The mottled gene is the mouse homologue of the Menkes' disease gene. *Nature Genet.* 6:369–373.

Li, Y., Y. Togashi, S. Sato, T. Emoto, J.-H. Kang, N. Takeichi, H. Kobayashi, Y. Kojima, Y. Une, and J. Uchino. 1991. Spontaneous hepatic copper accumulation in Long-Evans cinnamon rats with hereditary hepatitis: A model of Wilson's disease. *J. Clin. Invest.* 87:1858–1861.

Lyon, M. F., and A. G. Searle. 1990. Mo locus, chromosome X, pp. 241–244. In *Genetic Variants and Strains of the Laboratory Mouse,* eds. M. F. Lyon and A. G. Searle. Oxford University Press, Oxford.

MacLennan, D. H., D. M. Clarke, T. W. Loo, and I. S. Skerjanc. 1992. Site-directed mutagenesis of the Ca^{2+} ATPase of sarcoplasmic reticulum. *Acta Physiol. Scand.* 146:141–150.

Mann, J., J. Camakaris, and D. M. Danks. 1979. Copper metabolism in mottled mouse mutants. Distribution of ^{64}Cu in brindled mice. *Biochem. J.* 180:613–619.

Mann, J., J. Camakaris, and D. M. Danks. 1980. Copper metabolism in mottled mouse mutants: Defective placental transfer of Cu-64 to foetal brindled (Mo^{br}) mice. *Biochem. J.* 186:629–631.

McArdle, H. J. 1992. The transport of copper across the cell membrane: Different mechanisms for different metals? *Proc. Nutr. Soc.* 51:199–209.

McArdle, H. J., S. M. Gross, and D. M. Danks. 1988. Uptake of copper by mouse hepatocytes. *J. Cell. Physiol.* 136:373–378.

Melchers, K., T. Weitzenegger, A. Buhmann, W. Steinhilber, G. Sachs, and K. P. Schafer. 1996. Cloning and membrane topology of a P-type ATPase from *Helicobacter pylori. J. Biol. Chem.* 271:446–457.

Menkes, J. H., M. Alter, G. K. Stegleder, D. R. Weakley, and J. H. Sung. 1962. A sex-linked recessive disorder with retardation of growth, peculiar hair, and focal cerebral and cerebellar degeneration. *Pediatrics* 29:764–779.

Mercer, J. F. B., A. Grimes, L. Ambrosini, P. Lockhart, J. A. Paynter, H. Dierick, and T. W. Glover. 1994. Mutations in the murine homologue of the Menkes' disease gene in dappled and blotchy mice. *Nature Genet.* 6:374–378.

Mercer, J. F. B., J. Livingston, B. K. Hall, J. A. Paynter, C. Begy, S. Chandrasekharappa, P. Lockhart, A. Grimes, M. Bhave, D. Siemenack, and T. W. Glover. 1993. Isolation of a partial candidate gene for Menkes' disease by positional cloning. *Nature Genet.* 3:20–25.

Muller, T., H. Feichtinger, H. Berger, and W. Muller. 1996. Endemic Tyrolean infantile cirrhosis: An ecogenetic disorder. *Lancet* 347:877–880.

Nederbragt, H. 1989. Effect of the glutathione-depleting agents diethylmaleate, phorone and buthionine sulfoximine on biliary copper excretion in rats. *Biochem. Pharmacol.* 38:3399–3406.

Odermatt, A., H. Suter, R. Krapf, and M. Solioz. 1993. Primary structure of two P-type ATPases involved in copper homeostasis in *Enterococcus hirae. J. Biol. Chem.* 268:12775–12779.

Owen, C. A. 1965. Metabolism of radiocopper (Cu64) in the rat. *Amer. J. Physiol.* 221:1722–1727.

Pandit, A., and S. Bhave. 1996. Present interpretation of the role of copper in Indian childhood cirrhosis. *Amer. J. Clin. Nutr.* 63:830S–835S.

Paynter, J. A., A. Grimes, P. Lockhart, and J. F. B. Mercer. 1994. Expression of the Menkes gene homologue in mouse tissues: Lack of effect of copper on the mRNA levels. *FEBS Lett.* 351:186–190.

Peltonen, L., H. Kuivaniemi, A. Palotie, N. Horn, I. Kaitila, and K. I. Kivirikko. 1983. Alterations in copper and collagen metabolism in the Menkes' syndrome and a new type of the Ehrlers-Danlos syndrome. *Biochem.* 22:6156–6163.

Petris, M. J., J. F. B. Mercer, J. G. Culvenor, P. Lockhart, P. A. Gleeson, and J. Camakaris. 1996. Ligand-regulated transport of the Menkes' copper P-type ATPase efflux pump from the Golgi apparatus to the plasma membrane; a novel mechanism of regulated trafficking. *EMBO J.* 15:6084–6095.

Petrukhim, K., S. Lutsenko, I. Chernov, B. M. Ross, J. H. Kaplan, and T. C. Gilliam. 1994. Characterization of the Wilson disease gene encoding a copper transporting ATPase: Genomic organization, alternative splicing, and structure/function predictions. *Hum. Mol. Genet.* 9:1647–1656.

Procopis, P., J. Camakaris, and D. M. Danks. 1981. A mild form of Menkes' syndrome. *J. Pediatr.* 98:97–100.

Rad, M. R., L. Kirchrath, and C. P. Hollenberg. 1994. A putative P-type Cu^{2+} transporting ATPase on chromosome II of *Saccharomyces cerevisiae. Yeast* 10:1217–1225.

Rauch, H. 1983. Toxic milk, a new mutation affecting copper metabolism in the mouse. *J. Hered.* 74:141–144.

Reed, Y., and Y. Boyd. 1997. Mutation analysis provides additional proof that *mottled* is the mouse homologue of Menkes' disease. *Hum. Molec. Genet.* 6:417–423.

Rowe, D. W., E. B. McGoodwin, G. R. Martin, and D. Grahn. 1977. Decreased lysyl oxidase activity in the aneurism-prone mottled mouse. *J. Biol. Chem.* 252:939–942.

Royce, P. M., J. Camakaris, and D. M. Danks. 1980. Reduced lysyl oxidase activity in skin fibroblasts from patients with Menkes' syndrome. *Biochem. J.* 192:579–586.

Royce, P. M., J. Camakaris, J. R. Mann, and D. M. Danks. 1982. Copper metabolism in mottled mouse mutants. *Biochem. J.* 202:369–371.

Sherwood, G., B. Sarkar, and A. S. Kortsak. 1989. Copper histidinate therapy in Menkes' disease: Prevention of progressive neurodegeneration. *J. Inher. Metab. Disease* 12:393–396.

Silver, S., G. Nucifora, L. Chu, and T. K. Misra. 1989. Bacterial resistance ATPases: Primary pumps for exporting toxic cations and anions. *Trends in Biochem. Sci.* 14:76–80.

Solioz, M., A. Odermatt, and R. Krapf. 1994. Copper pumping ATPases: Common concepts in bacteria and man. *FEBS Lett.* 346:44–47.

Starcher, B., J. A. Madaras, D. Fisk, E. F. Perry, and C. H. Hill. 1978. Abnormal cellular copper metabolism in the blotchy mouse. *J. Nutr.* 108:1229–1233.

Tanzi, R. E., K. Petrukhin, I. Chernov, J. L. Pellequer, W. Wasco, B. Ross, D. M. Romano, E. Parano, L. Pavone, L. M. Brzustowicz, M. Devoto, J. Peppercorn, A. I. Bush, I. Sternlieb, M. Pirastu, J. F. Gusella, O. Evgrafov, G. K. Penchaszadeh, B. Honig, I. S. Edelman, M. B. Soares, I. H. Scheinberg, and T. C. Gilliam. 1993. The Wilson's disease gene is a copper transporting ATPase with homology to the Menkes' disease gene. *Nature Genet.* 5:344–350.

Theophilos, M. B., D. W. Cox, and J. F. B. Mercer. 1996. The toxic milk mouse is a murine model of Wilson's disease. *Hum. Mol. Genet.* 5:1619–1624.

Thomas, G. R., J. R. Forbes, E. A. Roberts, J. M. Walshe, and D. W. Cox. 1995a. The Wilson's disease gene: Spectrum of mutations and their consequences. *Nature Genet.* 9:210–216.

Thomas, G. R., O. Jensson, G. Gudmundsson, L. Thorsteinsson, and D. W. Cox. 1995b. Wilson's disease in Iceland: A clinical and genetic study. *Amer. J. Hum. Genet.* 56:1140–1146.

Tønneson, T., A. Petterson, T. A. Kruse, A.-M. Gerdes, and N. Horn. 1992. Multi-point linkage analysis in Menkes' disease. *Amer. J. Hum. Genet.* 50:668–672.

Tümer, Z., N. Horn, Tønneson, J. Christodoulou, J. T. R. Clarke, and B. Sarkar. 1995a. Early copper-histidine treatment for Menkes' disease. *Nature Genet.* 12:11–13.

Tümer, Z., B. Vural, T. Tønnesen, J. Chelly, A. P. Monaco, and N. Horn. 1995b. Characterization of the exon structure of the Menkes' disease gene using vectorette PCR. *Genomics* 26:437–442.

Tümer, Z., C. Lund, J. Tolshave, B. Vural, T. Tønnesen, and N. Horn. 1997. *Am. J. Hum. Genet.* 60:63–71.

Van Campen, D. R. 1971. Absorption of copper from the gastrointestinal tract, pp. 211–227. In *Intestinal Absorption of Metal Ions, Trace Elements and Radionuclides*, eds. S. C. Skoryna and D. Waldron-Edwards. Pergamon Press, Oxford.

Verga, V., B. K. Hall, S. Wang, S. Johnson, J. V. Higgins, and T. W. Glover. 1991. Localization of the translocation breakpoint in a female with Menkes' syndrome to Xq13.2–q13.3 proximal to PGK-1. *Amer. J. Hum. Genet.* 48:1133–1138.

Vulpe, C., B. Levinson, S. Whitney, S. Packman, and J. Gitschier. 1993a. (Correction) Isolation of a candidate gene for Menkes' disease and evidence that it encodes a copper-transporting ATPase. *Nature Genet.* 3:273.

Vulpe, C., B. Levinson, S. Whitney, S. Packman, and J. Gitschier. 1993b. Isolation of a candidate gene for Menkes' disease and evidence that it encodes a copper-transporting ATPase. *Nature Genet.* 3:7–13.

Walsh, J. M. 1973. Copper chelation in patients with Wilson's disease. *Quarterly J. Med.* 42:441–450.

Weiss, K. C., and M. C. Linder. 1985. Copper transport in rats involving a new transport protein. *Amer. J. Physiol.* 249:E77–E88.

Wu, J., J. R. Forbes, H. S. Chen, and D. W. Cox. 1994. The LEC rat has a deletion in the copper transporting ATPase homologous to the Wilson disease gene. *Nature Genet.* 7:541–545.

Xu, G. Q., T. Yamano, and M. Shimada. 1994. Copper distribution in fetus and placenta of the macular mutant mouse as a model of Menkes' kinky hair disease. *Biol. Neonate* 66:302–310.

Yamaguchi, Y., M. E. Heiny, and J. D. Gitlin. 1993. Isolation and characterization of a human liver cDNA as a candidate gene for Wilson's disease. *Biochem. Biophys. Res. Commun.* 197:271–277.

Yamaguchi, Y., M. E. Heiny, M. Suzuki, and J. D. Gitlin. 1996. Biochemical characterization and intracellular localization of the Menkes' disease protein. *Proc. Natl. Acad. Sci. U.S.A.* 93:14030–14035.

Yoshida, M. C., R. Masuda, M. Sasaki, N. Takeichi, H. Kobayashi, K. Dempo, and M. Mori. 1987. New mutation causing hereditary hepatitis in the laboratory rat. *J. Hered.* 78:361–365.

Yoshimura, N. 1994. Histochemical localization of copper in various organs of brindled mice. *Pathol. Int.* 44:14–19.

Yuzbasiyan, G. V., S. Wagnitz, S. H. Blanton, and G. J. Brewer. 1993. Linkage studies of the esterase D and retinoblastoma genes to canine copper toxicosis: A model for Wilson's disease. *Genomics* 15:86–90.

III

LOWER EUKARYOTES AND PLANTS

11

Metal Ion Stress in Yeast

Dennis R. Winge, Andrew K. Sewell, Wei Yu,
Joanne L. Thorvaldsen, and Rohan Farrell

1. Physiological Responses to Stress Induced by Metal Ions in Yeast

It is well known that metal ions are essential for normal physiology. At least 16 metal ions are known to be essential for life. The implication of being essential is that an optimal, intracellular concentration of each particular metal ion is required for homeostasis. Reduced concentrations cause reduced growth as a result of metal ion deficiency, and excess concentrations can cause metal-induced toxicity (Figure 11.1). The key issue is the intracellular concentration of available metal ions, which are likely to be present as metal complexes with dissociable ligands. Concentrations of available metal ions above or below a threshold value may evoke a variety of coordinated physiological responses that compensate for environmental changes to inadequate or excess metal ion concentrations. Although homeostasis can be maintained over a limited range of metal ion concentrations, extreme conditions of low or high concentrations in the environment or diet can result in impaired physiology (Figure 11.1). Deficiency of an essential metal ion may impair physiology through loss of critical enzymatic functions of metalloenzymes. Excessive metal concentrations may affect cellular metabolism in many ways, including disruption of cell membrane integrity, alteration of the cellular redox state, and inhibition of respiration. DNA replication, or protein synthesis (Ochai 1987; Gadd and White 1989; Agarwal et al. 1989). These toxic effects may arise from direct inhibition of protein function or indirect mechanisms such as metal-induced initiation of free radical reactions (Halliwell et al. 1988).

Cell survival depends on the ability to regulate the intracellular concentration of both essential and nonessential metal ions. Nonessential metal ions, such as Cd(II), can accumulate within cells because of an inability of membrane transport-

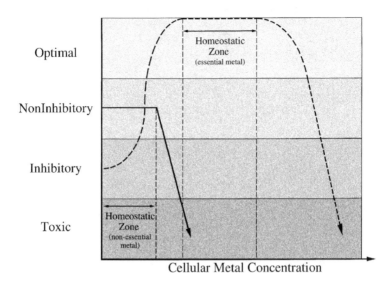

Cellular Metal Concentration

Figure 11.1 Metal ion homeostasis maintained over a limited range of metal ion concentrations. The concentration range in which homeostasis is achieved varies with the particular metal ion.

ers to discriminate between chemically related essential and nonessential metal ions.

The threshold level separating deficiency from toxicity of essential metals and toxicity of nonessential metal ions varies from metal to metal. These differences are related to the available homeostatic mechanisms, their efficiencies, and the toxic potential of a given metal ion. Considerable effort is being made to elucidate mechanisms by which cells sense intracellular metal concentrations. However, it is unclear what cells sense when the concentration of a particular metal ion is low. Metal ion deficiency leads to stress responses that are poorly understood for many metal ions. For example, iron availability is a significant biological problem. Iron deficiency in many species results in enhanced synthesis and secretion of siderophores (Crosa 1989; Mei and Leong 1994). Iron deficiency in yeast is also known to increase the Fe(II) transport system (Dancis et al. 1990a,b; Askwith et al. 1994). Iron-deficient barley roots were shown to have enhanced levels of metallothionein mRNA (Okumura et al. 1991). Studies on the effects of metal ion insufficiency are complicated by the technical problems of reducing the concentration of a given metal ion in culture media to low enough levels to cause deficiency.

This review will focus on progress made in determining the physiological responses when metal ions are present in excess. In addition to specific induction responses made to a given metal ion, a more general response to the metal-induced damage occurs. Together these responses protect the cell against metal-induced toxicity. Our focus will be yeast systems, particularly *Saccharomyces cerevisiae,* and we will examine the responses to one essential heavy metal ion, copper, and one nonessential metal ion, cadmium.

2. Cell Responses Specific to Copper Stress

Yeast cells require copper ions for a variety of enzymes including cytoplasmic superoxide dismutase (SOD), cytochrome oxidase, and Fet3. Most yeast strains can grow in culture medium containing copper salts in excess of 0.1 mM. Laboratory strains able to grow in medium containing between 0.3 and 1 mM Cu(II) are designated copper resistant (Cu^r). Cells unable to grow in medium containing this range of copper concentrations are designated copper sensitive (Cu^s). One discriminating factor between copper-sensitive and copper-resistant cells is the metallothionein (MT) locus. Regulation of MT expression is a cell response specific to copper stress in yeast, but a cell response common to a number of metal ions in animal cells. The repertoire of induced responses extends to other pathways, only some of which are well characterized.

2.1 Metallothionein

Genetic crosses between Cu^r and Cu^s strains demonstrated that copper resistance was mediated by a single genetic element, designated *CUP1,* later shown to encode MT (Brenes-Pomales et al. 1955; Fogel and Welch 1982; Fogel et al. 1983; Butt et al. 1984; Karin et al. 1984). The genetic analysis of copper resistance was followed by the targeted disruption of a single *CUP1* gene that reduced copper resistance from 100 µM to less than 10 µM (Hamer et al. 1985; Ecker et al. 1986). In addition, tandem gene amplification of *CUP1* results in copper resistance (Fogel and Welch 1982). The repeating DNA element within an amplified *CUP1* locus is a 2-kb sequence and contains two transcription units, one of which encodes MT (Karin et al. 1984). The second open reading frame is nonessential for the copper resistance phenotype (Karin et al. 1984; Fogel and Welch 1983). In strains examined, sensitive (Cu^s) cells contained a single 2-kb unit, whereas Cu^r strains contained multiple tandemly iterated units, from 2 to 14 copies of *CUP1* on chromosome VIII (Karin et al. 1984; Fogel et al. 1983). Thus, copper resistance in *S. cerevisiae* is related to the gene copy number of *CUP1.* It should be noted that additional *CUP1* sequences have also been observed at a locus on chromosome XVI (Naumov et al. 1992).

   S. cerevisiae contains a second metallothionein locus designated *CRS5* (Culotta et al. 1994). Disruption of *CRS5* resulted in only a modest copper sensitivity in strains containing a wild-type *CUP1* locus (Culotta et al. 1994). One interpretation of this study was that *CUP1* was dominant over *CRS5* in copper tolerance. A direct comparison between *CUP1* and *CRS5* was not possible in the first study as the disruption was made in a strain containing an amplified *CUP1* locus. To evaluate the role of *CRS5* in copper homeostasis, we compared the effectiveness of MTs encoded by *CRS5* and *CUP1* in conferring copper tolerance by swapping the promoters of the two MT genes and Cu-binding properties of the two molecules (Jensen et al. 1996). *CUP1* was found to be dominant over *CRS5* in conferring copper tolerance and it was found that this *CUP1* dominance arises predominantly from differences in the promoters. A second contributing factor in the dominance of *CUP1* MT is that Cu(I) ions associated with *CRS5* are more exchange labile compared to *CUP1* (Jensen et al. 1996).

MT genes are the dominant determinant of copper resistance in other species. The imperfect yeast, *C. glabrata,* contains three distinct chromosomal MT loci, one of which is present as tandemly duplicated genes (Mehra et al. 1990). Targeted disruption of the amplified MT locus and a second MT gene conferred a copper-sensitive phenotype (Mehra et al. 1992). Targeted MT disruptions have also been reported in cyanobacteria and the mouse, and in each case the only observed phenotype was metal sensitivity (Mehra et al. 1992; Michalska and Choo 1993; Masters et al. 1994; Turner et al. 1993).

MT confers copper resistance in yeast by buffering the intracellular Cu(I) concentration. Cu(I) ions bind to *S. cerevisiae* Cup1 within a single Cu_7S_{10} complex with cysteinyl thiolate ligands (Narula et al. 1991). The polycopper cluster exhibits mixed coordination geometries with 5 Cu(I) ions showing trigonal geometry and 2 Cu(I) ions with diagonal thiolate coordination (Narula et al. 1991). Thus, the CuCup1MT complex is distinct from known CdMT complexes that bind Cd(II) ions in tetrahedral geometry (Robbins et al. 1991). The CuCrs5 complex consists of 11–12 Cu(I) ions bound (Jensen et al. 1996). The presence of 19 cysteinyl thiolates may result in a CuCrs5 structure consisting of two domains, each enfolding a Cu(I)-thiolate cluster analogous to mammalian MTs.

There are two responses in the yeast MT loci to copper stress. One response involves further amplification of tandemly arrayed MT genes (Fogel et al. 1983; Mehra et al. 1990). The increase in the *CUP1* copy number arises by nonreciprocal recombination or gene conversion (Fogel and Welch 1982, 1983; Fogel et al. 1983). In *S. cerevisiae* amplification of *CUP1* can proceed to upwards of 14 iterated genes (Fogel and Welch 1982; Fogel et al. 1983). In *C. glabrata* amplification of *MTIIa* was shown to increase from three tandemly arrayed genes in the wild-type strain to near 30 copies in Cu-stressed cells (Mehra et al. 1990). It is not clear what limits the extent of amplification of *CUP1* near 14 copies and allows amplification of *MTIIa* in *C. glabrata* to proceed to 30 repetitions.

The second response in the yeast MT loci to copper stress is copper-induced transcriptional activation of MT gene expression (Figure 11.2) (Hamer et al.

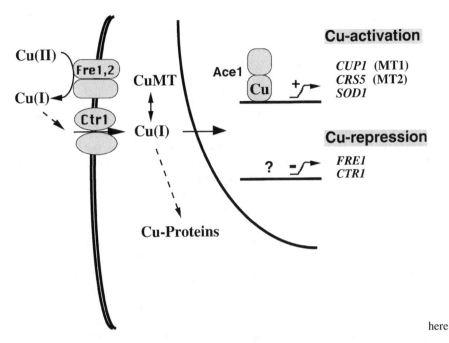

here

Figure 11.2 Cellular responses to copper ions in yeast. Cu(I) ions taken up by yeast cells are buffered by metallothioneins. Above a threshold concentration Cu(I) ions mediate the repression of at least two genes, one encoding the Fre1 metalloreductase and one encoding Ctr1, the high-affinity copper transporter. Cu(I) ions also mediate the transcriptional activation of three genes, two of which encode metallothioneins. The Cu-activation occurs through the Ace1. Binding of Cu(I) ions to Ace1 induces a DNA-binding conformation.

1995). The extent of Cu-induced transcription of MT genes is highly dependent on basal expression, but induction ratios in excess of 15 have been reported for *CUP1* (Hamer et al. 1985; Furst et al. 1988). The response of MT transcriptional activation is restricted to copper stress in yeast (Furst et al. 1988). This is in contrast to mammalian cells in which multiple metal ions mediate MT induction (Palmiter 1987). Yeast MT will confer resistance to metals other than copper when chimeric yeast MTs are constructed in which *CUP1* is placed under a constructive promoter (Ecker et al. 1986; Yu et al. 1994). *CUP1* MT is capable of binding Cd(II) ions, yet *CUP1* does not impart cadmium resistance unless the gene can be induced (Yu et al. 1994). MT confers resistance only to those metal ions capable of inducing the transcriptional activation of MT genes (Furst et al. 1988). In *S. cerevisiae* and *C. glabrata*, Ag(I) is the only other metal ion known to induce MT expression (Narula et al. 1991). Induction by Cu(I) and Ag(I) is expected as these ions exhibit similar coordination properties (Dance 1986). The mechanism for metal selectivity is discussed in Section 2.2.

2.2 Metallothionein Regulatory Molecules

Metal ion induction of MT expression was shown to reside in DNA sequences in the 5' flanking region to the MT open reading frame (Thiele and Hamer 1986). Deletion analysis of the 5' flanking region of *CUP1* led to the identification of a 39-bp DNA sequence that conferred limited copper regulation on a heterologous promoter when present in two tandem copies (Thiele and Hamer 1986). The DNA sequence has come to be designated UAS_c, for Cu-induced, upstream activation sequence (UAS) (Furst et al. 1988). Chromosomal footprinting detected a Cu-dependent DNaseI footprint of this UAS_c DNA sequence (Huibregtse et al. 1989). The implication from this work was that one or more cellular proteins was binding to UAS_c in a copper-dependent manner. This prediction was confirmed by cloning of the responsible *trans*-acting factor (Thiele 1988).

The *trans*-acting factor mediating Cu-dependent regulation of *CUP1* has been designated Ace1 (Figure 11.2) (Furst et al. 1988; Thiele 1988; Welch et al. 1989). The Cu-dependent footprint of UAS_c was not observed in *ace1-1* mutant cells and overexpression of *ACE1* episomally resulted in two additional regions of protection to DNaseI cleavage outside of UAS_c (Huibregtse et al. 1989). The UAS_c region extends from -107 to -140 upstream of the *CUP1* transcription start site and contains an inverted repeat. Only the upstream section of this repeat is fully protected in Cu-treated cells with one copy of *ACE1* (Huibregtse et al. 1989).

Using a cell-free transcription initiation system, it was shown that efficient transcription of *CUP1* required the *CUP1* DNA template, Ace1, copper ions, a mouse nuclear extract to provide general transcription factors, and RNA polymerase (Culotta et al. 1989; Buchman et al. 1990). A 45-bp oligonucleotide containing UAS_c could functionally replace the *CUP1* promoter and yielded transcription at nearly 40% of the efficiency of the complete *CUP1* promoter (Culotta et al. 1989). Thus, control sequences other than UAS_c are likely to be important for efficient Cu-induced expression of *CUP1*. Two additional candidate UAS elements exist in the 5' *CUP1* sequences as single elements that resemble the left half of the UAS_c palindrome, but the relative significance of these elements has not been addressed (Macreadie et al. 1994).

Mutations within UAS_c or Ace1 diminished Cu-induced expression of *CUP1* (Furst et al. 1988; Buchman et al. 1989). Mutations in UAS_c that abolish function are clustered in the 5' half of the palindrome (Furst et al. 1988). The critical nucleotides span a 16-nucleotide segment. Ace1 contacts DNA over one and one-half turn of the DNA helix and makes minor groove interactions between two major groove interaction sites (Buchman et al. 1990; Dobi et al. 1995)

(Figure 11.3). A mutation at codon 11 of Ace1 (*ace1-1*) results in a decrease in the apparent affinity of Ace1 to UAS_c and loss of the 5' distal major groove interaction sites (Buchman et al. 1990) (Figure 11.3).

The 225-residue Ace1 polypeptide consists of two regions with distinct functions (Furst et al. 1988; Szczypka and Thiele 1989). The binding of Ace1 to UAS_c maps to the N-terminal half of Ace1, and this segment of the polypeptide contains multiple cysteinyl residues in Cys-x-Cys or Cys-x-x-Cys sequence motifs (Furst et al. 1988; Buchman et al. 1989). Eleven cysteinyl residues are critical for the Ace1 function of Cu-induced expression of *CUP1* (Hu et al. 1990). The C-terminal segment of Ace1 is similar to many fungal transcription factors in the abundance of acidic residues and is believed to be the transactivation domain. This domain is critical for Cu-induced *CUP1* expression (Thiele 1988).

Cu-induced expression of MT genes in *Candida glabrata* is mediated by Amt1. Amt1 is 50% identical to Ace1 of *S. cerevisiae* in the DNA binding N-terminal segment of the polypeptide (Zhou and Thiele 1991). The 11 critical cysteinyl residues in Ace1 are conserved in Amt1. Multiple Amt1 binding sites exist within the 5' flanking sequences of the three MT genes in *C. glabrata* (Zhou et al. 1992). As expected, the various Amt1 binding sites are related in sequence. Within an Amt1 binding site a tetranucleotide core region exists that appears in Ace1 binding sites (Zhou et al. 1992) (Figure 11.3). Amt1 can functionally replace Ace1 in *S. cerevisiae* and confer Cu-induced expression on *CUP1* (Thorvaldsen et al. 1993). However, Ace1 expressed episomally in *C. glabrata* was unable to functionally replace Amt1 in *amt1-1 C. glabrata*, whereas episomally expressed *AMT1* restored copper tolerance (Thorvaldsen et al. 1995).

The mechanism of Cu-induced expression of yeast MT genes lies in the Cu-activation of Ace1 and Amt1. We have shown that activated Ace1 and Amt1 contain a tetracopper center with cysteinyl thiolates as ligands (Thorvaldsen et

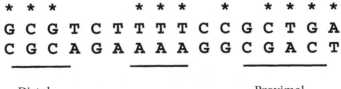

Distal   Proximal
Major Groove   Minor Groove   Major Groove

Figure 11.3 Cu(I)-induced upstream activation sequence of *CUP1*. The sequence shown is the 5' half of the palindromic UAS_c. CuAce1 makes base-specific contacts within two major groove regions separated by an A/T-rich region in which minor groove contacts are made (Buchman et al. 1990; Dobi et al. 1995). The asterisks above the sequence specify bases that were shown to be essential for Cu-induced expression (Furst et al. 1988).

al. 1994; Farrell et al. 1996). Expression of the N-terminal segments of Ace1 and Amt1 molecules in bacteria grown in the presence of $CuSO_4$ results in the isolation of each protein as a Cu, Zn-protein complex (Thorvaldsen et al. 1994; Farrell et al. 1996). Chemical analysis revealed 4 and 1 mol eq. copper and zinc ions bound, respectively, for each protein (Thorvaldsen et al. 1994; Farrell et al. 1996). Electrospray mass spectrometry was used to verify that a uniform species was present with 4 Cu(I) ions and 1 Zn(II) ion bound per Amt1 molecule (Thorvaldsen et al. 1994).

The observation of Zn(II) in bacterially expressed Ace1 and Amt1 raises the question of whether these Cu(I) sensors in yeast would, likewise, contain bound Zn(II). A number of observations suggest that Ace1 and Amt1 may indeed contain Zn(II) in yeast. First, the proteins were invariably isolated with a bound Zn(II) ion from bacteria cultured in the presence of 1.4 mM $CuSO_4$. Second, Cu(I) reconstitution studies of Amt1 did not reveal Cu(I) ions bound in excess of 4 mol eq. This is in contrast to Ace1 in which Cu(I) reconstitution studies showed binding of 6 mol eq (Dameron et al. 1991). Exogenous Cu(I) can displace the bound Zn(II) ion in Ace1, but not in Amt1. Third, a substitution of an Asp for a Cys codon which serves as a Zn(II) ligand did not totally abolish Ace1 function, which would be expected if that cysteinyl residue was a Cu(I) ligand (Farrell et al. 1996). The N-terminal 40 residues in Ace1 and Amt1 have been identified as the Zn(II) binding segment, and the Cys→Asp substitution was made at codon 11 within the Zn module (Farrell et al. 1996). Asp is a common Zn(II) ligand, whereas it is not expected to serve as a Cu(I) ligand.

The properties of the tetracopper center in Ace1 and Amt1 are similar to Cu(I) clusters in MT such as Cu(I)-thiolate coordination, trigonal coordination, and polycopper clusters (Thorvaldsen et al. 1994; Winge et al. 1994). The tetracopper center in Amt1, and presumably Ace1, forms in an all-or-nothing manner (Thorvaldson et al. 1994). The all-or-nothing formation of the cluster to activate Ace1 and Amt1 may enable these factors to respond rapidly to copper ion stress.

As mentioned, the Zn(II) site in Ace1 and Amt1 is formed by the N-terminal 40 residues of each protein (Figure 11.4). The Zn(II) ion is coordinated by three cysteinyl thiolates and a single histidyl imidazole (Posewitz et al. 1996). The Zn(II) module is an independently folded domain, and this domain is conserved in three known fungal proteins and three candidate fungal open reading frames (Farrell et al. 1996). This Zn domain occurs in *S. cerevisiae* Mac1, a factor responsible for basal transcription of *FRE1* and *CRT1* (Dancis et al. 1994a; Hassett and Kosman 1995; Jungmann et al. 1993). A third *S. cerevisiae* factor containing the conserved Zn domain is a molecule with high homology to Ace1 (Lpz8p in Figure 11.5). This ORF is an expressed protein and functions in an undefined aspect of copper homeostasis (Martins and Winge, unpublished observation).

The tetracopper center in Ace1 and Amt1 is enfolded by residues 41–110

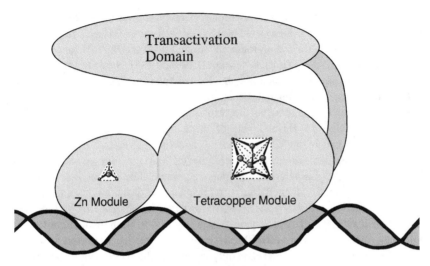

xGCGxxxxTxxxGTGC Ace1 sites

xxxxxxxxTxxxGTGC Amt1 sites

Figure 11.4 Model of CuZnAce1 and CuZnAmt1 interactions with DNA. The Zn module of each protein consists of the N-terminal 40 residues (Farrell et al. 1996). The tetracopper module is enfolded by residues 41–110, and this is followed in each case by the transactivation domain. The Zn module in Ace1 is believed to contact the 5′ distal major groove sequence in UASc (Buchman et al. 1990; Dobi et al. 1995). The tetracopper molecule in each molecule is believed, therefore, to contact the core consensus sequence, GTGC.

(Farrell et al. 1996) (Figure 11.4). We have demonstrated that expression of a polypeptide consisting of residues 35–110 of Amt1 results in the isolation of a Cu(I) protein with 4 Cu(I) ions bound. The Cu(I) ions are clusters as in the intact molecule (Graden et al. 1996).

Thus, the Cu_4Zn_1 complexes of Ace1 and Amt1 consist of a Cu_4S_8 cage cluster and a distinct ZnS_3N_1 site. A general model for the structure is shown in Figure 11.4 It is likely that the polycopper cluster stabilizes a tertiary fold that is competent to bind DNA. As mentioned, the UAS binding sites for Cu_4Zn_1Ace1 span one and one-half turns of B-form DNA helix. It appears that the Zn domain and tetracopper domains of Ace1 make base-specific contacts within the major groove at the two ends of the UAS sequence and span the minor groove near the middle of the UAS element (Dobi et al. 1995) (Figures 11.3 and 11.4). The contact of Ace1 within the A/T-rich minor groove may arise from residues 36–39 in the Zn module, sequence RGRP (Farrell et al. 1996) (Figure 11.5). This sequence is homologous to a DNA minor groove binding motif found in high-mobility-group proteins (Geierstanger et al. 1994).

It is unlikely that the Zn domain in Amt1 contacts DNA in an analogous manner to that in Ace1 (Thorvaldsen et al. 1994, 1995). Substitutions at codon 11 in Amt1 do not result in the same loss-of-function phenotype as in Ace1 (Simon and Winge, unpublished observation).

Ace1 and Amt1 appear to be the primary sensors of the intracellular Cu(I) concentration in yeast. Basal levels of these proteins must exist in a stable, unactivated state. It is unclear whether the unactivated state of Ace1 and Amt1 is the metal-free molecules or Zn_4 complexes, but the high avidity of these molecules for metal ions suggests that the unactivated state may be Zn_4 complexes (Ace1 binds maximally 4 Zn(II) ions). As such, activation would occur by metal exchange reactions. Zn(II) ions in Ace1 are rapidly displaced by Cu(I) ions (Dameron et al. 1993). The ability of Ace1 and Amt1 to form distinct conformers with Cu(I) and Zn(II) may form the basis of the exquisite metal ion specificity in the function of these proteins. The only metal ions that yield active conformers, Cu(I) and Ag(I), are ions that form structurally homologous metal:thiolate cages (Dance 1978).

Amt1 and Ace1 are activated at extracellular copper concentrations in excess of 5 μM (Thorvaldsen et al. 1993). The all-or-nothing formation of the tetracopper center in Amt1 and presumably Ace1 may be important in making the system responsive to low concentrations of copper. Whereas *ACE1* is expressed constitutively, *AMT1* is Cu induced. The Amt1 gene was shown to contain a single UAS element in the 5′ flanking sequences (Zhou and Thiele 1993). Mutations within this UAS element not only abolished copper regulation but also reduced the copper resistance of *C. glabrata* cells harboring these mutations. The autoregulation of Amt1 is rapid and provides an efficient way for cells to respond to high environmental copper concentrations (Zhou and Thiele 1993). The efficient coupling of external copper levels to the concentration of active Cu_4Zn_1AMT1 allows *C. glabrata* cells to stimulate high-level expression of the three MT genes, thereby minimizing Cu-induced toxicity.

Maximal expression of many genes requires displacement of nucleosomes. This displacement is a component of the activation process of certain inducible genes. Nucleosome loss results in a pronounced activation on an episomal *CUP1/lacZ* fusion but no effect on the chromosomal *CUP1* locus (Durrin et al. 1992; Kim et al. 1988). The effect was restricted to an episomally carried construct, so its relevance to the *in vivo* situation may be limited.

Posttranscriptional regulation deserves mention. The half-life of the *CUP1* mRNA is much shorter (6 minutes) than the average yeast mRNA (22 minutes) (Durrin et al. 1992; Chia and McLughlin 1979). Regulation of *CUP1* mRNA stability is another candidate mechanism for the acquisition of metal tolerance, although there is no evidence that this occurs in yeast. Posttranscriptional regulation of MT may occur in animal cells (Sadhu and Gedamu 1989).

2.3 Other Components Involved in Copper Resistance

Genetic analysis of copper resistance in *S. cerevisiae* revealed that multiple genes were important for this phenotype (Welch et al. 1989). At least 12 complementation groups have been identified and five genes that affect copper resistance have been mapped on the yeast genome (Fogel et al. 1983; Welch et al. 1989). Progress is being made to identify genes other than *CUP1* and *ACE1* that are critical for the copper resistance phenotype. Some of these genes will be discussed in the following sections. Some of the genes essential for copper tolerance encode components involved in metal-induced cell responses, whereas others likely encode components involved in general stress responses.

VACUOLES

The gene complementing one of the Cu^s mutants, cup^5, was recently cloned (Eide et al. 1993). The *CUP5* gene was found to be identical to *VMA3*, the gene encoding the proteolipid subunit of the H^+-ATPase in vacuoles (Eide et al. 1993). Mutations in *VMA3* are known to diminish the activity of the ATPase proton pump (Noumi et al. 1991). Mutations in *VMA3* result in a phenotype of hypersensitivity to multiple metal ions, including Cu(II), Ca(II), Mn(II), and Zn(II). Yeast vacuoles are known to be important in metal ion homeostasis, but their direct role remains unclear (Gadd and White 1989). The ATPase has been implicated in metal ion transport by the vacuole (Nelson and Nelson 1990). One obvious model of the importance of vacuoles in copper resistance is the vacuolar accumulation of CuMT complexes or accumulation of Cu(I) ions donated by CuMT. Vacuoles are known to be important for the compartmentalization of various metal ions (Gadd and White 1989), but there is no indication that CuMT becomes localized within vacuoles. Yeast CuMT appears to be predominantly cytoplasmic based on immunocytochemistry (Wright et al. 1987).

TRANSPORT

The high-affinity copper transporter in *S. cerevisiae* was recently cloned (Figure 11.2) (Dancis et al. 1994a,b). The transporter, Ctr1, appears specific for copper transport as other divalent metal ions did not inhibit copper transport (Dancis et al. 1994b). Disruption of the *CTR1* locus confers a copper resistance phenotype and reduces copper uptake. The Ctr1 polypeptide is localized within the plasma membrane. The CRT1 polypeptide appears to consist of distinct domains. The candidate ectodomain is characterized by an unusual abundance of Met and Ser residues (Dancis et al. 1994b). A Met-x-x-Met sequence motif is repeated 11 times. The central domain contains two long hydrophobic stretches that are candidate transmembrane segments. The C-terminal region of the polypeptide is the candidate third domain and is characterized by two Cys-x-Cys sequence repeats (Dancis et al. 1994b).

The Met-rich ectodomain of Ctrl may function as an initial copper receptor by coordinating Cu(II) or Cu(I) by the thioether groups of the multiple methionines. Transport across the lipid bilayer may be facilitated in part by binding of Cu ions to the Cys-x-Cys sequence motifs within the cytoplasmic domain. Previous studies on copper uptake in yeast indicated that the process was energy dependent (Lin and Kosman 1990). Details of the transport process must await further studies.

Because unregulated copper transport may be expected to result in Cu-induced toxicity, homeostatic regulation of transport is likely. Excess Cu was reported to down-regulate *CTR1* mRNA levels (Dancis et al. 1994a) (Figure 11.2). The down-regulation of *CTR1* may limit copper transport at low μM extracellular concentrations, but copper transport may persist at higher copper concentrations via putative low-affinity transporters. Such transporters must exist because *ctr1* cells gain only limited copper resistance (Dancis et al. 1994b). Limited copper resistance in *ctr1* cells may also arise from the toxicity of extracellular Cu(II) ions.

A second copper transporter has been identified in *S. cerevisiae* and designated Ctr2. Ctr2 bears no striking sequence similarity to Ctrl. Disruption of *CTR2* conferred limited copper resistance and overexpression of *CTR2* resulted in an increased sensitivity to copper toxicity.[71] Therefore, Ctr2 is believed to be a low affinity copper ion transporter.

Yeast contain a two metallo-reductase, *FRE1* and *FRE2*, capable of ferric and cupric ion reduction.[7,72] Fre1 has been implicated in copper transport from a study of a *FRE1* regulator.[56] Cells harboring a dominant gain-of-function mutation in the *MAC1* locus are copper hypersensitive and exhibit elevated Fre1 reductase levels.[55,56] Cells with a disrupted *mac1* locus are defective in the reductase. Thus, Mac1 appears to be a cellular regulator of *FRE1*. Mac1 is a cysteine-rich protein that exhibits sequence homology to the N-terminal segments of both Ace1 and Amt1.[56] (Fig. 11.5) Since the homology maps to the DNA binding Zn domain of Ace1, one implication is that Mac1 may further resemble Ace1 in being a metal-regulated protein.

CUP9

Another genetic locus has been described that is important in copper resistance.[73] The locus, *CUP9*, confers copper resistance to cells grown on non-fermentable carbon sources. The effect of *LOC1* is specific for copper and the

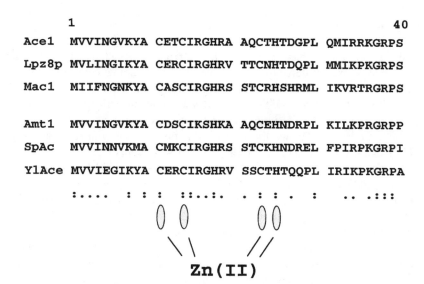

```
            1                                              40
Ace1    MVVINGVKYA  CETCIRGHRA  AQCTHTDGPL  QMIRRKGRPS

Lpz8p   MVLINGIKYA  CERCIRGHRV  TTCNHTDQPL  MMIKPKGRPS

Mac1    MIIFNGNKYA  CASCIRGHRS  STCRHSHRML  IKVRTRGRPS


Amt1    MVVINGVKYA  CDSCIKSHKA  AQCEHNDRPL  KILKPRGRPP

SpAc    MVVINNVKMA  CMKCIRGHRS  STCKHNDREL  FPIRPKGRPI

YlAce   MVVIEGIKYA  CERCIRGHRV  SSCTHTQQPL  IRIKPKGRPA
```

Zn(II)

Figure 11.5 The Zn module in Ace1 and Amt1 is a conserved domain. The module has been found in three proteins from *S. cerevisiae*. These are Ace1, Mac1, and a sequence Lpz8p, which we have recently demonstrated to be an expressed protein that functions in copper homeostasis. The Zn module is also found in sequences within the *Schizosaccharomyces pombe* (SpAc) and *Yarrowia lipolytica* (YlAce) genomes (Farrell et al. 1996). The Zn(II) ligands (indicated by the shaded ovals) are Cys11, Cys14, Cys23, and His25 in Amt1 and Ace1.

observed resistance is clearly independent of *CUP1* and *ACE1*.[73] There is no indication whether copper resistance imparted by Cup9 occurs through altered copper transport, intracellular copper sequestration in a non-MT molecule, compartmentalization with an organelle or efflux. The presence of a homodomain in Cup9 is consistent with it being a transcription factor.[73]

BSD2

A gene that suppresses the oxygen toxicity of *S. cerevisiae* cells lacking superoxide dismutase *SOD1* was shown to contribute to copper homeostasis (Georgatsou et al. 1995; Knight et al. 1994; Liu and Culotta 1994). A loss-of-function mutation in *BSD2* was found to exhibit a copper hypersensitivity phenotype. The cells accumulated more copper ions than wild-type cells. The significance of this molecule in copper homeostasis is unclear as the effect of the *BSD2* mutation is not restricted to copper ions (Liu and Culotta 1994).

GLUTATHIONE

Glutathione (GSH) has been called the initial line of defense against metal toxicity. Most studies demonstrating a role of GSH in animal cells have utilized buthionine sulfoximine to inhibit the γ-glutamylcysteine synthase (Singhal et al. 1987). The availability of GSH-deficient yeast mutants will enable definitive studies on the role of GSH in metal homeostasis (Kistler et al. 1990; Ohtake et al. 1990). Cells *gsh1* (lacking functional γ-glutamylcysteine synthetase) and *gsh2* cells (lacking GSH synthetase) are hypersensitive to copper salts in the growth medium and have an attenuated Cu-induced expression of a *CUP1/lacZ* fusion gene (unpublished observation). The effect of GSH depletion appears to be specific for induction from the *CUP1* promoter as no impaired *lacZ* expression was observed in cells containing a *HIS4/lacZ* fusion gene. Thus, GSH appears to be important for maximal *CUP1* expression.

The role of GSH in *CUP1* expression may be as a Cu(I) buffer. Alternatively, a Cu(I):GSH complex may be involved in the presentation of Cu(I) ions to Ace1 and Amt1. GSH is not an obligatory molecule in Cu-activation of Ace1 as expression of *CUP1/lacZ* shows only limited inhibition in *gsh1* cells (unpublished observation).

INTRACELLULAR COPPER TRANSPORTERS

Recently, two genes have been cloned that are responsible for copper homeostatic abnormalities in Menkes' and Wilson's patients (Chelly et al. 1993; Vulpe et al. 1993; Mercer et al. 1993; Bull et al. 1993; Petruhkin et al. 1993; Hamer 1993). The two genes appear to encode copper transport ATPases. Menkes' mutations in humans and the mouse result in excess copper accumulation in many tissues. The details of how these gene products function in copper homeostasis in animal cells remain to be elucidated. *S. cerevisiae* contains two homologs to these mammalian genes (Fu et al. 1995; Yuan et al. 1995; Rud et al. 1994). One of these genes, designated CCC2, functions in the export of copper ions into an extracytosolic compartment in which Cu ions are incorporated into Fet3, a protein required for iron transport in yeast (Yuan et al. 1995; Askaith et al. 1994). Fet3 is homologous to mammalian ceruloplasmin, which is a Cu-dependent ferrooxidase (Askwith et al. 1994). The activity site of Fet3 is expected to be a trinuclear copper center based on the homology with ceruloplasmin and ascorbate oxidase (Messerschmidt et al. 1992). The importance of Fet3 in iron transport establishes a link between iron and copper metabolism (Askwith et al. 1994). This link was predicted from early studies demonstrating that copper-deficient pigs develop iron deficiency anemia (Lee et al. 1968). Yeast cells lacking Ccc2 exhibit a respiratory defect that correlates with limited iron uptake (Yuan et al. 1995).

Yeast lacking *PCA1*, the second Wilson's homolog, do not display any obvious growth phenotype under standard conditions. However, these *PCA1* null cells grew more slowly in medium containing high copper concentrations (Rad et al. 1994).

BIOMINERALIZATION OF COPPER ON CELL SURFACE

A new yeast gene, designated *SLF1*, was identified as a multicopy suppressor of a *cup14* mutation (Yu et al. 1996). Slf1 appears to be a determining factor for (CuS)x biomineralization on the surface of yeast. Cells lacking a functional Slf1 remain white colored on the medium containing $CuSo_4$ (Yu et al. 1996). In contrast, wild-type cells exhibit a brownish hue on such a medium. This coloration directly correlates with their ability to deplete copper ions from the growth medium. Wild-type cells that show a brown coloration in Cu-containing medium have a greater ability to deplete medium copper than the Δ*slf1* cells that develop no coloration in Cu-containing medium. Moreover, an overproduction of Slf1 enhances both the brownish hue in cells grown in the presence of copper salts and depletion of copper from the medium. These results confirm the direct involvement of Slf1 in the physiology of coloration in Cu-treated yeast cells.

It has been long known that *S. cerevisiae* cells develop a brownish color in the presence of copper salts because of the formation of CuS mineral complexes on the cell surface (Ashida et al. 1963; Kikuchi; 1965; Naiki 1957). Electron-microscopic studies established that the electron dense particles present in Cu-treated yeast cells locate mainly inside the yeast cell wall. X-ray powder analysis confirmed that the electron dense particles contain copper and sulfur (Ashida et al. 1963). The sulfur is likely present as sulfide anions.

The formation of CuS biomineralization in yeast is dependent on Slf1. This pathway is clearly of importance in copper detoxification, as disruption of *SLF1* results in limited Cu sensitivity (Yu et al. 1996). In addition, overexpression of *SLF1* confers limited resistance and super-resistance to copper salts to *cup1* cells and the wild-type cells, respectively. Overexpression or disruption of *SLF1* cells has no effect on the tolerance of cells to cadmium salts.

There are indications that the Slf1-dependent sulfide generation pathway is copper dependent. No depletion of Zn(II) or Cd(II) ions occurs in the medium of cells overexpressing *SLF1* (Yu et al. 1996).

3. Cadmium-induced Stress Responses

Cadmium tolerance is conferred in part by sequestration, yet MT is not a dominant factor in cadmium buffering in yeast. The regulation of the pathways leading to cadmium sequestration is not well characterized to date, so it remains unclear whether these responses are cadmium induced or cadmium specific.

3.1 Glutathione-related Isopeptides

Cadmium-stressed yeast accumulate intracellular Cd(II) complexes with gluta-thione-related isopeptides of general structure (γGlu-Cys)$_n$Gly (Rauser 1990). The Cd(II):isopeptide complexes can incorporate sulfide anions to generate cad-mium:sulfide crystallites coated with the γGlu-Cys peptides. Mutants of *Schizo-saccharomyces pombe* have been described that are incapable of biosynthesizing these isopeptides (Mutoh and Hayashi 1988). The observed phenotype of the mutants is cadmium hypersensitivity.

Another class of cadmium-hypersensitive mutants has been described in *S. pombe*. These mutants are deficient in the accumulation of the sulfide form of Cd:(γGlu-Cys)$_n$Gly complexes (Ortiz et al. 1992). The gene that complements this mutation, *HMT1*, encodes a polypeptide associated with the vacuolar mem-brane. The implication of the *hmt1* cells is that sulfide incorporation into Cd(II):isopeptide complexes is critical for cells to be tolerant of cadmium ions. *S. cerevisiae* and *C. glabrata* accumulate Cd(II):(γGlu-Cys)$_n$Gly complexes in Cd-stressed cells, although Cd(II) complexes in *S. cerevisiae* involve predomi-nantly GSH (Yu et al. 1994; Kneer et al. 1992; Mehra et al. 1988).

3.2 Other Cd-resistant Loci

Cadmium tolerance in yeast cells is conferred by more than simple intracellular sequestration in stable complexes. Other genes have been described that confer cadmium tolerance. These include *YAP1, CAD1, CAD2, YCF1, ZCR1,* and *ADH1* (Wu et al. 1993; Yu et al. 1991; Tohoyama et al. 1990; Szczypka et al. 1994; Kamizono et al. 1989). Overexpression of each of these genes imparts metal resistance. This phenotype does not directly imply a physiological function for these gene products. High copy expression of one protein may activate one or more target genes by virtue of low affinity for promoter elements. Proof of a direct involvement in metal homeostasis requires the opposite phenotype (metal hypersensitivity) when the gene is disrupted. One good example of this is the homologous pair of proteins, yAP1 and Cad1. Overexpression of each gene confers metal resistance; however, disruption of *YAP1*, but not *CAD1*, results in cadmium hypersensitivity (Wu et al. 1993). In high copy the Cad1 may activate genes normally regulated only by yAP1. One candidate gene regulated by yAP1 that may affect cadmium tolerance is *GSH1*. This glutathione biosynthetic gene has a candidate yAP1 response element and is modulated by *YAP1* (Wu et al. 1993).

High copy expression of *ADH1* confers cadmium resistance by virtue of Cd(II) binding in the Zn(II) sites of alcohol dehydrogenase (Yu et al. 1991). It is possible that overexpression of other Zn(II)-containing enzymes may also permit limited Cd(II) buffering within the cell.

Overexpression of two candidate membrane transporters, *YCF1* and *ZCR1*, imparts cadmium resistance (Szczypka et al. 1994; Kamizono et al. 1989). *ZCR1* was shown to confer Cd and Zn tolerance when overexpressed, but not tolerance to copper salts, heat shock, UV irradiation, or alkylating agents (Kamizono et al. 1989). Disruption of the *ZCR1* locus confers hypersensitivity to Zn(II) and Cd(II) ions. The mechanism by which *ZCR1* confers metal tolerance is unknown.

4. General Responses to Metal Stress

Our discussion has so far focused on mechanisms of metal transport and sequestration of intracellular metal ions. In addition to these mechanisms, cells respond to stress induced by the deleterious action of metal ions. Damage induced by metal stress is dealt with by a set of coordinated stress responses that serve to either repair or replace function.

A number of common regulated stress responses exist in cells. All cells are capable of coping with changes in their environment, such as exposure to elevated temperatures, toxins, and oxidants (Morimoto 1993). First, in response to certain stress conditions, activation of stress gene expression occurs, resulting in an elevated synthesis of stress proteins. These stress proteins, commonly called heat shock proteins (Hsp), ensure survival under certain stressful conditions (Morimoto et al. 1992). A subset of heat shock proteins are the functional chaperone molecules that facilitate protein folding and refolding of misfolded proteins (Hendrick and Hartl 1993). A second response is facilitated protein degradation (Ciechanover et al. 1994). Third, DNA repair response occurs in response to agents or conditions that damage DNA. It is now clear that metal ions may elicit some or all of these stress responses (Figure 11.6). Thus, the repertoire of cell responses to metal stress appears to involve specific responses; i.e. metal-induced expression of MT genes, as well as the previously mentioned general stress responses. The general stress responses are likely to be more important for metal ions that cells are unable to efficiently sequester. The first two general responses will now be discussed separately.

4.1 Stress Protein Response

HSP(S)

Heat shock proteins (Hsp) are induced by a variety of environmental stresses including heavy metal ions, heat, amino acid analogs, and oxidants (Morimoto

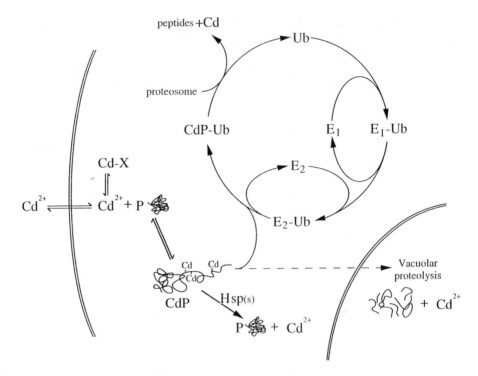

Figure 11.6 Potential metal-induced stress responses in cells. Evidence is presented in the text that metal ions can cause the misfolding of proteins, and the resulting abnormal molecules may be refolded via Hsp chaperones or removed by proteolysis. Two pathways of proteolytic degradation are shown, one the ubiquitin-dependent pathway (Ub) and the second a vacuolar process.

1993; Morimoto et al. 1992). That Hsp genes are induced by a variety of stress conditions implies that hsp proteins have broadly protective functions. Hsp families include constitutive as well as inducible proteins (Morimoto et al. 1992). A major question concerns how cells sense physiological stress. Most conditions that induce Hsp genes are known or believed to cause denaturation of preexisting or misfolding of nascent polypeptides, so one likely signal for Hsp gene induction is protein denaturation (Morimoto et al. 1992; Goff and Goldberg 1985; Pelham 1986; Ananthan et al. 1986; Parsell and Lindquist 1993; Baler et al. 1992). Intermediates in the protein folding pathway are susceptible to off-pathway interactions that may lead to aggregation. Denaturation of one protein molecule may induce coaggregation with other proteins leading to cell dysfunction (Parsell and Lindquist 1993). Injection of denatured, but not native, bovine albumin or β-lactoglobulin in *Xenopus* oocytes is known to cause a heat shock response. Thus, proteins with no known relationship to heat shock response and of different origins stimulate expression of Hsp(s) (Ananthus et al. 1986). This implies that an important function of hsp(s) is in the protection of proteins from denaturation and aggregation. Protein aggregation rather than simple denaturation may be critical for stress response induction (Mifflin and Cohen 1994a).

There are two ways in which Hsp(s) may alleviate stress-induced protein denaturation or aggregation (Parsell and Lindquist 1993). Hsp molecules may either renature proteins or promote their degradation. Evidence suggests that Hsp(s) function in both pathways.

A number of Hsp molecules are known to bind to and stabilize unstable conformers of proteins. By binding and releasing polypeptides that are in nonnative conformations, these Hsp(s) facilitate correct protein folding (Hendrick and Hartl 1993). Hsp molecules that function in protein folding are classified as chaperones. Two prominent classes of chaperones are the Hsp70 and Hsp60 families. Hsp70 proteins exist in multiple cell compartments and participate in a variety of functions other than protein folding such as unfolding or disassembly of certain proteins and membrane translocation (Parsell and Lindquist 1993). The chaperone-assisted folding of polypeptides is achieved through functional cooperation of different chaperones (Hendrick and Hartl 1993; Caplan et al. 1993). For example, the function of Hsp70, a well-established chaperone, is regulated by DnaJ-like molecules (Parsell and Lindquist 1993; Caplan et al. 1993; Georgopoulos 1992). *S. cerevisiae* contains at least eight different Hsp70 molecules and eight different molecules that regulate Hsp70 (Caplan et al. 1993; Craig et al. 1993; Werner-Washburne et al. 1987). Hsp(s) in the Hsp26, Hsp70, Hsp90, and Hsp100 families are induced by heat stress in yeast (Parsell and Lindquist 1993). There is less information available on metal-induced stress proteins. Metal ions are reported to induce expression of a subset of Hsp genes, including one or more members of the Hsp70 family and Hsp104 (Morimoto et al. 1992; Karin et al. 1981; Sanchez et al. 1992; Hatayama et al. 1991). The metal induction as with other stress responses is rapid, occurring within minutes (Morimoto et al. 1992).

In addition to the role of chaperones in preventing the aggregation of partially folded nascent and newly completed polypeptides, chaperones can also facilitate the refolding of previously misfolded proteins (Georgopoulos 1992). The bacterial DnaK can reactivate thermally denatured and aggregated RNA polymerase in a process dependent on ATP hydrolysis (Skowyra et al. 1990). Many cellular proteins can be coprecipitated with a monoclonal antibody to Hsp72 from cells exposed to heat shock or amino acid analogs (Beckmann et al. 1992). Thus, chaperones appear to have a repair function for proteins damaged by certain stresses in addition to a role in protein folding. It is unclear whether the function of chaperones in stress response is primarily at the level of assisted folding of nascent polypeptide chains or renaturation of misfolded proteins.

Hsp protein induction is often associated with increased tolerance, both to the immediate inducing agent and to other types of stress. Prior exposure of cells to stressful concentrations of cadmium or copper gives cross tolerance to heat (Sanchez et al. 1992). Cross tolerance does not extend to all types of stress. Disruption of yeast *HSP104* impairs the ability of cells to survive extreme temperatures, yet has no effect on the ability of cells to survive the amino acid analog, canavanine, or metal ions (copper and cadmium) (Parsell and Lindquist 1993; Sanchez et al. 1992). It is unclear which Hsp is responsible for metal tolerance. Hsp molecules appear to have a degree of specialization; different Hsp molecules are critical for different types of stress (Parsell and Lindquist 1993).

The induction of Hsp(s) occurs at the level of transcription (Parker and Topol 1984; Sorger and Pelham 1987). Genes encoding the various Hsp molecules contain a conserved promoter element, designated a heat shock element (HSE) (Morimoto et al. 1992; Pelham 1982). The induction of Hsp70 in animal cells by heat or metal ions requires only the HSE element in the promoter (Mosser et al. 1988; Williams and Morimoto 1990). Thus, HSEs are common stress elements. HSEs contain multiple 5 bp inverted repeats of the sequence nGAAn (Pelham 1982). The number of 5-bp boxes may range from 2 to 6. A perfect consensus array of three boxes would be the sequence: 5'-nGAAnnTTCnnGAAn-3', although not all HSEs have perfect inverted repeats (Lindquist 1986). Some sequence variation is observed, but it appears that HSEs have at least two perfect nGAAn boxes (Lindquist 1986). In addition to the variation in the number of 5-bp boxes within a given HSE, a heat shock promoter may have several HSEs (Xiao et al. 1991; Harrison et al. 1994).

HSF

Transcriptional activation of genes containing HSE promoter elements is mediated by the heat shock factor (Hsf). In yeast Hsf1 is a trimeric protein that binds HSE sequences constitutively at low temperature (Jakobsen and Pelham 1988; Sorger and Nelson 1989). Within the N-terminal region of the 833 residue yeast Hsf1 is a conserved sequence of 89 residues that is important in binding to the 5-bp HSE boxes (Harrison et al. 1994; Flick et al. 1994). It is likely that each subunit of the trimeric Hsf1 contacts a separate 5-bp box within an HSE, but the actual Hsf1:HSE complex appears to contain multiple Hsf1 trimers (Drees et al. 1994).

In yeast, the bound Hsf1:HSE complex is transcriptionally silent until stress activation (Jakobsen and Pelham 1991; Flick et al. 1994; Sorger et al. 1987; Sorger and Pelham 1988). However, the mechanism of Hsf1 activation remains unclear in yeast. In contrast, the activation process in animal cells involves oligomerization to the trimer state (Zamarino and Wu 1987).

Yeast Hsf1 contains domains that function as constitutive transcription activation domains when fused to heterologous DNA binding domains (Jakobsen and Pelham 1991; Nieto-Sotelo et al. 1990; Sorger 1990). These transactivation domains are not constitutive in Hsf1 at low temperature, thus it appears that the normal mode of action of Hsf1 is to hinder the effectiveness of these domains. This hindrance is relieved upon change to stress conditions (Jakobsen and Pelham 1991; Bonner et al. 1992). Constitutive activity is observed in *HSF1* mutations within the DNA binding domain (residues 167–256), deletions within the oligomerization domain (residues 350–402), or deletions within a C-terminal conserved region (residues 535–551) (Jakobsen and Pelham 1991; Sorger 1990; Bonner et al. 1992). A mutation at codon 232 in the DNA binding domain was shown to yield a 200-fold increase in activity at 26°C (Bonner et al. 1992). One model is that such mutations do not alter DNA binding but rather derepress the activation domain. Repression must involve the DNA binding domain, the trimerization domain, and the C-terminal conserved sequence.

Hsf1 activation may involve an autoregulatory loop in which one or more members of the hsp70 protein family is a negative regulator. Hsp70 has been found in a complex with activated Hsf1, and it is postulated that Hsp70 represses Hsf1 under conditions of low levels of abnormal proteins (Baler et al. 1992; Abravaya et al. 1992; Craig et al. 1994). Enhanced levels of Hsp70 attenuate the thermal stress response (Craig et al. 1994; Mifflin and Cohen 1994b). The appearance of denatured and aggregated proteins during stress may create a large pool of protein substrates that compete with Hsf1 for Hsp70 (Morimoto et al. 1992; Ananthan et al. 1986; Craig et al. 1994). Release of Hsp70 from Hsf1 by the presence of damaged protein substrates may permit a conformational change in Hsf1 to the active state. Consistent with this model is evidence in animal and bacteria cells that the heat shock response is initiated by accumulation of misfolded proteins (Goff and Goldberg 1985, Ananthan et al. 1986; Mifflin and Cohen 1994a; Parsell and Sauer 1989).

Hsf1 activation is postulated to confer metal tolerance by refolding of proteins via Hsp(s) (Figure 11.6). In addition, metal tolerance may be achieved by Hsf1-dependent expression of *CUP1*. From the limited data on the responsiveness of *CUP1* to heat stress, it is likely that heat stress results in only a limited increase in *CUP1* expression (Yang et al. 1991; Sewell et al. 1995). Hsf1 imparts metal tolerance in *cup1* cells, suggesting that metal tolerance is imparted by activation of the stress response (Sewell et al. 1995). Mutations in *HSF1* can result in substantial increases in *CUP1* expression. Mutation of *HSF1* at codon 203 (V203A) was shown to increase *CUP1* mRNA levels 10- to 20-fold (Yang et al. 1991; Silar et al. 1991). Wild-type Hsf1 can enhance *CUP1* expression when its gene is present on a high-copy plasmid (Sewell et al. 1995; Silar et al. 1991). This implies that both the wild-type Hsf1 and mutant Hsf1 molecules can activate transcription of *CUP1*. The mutant Hsf1 molecules exhibit an increased affinity for the *CUP1* promoter compared to the wild-type Hsf1 (Yang et al. 1991; Sewell et al. 1995). Although the mutant Hsf1 has enhanced activity on transcription of *CUP1*, no significant effect was observed with the Hsp70 gene, *SSA3*, and no enhanced thermal tolerance was observed (Silar et al. 1991).

The *CUP1* promoter region contains an HSE element, but two 5-bp consensus boxes are separated from a third nonconsensus box by a 5-bp spacer (Sewell et al. 1995; Tamai et al. 1994). The *CUP1* HSE is, therefore, a minimal HSE (Chen and Peterson 1993).

Cadmium-resistant *S. cerevisiae* strain 301N exhibits high basal as well as cadmium-induced expression of *CUP1* metallothionein. The gene responsible for the observed phenotype is a spontaneously mutated *HSF1* with substitutions at codons 206 and 256 within the DNA-binding domain of Hsf1 (Sewell et al. 1995). The mutant *HSF1* is responsible for the Cd-induced *CUP1* expression.

4.2 Protein Degradation

UBIQUITIN-DEPENDENT PROTEOLYSIS

Effective control over the level of abnormal proteins in a cell requires the cooperation of chaperone-assisted folding or refolding and protein degradation. Both processes deal with the similar problem of accumulation of denatured proteins in heat-stressed cells (Parsell and Lindquist 1993). A subset of Hsp(s) are components in the ubiquitin-dependent degradation pathway (Parsell and Lindquist 1993; Lindquist 1986). The ubiquitin pathway involves a series of proteins that transfer the 8.5-kDa ubiquitin polypeptide chain to protein substrates for ATP-dependent proteolysis. Ubiquitin-dependent degradation is the pathway responsible for the turnover of most naturally unstable proteins as well as abnormal proteins (Parsell and Lindquist 1993; Sadis et al. 1994; Jentsch 1992a).

Ubiquitin conjugation is a multistepped process requiring the function of a ubiquitin-activating enzyme, E1, a family of ubiquitin-conjugating E2 enzymes, and in some cases auxiliary substrate recognition proteins, E3 (Ciechanover et al. 1984; Finley et al. 1984; Jentsch 1992a) (Figure 11.6). Ubiquitin-protein conjugates are degraded by a large oligomeric complex, the proteosome (Ciechanover et al. 1984; Jentsch 1992a). *S. cerevisiae* contains four genes encoding ubiquitin (*UBI1-4*), at least ten genes coding for E2 ubiquitin-conjugating enzymes (UBC genes), and one E3 gene (*UBR1*) (Ciechanover et al. 1984; Jentsch 1992a; Finley et al. 1987). The ubiquitin pathway appears to be responsible for the turnover of many stress-damaged proteins (Parsell and Lindquist 1993). This pathway is expected to be important in proteins damaged by metal ion stress. Ubiquitin conjugation to stress-damaged polypeptides does not appear to involve the E3 component (Parsell and Lindquist 1993; Jentsch 1992a). As E3 is important for turnover of proteins by the N-end rule pathway, the ubiquitin-dependent catabolism of stress-damaged proteins may not involve N-end recognition.

Cell viability requires a functional ubiquitin proteolytic pathway. Cells with mutations in the E1 gene (*UBA1*), two E2 genes (*UBC3* and *UBC9*), or in *PRE1* encoding a proteosome subunit are nonviable (Parsell and Lindquist 1993; Jentsch 1992a,b). Protein degradation is required to maintain protein levels and to eliminate abnormal proteins generated under normal and stress conditions (Jentsch 1992a, Finley et al. 1987). The stress response is greatly enhanced in cell conditions that block ubiquitin conjugation, supporting the notion that stress response and ubiquitin-dependent catabolism are closely coupled (Parsell and Lindquist 1993; Hochstrasser 1992). The coordinate function of these two pathways may mitigate any deleterious effects of overload of either pathway (Hochstrasser 1992).

Stress conditions induce the expression of several proteins within the ubiquitin pathway, whereas other components are constitutive. The *UBI4* locus encoding polyubiquitin is induced by a variety of stress conditions (Jentsch 1992a,b). *UBI4* expression appears to provide ubiquitin under conditions of stress. Two E2 ubiquitin-conjugating enzymes, UBC4 and UBC5, are induced by stress conditions of heat shock, metal ions, and amino acid analogs (Jentsch 1992a,b; Seufert and Jentsch 1990). As expected, *ubc4* cells are hypersensitive to Cd(II) and amino acid analogs. It is likely that UBC4 and UBC5 function in the degradation of abnormal proteins, including metal-denatured proteins, in yeast (Sadis et al. 1994; Seufert and Jentsch 1990). Although *UBC7* is not heat shock induced, it is cadmium inducible. Cells that are *ubc7* are Cd(II) hypersensitive, but not hypersensitive to heat shock. DNA-damaging agents, or amino acid analogs (Jungmann et al. 1993). Mutant *ubc7* cells are not hypersensitive to copper, presumably because of the presence of the functioning *CUP1* pathway. Cells harboring a missense mutation within *PRE1*, which encodes a subunit of the proteosome, are hypersensitive to Cd(II), heat shock, and amino acid analogs (Jungmann et al. 1993). The implication is that cadmium resistance is mediated by ubiquitin-dependent proteolysis involving at least three E2 enzymes, UBC4, UBC5, and UBC7 and the proteosome.

Chaperones and proteases are components of a common stress response in cells. The ubiquitin pathway may be essential to stress responsiveness in two ways. First, ubiquitin-dependent proteolysis and chaperones may act in a coordinated manner to rid cells of abnormal proteins generated by stress. Proteins not renatured by chaperones may be substrates for proteolysis (Figure 11.6). Chaperones may also be directly involved in proteolytic turnover (Parsell and Lindquist 1993; Wagner et al. 1994; Goldberg et al. 1994). Accumulation of abnormal proteins is believed to be deleterious to cell viability. The cadmium sensitivity of *ubc7* may arise from accumulation of Cd-denatured proteins.

The fact that *UBC7* is cadmium inducible but not heat inducible implies that cadmium does not elicit a general stress response. Likewise, hsp104, which protects cells against heat shock is without effect with Cd(II). Lesions induced by Cd(II) differ in some still unknown way from those induced by heat (Jungmann et al. 1993). Cd(II) does not appear to induce Hsp(s) in all cells, so it is possible that a simple relationship between heat-induced stress and metal-induced stress does not exist. Alternatively, it is conceivable that metals do induce a stress response, but with the redundancy in stress responsiveness, heat-denatured proteins are repaired or removed by a different subset of proteins than those that repair or remove Cd-denatured or Cd-aggregated proteins. The various ubiquitin E2 enzymes are thought to interact with distinct proteins (Parsell and Lindquist 1993; Jentsch 1992a). The lack of hypersensitivity of *ubc7* to copper salts (Jungmann et al. 1993) may arise from the specificity of E2 enzymes, or loss of the ubiquitin system may not have a pronounced phenotype in cells having a functional Cu-induced MT pathway.

As mentioned above, ubiquitin-dependent proteolysis may contribute to cadmium tolerance in removing Cd-denatured proteins. The question arises of whether metal ions can denature proteins. The free energy of stabilization of protein structures is typically near 10 kcal/mol. This ΔG is equivalent to the energy of only a few hydrogen bonds. Thus, the balance between the folded and unfolded states is small. Conditions that stabilize the unfolded state or destabilize the folded state may shift the equilibrium toward the unfolded state. Heat is one condition capable of denaturation. Although there is little direct evidence for metal-induced protein denaturation, there are a host of proteins inhibited in function by Cd(II), but it is not known whether inhibition results from metal binding to active site residues or metal-induced denaturation (Vallee and Ulmer 1972).

It is clear that abnormal proteins can induce stress responsiveness. If metal stress increases the level of misfolded or aggregated proteins, a cooperative response of chaperone induction and ubiquitin-dependent proteolysis may occur. It is also possible that certain stress-damaged proteins are preferentially renatured by chaperones, whereas others are preferentially cleared by proteolysis. The combined processes may reduce levels of damaged proteins.

PROTEOLYSIS WITHIN ORGANELLES

Cellular proteolysis occurs via the cytoplasmic ubiquitin pathway as well as within lysosomes (Ciechanover et al. 1984). Short-lived proteins tend to be degraded by the ubiquitin system, whereas the long-lived cellular proteins are degraded by both proteolytic systems (Parsell and Lindquist 1993; Dice 1990). Proteins containing the sequence KFERQ are selectively degraded by lysosomes (Dice 1990). Both proteolytic systems may be important during stress. Proteins damaged by oxidation are known to accumulate in lysosomes (Dean et al. 1993). The two proteolytic systems may actually be functionally interrelated (Ciechanover et al. 1984; Finley et al. 1984). Proteins with the KFERQ motif are recognized by a member of the Hsp70 family, which may facilitate transport across the lysosomal membrane (Chiang et al. 1989). Degradation of long-lived proteins within lysosomes appears to require a functional E1 of the ubiquitin pathway (Ciechanover et al. 1984; Finley et al. 1984). Thus, ubiquitin-conjugation may occur in lysosomes as well as within the cytoplasm.

The vacuole in yeast is a complex organelle with multiple functions (Klionsky et al. 1990). In addition to a potential role in ubiquitin-dependent proteolysis, the vacuole is an organelle capable of ubiquitin-independent proteolysis as numerous proteolytic enzymes are present within vacuoles (Klionsky et al. 1990). The yeast vacuole is analogous to the mammalian lysosome in its ability to carry out proteolysis. Yeasts containing defective vacuoles are sensitive to copper salts as well as to several other cations (Eide et al. 1993). It is conceivable that proteolysis within vacuoles as well as within the cytoplasm is required for optimal metal tolerance in yeast (Figure 11.6).

5. Summary Model of Metal-induced Stress Response

Metal stress is likely to induce a set of coordinate responses that function to maintain normal physiology. At least three distinct levels of defense appear to exist in metal homeostasis. The efficiencies of these three levels determine the range of metal ion concentrations tolerated for homeostasis. Homeostasis ranges for essential and nonessential metal ions are shown in Figure 11.1. A threshold concentration may exist for each metal ion at which deleterious effects begin to be observed.

The first level of defense consists of largely unregulated processes. The affinity of metal ions for yeast cell wall constituents represents an initial metal buffer (Kosman 1994). Cells generating sulfide ions are capable of precipitating metal ions on the cell surface (Yu et al. 1996). An advantage may be gained by sulfide-producing cells under conditions of metal stress, although sulfide ions may also precipitate essential metal ions. Within the cell the high concentration of GSH represents an initial intracellular metal buffer. Cells capable of undergoing polymerization of GSH to $(\gamma EC)_n G$ isopeptides may gain an additional advantage. Metal ions may also be compartmentalized within the vacuole in a nonavailable form (Gadd and White 1989).

A second level is the metal-induced responses to regulate the intracellular concentration of a metal ion. The known Cu-induced responses are the Cu-induced repression of *CTR1* expression reducing copper transport and the Cu-induced expression of *SOD1* and MT genes, *CUP1* and *CRS5*. Expression of *CUP1* and *CRS5* leads to the accumulation of MTs for direct sequestration of Cu(I) ions in stable protein complexes. These responses are clearly the dominant processes in copper homeostasis in yeast. Not all MT genes are equally effective in conferring metal tolerance. In *S. cerevisiae* and *C. glabrata* only one MT isoform class is dominant in copper homeostasis. Defects in expression of those dominant MT genes result in a marked copper hypersensitivity. The magnitude of homeostatic effects of all other processes is lower.

The third level is the coordinate induction of Hsp(s) and protein catabolism activities. These responses are likely to be quite important in the detoxification of nonessential metal ions such as Cd(II). Because overexpression of *HSF1* enhances cadmium and copper tolerance in *CUP1* cells, it is likely that these stress response pathways confer some metal tolerance. The deleterious action of excess metal ions on protein stability may initiate these third-level responses. They may be initiated when a threshold concentration of a metal ion is exceeded such that the second-level metal-regulated responses are saturated. Although the exact mechanism of metal-induced heat shock response is unresolved, the model of metal-denatured protein accumulation is consistent with the data. The general stress response may also include enhanced expression of MT genes.

The third level of stress responses may also result in other protective action. We have preliminary evidence that metal ions, like heat shock, yield a G_1 arrest in the cell cycle. Heat shock lowers the levels of G_1 cyclins, resulting in an arrest in the cell cycle at G_1 (Rowley et al. 1993). As arrest at G_1 is known to facilitate gene amplification (Kirschner 1992), tandem amplification of MT genes during G_1 arrest will enhance cellular protection against copper stress in yeast.

A redundancy may exist within the multilevel defense system for metal-induced stress. Each metal ion may express a different set of pathways to deal with the stress. It is likely that the redundancy and efficiency of the pathways may dictate the dynamic range of metal concentrations in which a cell may propagate.

References

Abravaya, K., M. P. Myers, S. P. Murphy, and R. I. Morimoto. 1992. The human heat shock protein *hsp70* interacts with HSF, the transcription factor that regulates heat shock gene expression. *Genes Dev.* 6:1153–1164.

Agarwal, K., A. Sharma, and G. Talukder. 1989. Effects of copper on mammalian cell components. *Chem. Biol. Interactions.* 69:1–16.

Ananthan, A., A. L. Goldberg, and R. Voellmy. 1986. Abnormal proteins serve as eukaryotic stress signals and trigger the activation of heat shock genes. *Science* 232:522–524.

Ashida, J., N. Higashi, and T. Kikuchi. 1963. An electronmicroscopic study on copper precipitation by copper-resistant yeast cells. *Protoplasma* 57:27–32.

Askwith, C., D. Eide, A. Van Ho, P. S. Bernard, L. Li, S. Davis-Kaplan, D. M. Sipe, and J. Kaplan. 1994. The *FET3* gene of *S. cerevisiae* encodes a multicopper oxidase required for ferrous iron uptake. *Cell* 76:403–410.

Baler, R., W. J. Welch, and R. Voellmy. 1992. Heat shock gene regulation by nascent polypeptides and denatured proteins: *hsp70* as a potential autoregulatory factor. *J. Cell Biol.* 117:1151–1159.

Beckmann, R. P., M. Lovett, and W. J. Welch. 1992. Examining the function and regulation of *hsp70* in cells subjected to metabolic stress. *J. Cell Biol.* 117:1137–1150.

Bonner, J. J., S. Heyward, and D. L. Fackenthal. 1992. Temperature-dependent regulation of heterologous transcriptional activation domain fused to yeast heat shock transcription factor. *Mol. Cell. Biol.* 12:1021–1030.

Brenes-Pomales, A., G. Lindegren, and C. C. Lindegren. 1955. Gene control of copper sensitivity in *Saccharomyces. Nature* 176:841–842.

Buchman, C., P. Skroch, J. Welch, S. Fogel, and M. Karin. 1989. The CUP2 gene product, regulator of yeast metallothionein expression, is a copper-activated DNA-binding protein. *Mol. Cell. Biol.* 9:4091–4095.

Buchman, C., P. Skroch, W. Dixon, T. D. Tullius, and M. Karin. 1990. A single amino acid change in CUP2 alters its mode of DNA binding. *Mol. Cell. Biol.* 10:4778–4787.

Bull, P. C., G. R. Thomas, J. M. Rommens, J. R. Forbes, and D. Wilson Cox. 1993. The Wilson disease gene is a putative copper transporting P-type ATPase similar to the Menkes gene. *Nature Genet.* 5:327–337.

Butt, T. R., E. J. Sternberg, J. A. Gorman, P. Clark, D. Hamer, M. Rosenberg, and S. T. Crooke. 1984. Copper metallothionein of yeast, structure of the gene, and regulation of expression. *Proc. Natl. Acad. Sci. U.S.A.* 81:3332–3336.

Caplan, A. J., D. M. Cyr, and M. G. Douglas. 1993. Eukaryotic homologues of *Escherichia coli DnaJ:* A diverse protein family that functions with hsp70 stress proteins. *Molec. Biol. Cell* 4:555–563.

Chelly, J., Z. Tumer, T. Tonnesen, A. Petterson, Y. Ishikawa-Brush, N. Tommerup, N. Horn, and A. P. Monaco. 1993. Isolation of a candidate gene for Menkes disease that encodes a potential heavy metal binding protein. *Nature Genet.* 3:14–19.

Chen, J., and D. S. Peterson. 1993. A distal heat shock element promotes the rapid response to heat shock of the *hsp26* gene in the yeast *Saccharomyces cerevisiae. J. Biol. Chem.* 268:7442–7448.

Chia, L.-L., and C. McLughlin. 1979. The half-life of mRNA in *Saccharomyces cerevisiae. Mol. Gen. Genet.* 170:137–144.

Chiang, H.-L., S. R. Terleckey, C. P. Plant, and J. F. Dice. 1989. A role for a 70-kilodalton heat shock protein in lysosomal degradation of intracellular proteins. *Science* 246:382–385.

Ciechanover, A., D. Finley, and A. Varshavsky. 1984. Ubiquitin dependence of selective protein degradation demonstrated in the mammalian cell cycle mutant *ts85. Cell* 37:57–66.

Craig, E. A., D. B. Gambill, and J. R. Nelson. 1993. Heat shock proteins: Molecular chaperones of protein biogenesis. *Microbiol. Rev.* 57:402–414.

Craig, E. A., J. S. Weissman, and A. L. Horwich. 1994. Heat shock proteins and molecular chaperones: Mediators of protein conformation and turnover in the cell. *Cell* 78:365–372.

Crosa, J. H. 1989. Genetics and molecular biology of siderophore-mediated iron transport in bacteria. *Microbiol. Rev.* 53:517–530.

Culotta, V. C., W. R. Howard, and X. F. Liu. 1994. *CRS5* encodes a metallothionein-like protein in *Saccharomyces cerevisiae*. *J. Biol. Chem.* 269:25295–25302.

Culotta, V. C., T. Hsu, S. Hu, P. Furst, and D. Hamer. 1989. Copper and the ACE1 regulatory protein reversibly induce yeast metallothionein gene transcription in a mouse extract. *Proc. Natl. Acad. Sci. U.S.A.* 86:8377–8381.

Dameron, C. T., G. N. George, P. Arnold, V. Santhanagopalan, and D. R. Winge. 1993. Distinct metal binding configurations in ACE1. *Biochem.* 32:7294–7301.

Dameron, C. T., D. R. Winge, G. N. George, M. Sansone, S. Hu, and D. Hamer. 1991. A copper-thiolate polynuclear cluster in the ACE1 transcription factor. *Proc. Natl. Acad. Sci. U.S.A.* 88:6127–6131.

Dance, I. G. 1978. The hepta (μ-benzenethiolato) pentametallate(I) dianions of copper and silver: Formation and crystal structures. *Aust. J. Chem.* 31:2195–2206.

Dance, I. G. 1986. The structural chemistry of metal thiolate complexes. *Polyhedron* 5:1037–1104.

Dancis, A., D. Haile, D. S. Yuan, and R. D. Klausner. 1994a. The *Saccharomyces cerevisiae* copper transport protein (Ctr1p). *J. Biol. Chem.* 269:25660–25667.

Dancis, A., R. D. Klausner, A. G. Hinnebusch, and J. G. Barriocanal. 1990a. Genetic evidence that ferric reductase is required for iron uptake in *Saccharomyces cerevisiae*. *Mol. Cell. Biol.* 10:2294–2301.

Dancis, A., D. G. Roman, G. J. Anderson, A. G. Hinnebush and R. D. Klausner. 1990b. Ferric reductase of *Saccharomyces cerevisiae*: Molecular characterization, role in iron uptake, and transcriptional control by iron. *Proc. Natl. Acad. Sci. U.S.A.* 89:3869–3873.

Dancis, A., D. S. Yuan, D. Halle, C. Askwith, D. Eide, C. Moehle, J. Kaplan, and R. D. Klausner. 1994b. Molecular characterization of a copper transport protein in *S. cerevisiae*: An unexpected role for copper in iron transport. *Cell* 76:393–402.

Dean, R. T., S. Gieseg, and M. J. Davies. 1993. Reactive species and their accumulation on radical-damaged proteins. *Trends in Biochem. Sci.* 18:437–441.

Dice, J. F. 1990. Peptide sequences that target cytosolic proteins for lysosomal proteolysis. *Trends in Biochem. Sci.* 15:305–309.

Dobi, A., C. T. Dameron, S. Hu, D. Hamer, and D. R. Winge. 1995. Distinct regions of Cu(I):ACE1 contact two spatially resolved DNA major groove sites. *J. Biol. Chem.* 270:10171–10178.

Drees, B., K. Flick, E. Grotkopp, C. Harrison, S. Hubl, P. Peteranderl, and H. C. M. Nelson. 1994. Structure and function of the DNA binding and trimerization domains of the heat shock transcription factor. In *Biology of Heat Shock Proteins and Molecular Chaperones*. 1994 Cold Spring Harbor meeting, p. 5.

Durrin, L. K., R. K. Mann, and M. Grunstein. 1992. Nucleosome loss activates *CUP1* and *HIS3* promoters to fully induced levels in the yeast *Saccharomyces cerevisiae*. *Mol. Cell. Biol.* 12:1621–1629.

Ecker, D. J., T. R. Butt, E. J. Sternberg, M. P. Neeper, C. Debouck, J. A. Gorman, and

S. T. Crooke. 1986. Yeast metallothionein function in metal ion and detoxification. *J. Biol. Chem.* 261:16895–16900.

Eide, D. J., J. T. Bridgham, Z. Zhao, and J. R. Mattoon. 1993. The vacuolar H⁺-ATPase of *Saccharomyces cerevisiae* is required for efficient copper detoxification, mitochondrial function, and iron metabolism. *Mol. Gen. Genet.* 241:447–456.

Farrell, R. A., J. L. Thorvaldsen, and D. R. Winge. 1996. Identification of the Zn(II) site in the copper-responsive transcription factor AMT1: A conserved Zn module. *Biochem.* (in press).

Finley, D., A. Ciechanover, and A. Varshavsky. 1984. Thermolability of ubiquitin-activating enzyme from the mammalian cell cycle mutant *ts85*. *Cell* 37:43–55.

Finley, D., E. Ozkaynak, and A. Varshavsky. 1987. The yeast polyubiquitin gene is essential for resistance to high temperatures, starvation and other stresses. *Cell* 48:1035–1046.

Flick, K. E., L. Gonzalez, C. J. Harrison, and H. C. M. Nelson. 1994. Yeast heat shock transcription factor contains a flexible linker between the DNA-binding and trimerization domains. *J. Biol. Chem.* 269:12475–12481.

Fogel, S., and J. W. Welch. 1982. Tandem gene amplification mediates copper resistance in yeast. *Proc. Natl. Acad. Sci. U.S.A.* 79:5342–5346.

Fogel, S., and J. Welch. 1983. A recombinant DNA strategy for characterizing industrial yeast strains. In *Genetics: New Frontiers: Proceedings of the XV International Congress of Genetics,* eds. V. L. Chopra et al. pp. 133–142.

Fogel, S., J. W. Welch, G. Cathala, and M. Karin. 1983. Gene amplification in yeast: *CUP1* copy number regulates copper resistance. *Current Genet.* 7:347–355.

Fu, D., T. J. Beeler, and T. M. Dunn. 1995. Sequence, mapping and disruption of *CCC2,* a gene that cross-complements the Ca(II)-sensitive phenotype of *csg1* mutants and encodes a P-type ATPase belonging to the Cu(II)-ATPase subfamily. *Yeast* 11:283–292.

Furst, P., S. Hu, R. Hackett, and D. Hamer. 1988. Copper activates metallothionein gene transcription by altering the conformation of a specific DNA binding protein. *Cell* 55:705–717.

Gadd, G. M., and C. White. 1989. Heavy metal and radionuclide accumulation and toxicity in fungi and yeasts. In *Metal-Microbe Interactions,* eds. R. K. Poole and G. M. Gadd. pp. 19–38. IRL Press, Oxford.

Geierstanger, B.H., B. F. Volkman, W. Kremer, and D. E. Wemmer. 1994. Short peptide fragments derived from HMG-I/Y proteins bind specifically to the minor groove of DNA. *Biochem.* 33:5347–5355.

Georgatsou, E., L. Mavrogiannis, G. Frangiadakis, A. Klinakis, and D. Alexandraki. 1995. The *Saccharomyces cerevisiae* ferric reductase genes *FRE1* and *FRE2* are distinctly regulated by iron, copper and stress related factors. *17th International Conference on Yeast Genetics and Molecular Biology,* abstract, S155.

Georgopoulos, C. 1992. The emergence of the chaperone machines. *Trends in Biochem. Sci.* 17:295–299.

Goff, S. A., and A. L. Goldberg. 1985. Production of abnormal proteins in *E. coli* stimulates transcription of *Lon* and other heat shock genes. *Cell* 41:587–595.

Goldberg, A. L., D. H. Lee, O. Kandror, and M. Sherman. 1994. Involvement of molecular chaperones in degradation of abnormal proteins in *E. coli* and yeast. In *Biology of Heat Shock Proteins and Molecular Chaperones.* 1994 Cold Spring Harbor meeting, p. 231.

Graden, J. A., M. C. Posewitz, J. R. Simon, G. N. George, J. J. Pickering and I. R. Winge. 1996. In Presence of a Copper (II)-Thiolate Regulatory Domain in the Copper-Activated Transcription Factor Amt1. *Biochemistry* 35:14583–14589.

Gralla, E. B., D. J. Thiele, P. Silar, and J. S. Valentine. 1991. ACE1, a copper-dependent transcription factor, activates expression of the yeast copper, zinc superoxide dismutase. *Proc. Natl. Acad. Sci. U.S.A.* 88:8558–8562.

Halliwell, B., J. R. Hoult, and D. R. Blake. 1988. Oxidants, inflammation, and anti-inflammatory drugs. *FASEB J.* 2:2867–2873.

Hamer, D. H. 1993. "Kinky hair" disease sheds light on copper metabolism. *Nature Genet.* 3:3–4.

Hamer, D. H., D. J. Thiele, and J. E. Lemontt. 1985. Function and autoregulation of yeast copperthionein. *Science* 228:685–690.

Harrison, C. J., A. A. Bohm, and H. C. M. Nelson. 1994. Crystal structure of the DNA binding domain of the heat shock transcription factor. *Science* 263:224–227.

Hassett, R., and D. J. Kosman. 1995. Evidence for Cu(II) reduction as a component of copper uptake by *Saccharomyces cerevisiae. J. Biol. Chem.* 270:128–134.

Hatayama, R., Y. Tsukimi, T. Wakatsuki, T. Kitamura, and H. Imahara. 1991. Different induction of 70,000-Da heat shock protein and metallothionein in HeLa cells by copper. *J. Biochem.* 110:726–731.

Hendrick, J. P., and F.-U. Hartl. 1993. Molecular chaperone functions of heat-shock proteins. *Ann. Rev. Biochem.* 62:349–384.

Hochstrasser, M. 1992. Ubiquitin and intracellular protein degradation. *Curr. Opin. Cell Biol.* 4:1024–1031.

Hu, S., P. Furst, and D. Hamer. 1990. The DNA and Cu binding functions of ACE1 are interdigitated within a single domain. *New Biol.* 2:544–555.

Huibregtse, J. M., D. R. Engelke, and D. J. Thiele. 1989. Copper-induced binding of cellular factors to yeast metallothionein upstream activation sequences. *Proc. Natl. Acad. Sci. U.S.A.* 86:65–69.

Jakobsen, B. K., and H. R. B. Pelham. 1988. Constitutive binding of yeast heat shock factor to DNA in vivo. *Mol. Cell. Biol.* 8:5040–5042.

Jakobsen, B. K., and H. R. B. Pelham. 1991. A conserved heptapeptide restrains the activity of the yeast heat shock transcription factor. *EMBO J.* 10:369–375.

Jensen, L. T., W. R. Howard, D. R. Winge, and V. C. Culotta. 1996. Enhanced effectiveness of *CUP1* metallothionein compared to *CRS5* metallothionein in copper iron buffering in *Saccharomyces cerevisiae. J. Biol. Chem.* 271:18514–18519.

Jentsch, S. 1992a. The ubiquitin-conjugation system. *Ann. Rev. Genet.* 26:179–207.

Jentsch, S. 1992b. Ubiquitin-dependent protein degradation: A cellular perspective. *Trends Cell Biol.* 2:98–103.

Jungmann, J., H. A. Reins, J. Lee, A. Romeo, R. Hassett, D. Kosman, and S. Jentsch. 1993. MAC1, a nuclear regulatory protein related to Cu-dependent transcription factors is involved in Cu/Fe utilization and stress resistance in yeast. *EMBO J.* 12:5061–5066.

Jungmann, J., H.-A. Reins, C. Schobert, and S. Jentsch. 1993. Resistance to cadmium mediated by ubiquitin-dependent proteolysis. *Nature* 361:369–371.

Kamizono, A., M. Nishizawa, Y. Teranishi, K. Murata, and A. Kimura. 1989. Identification of a gene conferring resistance to zinc and cadmium ions in the yeast *Saccharomyces cerevisiae. Mol. Gen. Genet.* 219:161–167.

Kampfenkel, K., S. Kushnir, E. Babiychuk, D. Inze, and M. V. Montagu. 1995. Molecular characterization of a putative *Arabidopsis thaliana* copper transporter and its yeast homolog. *J. Biol. Chem.* 270:28479–28486.

Karin, M., R. Najarian, A. Haslinger, P. Valenzuela, J. Welch, and S. Fogel. 1984. Primary structure and transcription of an amplified genetic locus: The *CUP1* locus of yeast. *Proc. Natl. Acad. Sci. U.S.A.* 81:337–341.

Karin, M., E. P. Slater, and H. R. Herschman. 1981. Regulation of metallothionein synthesis in HeLa cells by heavy metals and glucocorticoids. *J. Cell Physiol.* 106:63–74.

Kikuchi, T. 1965. Studies on the pathway of sulfide production in a copper-adapted yeast. *Plant Cell. Physiol.* 68:195–210.

Kim, U. J., M. Han, P. Kayne, and M. Grunstein. 1988. Effects of histone H4 depletion on the cell cycle and transcription of *Saccharomyces cerevisiae. EMBO J.* 7:2211–2219.

Kirschner, M. 1992. The cell cycle then and now. *Trends in Biochem. Sci.* 17:281–285.

Kistler, M., K. Maier, and F. Eckardt-Schupp. 1990. Genetic and biochemical analysis of glutathione-deficient mutants of *Saccharomyces cerevisiae. Mutagenesis* 5:39–44.

Klionsksy, D. J., P. K. Herman, and S. D. Emr. 1990. The fungal vacuole: Composition, function and biogenesis, *Microbiol. Rev.* 54:266–292.

Kneer, R., T. M. Kutchan, A. Hochberger, and M. H. Zenk. 1992. *Saccharomyces cerevisiae* and *Neurospora crassa* contain heavy metal sequestering phytochelatin. *Arch. Microbiol.* 157:305–310.

Knight, S. A. B., K. T. Tamai, D. J. Kosman, and D. J. Thiele. 1994. Identification and analysis of a *Saccharomyces cerevisiae* copper homeostasis gene encoding a homeodomain protein. *Mol. Cell. Biol.* 14:7792–7804.

Kosman, D. J. 1994. Transition metal ion uptake in yeasts and filamentous fungi. In *Metal Ions in Fungi,* eds. G. Winkelmann and D. R. Winge. pp. 1–38. Marcel Dekker, New York.

Lee, R. L., S. Nacht, J. H. Lukens, and G. E. Cartwight. 1968. Iron metabolism in the copper-deficient swine. *J. Clin. Invest.* 47:2058–2069.

Lin, C. M., and D. J. Kosman. 1990. Copper uptake in wild-type and copper metallothionein-deficient *Saccharomyces cerevisiae. J. Biol. Chem.* 265:9194–9200.

Lindquist, S. 1986. The heat-shock response. *Ann. Rev. Biochem.* 55:1151–1191.

Liu, X. F., and V. C. Culotta. 1994. The requirement for yeast superoxide dismutase is bypassed through mutations in *BSD2*, a novel metal homeostasis gene. *Mol. Cell. Biol.* 14:7037–7045.

Macreadie, I. G., A. K. Sewell, and D. R. Winge. 1994. Metal ion resistance and the role of metallothionein in yeast. In *Metal Ions in Fungi*, eds. G. Winkelmann and D. R. Winge. pp. 279–310. Marcel Dekker, New York.

Masters, B. A., E. J. Kelly, C. J. Quaife, R. L. Brinster, and R. D. Palmiter. 1994. Targeted disruption of metallothionein I and II genes increases sensitivity to cadmium. *Proc. Natl. Acad. Sci. U.S.A.* 91:584–588.

Mehra, R. K., J. R. Garey, and D. R. Winge. 1990. Selective and tandem amplification of a member of the metallothionein gene family in *Candida glabrata. J. Biol. Chem.* 265:6369–6375.

Mehra, R. K., E. B. Tarbet, W. R. Gray, and D. R. Winge. 1988. Metal-specific synthesis of two metallothioneins and gamma-glutamyl peptides in *Candida glabrata. Proc. Natl. Acad. Sci. U.S.A.* 85:8815–8819.

Mehra, R. K., J. L. Thorvaldsen, I. G. Macreadie, and D. R. Winge. 1992. Disruption analysis of metallothionein-encoding genes of *Candida glabrata. Gene* 114:75–80.

Mei, B., and S. A. Leong. 1994. Molecular biology of iron transport in fungi. In *Metal Ions in Fungi*, eds. G. Winkelmann and D. R. Winge. pp. 117–148. Marcel Dekker, New York.

Mercer, J. F. B., J. Livingston, B. Hall, J. A. Paynter, C. Begy, S. Chandrasenkharappa, P. Lockhart, A. Grimes, M. Bhave, D. Siemieniak, and T. W. Glover. 1993. Isolation of a partial candidate gene for Menkes disease by positional cloning. *Nature Genet.* 3:20–25.

Messerschmidt, A., R. Ladenstein, R. Huber, M. Bolognesi, L. Avigliano, R. Petruzzelli, A. Rossi, and A. Finazzi-Agro. 1992. Refined crystal structure of ascorbate oxidase at 1.9 Å resolution. *J. Mol. Biol.* 224:179–205.

Michalska, A. E., and K. H. A. Choo. 1993. Targeting and germ-line transmission of a null mutation at the metallothionein I and II loci in mouse. *Proc. Natl. Acad. Sci. U.S.A.* 90:8088–8092.

Mifflin, L. C., and R. E. Cohen. 1994a. Characterization of denatured protein inducers of the heat shock (stress) response in *Xenopus laevis* oocytes. *J. Biol. Chem.* 269:15710–15717.

Mifflin, L. C., and R. E. Cohen. 1994b. *Hsc70* moderates the heat shock (stress) response in *Xenopus laevis* oocytes and binds to denatured protein inducers. *J. Biol. Chem.* 269:15718–15723.

Morimoto, R. I. 1993. Cells in stress: Transcriptional activation of heat shock genes. *Science* 259:1409–1410.

Morimoto, R. I., K. D. Sarge, and K. Abavaya. 1992. Transcriptional regulation of heat shock genes. *J. Biol. Chem.* 267:21987–21990.

Mosser, D. D., N. G. Theodorakis, and R. I. Morimoto. 1988. Coordinate changes in heat shock element-binding activity and *HSP70* gene transcription rates in human cells. *Mol. Cell Biol.* 8:4736–4744.

Mutoh, N., and Y. Hayashi. 1988. Isolation of mutants of *Schizosaccharomyces pombe* unable to synthesize cadystin, small cadmium-binding peptides. *Biochem. Biophys. Res. Comm.* 151:32–39.

Naiki, N. 1957. Studies on the adaptation of yeast to copper XVIII. Copper-binding sulfur substrates of the copper-resistant substrains. *Mem. Coll. Sci. Univ. Kyoto.* 24:243–248.

Narula, S. S., R. K. Mehra, D. R. Winge, and I. M. Armitage. 1991. Establishment of the metal-to-cysteine connectivities in silver-substituted yeast metallothionein. *J. Am. Chem. Soc.* 113:9354–9358.

Naumov, G. I., E. S. Naumova, H. Turakainen, and M. Korhola. 1992. A new family of polymorphic metallothionein-encoding genes *MTH1* (*CUP1*) and *MTH2* in *Saccharomyces cerevisiae*. *Gene* 119:65–74.

Nelson, H., and N. Nelson. 1990. Disruption of genes encoding subunits of yeast vacuolar H⁺-ATPase causes conditional lethality. *Proc. Natl. Acad. Sci. U.S.A.* 87:3503–3507.

Nieto-Sotelo, J., G. Wiederrecht, A. Okuda, and C. S. Parker. 1990. The yeast heat shock transcription factor contains a transcriptional activation domain whose activity is repressed under nonshock conditions. *Cell* 62:807–817.

Noumi, T., C. Beltran, H. Nelson, and N. Nelson. 1991. Mutational analysis of yeast vacuolar H⁺-ATPase. *Proc. Natl. Acad. Sci. U.S.A.* 88:1938–1942.

Ochai, E. I. 1987. In *General Principles of Biochemistry of the Elements,* Plenum Press, New York.

Ohtake, Y., A. Satou, and S. Yabuuchi. 1990. Isolation and characterization of glutathione biosynthesis-deficient mutants in *Saccharomyces cerevisiae*. *Agric. Biol. Chem.* 54:3145–3150.

Okumura, N., N. K. Nishizawa, Y. Umehara, and S. Mori. 1991. An iron deficiency-specific cDNA from barley roots having two homologous cysteine-rich MT domains. *Plant Mol. Biol.* 17:531–533.

Ortiz, D. F., L. Kreppel, D. M. Speiser, G. Scheel, G. McDonald, and D. W. Ow. 1992. Heavy metal tolerance in the fission yeast requires an ATP-binding cassette-type vacuolar membrane transporter. *EMBO J.* 11:3491–3499.

Palmiter, R. D. 1987. Molecular biology of metallothionein gene expression. *Experentia Suppl.* 52:63–80.

Parker, C. S., and J. Topol. 1984. A *Drosophila* RNA polymerase II transcription factor binds to the regulatory site of an *hsp70* gene. *Cell* 37:273–283.

Parsell, D. A., and S. Lindquist. 1993. The function of heat-shock proteins in stress tolerance: Degradation and reactivation of damaged proteins. *Ann. Rev. Genet.* 27:437–496.

Parsell, D. A., and R. T. Sauer. 1989. Induction of a heat shock-like response by unfolded protein in *Escherichia coli:* Dependence on protein degradation. *Genes Dev.* 3:1226–1232.

Pelham, H. R. 1986. Speculation on the functions of the major heat shock and glucose-regulated proteins. *Cell* 46:959–961.

Pelham, H. R. B. 1982. A regulatory upstream promoter element in the *Drosophila Hsp70* heat-shock gene. *Cell* 30:517–528.

Petrukhin, K., S. G. Fisher, M. Pirastu, R. E. Tanzi, I. Chernov, M. Devoto, et al. 1993. Mapping, cloning and genetic characterization of the region containing the Wilson disease gene. *Nature Genet.* 5:338–343.

Posewitz, M. C., J. R. Simon, R. A. Farrell, and D. R. Winge. 1996. Role of the conserved histidines in the Zn module of the copper-activated transcription factors in yeast. *J. Bioinorg. Chem.* 1:560–566.

Rad, M. R., L. Kirchrath, and C. P. Hollenberg. 1994. A putative P-type Cu(II)-transporting ATPase gene on chromosome II of *Saccharomyces cerevisiae. Yeast* 10:1217–1225.

Rauser, W. E. 1990. Phytochelatins. *Ann. Rev. Biochem.* 59:61–86.

Robbins, A. H., D. E. McRee, M. Williamson, S. A. Collett, N. H. Xuong, W. F. Furey, B. C. Wang, and C. D. Stout. 1991. Refined crystal structure of Cd,Zn metallothionein at 2.0 Å resolution. *J. Mol. Biol.* 221:1269–1293.

Rowley, A., G. C. Johnston, B. Butler, M. Werner-Washburne, and R. A. Singer. 1993. Heat shock-mediated cell cycle blockage and G1 cyclin expression in the yeast *Saccharomyces cerevisiae. Mol. Cell Biol.* 13:1034–1041.

Sadhu, C., and L. Gedamu. 1989. Metal-specific posttranscriptional control of human metallothionein gene. *Mol. Cell. Biol.* 9:5738–5741.

Sadis, S., C. Atienza, and D. Finley. 1994. Short hydrophobic amino acid sequences target proteins for ubiquitin-dependent degradation. In *Biology of Heat Shock Proteins and Molecular Chaperones.* 1994 Cold Spring Harbor meeting, p. 264.

Sanchez, Y., J. Taulien, K. A. Borkovich, and S. Lindquist. 1992. Hsp104 is required for tolerance to many forms of stress. *EMBO J.* 11:2357–2364.

Seufert, W., and S. Jentsch. 1990. Ubiquitin-conjugatin enzymes UBC4 and UBC5 mediate selective degradation of short-lived and abnormal proteins. *EMBO J.* 9:543–550.

Sewell, A. K., F. Yokoya, W. Yu, R. Miyagawa, T. Murayama, and D. R. Winge. 1995. Mutated yeast heat shock transcription factor exhibits elevated basal transcriptional activation and confers metal resistance. *J. Biol. Chem.* 270:25079–25086.

Silar, P., G. Butler, and D. J. Thiele. 1991. Heat shock transcription factor activates transcription of the yeast metallothionein gene. *Mol. Cell. Biol.* 11:1232–1238.

Singhal, R. K., M. E. Anderson, and A. Meister. 1987. Glutathione, a first line of defense against cadmium toxicity. *FASEB J.* 1:220–223.

Skowyra, D., C. Georgopoulos, and M. Zylicz. 1990. The *E. coli DnaK* gene product, the *hsp70* homolog, can reactivate heat-inactivated RNA polymerase in an ATP hydrolysis-dependent manner. *Cell* 62:939–944.

Sorger, P. K. 1990. Heat shock factor and the heat shock response. *Cell* 65:363–366.

Sorger, P., and H. R. B. Pelham. 1988. Yeast heat shock factor is an essential DNA-binding protein that exhibits temperature-dependent phosphorylation. *Cell* 54:855–864.

Sorger, P. K., M. Lewis, and H. R. B. Pelham. 1987. Heat shock factor is regulated differently in yeast and HeLa cells. *Nature* 329:81–84.

Sorger, P. K., and H. C. M. Nelson. 1989. Trimerization of a yeast transcriptional activator via a coiled-coil motif. *Cell* 59:807–813.

Sorger, P. K., and H. R. B. Pelham. 1987. Purification and characterization of a heat shock element binding protein from yeast. *EMBO J.* 6:3035–3041.

Szczypka, M., and D. J. Thiele. 1989. A cysteine-rich nuclear protein activates yeast metallothionein gene transcription. *Mol. Cell. Biol.* 9:421–429.

Szczypka, M. S., J. A. Wemmie, W. S. Moye-Rowley, and D. J. Thiele. 1994. A yeast metal resistance protein similar to human cystic fibrosis transmembrane conductance regulator (CFTR) and multidrug resistance. *J. Biol. Chem.* 269:22853–22857.

Tamai, K. T., X. Liu, P. Silar, T. Sosinowski, and D. J. Thiele. 1994. Heat shock transcription factor activates yeast metallothionein gene expression in response to heat and glucose starvation via distinct signalling pathways. *Mol. Cell. Biol.* 14:8155–8165.

Thiele, D. J. 1988. ACE1 regulates expression of the *Saccharomyces cerevisiae* metallothionein gene. *Mol. Cell. Biol.* 8:2745–2752.

Thiele, D. J., and D. H. Hamer. 1986. Tandemly duplicated upstream control sequences mediate copper-induced transcription of the *Saccharomyces cerevisiae* copper-metallothionein gene. *Mol. Cell. Biol.* 6:1158–1163.

Thorvaldsen, J. L., R. K. Mehra, W. Yu, A. K. Sewell, and D. R. Winge. 1995. Analysis of copper-induced metallothionein expression using autonomously replicating plasmids in *Candida glabrata. Yeast* 11:1501–1511.

Thorvaldsen, J. L., A. K. Sewell, C. L. McCowen, and D. R. Winge. 1993. Regulation of metallothionein genes by the ACE1 and AMT1 transcription factors. *J. Biol. Chem.* 268:12512–12518.

Thorvaldsen, J. L., A. K. Sewell, A. M. Tanner, J. M. Peltier, I. J. Pickering, G. N. George, and D. R. Winge. 1994. Mixed Cu(I), Zn(II) coordination in the DNA binding domain of AMT1 transcription factor from *Candida glabrata. Biochem.* 33:9566–9577.

Tohoyama, H., M. Inouhe, M. Joho, and T. Murayama. 1990. Resistance to cadmium is under the control of the CAD2 gene in the yeast *Saccharomyces cerevisiae. Current Genet.* 18:181–185.

Turner, J. S., A. P. Morby, B. A. Whitton, A. Gupta, and N. J. Robinson. 1993. Construction of Zn(II)/Cd(II) hypersensitive cyanobacterial mutants lacking a functional metallothionein locus. *J. Biol. Chem.* 268:4494–4498.

Vallee, B. L., and D. D. Ulmer. 1972. Biochemical effects of mercury, cadmium and lead. *Ann. Rev. Biochem.* 41:91–118.

Vulpe, C., B. Levinson, S. Whitney, S. Packman, and J. Gitschier. 1993. Isolation of a candidate gene for Menkes' disease and evidence that it encodes a copper-transporting ATPase. *Nature Genet.* 3:7–13.

Wagner, I., W. Neupert, and T. Langer. 1994. Functions of chaperone proteins in degradation of mitochondrial proteins. In *Biology of Heat Shock Proteins and Molecular Chaperones.* 1994 Cold Spring Harbor meeting, p. 232.

Welch, J., S. Fogel, C. Buchman, and M. Karin. 1989. The *CUP2* gene product regulates the expression of the *CUP1* gene, coding for yeast metallothionein. *EMBO J.* 8:255–260.

Werner-Washburne, M., D. E. Stone, and E. A. Craig. 1987. Complex interactions among members of an essential subfamily of *hsp70* genes in *Saccharomyces cerevisiae. Mol. Cell. Biol.* 7:2568–2577.

Williams, G. T., and R. I. Morimoto. 1990. Maximal stress-induced transcription from the human *HSP70* promoter requires interactions with the basal promoter elements independent of rotational alignment. *Mol. Cell. Biol.* 10:3125–3136.

Winge, D. R., C. T. Dameron, and G. N. George. 1994. The metallothionein structural motif in gene expression. *Adv. Inorg. Biochem.* 10:1–48.

Wright, C. F., K. McKenney, D. H. Hamer, J. Byrd, and D. R. Winge. 1987. Structural and functional studies of the amino terminus of yeast metallothionein. *J. Biol. Chem.* 262:112912–112919.

Wu, A., J. A. Wemmie, N. P. Edgington, M. Goebl, J. L. Guevara, and W. S. Moye-Rowley. 1993. Yeast bZip proteins mediate pleitropic drug and metal resistance. *J. Biol. Chem.* 268:18850–18858.

Xiao, H., O. Perisic, and J. T. Lis. 1991. Cooperative binding of *Drosophila* heat shock factor to arrays of a conserved 5 bp unit. *Cell* 64:585–593.

Yang, W., W. Gahl, and D. Hamer. 1991. Role of heat shock transcription factor in yeast metallothionein gene expression. *Mol. Cell. Biol.* 11:3676–3681.

Yu, W., R. A. Farrell, D. J. Stillman, and D. R. Winge. 1996. Identification of *SLF1* as a new copper homeostasis gene in *Saccharomyces cerevisiae* involved in copper sulfide mineralization. *Mol. Cell. Biol.* (in press).

Yu, W., I. G. Macreadie, and D. R. Winge. 1991. Protection against cadmium toxicity in yeast by alcohol dehydrogenase. *J. Inorg. Biochem.* 44:155–161.

Yu, W., V. Santhanagopalan, A. K. Sewell, L. T. Jensen, and D. R. Winge. 1994. Dominance of metallothionein in metal ion buffering in yeast capable of synthesis of (γEC)$_n$G isopeptides. *J. Biol. Chem.* 269:21010–21015.

Yuan, D. S., R. Stearman, A. Dancis, T. Dunn, T. Beeler, and R. D. Klausner. 1995. The Menkes/Wilson disease gene homolog in yeast provides copper to a ceruloplasmin-like oxidase required for iron uptake. *Proc. Natl. Acad. Sci. U.S.A.* 92:2632–2636.

Zhou, P., M. S. Szczypka, T. Sosinowski, and D. J. Thiele. 1992. Expression of a yeast metallothionein gene family is activated by a single metalloregulatory transcription factor. *Mol. Cell. Biol.* 12:3766–3775.

Zhou, P., and D. J. Thiele. 1991. Isolation of a metal-activated transcription factor gene from *Candida glabrata* by complementation in *Saccharomyces cerevisiae. Proc. Natl. Acad. Sci. U.S.A.* 88:6112–6116.

Zhou, P., and D. J. Thiele. 1993. Rapid transcriptional autoregulation of a yeast metallo-regulatory transcription factor is essential for high-level copper detoxification. *Genes Dev.* 7:1824–1835.

Zimarino, V., and C. Wu. 1987. Induction of sequence-specific binding of *Drosophila* heat shock activator protein without protein synthesis. *Nature* 327:727–730.

12

Yeast Metallothionein Gene Regulation

Simon A. B. Knight, Keith A. Koch, and Dennis J. Thiele

1. Introduction

A yeast provides an excellent model system for studying the function and regulation of genes and proteins within eukaryotic cells. The research advantage of this organism lies in the ease in which both classical and molecular genetics can be conducted, permitting the rapid identification of genes, the proteins they encode, and the functions associated with the proteins. This approach has been used highly successfully in understanding how cells are able to regulate cellular copper, and how copper in turn is able to regulate the genes that are involved in its homeostasis.

Copper was conclusively shown to be an essential trace element by Hart et al. (1928), who demonstrated that copper was necessary for growth and the prevention of anemia in rats. The essentiality of copper can be attributed to its role as a cofactor in a number of enzymes, including copper/zinc superoxide dismutase (Cu,Zn-SOD), cytochrome oxidase, and lysyl oxidase, which play critical roles in the defense of oxidative stress, oxidative phosphorylation, and collagen biosynthesis in organisms ranging from yeast to human. Copper, however, has a second face, that of a toxic compound. Species vary in their susceptibility to copper toxicity. Sheep are particularly sensitive, whereas in humans copper toxicity is rare, although a number of toxicity cases have been reported in India in suicides (Mertz 1987). There is, however, a heritable disease in humans, Wilson's disease, in which mutations in a proposed copper transporter protein lead to the accumulation of copper in renal tubules, cornea, and central nervous system, resulting in damage to these structure (Petruhkin et al. 1993; Tanzi et al. 1993). The destructive potential of copper towards cellular components can be attributed to two properties of the metal. The first is its ability to change its redox state within the cell [i.e., from Cu(I) and Cu(II)]. This permits copper to take part in Fenton-like reactions in which Cu(I) reacts with hydrogen peroxide (H_2O_2) to form Cu(II) and the highly reactive and extremely deleterious hydroxyl radical (•OH) (Halliwell and Gutteridge 1984). The second property is the ability of copper to interact nonspecifically with amino acid side chains such as those derived from cysteine, methionine, and histidine, thus rendering proteins inactive. This has been well documented for the estrogen receptor, a hormone-responsive DNA binding transcription factor that normally coordinates two zinc atoms for its proper structural configuration (Hutchens et al. 1992; Predki and Sarkar 1992). The consequence of this two-faced nature of copper is that all cells have to possess homeostatic mechanisms to allow sufficient copper to act as a cofactor where required for enzymatic processes and at the same time prevent the accumulation of copper to toxic levels.

It is becoming apparent that there are a number of genes whose protein products are required for copper homeostasis in eukaryotes. Five putative eukaryotic copper transporter proteins have been identified. In humans, the Wilson's disease and Menkes' disease genes encode proteins with similarity to members of the P-type ATPases, which transport cations across membranes (Chelly et al. 1993; Mercer et al. 1993; Petruhkin et al 1993; Tanzi et al. 1993; Vulpe et al. 1993). In the bakers' yeast, *Saccharomyces cerevisiae,* CCC2, an intracellular copper transporter with homology to the Menkes'/Wilson's disease gene products has been identified (Fu et al. 1995; Yuan et al. 1995), and two high-affinity copper transporter proteins have been identified, CTR1 and CTR3, whose function is critical at low environmental copper levels (Dancis et al. 1994; Knight et al. 1996). Furthermore, proper *CUP1* trancriptional upregulation in response to copper requires the presence of a functional allele of either *CTR1* or *CTR3* (Knight et al. 1996).

It is believed that for yeast to import copper into the cell the copper must be in the Cu(I) form. Recent work by Hassett and Kosman (1995) has demonstrated that the plasma membrane iron reductase, FRE1, and an as yet unidentified copper reductase are important for the reduction of Cu(II), adding convincing evidence that copper enters yeast cells as a Cu(I) ion. The biological targets of copper are still being identified. In addition to the well-characterized copper-dependent enzymes Cu, Zn-SOD and cytochrome oxidase, a number of new proteins have been identified in yeast. The first of these is the multicopper oxidase, FET3, which is postulated to oxidize Fe(II) to Fe(III), and is a copper-dependent component of the iron entry pathway (Askwith et al. 1994). The intracellular copper transporter CCC2, which is required for delivery of copper to the FET3 protein, provides a critical second component of the intimate connection between copper and iron homeostasis (Fu et al. 1995; Yuan et al. 1995). Additionally, two putative transcription factors have been identified. The MAC1 protein, which displays limited amino-terminal homology to ACE1 in *S. cerevisiae*, and AMT1 in *C. glabrata* (discussed later in this chapter), may bind copper or iron. MAC1 regulates *FRE1* transcription through an as yet uncharacterized mechanism, and as such is another protein that provides a link between copper and iron homeostasis in yeast (Jungmann et al. 1993). CUP9, which does not appear to bind copper directly, is important for the survival of yeast cells at high environmental copper concentrations when cells are undergoing respiration exclusively (as opposed to fermentation, the preferred means of energy generation in yeast) (Knight et al. 1994). If high amounts of copper enter the yeast cell, then the CUP1 (metallothionein)/ ACE1 (the copper-dependent activator of CUP1 transcription) system is activated to chelate the excess copper. A second metallothionein gene, *CRS5*, is also transcriptionally induced in response to copper (Culotta et al. 1994). The regulation of metallothionein-based copper detoxification systems is the subject of this chapter, and the regulation of the multifunctional protein, metallothionein (MT), in the bakers' yeast *S. cerevisiae* and in the opportunistic pathogenic yeast *Candida glabrata* will be discussed.

2. Structure of Metallothionein

The *S. cerevisiae* MT, encoded by the *CUP1* locus, is a small, cysteine-rich protein that binds both monovalent [Cu(I), Ag(I)] and divalent [Zn(II), Cd(II)] metals. The size of MT purified from yeast is believed to be 53 amino acids, although the size predicted from the nucleotide sequence of the *CUP1* gene is 61 amino acids (Butt et al. 1984; Karin et al. 1984; Winge et al. 1985; Wright et al. 1987). This discrepancy may be the result of *in vivo* processing of the newly synthesized MT, in which the eight amino terminal amino acids are cleaved from the molecule. Alternatively, the discrepancy in size could simply be due to proteolysis of MT during purification. If the former theory is true, the eight amino acids could potentially act as a signal peptide directing the protein to its functional location. Analysis of an amino-terminus mutant form of metallothionein, in which the first eight amino acids were not coded for, only slightly compromised the copper resistance of yeast cells expressing this mutant MT (Wright et al. 1987).

The ability of MT to impart metal resistance stems from its structure. Of the 61 amino acids in MT, 12 are cysteines. Apart from the two carboxy-terminal cysteines, these are all arranged as Cys-x-Cys or Cys-x_2-Cys (Butt et al. 1984). These residues coordinate metal ions as a single cluster by thiolate bridges. The solution structure derived for Ag-MT and Cu-MT shows the folding of the polypeptide around a single cluster of seven metal ions (Peterson et al. 1996). Two large parallel peptide loops are formed with a cleft in the center, where the metal cluster resides, with all amino acid side chains other than those of the cysteines residing outside the cleft region (Peterson et al. 1996). It is possible that a second type of cluster structure is formed: a truncated form of metallothionein in which five (T48) carboxy-terminal amino acids have been removed, including two cysteines, still appears to coordinate 7–8 moles of Cu (Byrd et al. 1988; George et al. 1988). In this structure it is envisioned that some of the cysteinyl thiolates are triply bridged with Cu(I) (George et al. 1988). The lack of a difference in the copper resistance of yeast expressing mutant MT, in which cysteines 49 and 50 have been converted to serines, compared to yeast expressing wild-type MT, supports the notion that these two cysteines are not involved in the coordination and therefore detoxification of copper (Thrower et al. 1988; Wright et al. 1986).

The *CRS5* metallothionein gene was identified as a multicopy suppressor of a *cup1Δ* strain of *S. cerevisiae*. This gene encodes a 69-amino-acid polypeptide, which contains 19 cysteine residues arranged predominantly in either Cys-x-Cys or Cys-x-x-Cys metal binding configurations, and shares only 28% amino acid sequence homology with CUP1. The CRS5 metallothionein, unlike CUP1, can bind 11–12 Cu(I) ions (Culotta et al. 1994; Jensen et al. 1996). The copper bound by CRS5, however, has been shown to be more solvent accessible and kinetically labile than that bound by CUP1. Given the greater number of Cu(I) ions bound by CRS5 and the number of coordinating cysteines, and utilizing hydrodynamic data derived from gel filtration of purified CRS5 protein, it has been postulated that CRS5 may exist as a two-domain metallothionein much like higher eukaryotic metallothioneins (Jensen et al. 1996).

3. Functions of Yeast Metallothionein

A body of convincing evidence has recently been presented that strongly suggests that *S. cerevisiae* MTs have multiple roles in cellular metabolism, although pinpointing the exact functions and mechanisms of action within the roles remains an ongoing area of research. The most intensely studied of these roles is metal detoxification and homeostasis. The *CUP1* locus was first associated with copper resistance in yeast by Brenes-Pomales (Brenes-Pomales et al. 1955). Cloning and sequencing of the *CUP1* gene revealed that the derived amino acid sequence contained 12 cysteine residues arranged in the characteristic manner of mammalian MT and known to be involved in the binding of metals (Kagi and Kojima 1987). Direct evidence of a role for MT in copper detoxification was obtained independently by two groups. Hamer and coworkers (Hamer et al. 1985) deleted the endogenous *CUP1* gene; the *cup1Δ* strain grew on standard growth media, but not in the same media to which 75 μM CuSO$_4$ had been added, demonstrating that MT is nonessential for growth under standard laboratory conditions but is essential in copper detoxification. Ecker and coworkers (Ecker et al. 1986) extended these observations and expressed the MT gene under normal or constitutive regulatory promoters in a copper-sensitive yeast strain in which the *CUP1* locus was deleted. The normally regulated MT gene only conferred resistance to copper; however, if the MT gene was constitutively expressed at high levels, the cells were resistant to cadmium (Ecker et al. 1986), consistent with the ability of yeast MT to bind both metals *in vivo*. Together these results demonstrated that one of the roles of yeast MT is the detoxification of copper. In the wild-type *S. cerevisiae* strains examined, cadmium does not significantly induce the synthesis of *CUP1* mRNA; thus, the binding of this metal by MT and subsequent detoxification of cadmium is perhaps a fortuitous rather than an evolutionary-designed event.

Mammalian MT synthesis is induced by a number of stimuli: heavy metals, interferon, glucocorticoids, lipopolysaccharides, and a variety of other effector molecules. The commonality of these stimuli is that they are either direct environmental stresses or are secondary signals of stress (Bremner and Beattie 1990; Hamer 1986). It is likely, therefore, that yeast MT may also have additional stress-related functions. To address this possibility, Silar and coworkers (Silar et al. 1991) utilized a yeast strain in which the *ACE1* gene (the copper-dependent transcriptional activator of *CUP1*) was deleted, and which contained a *CUP1-lacZ* fusion gene integrated into the chromosome. Selection of spontaneous suppressors of the *ace1* deletion (cells that grow in 150 mM CuSO₄ and which expressed elevated levels of β-galactosidase activity) identified one isolate designated *ADS* (*ace1* deletion suppressor) which was identified as a semidominant *trans*-acting mutation. Sequencing of the *ADS* locus revealed complete DNA homology with the heat shock transcription factor (HSF), except for a single point mutation resulting in a valine-to-alanine change in the DNA binding domain of HSF. Furthermore, wild-type HSF could both bind to the *CUP1* promoter and activate *CUP1* transcription in response to heat shock (Silar et al. 1991; Yang et al. 1991). *HSF is an essential gene* in *S. cerevisiae*, that transiently activates the *hsp70* family of heat shock proteins in response to heat stress [reviewed in Mager and Moradas Ferreira (1993)]. What role, if any, MT might play in heat shock is not clear, *CUP1* might represent a member of a second class of genes activated by HSF in response to environmental stress, instead of being a heat shock protein *per se*.

Recent evidence suggests that the CUP1 protein functions as an antioxidant. In yeasts and higher eukaryotes, normal cellular respiration generates reactive oxygen radicals, namely the superoxide anion ($O_2^-\bullet$), hydrogen peroxide (H_2O_2), and the highly reactive hydroxyl radical ($\bullet OH$). These molecules damage cellular macromolecules, such as proteins, DNA, lipids, and sugars. In humans, oxidative damage is associated with a number of disease states, including rheumatoid arthritis and familial amyotrophic lateral sclerosis (Lou Gehrig's disease) (Halliwell and Gutteridge 1984; Rosen et al. 1993). Cells have a number of defenses against oxygen toxicity, one of the most well characterized being Cu,Zn-SOD, which catalyzes the dismutation of the superoxide anion to hydrogen peroxide and singlet oxygen (McCord and Fridovich 1969). In *S. cerevisiae*, Cu,Zn-SOD is encoded by the *SOD1* gene. Deletion of the *SOD1* gene generates cells that are oxygen sensitive, require cysteine, methionine, and lysine for growth, and are incapable of growth on nonfermentable carbon sources such as lactate (Chang et al. 1991; Gralla and Valentine 1991). Supplementation of lactate-based media, in which respiration is the sole means of energy generation, with 1 mM $CuSO_4$ restored growth of a *sod1* deletion strain and suppressed the requirement of cysteine and methionine (Tamai et al. 1993). Tamai and colleagues further demonstrated that the suppression of the *sod1* strains growth phenotype required *CUP1* and moreover that purified Cu(I)-MT, but not apo⁻, Ag(I)⁻, or Zn(II)-MT, possesses antioxidant activity. Measurement of the decay constants of $O_2^-\bullet$ and •OH in the presence of purified yeast MT also indicate that Cu(I)-MT is a potent antioxidant, increasing the rate of $O_2^-\bullet$ dismutation by two orders of magnitude. In contrast, apo-MT and reconstituted Zn(II)-MT had little or no effect on the rate of $O_2^-\bullet$ dismutation. Cu(I)-MT was also superior to apo-MT and Zn(II)-MT in the quenching of •OH (Felix et al. 1993). It remains to be determined if the increased rate of decay of $O_2^-\bullet$ is due to a dismutase activity of Cu(I)-MT of if it represents a scavenging of oxygen free radicals. Recent evidence for the presence of transient thiyl radicals on the cysteines of yeast MT, without the formation of a Cu(II) within the cluster, suggests a possible role for the cysteine residues in the catalytic turnover of $O_2^-\bullet$ (Deters et al. 1994). Thus, it is apparent that yeast MT is not simply a metal-chelating protein, but may in fact serve a number of cellular functions.

4. Regulation of *S. cerevisiae* Metallothionein Gene Expression

The majority of studies investigating the regulation of MT gene expression in yeast have addressed the aspect of regulation by metals, specifically copper, because metal detoxification was the first phenotype associated with MT (Brenes-Pomales et al. 1955; Fogel and Welch 1982; Hamer et al. 1985; Karin et al. 1984). However, as more functions of MT are discovered, so the number of factors involved in *CUP1* regulation appears to be increasing. This section will first address regulation of MT gene expression by copper and other metals and then discuss other regulators of MT.

4.1 Gene Amplification

Work from several labs (Butt et al. 1984; Fogel et al. 1983; Hamer et al. 1985; Karin et al. 1984) associated yeast copper resistance with three key cellular events: amplification of the *CUP1* locus, increased levels of copper-inducible *CUP1* mRNA, and increased levels of MT protein. Laboratory yeast strains typically contain 9 to 10 copies of a 2-kb repeat unit at the *CUP1* locus on chromosome VIII; these strains are referred to as *CUP1^R* and are resistant to about 2 mM copper. To date, no amplification of the *CRS5* locus has been demonstrated. Successive plating of *CUP1^R* strains onto growth medium with increasing copper concentrations generated a hyperresistant strain, analysis of which revealed disomy of chromosome VIII. Additionally, the two copies of chromosome VIII contained 9 and 14 repeats of the 2-kb unit. These results suggested that not only had disomy of chromosome VIII occurred, but that gene amplification involving unequal sister chromatid exchange and intrachromosomal gene conversion had taken place (Fogel and Welch 1982). Unequal gene conversion also occurs during meiosis, during which copies of the 2.0-kb repeat unit loop out, permitting both reciprocal and nonreciprocal changes in 2.0-kb copy number between sister chromosomes and generating nonparental *CUP1* repeat arrays in the recombinant spores (Welch et al. 1990).

4.2 Cis-acting Elements in Transcriptional Control

The transcriptional activation of *CUP1* has been intensely studied. The initial observation that *CUP1* mRNA levels are induced by exposing cells to copper suggested that more *CUP1* mRNA was being transcribed, and/or the *CUP1* mRNA was stabilized in the presence of copper (Fogel et al. 1983). Copper-dependent transcription activation was suggested by Butt and coworkers (Butt et al. 1984), who fused a 431-bp DNA fragment from the 5'- noncoding region of *CUP1* to the *E. coli galK* gene. This reporter plasmid was transformed into a *CUP1^R S. cerevisiae* strain, and galactokinase activity was measured after the yeast had been exposed to 20, 100, and 500 µM $CuSO_4$ for 6 hours. Galactokinase activity was elevated between 20- to 50-fold with the addition of copper; no increase in galactokinase activity was observed in yeast transformed with the control plasmid. These results suggested that copper activation of *CUP1* transcription is a major cause of the elevation of *CUP1* mRNA. To identify the DNA sequences directly responsible for copper-activity transcription of *CUP1* within the 431-bp promoter region, Thiele and Hamer (Thiele and Hamer 1986) constructed a series of 5' and 3' deletion mutations in this region and fused them to the *E. coli galK* gene. The reporter constructs were transformed into a *CUP^R gal1* strain, and galactokinase activity of the transformants was determined after exposure to 0.5 mM $CuSO_4$ for 30 minutes. From this analysis, Thiele and Hamer (Thiele and Hamer 1986) were able to demonstrate that copper activation of transcription was the principal means of elevating *CUP1* mRNA levels. They were also able to identify two related sequences; <u>u</u>pstream <u>a</u>ctivation <u>s</u>equence proximal (UAS$_P$) and <u>d</u>istal (UAS$_D$) between nucleotides -105 and -180 with respect to the transcriptional start; these sequences are essential for copper-induced expression. The importance of UAS$_P$ was confirmed by placing tandem copies of a synthetic oligonucleotide, corresponding to positions -108 to -139, adjacent to a truncated *CYC1* promoter. Copper-induced expression from this construct was similar to a control construct of a native *CUP1* upstream sequence from -91 to -393, emphasizing the importance of the UAS in copper-induced expression.

An exhaustive systematic point mutagenesis and functional analysis of an oligonucleotide corresponding to the *CUP1* promoter region -105 to -142 was conducted by Fürst and coworkers (Fürst et al. 1988) to identify individual bases responsible for copper-induced transcription. Using the *galK* reporter system as described above, a region corresponding to -127 to -142 was identified, in which 11 of the 16 point mutations rendered this element noninducible by copper. This analysis also identified two mutations at positions A-118 and T-122 that caused substantial increases in basal transcription. An increase in basal transcription was also observed by Thiele and Hamer in which all nucleotides 5' to -105 were deleted (Thiele and Hamer 1986). These results are indicative of the presence of one or more negative control elements involved in *CUP1* transcription, which to date have not been characterized.

4.3 Trans-*acting Factors Controlling Metallothionein Gene Transcription*

The identification of the *cis*-acting elements that are involved in copper-regulated expression of *CUP1*, although necessary for copper-activated gene expression, did not reveal how copper was able to be sensed by the cell and act through these elements. This question was determined by classical yeast genetics. It was reasoned that a mutation in a *CUP1* transcriptional activation factor would render a *CUP1^R* strain copper sensitive and be defective in *CUP1* mRNA accumulation (Thiele 1988). This proved to be the case; two ethylmethanesulfonate-generated mutants met this criteria. One of these isolates, *ace1-1* (activation of *CUP1* expression) was then used to clone the transcription factor by complementation with a yeast genomic DNA library. Mapping of *ACE1* located the gene to chromosome VII, potentially allelic to *CUP2*, which was cloned shortly thereafter (Welch et al. 1989). Subsequent sequence comparisons and complementation analysis demonstrated that *ACE1* and *CUP2* were identical (Buchman et al. 1989); thus, for simplicity, the nomenclature *ACE1* will be used in this review. Sequencing of *ACE1* revealed an open reading frame encoding a protein of 225 amino acids that can be divided into two domains. One is an amino-terminal domain of 108 amino acids, which is extremely rich in basic residues and also contains the 12 cysteine residues arranged in the Cys-x-Cys or Cys-x$_2$-Cys motifs characteristic of metal binding proteins (Fürst et al. 1988; Szczypka and Thiele

1989). The remaining carboxy-terminal domain is highly negatively charged, a feature observed in other yeast transcriptional activators (Ma and Ptashne 1987).

The binding of ACE1 to the *CUP1* promoter was initially demonstrated using *in vitro* synthesized ACE1 and truncated form of ACE1 consisting of the 122 amino-terminal amino acids. Both the wild-type and truncated ACE1 exhibited copper-dependent binding to a synthetic oligonucleotide corresponding to the *CUP1* upstream activation sequence (UAS$_{CUP1}$), in mobility shift assays, indicating that the amino-terminal half of ACE1 is responsible for both copper and DNA binding (Fürst et al. 1988). These and similar experiments with protein extracts of *E. coli* cells overexpressing ACE1 showed that silver (electronically similar to copper) as well as copper stimulated formation of the ACE1 UAS$_{CUP1}$ complex, but zinc, cadmium, nickel, iron, lead, and mercury had no effect (Fürst et al. 1988; Welch et al. 1989). The results from these experiments were supported by yeast chromosomal footprinting studies, in which copper stimulated binding of one or more cellular factors to the UAS$_{CUP1}$. This binding was observed in a yeast strain harboring multiple copies of the *ACE1* gene, and was not detectable in a yeast strain with a nonfunctional *ace1* locus (Huibregtse et al. 1989). The significance of the copper-dependent binding of ACE1 to the UAS$_{CUP1}$ is underscored by examination of *in vivo CUP1* expression in an *ACE1* wild-type or *ace1* mutant yeast strains. *CUP1* expression, determined from *CUP1-lacZ* fusion genes, was only increased by the addition of copper or silver in cells with a functional *ACE1* locus (Fürst et al. 1988; Thiele 1988). Taken together, these results demonstrate that ACE1 is the major transcriptional activator of *CUP1* in cells responding to high environmental copper levels.

ACE1 acts as a cellular copper sensor. Four to six copper ions are coordinated per molecule of ACE1, as assessed by atomic absorption spectroscopy (Buchman et al. 1989; Dameron et al. 1993; Farrell et al. 1996; Thorvaldsen et al. 1994). These coppers are thought to be coordinated in a similar manner to copper in MT: trigonal coordination with the thiol groups of 11 of the 12 cysteine residues in the amino terminus of the protein forms a polynuclear copper cluster (Dameron et al. 1991; Nakagawa et al. 1991). More recent studies have identified a zinc ion associated with the amino-terminus of ACE1 (and AMT1, to be discussed later in this chapter) (Farrell et al. 1996; Thorvaldsen et al. 1994). In a cell not exposed to copper it is thought that ACE1 exists as an apoprotein. When environmental concentrations of copper increase, excess copper is bound by ACE1 in a cooperative manner, ensuring the sensitivity of the ACE1 protein to detect small changes in cellular copper concentrations. The binding of copper causes an as yet uncharacterized change in the conformation of ACE1, thereby permitting it to bind to the UAS$_{CUP1}$ and activate transcription of *CUP1* 10–50-fold, and to the promoter of the *CRS5* gene, resulting in a modest 2–3 activation of this metallothionein gene (Culotta et al. 1994).

The identification of ACE1 as the copper-dependent activator of *CUP1* transcription permitted a more detailed analysis of the *CUP1* promoter. *In vivo* footprinting studies by Huibregtse and colleagues (Huibregtse et al. 1989) identified three regions in the *CUP1* promoter in yeast chromatin which were protected from DNaseI cleavage in copper-treated wild-type cells but were not protected in *ace1* deletion cells. One of these regions correspond to the -127 to -147 region identified previously by point mutation analysis (Fürst et al. 1988). More detailed footprinting using a trpE-ACE1 protein produced in *E. coli* identified an ACE1 binding consensus of 5'-TC(T)$_{4-6}$GCTG-3', which appears six times within the *CUP1* promoter (Evans et al. 1990) (Figure 12.1). Support of this consensus, of which four copies are located in DNaseI protected regions, comes from methylation interference studies in which the two G residues of 5'-GCTG-3' were shown to be in close contact with the ACE1 protein (Buchman et al. 1990). Additional

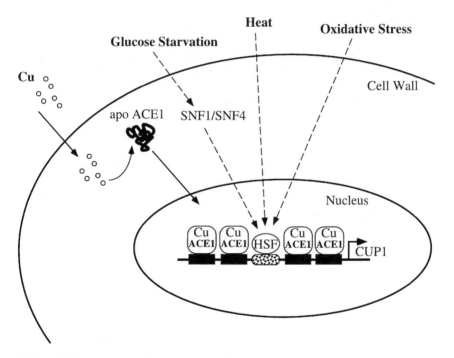

Figure 12.1 Environmental stress regulation of *CUP1* expression. Copper, heat, oxidative stress, and glucose starvation all induce *CUP1* expression. Induction by copper is mediated by the copper binding transcription factor ACE1, which when fully metalled binds to four sites in the *CUP1* promoter to activate transcription. The signals for heat shock, oxidative stress, and glucose starvation travel via different, as yet uncharacterized (dashed arrows), biochemical pathways. These three signals eventually converge on the same transcription factor, HSF, which activates *CUP1* transcription.

analysis with a combination of mobility shift, methylation interference, DNaseI, and hydroxyl radical footprinting methods, identified pseudo dyad symmetry in the region of ACE1 binding and suggested that two molecules of ACE1 bound to this region (Buchman et al. 1990). As discussed later in this chapter, the DNA binding consensus of AMT1, the Cu-activated transcription factor of C. glabrata, also contains AT-rich DNA followed by the 5'-GCTG-3' consensus sequence, underscoring the significance of this cis element in copper-activated gene transcription.

Once it is bound to the cis elements of the CUP1 promoter, the mechanism of how Cu(I)-ACE1 activates transcription has not been dissected. It is postulated that the highly acidic carboxyl-terminal domain might interact with the multicomponent cellular transcriptional machinery, as has been proposed for the GAL4 transcription factor with a similar negatively charged carboxy-terminal domain (Ma and Ptashne 1987). Equally little is known about how CUP1 transcription is turned off when copper concentrations decrease. The simplest model would be that as more MT is synthesized it chelates the excess cellular copper, depriving ACE1 from binding copper and rendering ACE1 incapable of binding to the UAS_{CUP1}. There is limited evidence of a negative regulatory element; in the analysis of the UAS_{CUP1} the CUP1-galK fusion construct Δ-105, which retained the TATA box but was devoid of all ACE1 binding sites, exhibited a basal level of expression that was 10-fold higher than the wild-type UAS_{CUP1} (Thiele and Hamer 1986). Additionally, two point mutations at A 118 and T 112 in the promoter of a CUP1-galK fusion construct gave expression levels that were 5- to 7-fold greater than that of the wild-type promoter (Fürst et al. 1988). The significance of these observations is not yet clear; however, there is evidence that nucleosome loss leads to activation of CUP1 transcription to levels similar to that observed by copper induction (Durrin et al. 1992). This might imply that the chromatin structure serves to maintain the CUP1 gene in an "inactive ground state," which can be overcome by Cu-ACE1 (Paranjape et al. 1994).

4.4 Nonmetal Regulators of Yeast Metallothionein Gene Expression

In accordance with the multiple functions of MT, there are factors other than metals that regulate MT biosynthesis. Aladjem and coworkers (Aladjem et al. 1988) have examined the effect of three carcinogens on CUP1 amplification. Treatment of yeast cells containing three copies of the CUP1 2-kb repeat with N-methyl-N'-nitro-N-nitrosoguanidine (MNNG), ethyl methanesulfonate (EMS), or 4-nitroquinoline N-oxide (4NQO) enhanced the frequency of cells resistant to 1 mM copper. The CUP1 copy number of these copper-resistant clones was increased 4- to 8-fold as ascertained from Southern blotting. EMS and MNNG modify bases directly, and 4NQO causes DNA damage by an oxidative mechanism; thus the gene amplification of the CUP1 locus in this instance correlates with the role of MT as an antioxidant (Tamai et al. 1993). Interestingly, the basal level of expression of both *CUP1* and *CRS5* metallothionein genes, mediated by ACE1, was aerobically repressed in the absence of added exogenous copper (Strain and Culotta 1996). Strain and Culotta noted that anaerobically grown cells are acutely sensitive to copper, because of an apparent 3–7-fold elevated level of copper associated with the cell compared to that of aerobically grown cells. It therefore appears that the regulation of the basal levels of metallothionein gene transcription in *S. cerevisiae* under aerobic and anaerobic conditions is a result of copper bioavailability. This is very reminiscent of *CUP1* regulation in a strain lacking functional alleles of the copper transporter genes *CTR1* and *CTR3*, as compaired to a strain harboring functional copper transporters (Knight et al. 1996).

As discussed earlier, the heat shock factor (HSF) activates *CUP1* transcription (Silar et al. 1991; Yang et al. 1991). HSF in *S. cerevisiae* is encoded by the essential gene *HSF1* and is thought to be constitutively bound as a trimer to heat shock elements (HSEs) in the promoters of heat-inducible genes (Sorger 1991). The heat shock consensus element consists of the palindromic sequence GAA-N_2-TTC. Mobility shift analysis has demonstrated that protein extract from yeast overexpressing HSF can bind to the wild-type *CUP1* promoter and that this binding can be completed with an unlabeled synthetic oligonucleotide containing the *CUP1* HSE. Not only does HSF bind to the UAS_{CUP1}, but *CUP1* expression is increased in wild-type or yeast cells with a nonfunctional *ace1* locus that are subjected to heat shock at 39°C. Together these results indicate that *CUP1* transcription can be activated by HSF and that the synthesis of CUP1 is a physiological response to heat shock.

DNaseI footprinting data identified a protected region from -146 to -171 within the UAS_{CUP1} containing the HSE consensus 5′-*n*TTC*nn*GAA*n* (Figure 12.1) (Tamai et al. 1994). Methylation interference indicated that the noncoding DNA strand G, at position -162, and the coding strand G, at position -159, were required for HSF-DNA complex formation. Mutation of these two bases abolished both HSF binding *in vitro*, and the induction of *CUP1* expression by heat shock, glucose starvation, and menadione (a pro-oxidant that generates superoxide anions in the cell) (Liu and Thiele 1996; Silar et al. 1991; Tamai et al. 1994). Genetic dissection of the activation pathway of these environmental stresses revealed that glucose starvation activation of *CUP1* expression required functional *SNF1* (encoding a serine-threonine protein kinase) and *SNF4* (encoding a SNF1 cofactor), which are essential for the activation of genes encoding key respiratory genes. In contrast, heat shock and menadione are able to induce *CUP1* expression independent of *SNF1* and *SNF4* (Figure 12.1) (Liu and Thiele 1996; Tamai et al. 1994). Furthermore, in support of at least three pathways for the activation of *CUP1* by HSF, Liu and Thiele demonstrated that HSF is differentially phosphorylated in response to heat shock or menadione, which is strong evidence that these pathways of HSF activation differ (Liu and Thiele 1996). Thus, in accordance with there being more than one function for MT, there are now two identified transcriptional activators of *CUP1* expression which can respond to four environmental stresses: copper toxicity, glucose starvation, oxidative stress, and heat shock.

5. *Candida glabrata*: The Metallothionein Genes and Proteins

The yeast *Candida glabrata* represents the closest-known relative, among the pathogenic members of the genus *Candida*, to *Saccharomyces cerevisiae* (Barns et al. 1991). Unlike *S. cerevisiae*, however, three distinct metallothionein (MT) genes have thus far been identified in *C. glabrata* (Mehra et al. 1989, 1992b). These genes are subdivided into classes: a single class I metallothionein gene, *MT-I*, and two class II metallothionein genes, *MT-IIa* and *MT-IIb*. The *MT-I* and *MT-IIa* genes were cloned by Mehra et al. (1989). The *MT-IIb* gene was not identified and cloned until 1992, after the deletion of the *MT-IIa* locus in strain 2001-L5 revealed the presence of another MT-II protein (Mehra et al. 1992b). The discovery of MT-IIb was impeded by the fact that the coding sequences for the *MT-IIa* and *MT-IIb* genes are identical, making their mRNAs and proteins indistinguishable. Outside of the coding region, however, significant differences are found in the 3' flanking DNA sequences as well as slight differences in 5' flanking sequences of the *MT-IIa* and *MT-IIb* genes (Mehra et al. 1992b). In fact, the copper-activated transcription from an *MT-IIa-lacZ* reporter gene was shown to be better than that of the *MT-IIb-lacZ* reporter gene, suggesting that the promoter differences correlate with differences in inducibility. Another major difference between the *MT-IIa* and *MT-IIb* genes is that only the *MT-IIa* gene has been found to be tandemly amplified in copper-resistant strains of *C. glabrata*. MT-I is a 6.4-kDa protein containing 18 cysteine residues that coordinate 11–12 mole equivalents of Cu(I); the MT-II proteins are 5.6 kDa and contain 16 cysteines that coordinate 10 mole equivalents of Cu(I) (Mehra et al. 1989). Because significant microheterogeneity of the amino acid sequence was detected during peptide sequencing of the MT-II protein (Mehra et al. 1989), as yet unidentified metallothioneins may exist in *C. glabrata*. Additionally, MT-II purified from cells grown in YTD media contained a minor amount of an amino-terminally truncated protein having Gln[7] at the amino-terminus (Mehra et al. 1989). Amino-terminal processing has also been observed for the purified *S. cerevisiae* metallothionein; however, the significance of this processing event for the function of MTs in either yeast is currently unknown (Wright et al. 1987). Unlike most higher eukaryotic metallothioneins, the two MTs from either *C. glabrata* or *S. cerevisiae* do not exhibit significant amino acid sequence homology with each other, or to other known MTs. In addition to the lack of amino acid sequence homology, the genes for MT-I and MT-II were shown to be localized to different chromosomes (Mehra et al. 1990). The lack of linkage of the *C. glabrata MT-I* and *MT-II* genes, like that of the *S. cerevisiae CUP1* and *CRS5* genes, is also distinct from that of mammals (West et al. 1990).

To date, the only factors reported to influence *C. glabrata* MT gene transcription are metal ions. *Candida glabrata* is thought to possess distinct mechanisms for coping with toxic levels of cadmium or copper. Cadmium resistance in *C. glabrata* has been proposed to be mediated by Cd binding γ-glutamyl peptides (Mehra et al. 1988), whereas copper resistance is mediated, in large part, by the transcriptional activation of the metallothionein genes and function of the encoded MTs. Activation of *C. glabrata* MT gene transcription in response to copper is carried out by the AMT1 (activator of metallothionein transcription 1) protein. The *AMT1* gene was cloned by Zhou and Thiele in 1991 using a complementation cloning strategy in *S. cerevisiae* (Zhou and Thiele 1991). AMT1, like ACE1, is only known to be activated to induce MT gene transcription in response to copper or silver. AMT1 shares 53% amino acid sequence identity, within the amino-terminal 115-amino-acid copper coordination-DNA binding domain, with the *S. cerevisiae* ACE1 protein. Within this region of sequence identity, the 11 cysteines that are proposed to coordinate copper and zinc in ACE1 are completely conserved in AMT1. Copper resistance in *C. glabrata*, as in *S. cerevisiae*, is fostered by elements of MT gene transcriptional regulation as well as MT gene amplification.

5.1 Transcriptional Activation of the Candida glabrata Metallothionein Genes

Metal-dependent transcription of the three *C. glabrata* MT genes is driven by AMT1. Disruption of the *AMT1* gene results in cells that are exquisitely sensitive to exogenous copper (Zhou et al. 1992). As AMT1 is the "master switch" governing activation of each of the three known MT isoform genes, it makes good sense that *amt1* deletion strains are more sensitive to copper than any single MT gene knock-out (Mehra et al. 1992b; Zhou et al. 1992). DNase I footprint analysis of the MT promoters, using AMT1 produced in *E. coli*, has established the presence of two AMT1 binding sites in the *MT-I* promoter and six AMT1 binding sites in the *MT-IIa* promoter (Zhou et al. 1992). Although footprinting of the *MT-IIb* promoter was not reported, the high degree of DNA sequence homology between the 5' regions of the two *MT-II* genes allows us to designate six putative AMT1 binding sites in the *MT-IIb* promoter, at analogous positions to the sites within the *MT-IIa* promoter. From DNase I footprinting, methylation and ethylation interference assays, and missing nucleoside experiments, a consensus AMT1 metal responsive element (MRE) can be derived (*w*T*whn*GCTG). The consensus AMT1 MRE contains an invariant GCTG core sequence and a highly variable

5′ AT-rich region. A comparison of the reported AMT1 MREs in Figure 12.2, along with the known ACE1 MREs, emphasizes the similarity in target sequences these highly homologous proteins share and the variable nature of the 5′ AT-rich region of their respective MREs. Experiments using the full-length and truncated AMT1, containing residues 1–122 encompassing the DNA binding domain, in electrophoretic mobility shift assays have provided evidence supporting monomeric binding of AMT1 to a single MRE (Zhou et al. 1992).

Zhou and Thiele (1993) identified a feature of *C. glabrata* metal responsive gene transcription distinct from that of *S. cerevisiae*. Unlike *ACE1* in *S. cerevisiae*,

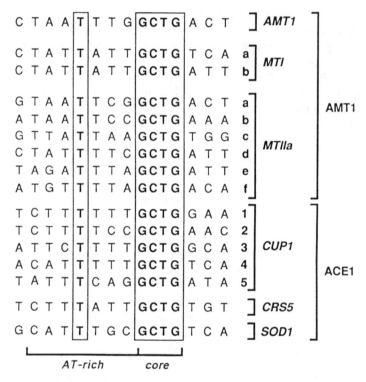

Figure 12.2 Sequences of the known AMT1 and ACE1 metal responsive elements. Each of the known AMT1 and ACE1 DNA binding sites is represented to emphasize the completely conserved nature of the GCTG core sequence and the variability in DNA sequence associated with the 5′ AT-rich region within each binding site. Completely conserved residues are indicated in bold. The genes in which the MREs are found are indicated directly to the right of the table. The corresponding metalloregulatory transcription factor that utilizes the MRE is indicated at the far right side of the table (Culotta et al. 1994; Evans et al. 1990; Gralla et al. 1991; Huibregtse et al. 1989; Zhou et al. 1992).

AMT1 mRNA levels increase 16-fold in response to exogenous copper. Subsequent DNase I footprinting of the *AMT1* promoter revealed a single extended AMT1 binding site, which conforms to the established consensus AMT1 MRE (Figure 12.2). When both G residues in the GCTG conserved cord binding site were changed to As, AMT1 binding was abolished *in vitro*, and the metal responsiveness of the *AMT1* promoter was abolished *in vivo*. These findings, and other evidence, describe an autoregulatory phenomenon whereby AMT1 regulates its own transcriptional activation in response to copper (Figure 12.3). Furthermore, reconstitution of the *AMT1* promoter mutant described above with the AMT1 structural gene, demonstrated that autoregulation of the *AMT1* gene is essential for cell survival at elevated copper levels (Zhou and Thiele 1993). Basal *AMT1* gene expression, and perhaps other as yet unidentified factors, are able to confer resistance up to 400 µM $CuSO_4$ in strain AMT1 (m)::URA3, which has an *AMT1* gene with a nonfunctional AMT1 binding site integrated at the URA3 locus. Given that the parental strain 85/038 can grow to 1.5 mM $CuSO_4$ on synthetic

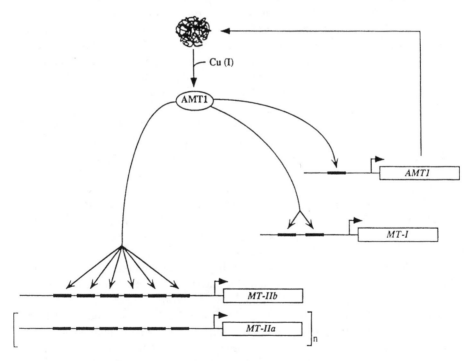

Figure 12.3 Copper activation of AMT1. The binding of Cu(I) by the AMT1 protein results in a conformational change that makes AMT1 DNA binding active. The activation of AMT1 results in subsequent transcriptional autoactivation of the *AMT1* gene and AMT1-dependent transcriptional activation of the *MT-I, MT-IIa,* and *MT-IIb* genes.

media, and an isogenic *AMT1* deletion strain can grow < 25 μM CuSO₄, these data suggest that AMT1 does not play a significant role in basal transcription from the *AMT1* promoter under standard laboratory growth conditions. In contrast to *AMT1* basal transcription, basal *MT-I* and *MT-II* mRNA levels decrease significantly in the absence of AMT1 (Zhou et al. 1992). The finding that a large part of *MT-I* and *MT-II* basal transcription is AMT1 dependent was surprising, because analysis of the kinetics of mRNA accumulation in response to copper has shown the order of maximal mRNA accumulation to be *AMT1* > *MT-I* > *MT-II* (Zhou and Thiele 1993). *MT-I* and *MT-II* mRNA levels reach their maximal levels much later than *AMT1*, yet their basal mRNA levels are regulated either transcriptionally or posttranscriptionally by an AMT1-dependent process. Interestingly, cadmium has been shown to decrease both basal *MT-I* and *MT-II* mRNA levels (Mehra et al. 1989), but has no discernible effect on basal *AMT1* mRNA levels (Zhou and Thiele 1993). One interpretation of this data is that cadmium is able to compete with copper for AMT1, and that Cd-AMT1 is a dead-end complex that is unable to bind MREs. Given that AMT1 supports a significant level of basal *MT-I* and *MT-II* mRNA accumulation, formation of this dead-end complex would result in deceased basal mRNA levels if AMT1 supports basal MT gene transcription. The lack of an effect of cadmium on *AMT1* basal mRNA levels strongly supports the hypothesis that basal transcription from the *AMT1* promoter is AMT1 independent. The role of transcription and posttranscriptional regulatory mechanisms in both basal and metal-induced expression of these genes is an area of future interest.

The absolute requirement for the transcriptional autoactivation of the *AMT1* gene, taken together with the fact that autoregulation occurs through a single MRE, allowed for a more detailed analysis of the interaction of AMT1 with the *AMT1* promoter MRE. The data from electrophoretic mobility shift assays, missing nucleoside experiments, and ethylation interference analysis describe the binding of monomeric AMT1 to one face of the double helix at adjacent major and minor grooves within the MRE (Koch and Thiele 1996). Furthermore, mutations that introduce a G or C at the single conserved T:A base pair within the AT-rich region of the MRE dramatically reduce the binding of AMT1 to the MRE, and correspondingly reduce the ability of AMT1 to activate transcription through an MRE harboring these mutations. Methylation interference and ethylation interference studies indicated that this region of the binding site was contacted by AMT1 in the minor groove (Koch and Thiele 1996; Zhou and Thiele 1993). Analysis of the primary amino acid sequence of the DNA binding domains of AMT1 and ACE1 revealed the presence of a completely conserved region, containing a GRP tripeptide, with homology to the minor groove binding peptide from the HMG-I/Y proteins. Mutation of the AMT1 coding sequence to convert arginine 38 to lysine resulted in a marked reduction in DNA binding affinity of the purified protein, and a severe defect in the ability of *C. glabrata* to mount a protective response against copper toxicity (Koch and Thiele 1996).

Upstream of the *AMT1* promoter MRE resides a homopolymeric stretch of 16 contiguous adenosine residues (A16). Zhu and Thiele examined the requirement of this A16 for rapid transcriptional autoactivation of the *AMT1* gene *in vivo* using mutant *AMT1* promoters that either had the A16 deleted or had a scrambled nucleotide sequence in its place (Zhu and Thiele 1996). The results of these experiments demonstrate the absolute requirement of the A16 for rapid transcriptional autoactivation of the *AMT1* gene. Further experiments demonstrated that both the A16 and the MRE were within the boundaries of a single nucleosome on the *AMT1* promoter. The observation that DNase I hypersensitive sites flanked the A16 *in vivo*, suggested that these regions were distorted. These distorted regions were subsequently shown to provide ready access to restriction endonucleases *in vivo*, and dimethylsulfate protection footprinting also revealed a strong correlation between MRE occupancy in early timepoints after copper addition and the presence of the A16. These data taken together describe the *AMT1* promoter in a novel class of specialized nucleosomal structures where a homopolymeric (dA-dT) element creates a distorted nucleosome with "access windows" that flank the element, providing access to transcription factors that would otherwise be hindered in the binding to their cognate site within a normal nucleosomal structure in chromatin (Zhu and Thiele 1996).

5.2 Candida glabrata *Metallothionein Gene Amplification*

High-level copper resistance is seen in many wild-type *Candida glabrata* strains. In large part, this is due to several wild-type strains having an amplified *MT-IIa* locus (Mehra et al. 1990). This selective amplification of a single MT gene in organisms harboring multiple MT genes is rare. *Drosophila melanogaster* is the only other organism known to selectively amplify a single member of its MT gene family (Mokdad et al. 1987; Otto et al. 1986). Mehra et al. have shown that the repeating unit of the *MT-IIa* gene is contained within 1.25 kb of DNA (Mehra et al. 1990), and the intensity of this fragment detected by Southern blotting, as has previously been demonstrated for *S. cerevisiae* (Fogel et al. 1983), directly correlates to the number of tandemly amplified *MT-IIa* genes. Through selective enrichment for highly copper-resistant mutants of the wild-type strains 2001, 67, CAP, and CDC, strains were identified which had tandemly amplified the *MT-IIa* locus. The parental strains 2001 and 67 display 50% inhibition of growth (IC_{50}) at 800 μM $CuSO_4$. The isogenic mutants, derived from the selective enrichment process, 2001-Cu^R-II and 67^R, display IC_{50} values of 7 mM and 3 mM, respectively.

Quantitation of the amount of the repeating 1.25-kb unit, as well as the size of a genomic DNA fragment encompassing all the tandemly amplified *MT-IIa* genes, indicates that the wild-type strain 2001 has three copies of *MT-IIa* and strain 67 has nine copies of *MT-IIa*. Both of these strains exhibit the same level of copper tolerance, so it is clear that other factors must also play a role in fostering high-level copper resistance in *C. glabrata*. There is a clearer correlation between the extent of *MT-IIa* gene amplification and copper resistance in strains having a highly amplified *MTIIa* locus. This point is illustrated by the fact that the strain 2001-Cu^R-II, which has > 29 copies of *MT-IIa* has an IC$_{50}$ of 7 mM CuSO$_4$, whereas the strain 67^R having 18 copies of *MT-IIa* has a much lower IC$_{50}$ of 3 mM. Interestingly, the mutant strain 67^R appears to have two completely independent sets of nine tandem copies of *MT-IIa*. The *MT-IIa* gene arrangement in 67^R may have resulted from chromosomal disomy, a phenomenon that has been observed in *S. cerevisiae* when amplification of the *CUP1* gene results from chromosome VIII duplication (Welch et al. 1983).

A characteristic of *Candida glabrata* strains having a highly amplified *MT-IIa* locus is that they have higher basal levels of *MT-IIa* mRNA, and the magnitude of *MT-IIa* gene induction in response to copper is reduced. This point is illustrated by the finding that wild-type strain 2001 induces its *MT-II* mRNA 60-fold over basal levels in response to 100 μM CuSO$_4$, whereas the isogenic strain 2001-Cu^R-II containing > 29 copies of *MT-IIa* elicits a 10-fold induction of *MT-II* mRNA in response to 1 mM CuSO$_4$ (Mehra et al. 1990). The mechanism utilized by *C. glabrata* to amplify the *MT-IIa* gene is currently unknown; however, putative autonomous replication sequences (*ARS*) have been identified in the 3′ untranslated region of the *MT-IIa* gene (Mehra et al. 1992a) that bear some resemblance to the *ARS* elements of both *S. cerevisiae* and *C. albicans* (Cannon et al. 1990). These *ARS*-like elements may play a role in *MT-IIa* gene amplification, which would explain their absence from the 3′ sequences of the *MT-IIb* gene.

The fact that higher eukaryotic organisms contain multiple metallothioneins, which are often differentially regulated, suggests that metallothioneins have diverse functions. *Candida glabrata* offers a unique opportunity to address the functions of very distinct metallothioneins, as well as their temporal regulation by a common metalloregulatory transcription factor.

6. Summary

As additional functions for yeast MT are determined, so the unraveling of the regulation of these genes will become more complex. In *S. cerevisiae*, four distinct environmental stresses have been shown to activate *CUP1* expression: copper, heat, oxidative stress, and glucose starvation. These environmental signals are mediated by two transcription factors, ACE1 in the case of copper, and HSF in response to heat shock, oxidative stress, and glucose starvation. The pathways, however, through which heat shock and glucose starvation act to induce HSF-activated transcription of *CUP1* involve different signaling intermediates, which eventually converge on HSF. The intermediates involved have yet to be identified, but it is known that the glucose starvation pathway requires SNF1 and SNF4, which are not needed for heat shock or menadione-based oxidative stress activation of *CUP1*. The dissection of these signaling pathways is therefore a key area of research.

The opportunistic yeast *C. glabrata* offers a yet more complex system for study. This yeast possesses three MT genes; *MT-I, MT-IIa,* and *MT-IIb.* The expression of all three genes is inducible by copper acting through the copper-dependent activator of transcription AMT1. Although unlike the constitutive expression of ACE1 is *S. cerevisiae,* AMT1 expression is inducible by copper. Despite the presence of three MT genes, the only confirmed function for MT in *C. glabrata* is metal detoxification. In light of what is now known about the multiple roles of MT in *S. cerevisiae,* it is quite likely that additional cellular functions for MT in *C. glabrata* will be identified. Overall, the use of yeast in the study of MT has shed light on both the regulation and the function of this important and ubiquitous protein, which has still not revealed all of its properties.

Acknowledgments

We thank the members of the Thiele laboratory for discussions and suggestions during the writing of this review. We are also grateful to Richard Hassett and Daniel Kosman for sharing their preprints with us. Research in the Thiele laboratory investigating copper-regulated gene expression is supported by National Institutes of Health Grant GM41840. Simon A. B. Knight is supported by a postdoctoral PHS National Research Service Award F32 GM15662. Keith A. Koch is supported in part by the Cellular Biotechnology Training Program at the University of Michigan, National Institute of Health Grant GM08353. Dennis J. Thiele is a Burroughs Wellcome Toxicology Scholar.

References

Aladjem, M. I., Y. Koltin, and S. Lavi. 1988. Enhancement of copper resistance and CUP1 amplification in carcinogen-treated cells. *Mol. Gen. Genet.* 211:88–94.

Askwith, C., D. Eide, A. Van Ho, P. S. Bernard, L. Li, S. Davis-Kaplan, D. M. Sipe, and J. Kaplan. 1994. The *FET3* gene of *S. cerevisiae* encodes a multicopper oxidase required for ferrous iron uptake. *Cell* 76:403–410.

Barns, S. M., D. J. Lane, M. L. Sogin, C. Bibeau, and W. G. Weisburg. 1991. Evolutionary relationships among pathogenic *Candida* species and relatives. *J. Bacteriol.* 173:2250–2255.

Bremner, I., and J. H. Beattie. 1990. Metallothionein and the trace minerals. *Ann. Rev. Nutr.* 10:63–83.

Brenes-Pomales, A., G. Lindegren, and C. C. Lindegren. 1955. Gene control of copper-sensitivity in *Saccharomyces. Nature* 176:841–842,

Buchman, C., P. Skroch, W. Dixon, T. D. Tullius, and M. Karin. 1990. A single amino acid change in *CUP2* alters its mode of DNA binding. *Mol. Cell. Biol.* 10:4778–4787.

Buchman, C., P. Skroch, J. Welch, S. Fogel, and M. Karin. 1989. The *CUP2* gene product, regulator of yeast metallothionein expression, is a copper-activated DNA-binding protein. *Mol. Cell. Biol.* 9:4091–4095.

Butt, T. R., E. J. Sternberg, J. A. Gorman, P. Clarke, D. Hamer, M. Rosenberg, and S. T. Crooke. 1984. Copper metallothionein of yeast, structure of the gene and regulation of expression. *Proc. Natl. Acad. Sci. U.S.A.* 81:3332–3336.

Byrd, J., R. M. Bergers, D. R. McMillin, C. F. Wright, D. Hamer, and D. R. Winge. 1988. Characterization of the copper-thiolate cluster in yeast metallothionein and two truncated mutants. *J. Biol. Chem.* 263:6688–6694.

Cannon, R. D., H. F. Jenkinson, and M. G. Shepherd. 1990. Isolation and nucleotide sequence of an autonomously replicating sequence (*ARS*) element function in *Candida albicans* and *Saccharomyces cerevisiae. Mol. Gen. Genet.* 221:210–218.

Chang, E. C., B. F. Crawford, Z. Hong, T. Bilinski, and D. J. Kosman. 1991. Genetic and biochemical characterization of Cu,Zn-superoxide dismutase mutant in *Saccharomyces cerevisiae. J. Biol. Chem.* 266:4417–4424.

Chelly, J., Z. Tumer, T. Tonnesen, A. Petterson, Y. Ishikawa-Brush, N. Tommerup, and A. P. Monaco. 1993. Isolation of a candidate gene for Menkes' disease that encodes a potential heavy metal binding protein. *Nature Genet.* 3:14–19.

Culotta, V. C., W. R. Howard, and X. F. Liu. 1994. CRS5 encodes a metallothionein-like protein in *Saccharomyces cerevisiae. J. Biol. Chem.* 269:25295–25302.

Dameron, C. T., G. N. George, P. Arnold, V. Santhanagopalan, and D. R. Winge. 1993. Distinct metal binding configurations in ACE1. *Biochem.* 32:7294–7301.

Dameron, C. T., D. R. Winge, G. N. George, M. Sansone, S. Hu, and D. H. Hamer. 1991. A copper-thiolate polynuclear cluster in the ACE1 transcription factor. *Proc. Natl. Acad. Sci. U.S.A.* 88:6127–6131.

Dancis, A., D. S. Yuan, D. Haile, C. Askwith, D. Elde, C. Moehle, J. Kaplan, and R. D. Klausner. 1994. Molecular characterization of a copper transport protein in *S. cerevisiae*: An unexpected role for copper in iron transport. *Cell* 76:393–402.

Deters, D., H. J. Hartmann, and U. Weser. 1994. Transient thiyl radicals in yeast copper(I) thionein. *Biochim. Biophys. Acta* 1208:344–347.

Durrin, L. K., K. M. Randall, and M. Grunstein. 1992. Nucleosome loss activates *CUP1* and *HIS3* promoters to fully induced levels in the yeast *Saccharomyces cerevisiae. Mol. Cell. Biol.* 12:1621–1629.

Ecker, D. J., T. R. Butt, E. J. Sternberg, M. P. Neeper, C. Debouck, J. A. Gorman, and S. T. Crooke. 1986. Yeast metallothionein function in metal ion detoxification. *J. Biol. Chem.* 261:16895–16900.

Evans, C. F., D. R. Engelke, and D. J. Thiele. 1990. ACE1 transcription factor produced in

Escherichia coli binds multiple regions within yeast metallothionein upstream activation sequences. *Mol. Cell. Biol.* 10:426–429.

Farrell, R. A., J. L. Thorvaldsen, and D. R. Winge. 1996. Identification of the Zn(II) site in the copper-responsive yeast transcription factor, AMT1: A conserved Zn module. *Biochem.* 35:1571–1580.

Felix, K., E. Lengfelder, H.-J. Hartmann, and U. Weser. 1993. A pulse radiolytic study on the reaction of hydroyl and superoxide radicals with yeast Cu(I)-thionein. *Biochim. Biophys. Acta* 1203:104–108.

Fogel, S., and J. W. Welch. 1982. Tandem gene amplification mediates copper resistance in yeast. *Proc. Natl. Acad. Sci. U.S.A.* 79:5342–5346.

Fogel, S., J. W. Welch, G. Cathala, and M. Karin. 1983. Gene amplification in yeast: *CUP1* copy number regulates copper resistance. *Curr. Genet.* 222:304–310.

Fu, D., T. J. Beeler, and T. M. Dunn. 1995. Sequence, mapping and disruption of *CCC2*, a gene that cross-complements the Ca(2+)-sensitive phenotype of *csg1* mutants and encodes a P-type ATPase belonging to the Cu(2+)-ATPase subfamily. *Yeast* 11:283–292.

Fürst, P., S. Hu, R. Hackett, and D. Hamer. 1988. Copper activates metallothionein gene transcription by altering the conformation of a specific DNA binding protein. *Cell* 55:705–717.

George, G. N., J. Byrd, and D. R. Winge. 1988. X-ray absorption studies of yeast copper metallothionein. *J. Biol. Chem.* 263:8199–8203.

Gralla, E. B., D. J. Thiele, P. Silar, and J. S. Valentine. 1991. ACE1, a copper-dependent transcription factor, activates expression of the yeast copper, zinc superoxide dismutase gene. *Proc. Natl. Acad. Sci. U.S.A.* 88:8558–8562.

Gralla, E. B., and J. S. Valentine. 1991. Null mutants of *Saccharomyces cerevisiae* Cu,Zn-superoxide dismutase: Characterization and spontaneous mutation rates. *J. Bacteriol.* 173:5918–5920.

Halliwell, B., and J.M.C. Gutteridge. 1984. Oxygen toxicity, oxygen radicals, transition metals and disease. *Biochem. J.* 219:1–14.

Hamer, D. H. 1986. Metallothioneins. *Ann. Rev. Biochem.* 55:913–951.

Hamer, D. H., D. J. Thiele, and J. E. Lemontt. 1985. Function and autoregulation of yeast copperthionein. *Science* 228:685–690.

Hart, E. B., H. Steenbock, J. Waddell, and C. A. Elvehjem. 1928. Iron in nutrition. vii. Copper as a supplement to iron for hemoglobin building in the rat. *J. Biol. Chem.* 77:797–812.

Hassett, R., and D. J. Kosman. 1995. Evidence for Cu(II) reduction as a component of copper uptake by *Saccharomyces cerevisiae*. *J. Biol. Chem.* 270:128–134.

Huibregtse, J. M., D. R. Engelke, and D. J. Thiele. 1989. Copper-induced binding of cellular factors to yeast metallothionein upstream activation sequences. *Proc. Natl. Acad. Sci. U.S.A.* 86:65–69.

Hutchens, T. W., M. H. Allen, C. M. Li, and T. T. Yip. 1992. Occupancy of a C2-C2 type "zinc-finger" protein domain by copper. Direct observation by electrospray ionization mass spectroscopy. *FEBS Lett.* 309:170–174.

Jensen, L. T., W. R. Howard, J. J. Strain, D. R. Winge, and V. C. Culotta. 1996. Enhanced effectiveness of copper ion buffering by *CUP1* metallothionein compared with *CRS5* metallothionein in *Saccharomyces cerevisiae. J. Biol. Chem.* 271:18514–18519.

Jungmann, J., H.-A. Reins, J. Lee, A. Romeo, R. Hassett, D. Kosman, and S. Jentsch. 1993. MAC1, a nuclear regulatory protein related to Cu-dependent transcription factors, is involved in Cu/Fe utilization and stress resistance in yeast. *EMBO J.* 12:5051–5056.

Kagi, J.H.R., and Y. Kojima. 1987. Chemistry and biochemistry of metallothionein. In *Metallothionein II*, eds. J. H. R. Kagi and Y. Kojima. pp. 25–61. Birkhäuser Verlag, Basel.

Karin, M., R. Najarian, P. Haslinger, J. Valenzuela, J. Welch, and S. Fogel. 1984. Primary structure and transcription of an amplified genetic locus: The CUP1 locus of yeast. *Proc. Natl. Acad. Sci. U.S.A.* 81:337–341.

Knight, S.A.B., S. Labbé, L. F. Kwon, D. J. Kosman, and D. J. Thiele. 1996. A widespread transposable element masks expression of a yeast copper transport gene. *Genes Dev.* 10:1917–1926.

Knight, S. A., K. T. Tamai, D. J. Kosman, and D. J. Thiele. 1994. Identification and analysis of a *Saccharomyces cerevisiae* copper homeostasis gene encoding a homeodomain protein. *Mol. Cell. Biol.* 14:7792–7804.

Koch, K. A., and D. J. Thiele. 1996. Autoactivation by a *Candida glabrata* copper metalloregulatory transcription factor requires critical minor groove interactions. *Mol. Cell. Biol.* 16:724–734.

Liu, X. D., and D. J. Thiele. 1996. Oxidative stress induced heat shock factor phosphorylation and HSF-dependent activation of yeast metallothionein gene transcription. *Genes Dev.* 10:592–603.

Ma, J., and M. Ptashne. 1987. Deletion analysis of *GAL4* defines two transcriptional activating segments. *Cell* 48:847–853.

Mager, W. H., and P. Moradas Ferreira. 1993. Stress response of yeast. *Biochem. J.* 290:1–13.

McCord, J. M., and I. Fridovich. 1969. Superoxide dismutase. An enzymic function for erythrocuprein (hemocuprein). *J. Biol. Chem.* 244:6049–6055.

Mehra, R. K., J. R. Garey, T. R. Butt, W. R. Gray, and D. R. Winge. 1989. *Candida glabrata* metallothioneins. *J. Biol. Chem.* 264:19747–19753.

Mehra, R. K., J. R. Garey, and D. R. Winge. 1990. Selective and tandem amplification of a member of the metallothionein gene family in *Candida glabrata. J. Biol. Chem.* 265:6369–6375.

Mehra, R. K., E. B. Tarbet, W. R. Gray, and D. R. Winge. 1988. Metal-specific synthesis of two metallothioneins and γ-glutamyl peptides in *Candida glabrata. Proc. Natl. Acad. Sci. U.S.A.* 85:8815–8819.

Mehra, R. K., J. L. Thorvaldsen, I. G. Macreadie, and D. R. Winge. 1992a. Cloning a system for *Candida glabrata* using element of the metallothionein-II$_a$-encoding gene that confer autonomous replication. *Gene* 113:119–124.

Mehra, R. K., J. L. Thorvaldsen, I. G. Macreadie, and D. R. Winge. 1992b. Disruption analysis of metallothionein-encoding genes in *Candida glabrata*. *Gene* 114:75–80.

Mercer, J. F. B., J. Livingston, B. Hall, J. A. Paynter, C. Begy, S. Candrasekharappa, P. Lockhart, A. Grimes, M. Bhave, D. Siemieniak, and T. W. Glover. 1993. Isolation of a partial candidate gene for Menkes' disease by positional cloning. *Nature Genet.* 3:20–25.

Mertz, W. 1987. *Trace Elements in Human and Animal Nutrition*, fifth ed. Academic Press, New York.

Mokdad, R., A. Debec, and M. Wegnez. 1987. Metallothionein genes in *Drosophila melanogaster* constitute a dual system. *Proc. Natl. Acad. Sci. U.S.A.* 84:2658–2662.

Nakagawa, K. H., C. Inouye, B. Hedman, M. Karin, T. D. Tullius, and K. O. Hodgson. 1991. Evidence from EXAFS for a Copper Cluster in the Metalloregulatory Protein CUP2 From Yeast. *J. Am. Chem. Soc.* 113:3621–3623.

Otto, E., J. E. Young, and G. Maroni. 1986. Structure and expression of a tandem duplication of the *Drosophila* metallothionein gene. *Proc. Natl. Acad. Sci. U.S.A.* 83:6025–6029.

Paranjape, S. M., R. T. Kamakaka, and J. T. Kadonaga. 1994. Role of chromatin structure in the regulation of transcription by RNA polymerase II. *Ann. Rev. Biochem.* 63:265–297.

Peterson, C. W., S. S. Narula, and I. M. Armitage. 1996. 3D solution structure of copper and silver-substituted yeast metallothioneins. *FEBS Lett.* 379:85–93.

Petruhkin, K., S. G. Fisher, M. Perastu, R. E. Tanzi, I. Cernov, M. Devoto, L. M. Brzustowicz, E. Cayanis, E. Vitale, and J. J. Russo. 1993. Mapping, cloning and genetic characterization of the region containing the Wilson's disease gene. *Nature Genet.* 5:338–343.

Predki, P. F., and B. Sarkar. 1992. Effect of replacement of "zinc finger" zinc on estrogen receptor DNA interactions. *J. Biol. Chem.* 267:5842–5846.

Rosen, D. R., T. Siddique, D. Patterson, and D. A. Figlewicz. 1993. Mutations in Cu/Zn superoxide dismutase gene are associated with familial amyotrophic lateral sclerosis. *Nature* 362:59–62.

Silar, P., G. Butler, and D. J. Thiele. 1991. Heat shock transcription factor activates transcription of the yeast metallothionein gene. *Mol. Cell. Biol.* 11:1232–1238.

Sorger, P. K. 1991. Heat shock factor and the heat shock response. *Cell* 65:363–366.

Strain, J., and V. C. Culotta. 1996. Copper ions and the regulation of *Saccharomyces cerevisiae* metallothionein genes under aerobic and anaerobic conditions. *Mol. Gen. Genet.* 251:139–145.

Szczypka, M., and D. J. Thiele. 1989. A cysteine-rich nuclear protein activates yeast metallothionein gene transcription. *Mol. Cell. Biol.* 9:421–429.

Tamai, K. T., E. B. Gralla, L. M. Ellerby, J. S. Valentine, and D. J. Thiele. 1993.

Yeast and mammalian metallothioneins functionally substitute for yeast copper-zinc superoxide dismutase. *Proc. Natl. Acad. Sci. U.S.A.* 90:8013–8017.

Tamai, K. T., X. Liu, P. Silar, T. Sosinowski, and D. J. Thiele. 1994. Heat shock transcription factor activates yeast metallothionein gene expression in response to heat and glucose starvation via distinct signalling pathways. *Mol. Cell. Biol.* 14:8155–8165.

Tanzi, R. E., K. Petrukhin, I. Chernov, J. O. Pellequer, W. Wasco, B. Ross, D. M. Rorano, E. Parano, L. Pavone, and L. M. Brzustowicz. 1993. The Wilson's disease gene is a copper transporting ATPase with homology to the Menkes' disease gene. *Nature Genet.* 5:344–350.

Thiele, D. J. 1988. ACE1 regulates expression of the *Saccharomyces cerevisiae* metallothionein gene. *Mol. Cell. Biol.* 8:2745–2752.

Thiele, D. J., and D. H. Hamer. 1986. Tandemly duplicated upstream control sequences mediate copper-induced transcription of the *Saccharomyces cerevisiae* copper-metallothionein gene. *Mol. Cell. Biol.* 6:1158–1163.

Thorvaldsen, J. L., A. K. Sewell, A. M. Tanner, J. M. Peltier, I. J. Pickering, G. N. George, and D. R. Winge. 1994. Mixed Cu^{2+} and Zn^{2+} coordination in the DNA-binding domain of the AMT1 transcription factor from *Candida glabrata*. *Biochem.* 33:9566–9577.

Thrower, A. R., J. Byrd, E. B. Tarbet, R. K. Mehra, D. H. Hamer, and D. R. Winge. 1988. Effect of mutation of cysteinyl residues in yeast Cu-metallothionein. *J. Biol. Chem.* 263:7037–7042.

Vulpe, C., B. Levinson, S. Whitney, S. Packman, and J. Gitschier. 1993. Isolation of a candidate gene for Menkes' disease and evidence that it encodes a copper transporting ATPase. *Nature Genet.* 3:7–13.

Welch, J., S. Fogel, C. Buchman, and M. Karin. 1989. The *CUP2* gene product regulates the expression of the *CUP1* gene, coding for yeast metallothionein. *EMBO J.* 8:255–260.

Welch, J. W., S. Fogel, G. Cathala, and M. Karin. 1983. Industrial yeasts display tandem gene iteration at the *CUP1* region. *Mol. Cell. Biol.* 3:1353–1361.

Welch, J. W., D. H. Maloney, and S. Fogel. 1990. Unequal crossing-over and gene conversion at the amplified *CUP1* locus. *Mol. Gen. Genet.* 222:304–310.

West, A. K., R. Stallings, C. E. Hildebrand, R. Chili, M. Karin, and R. I. Richards. 1990. Human metallothionein genes: Structure of the functional locus at 16q13. *Genomics* 8:513–518.

Winge, D. R., K. B. Nielson, W. R. Gray, and D. H. Hamer. 1985. Yeast metallothionein. *J. Biol. Chem.* 260:14464–14470.

Wright, C. F., D. H. Hamer, and K. Mckenney. 1986. Chromogenic identification of oligonucleotide directed mutants. *Nucl. Acids Res.* 14:8489–8499.

Wright, C. F., K. McKenney, D. H. Hamer, J. Byrd, and D. R. Winge. 1987. Structure and functional studies of the amino terminus of yeast metallothionein, *J. Biol. Chem.* 262:12912–12919.

Yang, W., W. Gahl, and D. Hamer. 1991. Role of heat shock transcription factor in yeast metallothionein gene expression. *Mol. Cell. Biol.* 11:3676–3681.

Yuan, D. S., R. Stearman, A. Dancis, T. Dunn, T. Beeler, and R. D. Klausner. 1995. The Menkes'/Wilson's disease gene homologue in yeast provides copper to a ceruloplasmin-like oxidase required for iron uptake. *Proc. Natl. Acad. Sci. U.S.A.* 92:2632–2636.

Zhou, P., M. S. Szczypka, T. Sosinowski, and D. J. Thiele. 1992. Expression of a yeast metallothionein gene family is activated by a single metalloregulatory transcription factor. *Mol. Cell. Biol.* 12:3766–3775.

Zhou, P., and D. J. Thiele. 1991. Isolation of a metal-activated transcription factor gene from *Candida glabrata* by complementation in *Saccharomyces cerevisiae. Proc. Natl. Acad. Sci. U.S.A.* 88:6112–6116.

Zhou, P., and D. J. Thiele. 1993. Rapid transcriptional autoregulation of a yeast metallo-regulatory factor is essential for high-level copper detoxification. *Genes Dev.* 7:1824–1835.

Zhou, P., and D. J. Thiele. 1996. A Specialized Nucleosome Modulates Transcription Factor Access to a *C. glabrata* Metal Responsive Promoter. *Cell* 87:459–470.

13

The Molecular Biology of Iron and Zinc Uptake in *Saccharomyces cerevisiae*

David Eide

1. Introduction

Metals such as iron and zinc are essential nutrients because of the critical roles they play in a large number of biochemical processes. Iron, for example, readily donates and accepts electrons from substrates and can display a broad range of oxidation-reduction potentials depending on the ligand environment surrounding the metal cation. Because of this unique property, iron is an important cofactor of several metalloenzymes such as ribonucleotide reductase and aconitase. Moreover, iron is required for heme biosynthesis and the activity of many heme-containing enzymes such as catalase, the cytochromes of the electron transport chain, and hemoproteins involved in oxygen transport. Zinc, in contrast, has only one biologically relevant valence but is essential because it is an integral cofactor of over 300 different metalloenzymes and is indispensable to their catalytic activity and/or structural stability. Examples include alkaline phosphatase, alcohol dehydrogenase, aspartate transcarbamoylase, carbonic anhydrase, and several proteases. Zinc is also an important component of enzymes involved in transcription and of accessory transcription factors, the zinc-finger proteins, that regulate gene expression.

Over the past 10 years, our understanding of the molecular biology of how these metals are transported across the plasma membrane of eukaryotic cells has benefited greatly from studies of *Saccharomyces cerevisiae*. The rapid progress of this research is largely the consequence of the amenability of this organism to genetic, molecular, and biochemical analyses. Numerous genes involved in the process of metal cation uptake have been identified, and their biochemical roles are becoming increasingly clear. These genes are also providing the means to identify related genes from other eukaryotes that play similar roles in iron or zinc uptake. Moreover, studies of metal uptake in this yeast have provided insight into an important interaction between copper and iron that was observed many years ago in mammals but remained poorly understood until recently. Studies of iron metabolism in both mammals and *S. cerevisiae* have synergistically led us to a greater understanding of these processes in both types of organisms. Thus, *S. cerevisiae* is proving to be a useful model system for the study of the mechanism and regulation of metal cation uptake in all eukaryotic cells. This chapter describes the recent advancements in our understanding of the mechanism and regulation of these processes in *S. cerevisiae*.

2. General Features of Iron-uptake Systems

Although iron is critical to cellular biochemistry, its properties cause problems that an organism must surmount. First, despite iron's abundance in the environment, the oxidized Fe(III) form of the metal is extremely insoluble at all but the most acidic pHs. At pH 7, for example, the concentration of free Fe(III) in an aqueous solution is only about 10^{-17} M! Therefore, organisms have evolved with efficient mechanisms to obtain iron and store it intracellularly in a usable form. Second, excess iron can be extremely toxic. Free iron can convert molecular oxygen into a series of potent oxidizing agents, such as superoxide anion, hydrogen peroxide, and hydroxyl radical. These agents can damage cellular components like lipids, proteins, and DNA, and may ultimately cause cell death. Although the iron-storage protein ferritin and protective enzymes (e.g., superoxide dismutase, catalase) play important roles in limiting the levels of these oxidizing agents, a critical line of defense is the tightly regulated activity of iron uptake. By controlling iron uptake, cells maintain an adequate supply of the metal for metabolism while protecting themselves against its toxic effects.

Two basic strategies of iron uptake across cellular membranes have been identified in studies of many different organisms. One strategy involves Fe(III) binding compounds, called siderophores, that are secreted by some bacteria, certain yeast and other fungi, as well as some species of plants [for review, see Winkelmann et al. (1987)]. Siderophores bind Fe(III) and are then internalized by specific Fe(III)-siderophore transport systems. Siderophores have a very high affinity for Fe(III), and this characteristic allows efficient mobilization of the metal cation from otherwise insoluble sources. For example, desferrioxamine B, a siderophore produced by *Streptomyces pilosus*, binds Fe(III) with an association constant of 10^{33} M^{-1}.

A second commonly used strategy of iron uptake is a two-step process in which extracellular Fe(III) is reduced to the more soluble Fe(II) form by plasma membrane Fe(III) reductases. This reduction is also useful in releasing the iron from chelators that would otherwise interfere with uptake of the metal. Once reduced, the Fe(II) product is taken up by Fe(II)-specific transport systems. This strategy of iron uptake is found in the yeasts *S. cerevisiae, Schizosaccharomyces pombe,* and *Candida albicans* (Lesuisse and Labbe 1989; Dancis et al. 1990; Eide et al. 1992; Roman et al. 1993; Morrissey et al. 1996), as well as in some bacteria (Evans et al. 1986; Johnson et al. 1991), and many plant species (Romheld and Marschner 1983; Grusak et al. 1990; Yi et al. 1994; Fox et al. 1996; Yi and Guerinot 1996). Mammals may use a similar mechanism of uptake to bring iron across the mucosal membrane of the intestine (Raja et al. 1992; Han et al. 1994; Núñez et al. 1994) and in the uptake of "free" iron from blood plasma (Oshiro et al. 1993; Jordan and Kaplan 1994). Moreover, most mammalian cells acquire iron from transferrin, a serum iron binding protein; this system may represent a combination of the siderophore and reductive mechanisms described above. Fe(III)-transferrin complexes bind to the transferrin receptor on the cell surface. This receptor-ligand complex is taken up by endocytosis into an endosomal compartment, the iron is dissociated and transported across the endosomal membrane, and the apotransferrin-receptor complex is returned to the cell surface where the apotransferrin is released back into the bloodstream. Some studies have indicated that transferrin-delivered Fe(III) is reduced to Fe(II) in the endosome, and the Fe(II) is then transported into the cytoplasm by Fe(II)-specific transport systems (Thorstensen and Romslo 1988; Nunez et al. 1992; Watkins et al. 1992). Thus, Fe(II)-specific transporter systems are essential components of iron-uptake systems at all phylogenetic levels.

3. An Overview of Iron Uptake in *S. cerevisiae*

At this time, there is no evidence from studies of *S. cerevisiae* for siderophore production or uptake of iron as Fe(III). Rather, as described above, extracellular Fe(III) is reduced to Fe(II) by plasma membrane Fe(III) reductases. The Fe(II) product is then taken up by either of two separate Fe(II)-specific transport systems. One system has a high affinity for iron (apparent k_m of 0.15 µM) and is necessary for iron-limited growth. Both Fe(III) reductase activity and high-affinity Fe(II) uptake activity are highly regulated by iron availability; cells grown on iron-limiting media have greater than 30-fold higher rates of Fe(III) reduction and Fe(II) uptake than do cells grown on iron-supplemented media. Iron-replete cells obtain iron through a second, low-affinity uptake system with an apparent K_m of 30 µM. The low-affinity system is also induced by iron limitation, albeit more modestly (~4-fold) than the high-affinity system. Recent genetic studies have identified several genes that encode the components or the regulatory proteins

Table 13.1 *S. cerevesiae* genes implicated in iron and zinc homeostasis

Gene	Proposed Function	Reference
AFT1	Iron-responsive transcriptional regulator	Yamaguchi-Iwai et al. 1995
CCC2	Intracellular (post-Golgi) copper transporter	Yuan et al. 1995
CTR1	High-affinity Cu(I) transporter	Dancis et al. 1994
FET3	High-affinity system multicopper ferroxidase	Askwith et al. 1994
FET4	Low-affinity system Fe(II) transporter	Dix et al. 1994
FET5	Unknown (*FET3* homolog)	D. Eide, unpublished result
FRE1	Plasma membrane Fe(III) [Cu(II)] reductase	Dancis et al. 1990
FRE2	Plasma membrane Fe(III) [Cu(II)?] reductase	Georgatsou and Alexandraki 1994
FTH1	Unknown (*FTR1* homolog)	Stearman et al. 1996
FTR1	High-affinity system iron transporter	Stearman et al. 1996
PCA1	Unknown (*CCC2* homolog)	Rad et al. 1994
UTR1	Cytoplasmic Fe(III) [Cu(II)] reductase subunit	Anderson et al. 1994
COT1	Intracellular cobalt/zinc transporter	Conklin et al. 1992
ZRC1	Intracellular zinc transporter	Kamizono et al. 1989
ZRT1	High-affinity zinc transporter	Zhao and Eide 1996a
ZRT2	Low-affinity zinc transporter	Zhao and Eide 1996b

of these systems. These genes and their functions are listed in Table 13.1. The roles of those proteins located on the plasma membrane are depicted in Figure 13.1.

4. The Plasma Membrane Fe(III) Reductases

The ability of *S. cerevisiae* to reduce extracellular Fe(III) was first demonstrated by Crane et al. (1982) and Ramirez et al. (1984). Because the substrate used in these experiments, ferricyanide, is not highly permeable to lipid membranes, it was concluded that this reduction occurred on the extracellular surface of the plasma membrane. The physiological role of Fe(III) reduction in iron uptake was not apparent in these studies. The hypothesis that these reductases play a role in iron uptake was first proposed by Lesuisse et al. (1987), who demonstrated that this yeast was capable of accumulating iron supplied as Fe(II) at a faster rate than when it was supplied as Fe(III). They also observed that Fe(III) reductase activity was markedly increased by growing the cells in an iron-limiting medium prior to analysis. These studies established the first link between Fe(III) reduction and iron uptake in *S. cerevisiae*. This link was further supported by the observation that heme-deficient strains were defective for both Fe(III) reduction as well as accumulation of iron when supplied as Fe(III) but not when the metal cation was supplied as Fe(II) (Lesuisse and Labbe 1989). The substrate of this reductase

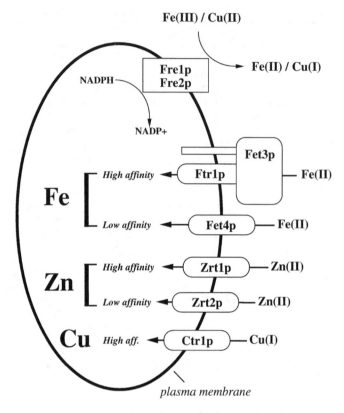

Figure 13.1 The metal cation uptake systems of *S. cerevisiae* and their plasma membrane components. The plasma membrane of the cell is depicted with the Fe(III) reductases Fre1p and Fre2p. The high-affinity iron-uptake system is composed of a complex of Fet3p and Ftr1p. Fet4p is the only known subunit of the low-affinity system for iron. High- and low-affinity zinc uptake is mediated by the Zrt1p and Zrt2p proteins, respectively. High-affinity copper uptake is mediated by Ctr1p. Additional uptake systems for all three metals exist but are not yet characterized.

activity is not limited to ferricyanide; it has also been found to be capable of reducing Fe(III) in a variety of chelates including Fe(III)-citrate, phosphate, NTA, EDTA, rhodotorulic acid, and desferrioxamine B.

Molecular insight into the reductive mechanism of iron uptake in *S. cerevisiae* was provided by the isolation of a mutant strain that was specifically defective in plasma membrane Fe(III) reductase activity (Dancis et al. 1990). The gene mutated in this strain, *FRE1*, probably encodes an Fe(III) reductase per se (or one subunit of a heteromeric complex, see below). Mutants defective for *FRE1* were unable to grow under iron-limiting conditions. Although defective for Fe(III) reductase activity and uptake of iron supplied as Fe(III), the *fre1* mutant strain retained its ability to accumulate iron supplied as Fe(II) (Dancis et al. 1990; Eide et al. 1992). Surprisingly, *fre1* mutants still possess a low level of iron-regulated Fe(III) reductase activity (Anderson et al. 1992; Dancis et al. 1992). This activity was later demonstrated to be the activity of the product of a second gene, *FRE2* (Georgatsou and Alexandraki 1994). *FRE2* encodes a protein related to Fre1p; they share 24.5% amino acid sequence identity over 693 amino acids. Depending on the strain and the growth conditions, either *FRE1* or *FRE2* expression can account for the majority of the total Fe(III) reductase activity in *S. cerevisiae* cells. Another gene, *frp1*, was identified in the distantly related yeast *Schizosaccharomyces pombe* and is thought to play a similar role in iron uptake by this organism (Roman et al. 1993). The *frp1* reductase shares 27% amino acid sequence identity with Fre1p.

A great deal of evidence supports the hypothesis that *FRE1* and *FRE2* of *S. cerevisiae* and *frp1* of *S. pombe* encode Fe(III) reductases. In addition to the effects of mutations in these three genes on Fe(III) reductase activity, it was found that the *FRE1, FRE2,* and *frp1* genes are regulated at the transcriptional level and that their mRNA levels closely correlated with Fe(III) reductase activity; i.e., mRNA levels were increased by iron-limited growth, as was Fe(III) reductase activity (Dancis et al. 1992; Roman et al. 1993; Georgatsou and Alexandraki 1994). Perhaps the strongest evidence comes from the DNA sequence analysis of these genes. The predicted products of each gene have potential signal sequences at their amino-termini and contain several hydrophobic domains of sufficient length (≥ 20 amino acids) to be transmembrane domains. Most important is the similarity that these proteins share with the gp91^{phox} subunit of cytochrome b$_{558}$, the phagocyte respiratory burst oxidase (Rotrosen et al. 1992). Cytochrome b$_{558}$ is a complex of proteins that transfers electrons across the phagosome membrane to molecular oxygen, thus reducing it to superoxide anion. This superoxide anion serves an antimicrobial function by killing microbes contained within the phagosome. The substrate of cytochrome b$_{558}$ is O_2, but the substrate of the yeast enzymes is Fe(III); Fre1p has little ability to reduce O_2 (Lesuisse et al. 1996; Shatwell et al. 1996). The similarity between the yeast proteins and gp91^{phox} is extensive. For example, the carboxy-terminal 402 amino acids of Fre1p share 17.9% identity and 62.2% similarity to gp91^{phox}. More specifically, all of these proteins share a high degree of sequence similarity in the region of gp91^{phox} implicated in flavin adenine dinucleotide (FAD) and NADPH binding.

S. cerevisiae has several NAD(P)H-dependent Fe(III) reductase activities that can be detected in different subcellular fractions (Lesuisse et al. 1990). A cytoplasmic flavoprotein was purified to homogeneity and found to utilize NADPH as the electron donor for Fe(III) reduction. Lesuisse et al. (1990) suggested that this enzyme was involved in mobilizing intracellular Fe(III). Attempts to purify an active Fe(III) reductase protein from plasma membranes were unsuccessful, perhaps because of the requirement for multiple subunits by these enzymes. However, a partially purified fraction of plasma membrane Fe(III) reductase was obtained and characterized biochemically. Consistent with the relationship of Fre1p and Fre2p with gp91[phox], these studies demonstrated that the yeast Fe(III) reductase requires NADPH and a flavin cofactor for activity. Two recent studies have supported the hypothesis that Fre1p is a plasma membrane flavocytochrome. Fre1p was found to be a glycosylated integral membrane protein enriched in plasma membrane fractions (Lesuisse et al. 1996). Plasma membrane fractions have an absorbance spectrum of a b-type cytochrome similar to that of cytochrome b_{558} (Lesuisse et al. 1996; Shatwell et al. 1996). The magnitude of this absorbance correlated with the level of Fe(III) reductase activity and the level of *FRE1* expression. Plasma membrane FAD levels also showed a correlation with Fre1p expression level (Shatwell et al. 1996).

The *UTR1* gene encodes a potential cytoplasmic factor involved in Fe(III) reductase activity (Anderson et al. 1994). This gene was identified in a genetic screen for Fe(III) reductase-deficient mutants. A strain bearing a *utr1* null mutation retains only 5% of the wild-type Fe(III) reductase activity. The predicted product of the *UTR1* gene is hydrophilic, suggesting that this protein is soluble and not integrally associated with the plasma membrane. A specific role for Utr1p in Fe(III) reductase activity was indicated by overexpression studies. Overexpression of either Fre1p or Utr1p alone failed to increase Fe(III) reductase activity, whereas simultaneous overexpression of both proteins resulted in a 5- to 7-fold increase in activity (Lesuisse et al. 1996). These observations indicated that both Fre1p and Utr1p are required for Fe(III) reductase activity. Little if any similarity exists between Utr1p and the known cytoplasmic subunits (p47 and p67) of cytochrome b_{558}. Therefore, the role of Utr1p in Fe(III) reductase activity is currently unknown.

Do Fre1p and Fre2p play different roles in *S. cerevisiae* iron uptake? Reductase activity generated by these two proteins shows different temporal regulation during batch culturing; Fre1p provides most of the activity measured early in culturing, whereas Fre2p provides the majority of the activity during the later stages of growth (Georgatsou and Alexandraki 1994). This pattern of regulation must be interpreted with caution because Fre1p activity was measured in a *fre2* mutant and Fre2p activity was assayed in a *fre1* mutant. Because Fre1p and Fre2p are both regulated by iron (see below) and either mutation will alter the iron status of the cell, the pattern of regulation observed in these mutant strains may not accurately reflect their wild-type regulation.

The Fe(III) reductase encoded by the *FRE1* gene has also been implicated in the uptake of copper. Fre1p reduces Cu(II) to Cu(I) prior to uptake of this metal cation (Hassett and Kosman 1995). Although *FRE2* has not been examined in this regard, it is possible that this gene is also important for copper uptake. *FRE1* is under the control of *MAC1*, a copper-responsive transcriptional activator that regulates copper uptake (Jungmann et al. 1993) as well as *AFT1*, an iron-responsive transcriptional activator (see below).

Fre1p and Fre2p are not the only sources of extracellular Fe(III) reductants; a *fre1 fre2* mutant still exhibited as much as 25% of the total Fe(III) reducing activity observed in a wild-type strain (Georgatsou and Alexandraki 1994). These residual reductants differed from the *FRE1*- and *FRE2*-dependent activities because they were excreted into the medium and could be detected in culture supernatants (Lesuisse et al. 1992; Georgatsou and Alexandraki 1994). It was suggested that these excreted reducing agents are anthranilate and/or 3-hydroxyanthranilate because their production correlated with the Fe(III) reducing properties of the culture medium (Lesuisse et al. 1992). Because excretion of these reductants was not iron regulated, it is unlikely that they are dedicated components of the iron-uptake system. Nonetheless, they may contribute to Fe(III) reduction under some growth conditions.

5. The High-affinity Fe(II) Uptake System and the Role of Copper in Iron Uptake

S. cerevisiae has multiple systems for the accumulation of iron, including two well-characterized systems that are specific for Fe(II). After reduction of Fe(III) by Fre1p or Fre2p, the Fe(II) product appears to be released from the cell surface. This hypothesis was suggested by the observation that there is no obligate coupling between Fe(III) reduction and uptake. Inhibition of reductase activity, either mutationally or by treating the cells with platinum, a potent inhibitor of reductase activity, could be bypassed by chemical reduction of Fe(III) to Fe(II) with a variety of reducing agents (Dancis et al. 1990; Eide et al. 1992). Following reduction, uptake of Fe(II) occurs through either the high- or low-affinity uptake systems. The existence of these two systems was initially suggested by analyses of the Michaelis-Menten kinetic properties of iron uptake (Dancis et al. 1990; Eide et al. 1992), which suggested the presence of two K_m values for uptake, a high-affinity component with an apparent K_m of 0.15 μM and a low-affinity component with an apparent K_m of 30 μM. Subsequent genetic studies demonstrated that these two systems are genetically separable; i.e., a mutation that disrupts one system fails to affect uptake by the other. These studies also indicated that uptake through the low- and high-affinity systems occur by different mechanisms.

The high-affinity system is exquisitely specific for Fe(II) over a wide variety of other metal cations (Eide et al. 1992). The first molecular insight into the mechanism of high-affinity uptake was provided by the isolation of a mutation in the *FET3* gene (Askwith et al. 1994). This mutation was isolated in a genetic selection using the hydroquinone antibiotic streptonigrin, which has been used previously to obtain mutants in iron metabolism in bacteria (Braun et al. 1983; Cohen et al. 1987). In the presence of intracellular Fe(II), streptonigrin generates free radicals that damage cellular components. A strain mutant in *FET3* was isolated because of its greater resistance to streptonigrin, perhaps because of a decreased intracellular iron content under the conditions used in this selection. Cells mutant in the *FET3* gene grow poorly on iron-limited media, suggesting a defect in the high-affinity uptake system. This defect was later confirmed by kinetic assays of Fe(II) accumulation. The *FET3* gene is required for activity of the high-affinity Fe(II) uptake system but is not required for low-affinity uptake activity (Askwith et al. 1994; Dix et al. 1994). This result indicated that the high- and low-affinity systems are separate and distinct pathways of iron accumulation.

The sequence of the *FET3* gene indicated that its predicted product is a protein of 636 amino acids with a molecular mass of 72 kDa (Askwith et al. 1994; De Silva et al. 1995). Surprisingly, Fet3p shares a remarkable degree of similarity not to transporter proteins but, rather, to a family of soluble enzymes known as the multicopper oxidases. This family includes ascorbate oxidase, laccase, and ceruloplasmin. These proteins carry out a variety of different functions including wound healing in plants (Butt 1980), lignin degradation by fungi (Ander and Eriksson 1976), and copper resistance in bacteria (Cha and Cooksey 1991). Despite this diversity of function, however, multicopper oxidases share the remarkable ability to carry out four single-electron oxidations of substrate followed by a four-electron reduction of O_2 to generate two molecules of H_2O. This is accomplished by 4–6 copper atoms that are bound by a group of highly conserved ligands present in these proteins. Homologous ligands are found in Fet3p. Fet3p has an amino-terminal signal sequence for transport across membranes that conforms well to a consensus sequence devised for such elements (von Heijne 1983) and was unique among multicopper oxidases in having an extended carboxy-terminal tail containing a potential transmembrane domain. These observations suggested that Fet3p is a type I integral membrane protein whose oxidase catalytic domain is located on the cell surface or in an extracytoplasmic intracellular compartment.

The predicted amino acid sequence of Fet3p and the effects of *FET3* mutations on the activity of the high-affinity system suggested that copper-dependent oxidation was an important step in the uptake of iron by the high-affinity system. This hypothesis was supported by many years of research on the mammalian multicopper oxidase, ceruloplasmin. In humans and other mammals, ceruloplasmin has been implicated in iron transport across cellular membranes. Ceruloplasmin is a serum protein that is capable of oxidizing Fe(II) to Fe(III), i.e., ferroxidase activity (Osaki et al. 1966). Many studies have indicated a relationship between ceruloplasmin, copper status, and iron metabolism in mammals. Insufficient dietary copper resulted in low levels of ceruloplasmin and a reduced ability to export iron from intestinal mucosal cells and macrophages into the blood plasma

(Lee et al. 1968). Plasma iron levels rose rapidly when either copper or ceruloplasmin was injected into these copper-deficient animals. More recently, a genetic defect in human patients has been identified in which a mutation in the ceruloplasmin gene itself results in the apparent absence of circulating ceruloplasmin (Harris et al. 1995; Yoshida et al. 1995). This mutation also causes alterations in iron homeostasis including anemia and low serum iron levels. Thus, the multicopper oxidase activities of ceruloplasmin and Fet3p have been implicated as important in the movement of iron across cellular membranes.

The relationship between Fet3p ferroxidase, copper status, and iron uptake in *S. cerevisiae* has been established in several ways. First, it was determined that copper-deficient cells also had low levels of high-affinity uptake activity (Askwith et al. 1994). Activity could be restored by including copper in the uptake assay in the presence or absence of the protein synthesis inhibitor cycloheximide. This result indicated that new protein synthesis was not necessary for reconstituting uptake activity. Second, high-affinity uptake was reduced in cells defective for copper uptake. The *CTR1* gene encodes the high-affinity copper transporter, and mutations in this gene caused defects in both copper and iron uptake (Dancis et al. 1994). Growth of the *ctr1* mutant in high copper restored iron uptake but did not restore copper uptake activity. Fet3p was demonstrated to have a copper-dependent oxidase activity *in vitro* by Yuan et al. (1995). Furthermore, Fet3p was required *in vivo* for a mitochondrial-independent O_2 consumption that was also dependent on Fe(II) (De Silva et al. 1995). This activity had a stoichiometry of 4 moles of Fe(II) oxidized to 1 mole of O_2 consumed as would be expected for Fe(II) oxidation mediated by a multicopper oxidase. The prediction that Fet3p is an integral membrane protein has also been confirmed. Furthermore, the catalytic domain is heavily *N*-glycosylated (Yuan et al. 1995). This domain was sensitive to trypsin digestion, and oxidase activity could be inhibited by monoclonal antibodies in spheroplasts, i.e., intact cells in which the cell wall has been removed (De Silva et al. 1995). These results demonstrated that the catalytic domain is exposed on the cell surface.

The important role of copper in iron uptake was also demonstrated by the analysis of the *CCC2* gene. *CCC2* encodes a P-type ATPase transporter with a high degree of similarity to the proteins encoded by the Menkes' (*MNK*) and Wilson's disease (*WD*) genes (Yuan et al. 1995). Both *MNK* and *WD* are thought to encode intracellular copper transporters that pass the metal cation from the cytoplasm into an unknown organelle of the secretory pathway. In the human liver, this copper transport is mediated by the *WD* protein (Bull et al. 1993; Tanzi et al. 1993). The copper is excreted in the bile or loaded into ceruloplasmin, which is then secreted into the blood plasma. The *MNK* protein is responsible for the export of copper from other tissues including the intestinal mucosa (Chelly et al. 1993; Vulpe et al. 1993; see Chapter 10 by Mercer and Camakaris). Thus, mutation in the *MNK* gene block the absorption of dietary copper. Ccc2p shares

31% amino acid identity over 604 residues to *WD* and 29% identity over 814 residues to *MNK*. In a role similar to that proposed for the *WD* and *MNK* proteins, Ccc2p appears to be an intracellular copper transporter that is responsible for delivering copper to Fet3p in a post-Golgi compartment (Yuan et al. 1995). Although *ccc2* mutants show no defect in plasma membrane copper uptake or in their intracellular level of copper, they are defective in iron uptake. As with *ctr1* mutants, the iron uptake defect of *ccc2* is copper suppressible. It is currently unclear if this copper suppression is due to the reconstitution of apo-Fet3p on the cell surface or due to the activity of additional intracellular copper transporters such as the product of *PCA1*, a gene homologous to *CCC2* (Rad et al. 1994).

5.1 The Oxidase/Permease Model of the High-affinity System

The observation that Fet3p contains only a single transmembrane domain suggested that this protein may be just one subunit of a heteromeric protein complex, i.e., one or more additional subunits might be required to form a functional transporter complex. This hypothesis was further suggested by the failure of cells overexpressing the *FET3* gene to exhibit increased Fe(II) uptake activity (Stearman et al. 1996). The identity of a second subunit of the high-affinity system was provided by the characterization of the *FTR1* gene (Stearman et al. 1996). *FTR1* is required for Fe(II) uptake by the high-affinity system. The iron uptake defect in an *ftr1* mutant was not suppressed by copper supplements as it was in the case of *ctr1* and *ccc2* mutants. Thus, the phenotype of the *ftr1* mutant is similar to that of the *fet3* mutant, suggesting that the *ftr1* defect lies downstream of the Fet3p copper binding step. *FTR1* encodes a protein of 404 amino acids with an amino-terminal signal sequence and six potential transmembrane domains, consistent with its proposed role as a transporter. Ftr1p shares little amino acid similarity to other known iron transporter proteins such as Fet4p of *S. cerevisiae* (Dix et al. 1994, see below), *feoB* of *E. coli* (Kammler et al. 1993), or *IRT1* from *A. thaliana* (Eide et al. 1996). Genes similar to *FTR1*, but of unknown function, were also identified in the genomes of *S. pombe* and *Bacillus subtilus* (Stearman et al. 1996).

Ftr1p is required for two aspects of the high-affinity Fe(II)-uptake system. First, Ftr1p is necessary for movement of Fet3 through the secretory pathway to the plasma membrane. This was first indicted by the observation that deletion of the *FTR1* gene markedly reduced Fet3p oxidase activity *in vitro* (Stearman et al. 1996). Oxidase activity could be restored by adding copper to the cell homogenization buffer, suggesting that the defect was in copper-loading Fet3p. The copper-reconstituted protein, however, had an aberrant pattern of glycosylation, indicating that in the absence of Ftr1p, Fet3p may be mislocalized within the secretory pathway. These results are consistent with a model in which Ftr1p and Fet3p aggregate in the endoplasmic reticulum, pass through the Golgi apparatus into a post-Golgi compartment where copper is delivered to Fet3p by Ccc2p, and finally move to the plasma membrane (Figure 13.2). Similarly, it was found that Fet3p is required for the appearance of Ftr1p in the plasma membrane. In the absence of Fet3p, Ftr1p is found in a perinuclear compartment, probably the endoplasmic reticulum. These results have suggested that the interaction between Fet3p and Ftr1p is required for either protein to proceed through the secretory pathway, during which time Fet3p acquires copper and becomes localized to the plasma membrane.

Ftr1p also appears to be the permease of the high-affinity system. Evidence of this role for Ftr1p was provided by deletion and point mutations introduced into the *FTR1* gene (Stearman et al. 1996). Specifically, mutations were generated that deleted a series of glutamates, in repeats of EXXE, located at the carboxyl-terminus. Also, either or both of the glutamates in a sequence found within a predicted transmembrane domain, REGLE, were replaced with alanine residues. The REGLE sequence is also conserved in the Ftr1p-related proteins from other

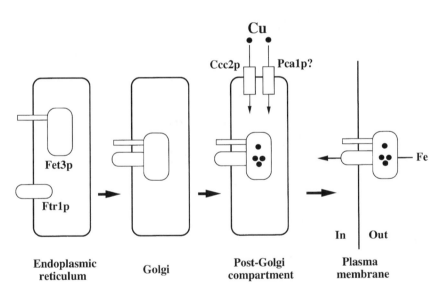

Figure 13.2 A model for assembly of the high-affinity oxidase/permease complex. The boxes depict the indicted organellar compartments along the secretory pathway. Fet3p and Ftr1p are both cotranslationally inserted into the membrane of the endoplasmic reticulum. These proteins form a complex, and this interaction allows the complex to migrate to the Golgi. Subsequently, the complex passes into a post-Golgi compartment where Cu (•) is bound by Fet3p. This copper is transported into the post-Golgi compartment by Ccc2p and possibly by Pca1p. The mature complex is then delivered to the plasma membrane where it is active in the uptake of iron.

organisms. This sequence was of interest because it is similar to a sequence, REGAE, found in the iron-storage protein, ferritin. Ferritin is a 24-subunit, spherical protein complex that is capable of taking up iron. The REGAE sequence forms a hydrophilic "spine" that liens a cation-selective pore in the protein that allows iron to pass into the hollow interior of the protein complex (Trikha et al. 1995). The parallel between this process and uptake of iron across the plasma membrane is striking and suggests that the conserved REGLE domain in Ftr1p may perform a related role in iron uptake by the high-affinity system. All of these mutations eliminated Fe(II) uptake activity. In marked contrast to the effects of an *ftr1* deletion mutation, however, Fet3p in these strains was correctly glycosylated and copper loaded, and had oxidase activity. These results implicated Ftr1p directly in the process of transporting iron across the membrane.

How might Fet3p and Ftr1p mediate iron uptake? One model for the mechanism of high-affinity Fe(II) uptake is described in Figure 13.3. In this model, Fe(II) produced by the Fe(III) reductases is oxidized to Fe(III) by the Fet3p multicopper

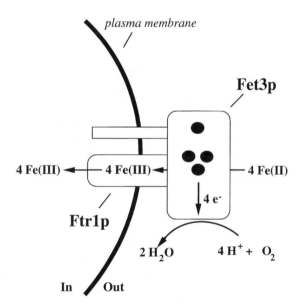

Figure 13.3 Proposed mechanism of iron uptake by the Fet3p/Ftr1p oxidase/permease complex. Fe(II) is generated by the activity of the Fe(III) reductases. This Fe(II) is oxidized to Fe(III) by the Fet3p ferroxidase subunit of the high-affinity transporter complex and then passed to the Ftr1p permease subunit for transport into the cell. Sequential oxidation of 4 Fe(II) atoms is followed by a single 4-electron reduction of O_2 to $2H_2O$. The manner in which the proteins are depicted is not meant to indicate any physical attributes of their association. The black circles indicate the copper atoms bound by Fet3p.

oxidase. The Fe(III) product is then transferred directly to an Fe(III) binding site on the Ftr1p permease and, following a conformational change in the permease, the Fe(III) is delivered to the cytoplasm of the cell.

The question then arises as to why opposing enzymatic reactions, i.e., Fe(III) reduction and Fe(II) oxidation, are both occurring on the extracellular surface of the plasma membrane prior to transport. One possible answer is that they provide substrate specificity to the uptake process. The high-affinity system is specific for iron and is not inhibited in the presence of high concentrations of other metals. This specificity may be the result of the sequential recognition of Fe(II) by Fet3p and Fe(III) by Ftr1p.

Why then can't Ftr1p utilize Fe(III) supplied directly in the medium? An answer to this question is provided by the solution chemistry of Fe(III). Fe(III) is extremely insoluble and readily forms $Fe(OH)_3$ precipitates. Thus, in aqueous solutions of moderate pH, soluble Fe(III) is almost entirely bound to chelators such as amino acids, citrate, and phosphate. It is hard to imagine how a transport system might be capable of utilizing a variety of different Fe(III) chelates and still maintain specificity for the metal cation. Siderophore-producing organisms have solved this problem by excreting their own high-affinity chelator which supplants the chelators to which iron in the medium is bound. In *S. cerevisiae*, reduction of Fe(III) to Fe(II) weakens the association between the chelator and the metal; the affinity of many chelators for Fe(II) is several orders of magnitude lower than for Fe(III). Fe(II) is then reoxidized by Fet3p, and the Fe(III) product is presented to the Ftr1p permease. Given the propensity of free Fe(III) to reform Fe(III)-chelator complexes or insoluble ferric hydroxides, it seems likely that Fe(II) oxidation is coupled directly to the delivery of the metal to the transporter. This hypothesis is supported by the evidence suggesting a physical association between Fet3p and Ftr1p.

It should be noted that the Fet3p-Ftr1p iron transporter is probably not the only iron transport oxidase/permease complex found in *S. cerevisiae*. Homologs of both *FET3* and *FTR1* [the *FET5* (D. Eide, unpublished result) and *FTH1* (Stearman et al. 1996) genes, respectively] are found in the *S. cerevisiae* genome. The function of these genes is not known, but a role in iron metabolism seems likely. Both *FET5* and *FTH1* genes are regulated by iron, and their mRNA levels increase several-fold in iron-limited cells (Yamaguchi-Iwai et al. 1996; D. Eide, unpublished result). The *FET5* gene was isolated in a genetic screen designed to identify yeast genes that, when overexpressed, suppress the iron-limited growth defect of an *fet3* mutant strain. Fet5p is 47% identical to Fet3p and contains both the amino-terminal signal sequence and the potential transmembrane domain. Like Fet3p, Fet5p is an integral membrane protein whose oxidase catalytic domain is located in an extracytoplasmic compartment. Assays of oxidase activity in membrane preparations from cells overexpressing *FET5* or bearing an *fet5* deletion mutation demonstrated that Fet5p has oxidase activity. *FET5* overexpression increased cell-associated iron levels and the rate of iron uptake, indicating that Fet5p can participate in iron transport. Moreover, Fet5p is highly enriched in purified vacuoles. Taken together, these results suggest that the Fet5p multicopper oxidase, perhaps in combination with the Fth1p permease, plays a role in vacuolar iron transport.

6. The Low-affinity Fe(II)-Uptake System

The high-affinity system is not active when iron is abundant in the medium. Iron-replete cells rely on a second, low-affinity system that also utilizes Fe(II) as its substrate. This low-affinity system has an apparent K_m for Fe(II) that is approximately 200-fold higher than that of the high-affinity system (Dix et al. 1994). Whereas the high-affinity system is extremely specific for Fe(II), the low-affinity system may be capable of transporting other metal cations as well. This was suggested by the observation that the low-affinity system is inhibited by high concentrations of Co(II), Cd(II), Ni(II), and Cu(I). The mechanism of this inhibition is not known. The transporter protein of the low-affinity system is encoded by the *FET4* gene. *FET4* was isolated in a genetic screen because its overexpression can suppress the iron-limited growth defect of an *fet3* mutant (Dix et al. 1994). Consistent with its proposed role as a transporter, overexpression of *FET4* increased low-affinity uptake activity, whereas *FET4* disruption eliminated that activity but did not reduce high-affinity uptake.

Sequence analysis of *FET4* indicated that this gene encodes a protein of 552 amino acids (Dix et al. 1994). More than 50% of these residues are hydrophobic and many of them are found in six potential transmembrane domains. More recently, it was demonstrated that Fet4p is an integral membrane protein and found in the plasma membrane; i.e., Fet4p colocalizes with plasma membrane marker proteins such as the plasma membrane H^+-ATPase and Ctr1p (Dix et al. 1997). Furthermore, mutations in *FET4* were isolated that alter the apparent K_m of the low-affinity system without greatly altering the V_{max}. For example, a mutation that substitutes an alanine for an aspartate residue (D271) located within one of the transmembrane domains increases the apparent K_m 30-fold with a less than 2-fold change in V_{max}. The isolation of mutant alleles such as this supports the hypothesis that Fet4p interacts directly with the Fe(II) substrate during uptake by the low-affinity system. Fet4p contains no ATP binding domains, nor does it share any significant sequence similarity with the ubiquitous P-type ATPase family of cation transport proteins. This observation suggests that iron uptake by the low-affinity system is driven by indirect coupling to energy metabolism. Given the positive charge of its substrate, the driving force for iron uptake by Fet4p may simply be the inside-negative electrochemical potential across the plasma membrane. Alternatively, uptake of iron may be coupled with the transport of an anion (e.g., Cl⁻, symport) or another cation (e.g., K⁺, antiport).

The relatively high apparent K_m displayed by the low-affinity system called into question whether or not this system could serve as a biologically relevant pathway of Fe(II) uptake. This relevance was established by examining the regulation of the high-affinity system in *fet4* mutant cells. The high-affinity system is induced under conditions of iron deprivation (Eide et al. 1992). In *fet4* mutants, high-affinity activity was increased, indicating that these cells are compensating for loss of the low-affinity system (Dix et al. 1994). Furthermore, the low-affinity system is itself regulated by iron availability. Growth of cells in iron-limiting conditions resulted in a 4-fold increase in Fet4p levels and uptake activity compared with iron-replete cells (Dix et al. 1997). Taken together, these results demonstrated that the low-affinity system is an important source of iron for cells, especially when grown under conditions in which iron is abundant and the high-affinity system is not expressed.

7. Additional Iron-uptake Pathways

In kinetic assays of iron uptake in *S. cerevisiae*, only the high- and low-affinity uptake systems were detectable (Dix et al. 1994). Thus, it was surprising to discover that an *fet3 fet4* double mutant that lacks both of these systems is still viable. This observation demonstrated that, despite our inability to detect them, one or more additional pathways of iron accumulation are present in these cells. One possible mechanism for this accumulation is that of fluid phase endocytosis of iron and subsequent delivery of the metal to the vacuole. The vacuole is known to have its own iron transport system(s) (Bode et al. 1995), which may then be responsible for transporting the iron from the organelle to the cytoplasm. Whatever the mechanism, it is clearly less efficient than either the low- or high-affinity systems because the *fet3 fet4* strain is extremely sensitive to iron limitation.

Outside of the laboratory environment, *S. cerevisiae* may also be able to obtain iron via a siderophore-mediated process. Although this yeast does not appear to synthesize its own siderophores (Schwyn and Neilands 1987), it is capable of utilizing siderophores produced by other microbes. It was demonstrated that *S. cerevisiae* can accumulate iron bound to desferrioxamine B, a hydroximate-type siderophore produced by *Streptomyces pilosus* (Lesuisse and Labbe 1989). These results indicated that this yeast has a specific transport system for the Fe(III)-siderophore complex and that this complex is taken up into the cell prior to dissociation of the iron. The ability of *S. cerevisiae* to utilize siderophores produced by other microbes may confer an advantage on these cells when growing in the presence of other microbes and competing with them for a limited supply of iron.

8. Transcriptional Regulation of Iron Uptake

Iron uptake in *S. cerevisiae* is tightly regulated in response to the availability of the metal cation in the medium as well as by the metabolic demand for iron. For example, transfer of growing cells from an iron-rich to an iron-limiting medium results in a greater than 30-fold increase in Fe(III) reductase and high-affinity uptake activities (Dancis et al. 1990; Eide et al. 1992). It is clear that the primary level of this control is transcriptional. Iron-limited growth increases the level of mRNA synthesized from the *CCC2, FET3, FET5, FRE1, FRE2, FTH1,* and *FTR1* genes (Dancis et al. 1990; Askwith et al. 1994; Georgatsou and Alexandraki 1994; Yamaguchi-Iwai et al. 1996; Eide et al. 1997). β-galactosidase reporter gene fusions constructed using elements of the *FRE1* and *FET3* promoters demonstrated that this regulation is at the transcriptional level for these two genes, and subsequent studies suggested that this is true for the other genes as well (Dancis et al. 1992; Yamaguchi-Iwai et al. 1996). Mutational studies of the *FET3* promoter and alignment of the promoter sequences of other iron-responsive genes have allowed the identification of a consensus iron-responsive UAS (upstream activation sequence) element, 5′-PyPuCACCCPu-3′ (Yamaguchi-Iwai et al. 1996). This sequence is found in the promoters of all of the iron-responsive genes characterized thus far in *S. cerevisiae*.

Presumably, the iron-responsive UAS is the binding site for a transcriptional activator protein that mediates this regulation. The *AFT1* gene is an excellent candidate for encoding this protein. This gene was identified through the isolation of a mutation, *AFT1-1^{up}*, that increases the expression of the *FRE1* promoter in iron-replete conditions (Yamaguchi-Iwai et al. 1995). The *AFT1-1^{up}* allele is dominant, suggesting a "gain-of-function" mutation. Cells carrying a disruption mutation, *aft1*, that eliminates Aft1p synthesis were unable to grow on iron-limiting media. Furthermore, these cells had no iron-regulated Fe(III) reductase activity and were unable to induce transcription of the *FRE1* gene in response to iron limitation. The allele-specific effects of the *AFT1-1^{up}* and *aft1* mutations were also observed with the other genes found to be induced by iron-limiting conditions, i.e., *CCC2, FET3, FRE2, FTH1,* and *FTR1* (Yamaguchi-Iwai et al. 1995; 1996). Regulation of *FET5* by Aft1p has not yet been examined. These results suggest that Aft1p is an important component of a transcriptional regulatory system that increases transcription of its target genes in response to iron-limiting conditions.

DNA sequence analysis of *AFT1* provided additional support for the role of Aft1p in the regulation of these genes. The product of *AFT1* is a protein of 690 amino acids with a predicted molecular mass of 78 kDa (Yamaguchi-Iwai et al. 1995). Although Aft1p bears no significant homology to other proteins found in the current databases, it has several features suggestive of its role as a transcriptional activator protein. A 140-amino-acid segment in the amino-terminal half of the protein is highly basic (pI = 10.6) and reminiscent of other DNA-binding proteins. It should be noted that Aft1p does not contain any of the common DNA binding motifs (e.g., zinc finger, helix-loop-helix, and leucine zipper). The carboxy-terminus of the protein is glutamine rich (25% over 60 amino acid), and such domains have been demonstrated to serve as transcription activation domains. A subsequent study demonstrated that Aft1p was capable of binding to DNA *in vitro* (Yamaguchi-Iwai et al. 1996). Moreover, this binding was specific to the iron-responsive UAS identified in the *CCC2, FET3, FRE1, FRE2, FTH1*, and *FTR1* promoters. These results indicated that Aft1p alone can recognize and bind to these target sites (although other proteins may still be required for activation).

An important and, as yet, unanswered question is: How does the cell sense intracellular iron levels and transduce that signal into increased or decreased transcriptional activity? *In vivo* footprinting indicated that Aft1p binds to the UASs in the *FET3* promoter in an iron-dependent fashion; i.e., the UAS is occupied by protein in iron-limited cells but not in iron-replete cells (Yamaguchi-Iwai et al. 1996). The simplest model is that Aft1p is itself an iron binding protein whose activity is directly altered by this interaction. How might iron binding control Aft1p activity? Several possibilities exist. For example, iron binding might induce a conformational change in the protein that lessens the affinity of the Aft1p DNA binding domain for its UAS. Alternatively, iron binding could induce degradation of the protein or prevent its movement into the nucleus.

Does Aft1p bind iron? An examination of the Aft1p amino acid sequence did not identify any defined metal binding motifs. The sequence does contain histidine-rich amino- and carboxy-terminal domains that could potentially play an iron binding role. An important clue to how Aft1p senses iron is provided by the *AFT1-1^up* allele. This allele contains a point mutation that changes a cysteine residue to phenylalanine (Yamaguchi-Iwai et al. 1995). This result was exciting because the altered residue is located in a Cys-x-Cys triplet. This Cys-x-Cys motif could bind iron and, perhaps, form part of an Fe-S cluster. An example of iron-responsive regulation involving an Fe-S cluster is the IRP1 protein in animal cells (see Chapter 8). IRP1 is an RNA binding protein that regulates ferritin and transferrin receptor synthesis by controlling either the rate of translation or mRNA degradation, respectively. RNA binding activity is controlled in this protein through a reversibly assembled [4Fe-4S] cluster. Aft1p might also bind intracellular iron in an Fe-S cluster, which could then inhibit its ability to activate transcription. In the absence of iron, this cluster would not form and Aft1p would be fully active. The Cys to Phe mutation could mimic iron deprivation by preventing formation of the Fe-S cluster and, thereby, cause constitutive transcription from Aft1p's target promoters. As this model predicts, the *AFT1-1^up* allele showed occupancy of the *FET3* iron-responsive UAS in iron-replete as well as iron-limited cells (Yamaguchi-Iwai et al. 1996).

9. Other Mechanisms of Regulating Iron Uptake

Given the complexity of iron uptake in *S. cerevisiae*, the potential exists for mechanisms of regulation in addition to the *AFT1*-dependent transcriptional control described above. For example, it was found that despite their coregulation by *AFT1*, the Fe(III) reductase and high-affinity Fe(II) uptake activities can be separately regulated (Eide et al. 1992). We now know that this differential regulation occurs because Fe(III) reductase activity is also regulated by general growth conditions as well as by the availability of other nutrients, namely copper. A consequence of this regulatory complexity is that, depending on the particular conditions of growth, the rate of iron uptake may be controlled by changes in either reductase or uptake activity (Eide et al. 1992). For example, when stationary phase cells are inoculated into fresh media, Fe(III) reductase activity increases rapidly such that iron accumulation is limited by Fe(II) uptake activity. In contrast, when growing cells enter the stationary phase, iron accumulation becomes limited by a rapid decrease in Fe(III) reductase activity.

Regulation of Fe(III) reductase in response to growth phase is apparently mediated by a cyclic AMP (cAMP)-dependent pathway. As a secondary messenger in *S. cerevisiae*, cAMP activates protein kinase A (PKA), which, in turn, regulates a variety of key metabolic processes in response to growth conditions. Cells mutant in the *RAS2* gene, which encodes a G-protein important in the GTP-dependent activation of adenylate cyclase, were unable to induce Fe(III) reductase activity in growing cells (Lesuisse et al. 1991). Moreover, mutations in either the *CDC35* or *CDC25* genes, both of which reduce intracellular cAMP levels, also caused low Fe(III) reductase activity. *CDC35* encodes adenylate cyclase, and *CDC25* encodes the guanine nucleotide exchange factor that is required to activate Ras2p. Furthermore, raising intracellular cAMP levels by either providing that compound in the medium or through mutations in *SRA5*, which encodes the cAMP phosphodiesterase, increased Fe(III) reductase activity. Thus, a cAMP-dependent activation of PKA may be necessary to activate the Fe(III) reductase activity in growing cells. These studies do not demonstrate that the plasma membrane reductase is itself phosphorylated by PKA, but they do suggest that a posttranscriptional regulatory mechanism exists to control iron uptake in response to general growth conditions. This growth control is dominant to the iron-responsive regulation.

A second difference in the regulation of Fe(III) reductase activity and Fe(II) uptake activity was observed in cells grown on different media. Iron accumulation by exponentially growing cells in some media was limited by the Fe(III) reductase activity, whereas in other media Fe(II) uptake activity was rate limiting (Eide et al. 1992). These surprising results were explained by the discovery that Fre1p plays a role in copper uptake. Fre1p mediates the reduction of Cu(II) to Cu(I), which is then taken up by Cu(I)-specific transporters such as Ctr1p (Hassett and

Kosman 1995). *FRE1* is regulated at the transcriptional level by copper as well as iron. The copper-dependent regulation of *FRE1* is controlled by the product of the *MAC1* gene (Jungmann et al. 1993). Thus, reductase activity is an indicator of copper status as well as iron status. To my knowledge, regulation of *FRE2* by copper has not been examined.

Finally, it should be noted that the low-affinity Fe(II) uptake system is also iron regulated. The V_{max} of the low-affinity system increases approximately 4-fold in iron-limited cells compared to iron-replete cells whereas there is no change in the apparent K_m (Dix et al. 1997). This result is consistent with an increased number of transporter proteins on the plasma membrane. The level of Fet4p shows a similar increase in iron-limited cells. Surprisingly, this regulation does not appear to be mediated by Aft1p. This conclusion is based on the observation that the *AFT1-1^{up}* allele does not increase low-affinity activity in iron-replete cells. The mechanism of this regulation is not yet known.

10. Mechanism and Regulation of Zinc Uptake in *S. cerevisiae*

Compared to iron, little is known about the uptake of zinc by any organism. However, recent studies of *S. cerevisiae* have provided new and exciting insights into the molecular biology of this process. In earlier studies, assays of zinc uptake indicated that this process is transporter-mediated; i.e., zinc uptake is time, temperature, and concentration dependent and requires metabolic energy (Rothstein et al. 1958; Fuhrmann and Rothstein 1968; Mowll and Gadd 1983; White and Gadd 1987). Analysis of the concentration dependence of zinc-uptake rate suggested the presence of two systems (Zhao and Eide 1996a). One system has a high affinity for zinc with an apparent K_m of 0.5–1 μM and is regulated by zinc availability. This high-affinity system is specific for zinc over other metal cations as its substrate (Zhao and Eide 1996b). The second system for zinc uptake has a lower affinity (apparent K_m of 10 μM), is active in zinc-replete cells, and also prefers zinc over other metals as its substrate (Zhao and Eide 1996b).

The *ZRT1* (for zinc regulated transporter) gene is required for high-affinity uptake. Our results suggest that *ZRT1* encodes the transporter protein of this system (Zhao and Eide 1996a). *ZRT1* first attracted attention because it is a member of a closely related family of transporter genes found in organisms as diverse as fungi, plants, nematodes, and humans. This family includes the *IRT1* gene from *Arabidopsis thaliana*, which was shown to encode an Fe(II) transporter by functional expression in an *fet3 fet4* strain defective for iron uptake (Eide et al. 1996). Like the other members of this family, Zrt1p is predicted to be an integral membrane protein with eight potential transmembrane domains. Given that *IRT1* encodes an Fe(II) transporter and *ZRT1* (and *ZRT2*, see below) encodes a zinc transporter, it is an exciting possibility that the other genes in this family encode metal cation transporters in the organisms in which they are found. The amino acid sequences of these proteins do not have extensive similarity to other known transport proteins, nor do they contain an ATP binding motif. Thus, a clue to what driving force is utilized by these proteins for metal uptake is not provided by their sequences. This driving force may simply be the inside-negative electrochemical potential of the plasma membrane.

It was found that the level of *ZRT1* expression correlated with activity of the high-affinity system. Overexpression of *ZRT1* increased high-affinity uptake, whereas disruption of the *ZRT1* gene eliminated high-affinity activity and resulted in poor growth of the mutant on zinc-limited media (Zhao and Eide 1996a). The high-affinity system was induced in activity greater than 100-fold in response to zinc-limiting growth conditions. This regulation is mediated at the transcriptional level. When cells were grown in media containing different zinc concentrations, high-affinity uptake and *ZRT1* mRNA levels were closely correlated, as was the β-galactosidase activity generated by a reporter gene in which the *ZRT1* promoter was fused to the *E. coli lacZ* gene. The *ZRT1-lacZ* fusion gene showed a similar pattern of regulation in response to cell-associated zinc levels in both wild-type and *zrt1* mutant cells despite the 75-fold higher extracellular zinc level required to down-regulate the promoter in the mutant. These results indicted that the activity of the high-affinity system is controlled, at least in part, by transcriptional regulation of the *ZRT1* gene in response to a pool of intracellular zinc.

Low-affinity uptake was unaffected by the *zrt1* mutation, suggesting that this system is a separate uptake pathway for zinc. The *ZRT2* gene appears to encode the transporter protein of the low-affinity system (Zhao and Eide 1996b). The amino acid sequence of Zrt2p, which also contains eight potential transmembrane domains, is 44% identical to Zrt1p and 35% identical to Irt1p. As was found for *ZRT1* and the high-affinity system, overexpressing *ZRT2* increased low-affinity uptake, whereas disrupting this gene eliminated that activity but had little effect on the high-affinity system. This result also supports the hypothesis that the high- and low-affinity systems are separate uptake pathways. The low-affinity system is also regulated by zinc. Cells grown in a medium supplemented with a high level of zinc had less than 10% of the uptake rate of cells grown in unsupplemented media. The mechanism of this regulation is not yet known but may involve transcriptional regulation of the *ZRT2* gene.

The observation that the low-affinity system is zinc regulated indicates that this system is a biologically relevant mechanism of zinc accumulation. This hypothesis is also supported by the analysis of the zinc levels required for growth of *zrt2* mutant strains as well as by the effects of the *zrt2* mutation on the regulation of the high-affinity system (Zhao and Eide 1996b). As was observed for iron uptake, disruption of the low-affinity zinc system increased the activity of the high-affinity system. Finally, it should be noted that one or more zinc uptake pathways in addition to the high- and low-affinity systems characterized thus far are present in *S. cerevisiae*. This is indicated by the fact that a *zrt1 zrt2* double mutant is viable, albeit very sensitive to zinc limitation. These additional pathways are not likely to be of major importance for supplying zinc for growth when it is present in moderate amounts but are probably responsible for zinc toxicity when cells are grown in zinc-rich conditions. This conclusion is based on two observations. First, the *zrt1 zrt2* double mutant requires an estimated 10^5- to 10^6-fold more zinc for growth than the wild-type strain. Second, this strain is not more resistant to toxic levels of zinc in the medium than the wild-type strain. Thus, in zinc-rich conditions, the high- and low-affinity systems may be down-regulated to a low level while the additional systems are still active.

The metabolism of intracellular zinc is still poorly understood. Two potential intracellular zinc transporters have been identified in *S. cerevisiae*, although their physiological roles are not known. These transporters are encoded by the *ZRC1* and *COT1* genes. *ZRC1* was isolated as a multicopy suppressor of zinc toxicity; i.e., overexpression of *ZRC1* from a multicopy plasmid results in zinc resistance (Kamizono et al. 1989). The *COT1* gene was isolated in a similar fashion as a suppressor of cobalt toxicity but was later found to confer zinc resistance as well (Conklin et al. 1992, 1994). Disruption of either *ZRC1* or *COT1* resulted in greater sensitivity to zinc. Zrc1p and Cot1p are closely related proteins (60% amino acid sequence identity) consisting of approximately 440 amino acids and 6–7 potential transmembrane domains. What role do these proteins play in zinc metabolism? Neither *ZRC1* nor *COT1* are essential genes, and a *zrc1 cot1* double mutant is also viable. Thus, these two genes do not redundantly provide a function essential for growth under standard laboratory conditions. The subcellular location of Zrc1p is not known, but Cot1p was found to be enriched in mitochondrial preparations. This result suggested that accumulation of zinc in the mitochondria may be important in its sequestration and detoxification. It should be noted that the mitochondrial preparations analyzed in this experiment were relatively crude and potentially contained other organelles. Thus, the true intracellular location of Cot1p is still unknown. One hypothesis that is consistent with these data is that these proteins are located in the membrane of an intracellular vesicular compartment and sequester cytoplasmic zinc within those vesicles. Alternatively, these proteins could be localized in the plasma membrane and mediate zinc efflux. These two hypotheses are supported by recent studies on ZnT-1 and ZnT-2, mammalian zinc transporters that are related to Zrc1p and Cot1p. ZnT-1 is a zinc efflux transporter in the plasma membrane (Palmiter and Findley 1995) whereas ZnT-2 is found on an intracellular vesicle and may play a role in zinc sequestration (Palmiter et al. 1996). Thus, the role of Zrc1p and/or Cot1p may be to detoxify excess zinc by either transporting the metal cation out of the cell or into an intracellular vesicular compartment. It is clear that the balance between zinc-uptake transporters like Zrt1p and Zrt2p and efflux/organellar transporters like Zrc1p and Cot1p will play an important role in controlling cellular zinc homeostasis.

11. Conclusions and Prospects

The main reason why many of us in the field of *S. cerevisiae* metal uptake initiated these studies was the expectation that this yeast would provide a useful model system for the understanding of metal metabolism in animals and plants. Over the past 10 years, that expectation has clearly been realized. The points of similarity in the biochemistry and molecular biology of metal cation uptake between *S. cerevisiae* and higher eukaryotes are as numerous as they are informative. For example, the identification of the *FET3* multicopper oxidase confirms and clarifies the role of ferroxidases in iron uptake by both *S. cerevisiae* and mammals. The *CCC2* copper-ATPase has provided useful information to our understanding of the mammalian *MNK* and *WD* copper-ATPases and the etiology of Menkes' and Wilson's disease. Studies of *S. cerevisiae* metal metabolism are also providing reagents useful in identifying functionally related genes from higher eukaryotes. The best example of this is the case of *IRT1*, an *Arabidopsis* Fe(II) transporter. IRT1 was identified by virtue of its ability, when expressed in *S. cerevisiae*, to suppress the iron-limited growth defect of an *fet3 fet4* mutant. Its isolation then led to the discovery of its family of related genes including *ZRT1* and *ZRT2*, the zinc transporters from *S. cerevisiae*. Members of this family may play roles in metal cation uptake in all eukaryotes. Moreover, preliminary studies using expression of *Arabidopsis* genes in the *zrt1 zrt2* mutant indicate that a similar approach may be useful in identifying zinc transporters from plants as well (D. Eide, unpublished results). Thus, studies of *S. cerevisiae* iron uptake have aided our understanding of both iron and zinc uptake in yeasts as well as in higher eukaryotes.

One general feature of metal cation uptake in *S. cerevisiae* is that the uptake of any single substrate is mediated by more than one system. This redundancy is especially clear in the case of iron and zinc, both of which have at least three different uptake systems. Copper uptake also appears to be mediated by at least two separate systems. These observations raise the question of why *S. cerevisiae* has multiple systems for the uptake of each metal cation. In the case of iron, the differences in the mechanisms of uptake used by the high- and low-affinity systems offer one possible explanation. The high-affinity system is dependent on the *FET3* multicopper oxidase and, therefore, requires O_2 for activity. This suggests that whereas the high-affinity system is essential in iron-limited aerobic conditions, the low-affinity system mediated by Fet4p may be essential for anaerobic growth. In the case of zinc uptake, an explanation for redundancy is less obvious. Given the amino acid similarity of *ZRT1* and *ZRT2*, it seems unlikely that these two transporters utilize greatly different mechanisms of uptake. Moreover, the high- and low-affinity systems show nearly identical substrate specificity for zinc. Why then does the cell have two systems for zinc uptake? The answer to this question may lie in the differential regulation of these two systems. The high-affinity system is undetectable in zinc-replete cells, whereas it is highly induced by zinc limitation. In contrast, the low-affinity system, although down-regulated by zinc excess, appears to be produced constitutively over a broad range of zinc concentrations. Thus, one could consider the low-affinity system as a "housekeeping system" for zinc accumulation and the high-affinity system as specialized for zinc-limited growth. Transport of many substrates in *S. cerevisiae* has been found to be mediated by redundant systems, including the high- and low-affinity K$^+$ transporters encoded by *TRK1* and *TRK2* (Ko and Gaber 1991) and the more than five different glucose transporter genes (Ko et al. 1993). Thus, the redundancy observed in metal uptake systems is not surprising.

Many important questions about iron and zinc uptake and their regulation remain unanswered. For example, it should be noted that Fet3p and ceruloplasmin, the mammalian ferroxidase, mobilize iron in opposite directions; i.e., Fet3p functions in transporting iron *into* cells, whereas ceruloplasmin promotes transport *out* of cells. Future studies will resolve this paradox. The explanation may lie in the interaction of the multicopper oxidase with the particular transporter protein that mediates movement of iron across the membrane. Fet3p is coupled with an Fe(III) transporter (Ftr1p), but ceruloplasmin may work in conjunction with an Fe(II) transporter. Another important question is: How does a metal-responsive transcriptional regulator like Aft1p sense intracellular iron levels and transduce that signal into differential control of gene expression? Furthermore, no mechanistic studies have been conducted to determine how transporters like Fet4p, Zrt1p, and Zrt2p work. These studies will clarify what the driving force is for uptake by these transporters. Finally, we understand little of the fate of metal cations once they enter the cell. Genetic, biochemical, and molecular studies on *S. cerevisiae* will no doubt prove as useful in the analysis of intracellular metal metabolism as they have for the dissection of uptake processes. The next decade of research in this field will be as exciting as the last one.

Acknowledgments

The research from the author's laboratory described in this chapter was funded by grants from the National Institutes of Health and the National Science Foundation. I thank A. Dancis and R. D. Klausner for information shared prior to publication.

References

Ander, P., and K.-E. Eriksson. 1976. The importance of phenol oxidase activity in lignin degradation by the white-rot fungus *Sporotrichum pulverulentum*. *Arch. Microbiol.* 109:1–8.

Anderson, G. J., A. Dancis, D. G. Roman, and R. D. Klausner. 1994. Ferric iron reduction and iron uptake in eucaryotes: Studies with the yeasts *Saccharomyces cerevisiae* and *Schizosaccharomyces pombe*. *Adv. Exp. Med. Biol.* 356:81–89.

Anderson, G. J., E. Lesuisse, A. Dancis, D. G. Roman, P. Labbe, and R. D. Klausner. 1992. Ferric iron reduction and iron assimilation in *Saccharomyces cerevisiae*. *J. Inorg. Biochem.* 47:249–255.

Askwith, C., D. Eide, A. Van Ho, P. S. Bernard, L. Li, S. Davis-Kaplan, D. M. Sipe, and J. Kaplan. 1994. The *FET3* gene of *S. cerevisiae* encodes a multicopper oxidase required for ferrous iron uptake. *Cell* 76:403–410.

Askwith, C. C., D. De Silva, and J. Kaplan. 1996. Molecular biology of iron acquisition in *Saccharomyces cerevisiae*. *Mol. Microbiol.* 20:27–34.

Bode, H. P., M. Dumschat, S. Garotti, and G. F. Fuhrmann. 1995. Iron sequestration by the yeast vacuole. *Eur. J. Biochem.* 228:337–342.

Braun, V., R. Gross, W. Koster, and L. Zimmermann. 1983. Plasmid and chromosomal mutants in the iron (III)-aerobactin transport system of *Escherichia coli*. Use of streptoni-grin for selection. *Mol. Gen. Genet.* 192:131–139.

Bull, P. C., G. R. Thomas, J. M. Rommens, J. R. Forbes, and D. W. Cox. 1993. The Wilson's disease gene is a putative copper transporting P-type ATPase similar to the Menkes' gene. *Nature Genet.* 5:327–336.

Butt, V. S. 1980. Direct oxidases and related enzymes, p. 81–123. In *The Biochemistry of Plants*, ed. D. D. Davies, Academic Press, New York.

Cha, J., and S. A. Cooksey. 1991. Copper resistance in *Pseudomonas syringae* mediated by periplasmic and outer membrane proteins. *Proc. Natl. Acad. Sci. U.S.A.* 88:8915–8919.

Chelly, J., Z. Tumer, T. Tonnesen, A. Petterson, Y. Ishikawa-Brush, N. Tommerup, N. Horn, and A. P. Monaco. 1993. Isolation of a candidate gene for Menkes' disease that encodes a potential heavy metal binding protein. *Nature Genet.* 3:14–19.,

Cohen, M. S., Y. Chai, B. E. Britigan, W. McKenna, J. Adams, T. Svendsen, K. Bean, D. J. Hasssett, and P. F. Sparling. 1987. Role of extracellular iron in the action of the quinone antibiotic streptonigrin: Mechanisms of killing and resistance of *Neisseria gonorrhoeae*. *Antimicrob. Agents Chemother.* 31:1507–1513.

Conklin, D. S., M. R. Culbertson, and C. Kung. 1994. Interactions between gene products involved in divalent cation transport in *Saccharomyces cerevisiae*. *Mol. Gen. Genet.* 244:303–311.

Conklin, D. S., J. A. McMaster, M. R. Culbertson, and C. Kung. 1992. *COT1*, a gene involved in cobalt accumulation in *Saccharomyces cerevisiae*. *Mol. Cell. Biol.* 12:3678–3688.

Crane, F. L., H. Roberts, A. W. Linnane, and H. Low. 1982. Transmembrane ferricyanide reduction by cell of the yeast *Saccharomyces cerevisiae*. *J. Bioenerg. Biomemb.* 14:191–205.

Dancis, A., R. D. Klausner, A. G. Hinnebusch, and J. G. Barriocanal. 1990. Genetic evidence that ferric reductase is required for iron uptake in *Saccharomyces cerevisiae*. *Mol. Cell. Biol.* 10:2294–2301.

Dancis, A., D. G. Roman, G. J. Anderson, A. G. Hinnebusch, and R. D. Klausner. 1992. Ferric reductase of *Saccharomyces cerevisiae*: Molecular characterization, role in iron uptake, and transcriptional control by iron. *Proc. Natl. Acad. Sci. U.S.A.* 89:3869–3873.

Dancis, A., D. S. Yuan, D. Haile, C. Askwith, D. Eide, C. Moehle, J. Kaplan, and R. D. Klausner. 1994. Molecular characterization of a copper transport protein in *S. cerevisiae*: An unexpected role for copper in iron transport. *Cell* 76:393–402.

De Silva, D. M., C. C. Askwith, D. Eide, and J. Kaplan. 1995. The *FET3* gene product required for high affinity iron transport in yeast is a cell surface ferroxidase. *J. Biol. Chem.* 270:1098–1101.

Dix, D. R., J. T. Bridgham, M. A. Broderius, C. A. Byersdorfer, and D. J. Eide. 1994. The *FET4* gene encodes the low affinity Fe(II) transport of *Saccharomyces cerevisiae*. *J. Biol. Chem.* 269:26092–26099.

Dix, D., J. Bridgham, M. Broderius, and D. Eide. 1997. Characterization of the Fet4p protein of yeast: Evidence for a direct role in the transport of iron. *J. Biol. Chem.* (in press).

Eide, D., M. Broderius, J. Fett, and M. L. Guerinot. 1996. A novel iron-regulated metal transporter from plants identified by functional expression in yeast. *Proc. Natl. Acad. Sci. U.S.A.* 93:5624–5628.

Eide, D., S. Davis-Kaplan, I. Jordan, D. Sipe, and J. Kaplan. 1992. Regulation of iron uptake in *Saccharomyces cerevisiae*: The ferrireductase and Fe(II) transporter are regulated independently. *J. Biol. Chem.* 267:20774–20781.

Evans, S. L., J.E.L. Arceneaux, B. R. Byers, M. E. Martin, and H. Aranha. 1986. Ferrous iron transport in *Streptococcus mutans*. *J. Bacteriol.* 168:1096–1099.

Fox, T. C., J. E. Shaff, M. A. Gruzak, W. A. Norvell, Y. Chen, R. L. Chaney, and L. V. Kochian. 1996. Direct measurement of ^{59}Fe-labeled Fe^{2+} influx in roots of *Pisum sativum* using a chelator buffer system to control free Fe^{2+} in solution. *Plant Physiol.* 111:93–100.

Fuhrmann, G. F., and A. Rothstein. 1968. The transport of Zn^{2+}, Co^{2+} and Ni^{2+} into yeast cells. *Biochim. Biophys. Acta* 163:325–330.

Georgatsou, E., and D. Alexandraki. 1994. Two distinctly regulated genes are required for ferric reduction, the first step of iron uptake in *Saccharomyces cerevisiae*. *Mol. Cell. Biol.* 14:3065–3073.

Grusak, M. A., R. M. Welch, and L. V. Kochian. 1990. Does iron deficiency in *Pisum sativum* enhance the activity of the root plasmalemma iron transport protein? *Plant Physiol.* 94:1353–1357.

Han, O., M. L. Failla, A. D. Hill, E. R. Morris, and J. C. Smith. 1994. Reduction of Fe(III) is required for uptake of nonheme iron by Caco-2 cells. *J. Nutr.* 125:1291–1299.

Harris, L., Y. Takahashi, H., Miyajima, M. Serizawa, R. T. A. MacGillivray, and J. D. Gitlin. 1995. Aceruloplasmin: Molecular characterization of this disorder of iron metabolism. *Proc. Natl. Acad. Sci. U.S.A.* 92:2539–2543.

Hassett, R., and D. J. Kosman. 1995. Evidence for Cu(II) reduction as a component of copper uptake by *Saccharomyces cerevisiae. J. Biol. Chem.* 270:128–134.

Johnson, W., L. Varner, and M. Poch. 1991. Acquisition of iron by *Legionella pneumophila*: Role of iron reductase. *Infect. Imm.* 59:2376–2381.

Jordan, I., and J. Kaplan. 1994. The mammalian transferrin-independent iron transport system may involve surface ferroxidase activity. *Biochem. J.* 302:875–879.

Jungmann, J., H. Reins, J. Lee, A. Romeo, R. Hassett, D. Kosman, and S. Jentsch. 1993. *MAC1*, a nuclear regulatory protein related to Cu-dependent transcription factors is involved in Cu/Fe utilization and stress resistance in yeast. *EMBO J.* 12:5051–5056.

Kamizono, A., M. Nishizawa, Y. Teranishi, K. Murata, and A. Kimura. 1989. Identification of a gene conferring resistance to zinc and cadmium ions in the yeast *Saccharomyces cerevisiae. Mol. Gen. Genet.* 219:161–167.

Kammler, M., C. Schon, and K. Hantke. 1993. Characterization of the ferrous iron uptake system of *Escherichia coli. J. Bacteriol.* 175:6212–6219.

Ko, C. H., and R. F. Gaber. 1991. *TRK1* and *TRK2* encode structurally related K⁺ transporters in *Saccharomyces cerevisiae. Mol Cell. Biol.* 11:4266–4273.

Ko, C. H., H. Liang, and R. F. Gaber. 1993. Roles of multiple glucose transporters in *Saccharomyces cerevisiae. Mol. Cell. Biol.* 13:638–648.

Lee, G. R., S. Nacht, J. N. Lukens, and G. E. Cartwright. 1968. Iron metabolism in copper deficient swine. *J. Clin. Invest.* 47:2058–2069.

Lesuisse, E., M. Casteras-Simon, and P. Labbe. 1996. Evidence for *Saccharomyces cerevisiae* ferrireductase system being a multicomponent electron transport chain. *J. Biol. Chem.* 271:13578–13583.

Lesuisse, E., R. R. Crichton, and P. Labbe. 1990. Iron-reductases in the yeast *Saccharomyces cerevisiae. Biochim. Biophys. Acta* 1038:253–259.

Lesuisse, E., B. Horion, P. Labbe, and F. Hilger. 1991. The plasma membrane ferrireductase activity of *Saccharomyces cerevisiae* is partially controlled by cyclic AMP. *Biochem. J.* 280:545–548.

Lesuisse, E., and P. Labbe. 1989. Reductive and nonreductive mechanisms of iron assimilation by the yeast *Saccharomyces cerevisiae. J. Gen. Microbiol.* 135:257–263.

Lesuisse, E., F. Raguzzi, and R. R. Crichton. 1987. Iron uptake by the yeast *Saccharomyces cerevisiae*: Involvement of a reduction step. *J. Gen. Microbiol.* 133:3229–3236.

Lesuisse, E., M. Simon, R. Klein, and P. Labbe. 1992. Excretion of anthranilate and 3-hydroxyanthranilate by *Saccharomyces cerevisiae*: Relationship to iron metabolism. *J. Gen. Microbiol.* 138:85–89.

Morrissey, J. A., P. H. Williams, and A. M. Cashmore. 1996. *Candida albicans* has a cell-associated ferric-reductase activity which is regulated in response to levels of iron and copper. *Microbiology* 142:485–492.

Mowll, J. L., and G. M. Gadd. 1983. Zinc uptake and toxicity in the yeast *Sporobolomyces roseus* and *Saccharomyces cerevisiae*. *J. Gen. Microbiol.* 129:3421–3425.

Núñez, M. T., X. Alvarez, M. Smith, V. Tapia, and J. Glass. 1994. Role of redox systems on Fe^{3+} uptake by transformed human intestinal epithelial (Caco-2) cells. *Amer. J. Physiol.* 267:C1582–C1588.

Núñez, M. T., A. Escobar, A. Ahumada, and M. Gonzalez-Sepulveda. 1992. Sealed reticulocyte ghosts. An experimental model for the study of Fe(II) transport. *J. Biol. Chem.* 267:11490–11494.

Osaki, S., D. A. Johnson, and E. Frieden. 1966. The possible significance of the ferrous oxidase activity of ceruloplasmin in normal human serum. *J. Biol. Chem.* 241:2746–2751.

Oshiro, S., H. Nakajima, T. Markello, D. Krasnewich, I. Bernardini, and W. A. Gahl. 1993. Redox, transferrin-independent, and receptor-mediated endocytosis iron uptake systems in cultured human fibroblasts. *J. Biol. Chem.* 268:21586–21591.

Palmiter, R. D., T. B. Cole, and S. D. Findley. 1996. ZnT-2, a mammalian protein that confers resistance to zinc by facilitating vesicular sequestration. *EMBO J.* 15:1784–1791.

Palmiter, R. D., and S. D. Findley. 1995. Cloning and functional characterization of a mammalian zinc transporter that confers resistance to zinc. *EMBO J.* 14:639–649.

Rad, M. R., L. Kirchrath, and C. P. Hollenberg. 1994. A putative Cu^{2+}-transporting ATPase gene of chromosome II of *Saccharomyces cerevisiae*. *Yeast.* 10:1217–1225.

Raja, K. B., R. J. Simpson, and T. J. Peters. 1992. Investigation of a role for reduction in ferric iron uptake by mouse duodenum. *Biochim. Biophys. Acta* 1135:141–146.

Ramirez, J. M., G. G. Gallego, and R. Serrano. 1984. Electron transfer constituents in plasma membrane fractions of *Avena sativa* and *Saccharomyces cerevisiae*. *Plant Sci. Lett.* 34:103–110.

Roman, D. G., A. Dancis, G. J. Anderson, and R. D. Klausner. 1993. The fission yeast ferric reductase gene *frp1+* is required for ferric iron uptake and encodes a protein that is homologous to the gp91-*phox* subunit of the human NADPH phagocyte oxidoreductase. *Mol. Cell. Biol.* 13:4342–4350.

Romheld, V., and H. Marschner. 1983. Mechanism of iron uptake by peanut plants. *Plant Physiol.* 71:949–954.

Rothstein, A., A. Hayes, D. Jennings, and D. Hooper. 1958. The active transport of Mg(II) and Mn(II) into the yeast cell. *J. Gen. Physiol.* 41:585–594.

Rotrosen, D., C. L. Yeung, T. L. Leto, H. L. Malech, and C. H. Kwong. 1992. Cytochrome b_{558}: The flavin-binding component of the phagocyte NADPH oxidase. *Science* 256:1459–1462.

Schwyn, B., and J. B. Neilands. 1987. Universal chemical assay for the detection and determination of siderophores. *Anal. Biochem.* 160:47–56.

Shatwell, K. P., A. Dancis, A. R. Cross, R. D. Klausner, and A. W. Segal. 1996. The

FRE1 ferric reductase of *Saccharomyces cerevisiae* is a cytochrome b similar to that of NADPH oxidase. *J. Biol. Chem.* 271:14240–14244.

Stearman, R., D. S. Yuan, Y. Yamaguchi-Iwai, R. D. Klausner, and A. Dancis. 1996. A permease-oxidase complex involved in high-affinity iron uptake in yeast. *Science* 271:1552–1557.

Tanzi, R. E., K. Petrukhin, I. Chernov, J. L. Pellequer, W. Wasco, B. Ross, D. M. Romano, E. Parano, L. Pavone, L. M. Brzustowicz, M. Devoto, J. Peppercorn, A. I. Bush, I. Sternlieb, M. Pirastu, J. F. Gusella, O. Evgrafov, G. K. Penchaszadeh, B. Honig, I. S. Edelman, M. B. Soares, I. H. Scheinberg, and T. C. Gilliam. 1993. The Wilson's disease gene is a copper transporting ATPase with homology to the Menkes' disease gene. *Nature Genet.* 5:344–350.

Thorstensen, K., and I. Romslo. 1988. Uptake of iron from transferrin by isolated rat hepatocytes. *J. Biol. Chem.* 263:8844–8850.

Trikha, J., E. C. Theil, and N. M. Allewell. 1995. High resolution crystal structures of amphibian red-cell L ferritin: Potential roles for structural plasticity and solvation in function. *J. Mol. Biol.* 248:949–967.

von Heijne, G. 1983. Patterns of amino acids near signal-sequence cleavage sites. *Eur. J. Biochem.* 133:17–21.

Vulpe, C., B. Levinson, S. Whitney, S. Packman, and J. Gitschier. 1993. Isolation of a candidate gene for Menkes' disease and evidence that it encodes a copper-transporting ATPase. *Nature Genet.* 3:7–13.

Watkins, J. A., J. D. Altazan, P. Elder, C. Y. Lin, M. T. Nunez, X. X. Cui, and J. Glass. 1992. Kinetic characterization of reductant dependent processes of iron mobilization from endocytic vesicles. *Biochem.* 31:5820–5830.

White, C., and G. M. Gadd. 1987. The uptake and cellular distribution of zinc in *Saccharomyces cerevisiae*. *J. Gen. Microbiol.* 133:727–737.

Winkelmann, G., D. van der Helm, and J. B. Neilands. (ed.). 1987. *Iron Transport in Microbes, Plants and Animals*. VCH Verlagsgesellschaft, New York.

Yamaguchi-Iwai, Y., A. Dancis, and R. D. Klausner. 1995. AFT1: A mediator of iron regulated transcriptional control in *Saccharomyces cerevisiae*. *EMBO J.* 14:1231–1239.

Yamaguchi-Iwai, Y., R. Stearman, A. Dancis, and R. D. Klausner. 1996. Iron-regulated DNA binding by the AFT1 protein controls the iron regulon in yeast. *EMBO J.* (in press).

Yi, Y., and M. L. Guerinot. 1996. Genetic evidence that induction of root Fe(III) chelate reductase activity is necessary for iron uptake under iron deficiency. *Plant J.* 15:3377–3384.

Yi, Y., J. Saleeba, and M. L. Guerinot. 1994. Iron uptake in *Arabidopsis thaliana*. In *Biochemistry of Metal Micronutrients in the Rhizosphere*, eds. J. Manthey, D. Luster, and D. E. Crowley. pp. 295–307. CRC Press, Boca Raton, FL.

Yoshida, K., K. Furihata, S. Takeda, A. Nakamura, K. Yamamoto, H. Morita, S. Hiyamuta, S. Ikeda, N. Shimizu, and N. Yanagisawa. 1995. A mutation in the ceruloplasmin gene is associated with systemic hemosiderosis in humans. *Nature Genet.* 9:267–272.

Yuan, D. S., R. Stearman, A. Dancis, T. Dunn, T. Beeler, and R. D. Klausner. 1995. The Menkes'/Wilson's disease gene homologue in yeast provides copper to a ceruloplasmin-like oxidase required for iron uptake. *Proc. Natl. Acad. Sci. U.S.A.* 92:2632–2636.

Zhao, H., and D. Eide. 1996a. The yeast *ZRT1* gene encodes the zinc transporter of a high affinity uptake system induced by zinc limitation. *Proc. Natl. Acad. Sci. U.S.A.* 93:2454–2458.

Zhao, H., and D. Eide. 1996b. The *ZRT2* gene encodes the low affinity zinc transporter in *Saccharomyces cerevisiae. J. Biol. Chem.* 271:23203–23210.

14

Metallothionein Gene Regulation in Cyanobacteria

Nigel J. Robinson, Amanda J. Bird, and Jennifer S. Turner

1. Introduction

Metal-induced synthesis of animal and yeast metallothioneins is regulated at the level of transcription via known *cis*-acting metal-regulatory elements. Reports describing the use of such elements to control the expression of foreign genes in transgenic animals (Brinster et al. 1982; Palmiter et al. 1982, 1983) are widely cited. *Trans*-acting metal-responsive activatory factors that interact with these elements have been cloned and characterized from both yeasts (Thiele, 1988; Welch et al. 1989; Dameron et al. 1991; Zhou and Thiele 1993) and animals (Radtke et al. 1993, 1995; Otsuka et al. 1994; Brugnera et al. 1994; Heuchel et al. 1994). A prokaryotic metallothionein gene sequence was first determined from *Synechococcus* PCC 6301 (Robinson et al. 1990). Regions containing *cis*-acting elements involved in the regulation of the related gene from *Synechococcus* PCC 7942 have been identified (Huckle et al. 1993; Morby et al. 1993; Erbe et al. 1995), and one metal-responsive *trans*-acting factor (a repressor) has been cloned and characterized (Huckle et al. 1993; Morby et al. 1993). The mechanisms involved in metalloregulation of this cyanobacterial metallothionein gene, which clearly differ from those in eukaryotes, are described here. The functions of cyanobacterial metallothionein are also discussed because this has obvious implications for understanding the nature of its regulation and, furthermore, it is hypothesized that SmtA may itself perform a regulatory role.

2. Isolation of Cyanobacterial Metallothionein Genes

The only published amino acid sequence of a prokaryotic metallothionein is from the cyanobacterium *Synechococcus* TX-20 (Olafson et al. 1988) (this and very closely related strains are also referred to as *Anacystis nidulans, Synechococcus* PCC 6301, *Synechococcus* PCC 7942, and *Synechococcus* R2). The *Synechococcus* sp. polypeptide has an abundance of cysteine residues arranged in characteristic cysteine-Xaa-cysteine motifs (where Xaa is an amino acid other than cysteine). There is otherwise little sequence similarity to other metallothioneins, and this protein is thus defined as a class II metallothionein [for nomenclature refer to Kojima (1991)]. A metal-thiolate cluster similar to that of eukaryotic metallothionein, but in a single domain, was suggested following spectroscopic studies (Olafson et al. 1988). However, an unusual feature of the cyanobacterial metallothionein is three histidine residues (two of which are located in glycine-histidine-threonine-glycine sequences) near the carboxy-terminus (Olafson et al. 1988; Huckle et al. 1993), and the possibility that these histidine residues are involved in metal coordination remains to be investigated.

Polymerase chain reaction products corresponding to part of the metallothionein gene were generated using template DNA from *Synechococcus* PCC 6301; the gene was called *smtA* (Robinson et al. 1990). These fragments were subsequently used as probes to isolate *smtA* and a divergently transcribed gene *smtB*, these two genes forming the metallothionein divergon *smt* (Huckle et al. 1993). The *smtA* and *smtB* protein coding sequences are separated by a 100-bp operator-promoter region containing divergent promoters (Figure 14.1). An open reading frame (ORF), designated *mtnA*, from *Synechococcus vulcanus* encoding a polypeptide with similarity to SmtA, has also been identified within the DNA flanking a previously characterized gene *psaC* for a photosystem-1 subunit (Shimizu et al. 1992).

3. Metal Sequestration by Cyanobacterial Metallothionein

3.1 Metal Binding Properties

Following exposure of cells to elevated concentrations of Cd^{2+} or Zn^{2+}, but not copper ions, metallothioneins accumulated within *Synechococcus* sp. cells (Olafson et al. 1980). The purified protein was associated with either Cd^{2+} or Zn^{2+} (dependent upon the metal administered) with copper as a minor component (Olafson et al. 1980, 1988). It was considered probable that SmtA may sequester, and hence detoxify, certain metal ions in cyanobacteria.

SmtA has been expressed in *Escherichia coli* as a recombinant fusion protein (Shi et al. 1992). The protein was associated with Zn^{2+}, Cd^{2+}, copper ions, and Hg^{2+} (all metals examined) following purification from cells grown in metal-supplemented media. Relative metal affinities were estimated from the pH at which 50% of the metal ions were displaced (Shi et al. 1992). Compared to equine metallothionein, recombinant SmtA has a greater relative affinity for Zn^{2+},

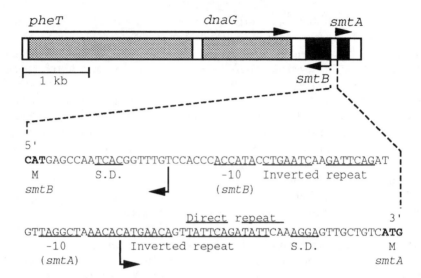

Figure 14.1 Gene organization in the metallothionein gene region. *smtA* and *smtB* (black rectangles coincide with ORFs) are divergently transcribed and separated by a 100-bp operator-promoter region (sequence shown in full). Consensus -10 promoter sequences (underlined), determined transcript start sites (bent arrows), Shine-Dalgarno sequences (S.D.), and hyphenated inverted/direct repeats (under/overlined) are shown. The *smtA* and *smtB* promoters have no region corresponding to an *E. coli* consensus -35 sequence; however, the *smtA* promoter has an "extended-10" sequence. *pheT* and *dnaG* (shaded rectangles coincide with ORFs) are cotranscribed.

but lesser relative (to equine metallothionein) affinities for Cd^{2+} and copper ions. Enhanced accumulation of Zn^{2+} is also observed in *E. coli* cells containing the *smtA*-fusion construct (in which SmtA production is not metalloregulated) compared to cells containing the vector alone, suggesting Zn^{2+} binding *in vivo* (Shi et al. 1992).

3.2 Phenotype of Metallothionein-deficient Mutants

Cyanobacterial (*Synechococcus* PCC 7942) *smt* mutants, with a nonfunctional metallothionein divergon (deficient in *smtA* and *smtB*), have been generated by insertional inactivation and partial gene deletion mediated by homologous recombination (Turner et al. 1993). The *smt* mutants are sensitive to Zn^{2+} (ca. 5-fold reduction in tolerance) and to a lesser degree Cd^{2+} (Turner et al. 1993). Normal tolerance to copper ions and Hg^{2+} is retained in these cells. Roles for the *smt* divergon in the homeostasis and metabolism of Zn^{2+} and detoxification of Cd^{2+} have thus been proposed (Turner et al. 1993; Turner and Robinson 1995). The *smt*-deficient cells accumulate slightly less Zn^{2+} than cells with an intact divergon (Turner et al. 1995), providing evidence that SmtA sequesters Zn^{2+} within cyanobacteria. Although *smt* causes some additional accumulation of Zn^{2+}, it is proposed that this is less than the total amount of Zn^{2+} sequestered. SmtA thereby effects a reduction in the "free" intracellular concentration of Zn^{2+} (Turner et al. 1995).

3.3 Amplification of the Metallothionein Gene in Metal-resistant Cells

Cyanobacterial mutants have been selected for metal (Cd^{2+}) tolerance by step-wise adaptation (Gupta et al. 1993). These mutants have an increased number of copies of *smtA* and a partial deletion of *smtB* (see later sections). This was interpreted as further support for a role for *smtA* in metal detoxification (Gupta et al. 1993). However, more recently the extent of the amplified unit of DNA in this metal-tolerant strain has been mapped (Figure 14.2) (Cranenburgh 1997). This has revealed that the amplified unit of DNA is much larger than originally anticipated. In addition to *smtA*, the amplified unit of DNA includes *dnaG*, a section (in excess of 20 kb) of DNA that has not been sequenced, and *moaA* (involved in molybdopterin biosynthesis). It is therefore possible that other genes within this region contribute to the metal-tolerant phenotype. Thus there is no formal proof that an increase in *smtA* copy number alone confers an increase in tolerance to Cd^{2+}. A copy of *moaA* is immediately adjacent to the 5' end of the second copy of *dnaG*, and *moaA* is also contained within the amplified unit of DNA (Figure 14.2). In wild-type cells, *pheT* is located at the 5' end of *dnaG*. Furthermore, the sequences at the 3' end of the second copy of *moaA* (Figure 14.2) are equivalent to those found in wild-type cells. Thus, there is only a single discontinuity in the sequence (compared to the wild-type sequence) located between *dnaG* and *moaA*. These observations thus support amplification via a single illegitimate recombination event during chromosome replication.

4. Induction of *smtA* Transcription in Response to Metal Ions

The abundance of *smtA* transcripts increases in response to elevated concentrations of metal ions, including Cd^{2+}, Zn^{2+}, copper ions, Hg^{2+}, Co^{2+}, and Ni^{2+}, but not heat shock (Huckle et al. 1993). There is no detectable effect of Cd^{2+} on transcript stability, consistent with *smtA* being regulated at the level of transcription. At maximum permissive concentrations, only Zn^{2+} and to a lesser extent Cd^{2+} and copper ions substantially increase expression of a reporter-gene (*lacZ*) driven by the *smtA* operator-promoter in *Synechococcus* PCC 7942 cells (Huckle

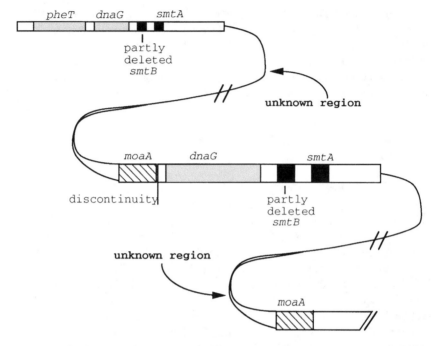

Figure 14.2 A diagrammatic representation of the amplification event in the metal-resistant cell line. The extent of the amplified unit of DNA is shown. The amplified unit of DNA includes *dnaG* (shaded rectangle), a partly deleted copy of *smtB* (filled rectangle), *smtA* (filled rectangle), an as yet uncharacterized region (in excess of 20 kb), and *moaA*, which is involved in molybdopterin biosynthesis (diagonal shading). There is only a single discontinuity in the sequence (compared to wild-type cells) located at the 5′ end of the amplified copy of *dnaG*.

et al. 1993). Reporter gene expression in response to increasing concentrations of Zn^{2+} is shown (Figure 14.3).

5. SmtB is a Zinc-responsive Repressor of *smtA* Transcription

5.1 Regulation of smtA in smtB-deficient Mutants

Reporter gene assays have shown that SmtB is a *trans*-acting repressor of expression from the *smtA* operator-promoter. In *smt* mutants devoid of *smtB*, highly elevated expression of *lacZ* (driven by the *smtA* operator-promoter) was detected even in the absence of added metal ions (Huckle et al. 1993). Repression and metal-dependent expression of *lacZ* were restored (at least in part) in cells

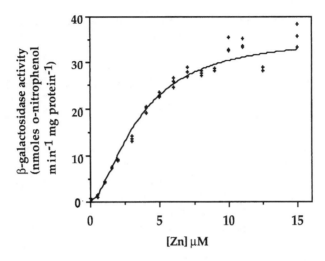

Figure 14.3 β-galactosidase activity in *Synechococcus* PCC 7942 as a function of Zn^{2+} concentration. β-galactosidase activity was measured in *Synechococcus* PCC 7942 cells containing plasmid-borne *lacZ* driven by the *smtA* operator-promoter. Cells were exposed to increasing concentrations of Zn^{2+} for 2 hours prior to the assay (J. W. Huckle, unpublished results).

containing plasmid-borne and/or chromosomal *smtB*. SmtB-dependent repression (in the absence of added metal ions) is most pronounced when *smtB* is associated with (and divergently transcribed from) the *smtA* operator-promoter. In common with many other known bacterial regulators (McFall 1986), SmtB acts more efficiently when synthesized adjacent to its target gene (Turner et al. 1995).

Loss of the SmtB repressor can be advantageous to cells continuously challenged with Cd^{2+} and/or Zn^{2+} (Gupta et al. 1993). Cells selected for Cd^{2+} tolerance had a deletion within all copies of *smtB* (Gupta et al. 1993). The deleted sequence was flanked by a highly iterated octameric palindrome that has been designed HIP1 (Gupta et al. 1993; Robinson et al. 1995). Cells devoid of functional repressor would be expected to show enhanced expression of *smtA*. Overaccumulation of *smtA* transcripts in nonmetal supplemented media has been confirmed in the *smtB*-deficient cells (Turner et al. 1995). In addition, loss of *smtB* has been shown to confer increased tolerance to Zn^{2+} and Cd^{2+} following the reconstruction of this genotype via insertional mutagenesis (Turner et al. 1995).

5.2 SmtB Binds to the smt Operator-Promoter

A protein complex (designated MAC1) that formed with the *smt* operator-promoter was identified by electrophoretic mobility shift assays using extracts from *Synechococcus* PCC 7942 (Morby et al. 1993). MAC1 was missing when extracts from *smt* mutants were used, whereas reintroduction of (plasmid-borne) *smtB* restored ability to form MAC1. It was therefore proposed that SmtB forms the protein component of MAC1 (Morby et al. 1993). MAC1 showed Zn^{2+}-dependent dissociation and treatment with a metal-chelator facilitated reassociation of MAC1 *in vitro* (Morby et al. 1993).

Recombinant SmtB has more recently been expressed in *E. coli* and confirmed to bind to the *smt* operator-promoter, forming a complex similar to MAC1 (at low protein concentrations) (Erbe et al. 1995). Furthermore, recombinant SmtB was shown to form additional, slower migrating (shown to be multimeric-SmtB) complexes with the *smt* operator-promoter when high concentrations of protein were used (Erbe et al. 1995). These complexes were similarly observed with lower concentrations of SmtB (stripped of metal ions) following treatment with a metal-chelator (Erbe et al. 1995). Analogous concentration-dependent complexes were also detected in electrophoretic mobility shift assays using crude protein extracts from cyanobacteria containing *smtB* (Erbe et al. 1995).

The mode of action of SmtB, a Zn^{2+}-inducible negative regulator of *smtA* transcription, can be contrasted with the metal-inducible positive regulation of eukaryotic metallothionein genes. Metal-induced expression of animal and yeast metallothionein genes involves metal-dependent binding of activatory proteins to *cis*-acting MREs (metal response elements) or UASs (upstream activation sequences), respectively (Labbé et al. 1991; Heuchel et al. 1994; Brugnera et al. 1994; Dameron et al. 1991; Radtke et al. 1995; Fürst et al. 1988; Zhou and Thiele 1993).

5.3 SmtB Binds Zn^{2+}

Binding of recombinant SmtB to the *smt* operator-promoter is enhanced by treatment with Zn^{2+}-chelators *in vitro* and is diminished with Zn^{2+} (Erbe et al. 1995), indicating a direct association of SmtB with Zn^{2+}. This has been confirmed using ^{65}Zn binding assays (Erbe et al. 1995). Binding of ^{65}Zn *in vitro* was enhanced following analysis of nonreduced protein compared to dithiothreitol-reduced protein (Turner et al. 1996). This indicates a coordination of Zn^{2+} by histidine (or cysteine and histidine) residues rather than purely cysteine residues (Barbosa et al. 1989).

6. A Family of Metal-responsive Repressors

The sequence of SmtB is similar to a large number of bacterial repressor proteins, most of which are known to be metal responsive (Huckle et al. 1993;

Morby et al. 1993) (Figure 14.4). It is apparent that SmtB is a member of a new family of metal-responsive transcription factors, each presumed to be capable of sensing a different spectrum of metal ions (Figure 14.5). These have been referred to as the arsR family of regulators (Shi et al. 1994).

There is a divergent ORF, which we have designated *mtnB*, within the sequences upstream of *mtnA* from *Synechococcus vulcanus* (Shimizu et al. 1992). By analogy to SmtB, the deduced product of *mtnB* is proposed to be a Zn^{2+}-responsive repressor of *mtnA*. A novel sequence has been derived from the *Synechocystis* PCC 6803 genome project, which we have designated PacR, as it is divergently transcribed from a gene encoding a predicted copper ATPase (*pacS*) (Kaneko et al. 1995). PacR is hypothesized to be a repressor that preferentially senses copper ions, however the metal cation transported by the ATPase and hence sensed by PacR remains to be established. At maximum permissive concentrations for growth, SmtB appears to preferentially sense Zn^{2+} (refer to preceding sections). ArsR from *E. coli* plasmid R773 (San Fransisco et al. 1990), *Staphylococcus xylosus* plasmid pSX267 (Rosenstein et al. 1994), *Staphylococcus aureus* plasmid pI258 (Ji and Silver 1992), the *E. coli* chromosome (Xu et al. 1996), and *Yersinia enterocolitica* Tn*2502* (Neyt et al. 1997) are all metal oxyanion-responsive repressors of the *ars* (arsenic-resistance) operons and dissociate from the *ars* promoters following binding to arsenite (or antimonite). CadC from *S. aureus* plasmid pI258 is a Cd^{2+}-responsive repressor of the *cadA* (Cd^{2+}-efflux ATPase) operon (Endo and Silver 1995). A similar function is proposed for CadC from *Listeria monocytogenes* pLm74 (Lebrun et al. 1994), *S. aureus* (chromosomal *cadC*; GenBank L10909) and *Bacillus firmus* strain OF4 (Ivey et al. 1992). Another gene product CadX is thought to be a Cd^{2+}-responsive repressor of a different Cd^{2+}-resistance determinant from *S. aureus* plasmid pOX4 (K. Dyke, personal communication). MerR from *Streptomyces lividans* is thought to be a Hg^{2+}-responsive regulator of the *S. lividans mer* operon (Sedlmeier and Altenbuchner 1992). Note that this is distinct from the MerR "transcriptional switch" proteins from other bacteria and is the only MerR within the family of SmtB-related proteins.

Other proteins with similarity to SmtB but not known to be involved in metal metabolism are NolR, a repressor of nodulation gene expression in *Rhizobium meliloti* (Kondorosi et al. 1991), and HlyU, an activator of haemolysin expression in *Vibrio cholerae* (Williams et al. 1993) (Figure 14.4).

7. What Determines SmtB-DNA Specificity?

7.1 Nucleotides Involved in Binding

The SmtB complex (MAC1) that associates with the *smt* operator-promoter at low concentrations of SmtB (Morby et al. 1993; Erbe et al. 1995) forms with

```
                                                            S*    S
                                                            |     |
Synechococcus PCC 7942     SmtB  MTKPVLQDG......ETVVCQG....THAAIASELQAIAPEVAQSLAEFFAVLADPNRLR
Synechococcystis PCC 6803  PacR  MSKSSLSKSQSCQNEEMPLCDQPLVHLEQVRQVPEVMSLDQAQQMAEFFSALADPSRLR
Synechococcus vulcanus     MtnB  MSIGYCLTSAYDEDMNTLDPHHPEALAQVSDRLLSTEKAQRMAQFFGLLADTNRVR
S. aureus pI258            CadC       MKKKDTCEIFCYDEEKVNRIQGDLQTVD.ISGVSQILKAIADENRAK
S. aureus                  CadC       MTKDMCEVTYIHEDKVNRAKKDLAKQN.PMDVAKVFKALSDDTRVK
B. firmus OF4              CadC       VNKKDTCEIFCYDEEKVNRIQGDLKTID.IVSVAQMLKAIADENRAK
L. monocytogenes pLm74     CadC       MTVDICEITCIDEEKVKRVKTGLETVE.VTTISQIFKILSDETRVK
S. aureus pOX4             CadX       MSYENTCDVICVHEDKVNNALSFLEDDK.SKKLLNILEKICDEKKLK
S. aureus pI258            ArsR                     MSYKELSTILKILSDSSRLE
E. coli R773               ArsR                     MLQLTPLQLFKNLSDETRLG
S. xylosus pSX267          ArsR                     MSYKELSTILKVLSDPSRLE
E. coli                    ArsR                     MSFLLPIQLFKILADETRLG
Y. enterocolitica Tn2502   ArsR                       MLQPVQLFKILSDETRLA
Strep. lividans            MerR          MKSPALAGSLATAEVPCTHPDTTARFFRALADPTRLK
R. meliloti                NolR        MNFRMEHTMQPLPPEKHEDAEIAAGFLSAMANPKRLL
V. cholerae                HlyU      MPYLKGAPMNLQEMEKNSAKAVVLLKAMANERRLQ
Consensus (8)                                            FK LADE RL


                                          S                           RR
                                          |                           ||
                                        +++++++++
Synechococcus PCC 7942     SmtB  LLSLLAR.SELCVGDLAQAIGVSESAVSHQLRSLRNLRIVSYRKQGRHVYYQLQDHHIVA
Synechococcystis PCC 6803  PacR  LMSALAR.QELCVCDLAAAMKVSESAVSHQLRLRSQRLVKYRRVGRNVYYSLADNHVMN
Synechococcus vulcanus     MtnB  IVALLAQ.GEFCVRDIAVALEST...(sequence is incomplete)
S. aureus pI258            CadC  ITYALCQDEELCVCDIANILGVTIANASHHLRTLYKQGVVNFRKEGKLALYSLGDEHIRQ
S. aureus                  CadC  IAYVLSLEGELCVCDVANIIESSTATASHHLRLLKNLGIAKYRKEGKLVYYSLDDEHVKQ
B. firmus OF4              CadC  ITYALCQDEESCVCDIANIIGITAANASHHLRTLHKQGIVRYRKEGKLAFYSLDDEHIRQ
L. monocytogenes pLm74     CadC  IVYALLTENELCVCDLANIVEATVAATSHHLRFLKKQGIANYRKDGKLVYYSLANERVRD
```

380

```
                              CadX  IILSLIKEDELCVCDISLILKMSVASTSHHLRLLYKNEVLDFYKDGKMAYYFIKDDEIRE
S. aureus pOX4
S. aureus pI258               ArsR  ILDLLS.CGELCACDLLEHFQFSQPTLSHHMKSLVDNELVTTRKDGNKHWYQL..NH..A
E. coli R773                  ArsR  IVLLLREMGELCVCDLCMALDQSPKISRHLAMLRESGILLDRKQGKWVHYRLS.PHIPS
S. xylosus pSX267             ArsR  ILDLLS.CGELCACDLLEHFQFSQPTLSHHMKSLVDNELVTTRKNGNKHMYQL..NH..E
E. coli                       ArsR  IVLLLSELGELCVCDLCTALDQSQPKISRHLALLRESGLLLDRKQGKWVHYRLS.PHIPA
Y. enterocolitica Tn2502      ArsR  IVMLLRESGEMCVCDLCGATSESQPKISRHMAILREAELVLDRREGKWVHYRLS.PHMPA
Strep. lividans               MerR  LLQFILR.GERTSAECVEHAGISQPRVSVHLSCLVDCGYVSARRDGKKLRYSVGDPRVAD
R. meliloti                   NolR  ILDSIVK.EEMAVGALAHKVGLSQSALSQHLSKLRAQNIVSTRRDAQTIYYSSSSDAVLK
V. cholerae                   HlyU  ILCMLLD.NELSVGELSSRLELSQSALSQHLAWLRRDGLVNTRKEAQTVFYTLSSTEVKA
Consensus (8)                       I    L   GELCVCDLA    SQ    SHHL   L    GLV   RK GK  V Y L    H

                                                         S
                                                         |

Synechococcus PCC 7942        SmtB  LYQNALDHLQECR
Synechococcystis PCC 6803     PacR  LYREVADHLQESD
S. aureus  pI258              CadC  IMMIALAHKKEVKVNV
S. aureus                     CadC  LVEKAFLHQREVASIG
B. firmus OF4                 CadC  IMMIVLEHKKEVNVNV
L. monocytogenes pLm74        CadC  RIKLILLNFEGVGV
S. aureus pOX4                CadX  FFSKNHEGF
S. aureus pI258               ArsR  ILDDIIQNLNIINTSNQRCVCKNVKSGDC
E. coli R773                  ArsR  WAAQIIEQAWLSQQDDVQVIARKLASVNCSGSSKAVCI
S. xylosus  pSX267            ArsR  FLDYINQNLDIINTSDQGCACKNMKSGEC
E. coli                       ArsR  WAAKIIDEAWRCEQEKVQAIVRNLARQNCSGDSKNICS
Y. enterocolitica Tn2502      ArsR  WAAETITTSWHCCGKMFVSGWINQRHHPAEMNRTHSFNHM
Strep. lividans               MerR  LVMLARCLAADNAAALDCCTRIPGEGEQR
R. meliloti                   NolR  ILGALSDIYGDDTDAVEEKPLVRKSA
V. cholerae                   HlyU  MIELLHRLYCQANQ
Consensus (8)
```

Figure 14.4 An alignment of proteins with similarity to SmtB. A consensus sequence with a plurality of 8 is shown. The putative helix-turn-helix DNA binding motif of SmtB is underlined. The locations of site-directed (and spontaneous, marked*) mutations that have been introduced into SmtB and phenotypically analyzed are indicated by the vertical lines, with the altered residue above.

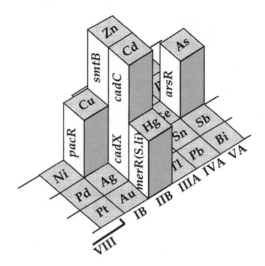

Figure 14.5 A representation of part of the periodic table showing the anticipated "preferred" metals sensed by SmtB-like repressors (after O'Halloran 1993, *Science* 261: front cover). The block heights are proportional to the number of different proteins anticipated to "preferentially" sense a metal (the proposed Zn^{2+}-sensor MtnB is not visible). *S. li* refers to *Streptomyces lividans*.

a 39-bp region of DNA immediately upstream of the ATG of *smtA* (nucleotides −15 to +24 relative to the mapped *smtA* transcription start site) (Morby et al. 1993). Methylation interference assays have identified a pair of nucleotides within an imperfect inverted repeat (5′-TGAACA-GT-TATTCA-3′, the mapped guanine and complement are underlined) in this region which interact with SmtB (Erbe et al. 1995). Substitutions have been introduced into this region of the *smt* operator-promoter to more extensively map the MAC1 binding site and examine the roles of nucleotides adjacent to the *smtA* transcription start site *in vivo*. These directed mutations conferred increased expression of an associated reporter gene consistent with loss of repression (Turner et al. 1996). Expression from the mutant operator-promoter regions indicated that nucleotides that form functional interactions with SmtB are not confined to the imperfect 6-2-6 inverted repeat as previously proposed (Morby et al. 1993; Erbe et al. 1995) but are contained within a larger region that includes an imperfect 12-2-12 inverted repeat (Figure 14.6). Furthermore, the *in vivo* studies revealed that functional interactions with nucleotides in both sides of this repeat are required for normal repression (Turner et al. 1996).

Methylation interference assays revealed that SmtB can form additional interactions with a second pair of nucleotides within an inverted repeat located between

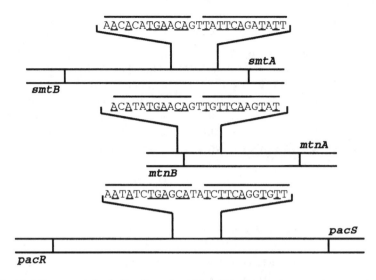

Figure 14.6 Conserved (underlined) nucleotides within the operator-promoter regions of the target genes of the cyanobacterial proteins SmtB, PacR, and MtnB. A similar hyphenated inverted repeat (overlined) is present within the operator-promoter regions of *smtA, mtnA,* and *pacS.*

the *smtA* and *smtB*-10 promoter sequences, generating the slower migrating complexes observed in electrophoretic mobility shift assays (Erbe et al. 1995) (Figure 14.8). These sites are located within a region that includes a perfect 7-2-7 inverted repeat and shows similarity to part of the MAC1 binding site (5'-ACCTGAATC-AA-GATTCAGAT-3', identical nucleotides underlined). SmtB binding to this second region is only observed with high concentrations of apo-SmtB, and it remains to be determined whether or not SmtB binding to this region is physiologically relevant. Deletion of nucleotides in this region (retaining only the two most 3' nucleotides, AT) from the *smtA* operator-promoter does not confer elevated expression of an associated reporter gene in cyanobacteria, and metal-dependent expression is retained (Morby et al. 1993), which is inconsistent with SmtB-mediating repression by binding to nucleotides within this region.

7.2 Amino Acid Residues Involved in Binding

SmtB contains a region that scores as a helix-turn-helix DNA binding motif on the Dodd and Egan (1990) prediction matrix (Huckle et al. 1993) (Figure 14.4). The second helix of the predicted helix-turn-helix is conserved within SmtB and PacR (Figure 14.4), and it is these residues that are predicted to be involved in functional interactions with DNA. These two proteins are, in consequence, predicted to interact with similar nucleotides within the promoters of their target genes. Within the promoter of *pacS* (the predicted target gene for PacR) is a similar degenerate 12-2-12 inverted repeat to that contained within the MAC1-binding site (Figure 14.6). Furthermore, a similar inverted repeat is present within the operator-promoter region of the predicted target gene, *mtnA*, of MtnB (Figure 14.6). However, the sequence of the DNA-recognition helix of MtnB remains to be determined. Similar amino acid-nucleotide interactions are proposed for all three of these cyanobacterial proteins.

8. Which Residues in SmtB Sense Zn²⁺?

SmtB has three cysteine and six histidine residues which are candidates for metal coordination (Figure 14.7). Within the aligned sequences of several of the family of related bacterial proteins are conserved cysteine residues that are associated with the amino-terminal end of putative helix-turn-helix DNA binding

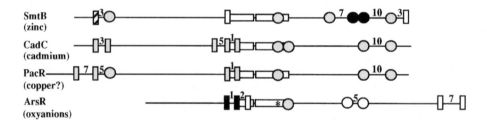

Figure 14.7 Arrangements of cysteine and histidine residues within SmtB-like proteins. A diagrammatic representation of the arrangements of cysteine (small boxes) and histidine (circles) residues in SmtB (from *Synechococcus* PCC 7942), CadC (from *S. aureus* plasmid pI258), PacR (from *Synechocystis* PCC 6803) and ArsR (from *E. coli* plasmid R773). The numbers of intervening residues are given when this value is ≤10. Putative helix-turn-helix DNA binding regions are indicated (rectangles). The "preferred" (predicted for PacR) sensed metal ions indicated for each protein refer to the most potent inducer (of expression from their target genes) *in vivo*. It is noted that other metal ions could act as inducers *in vitro* (although at lethal levels) or may bind more avidly *in vitro*. Residues shown to be essential (filled symbols), not essential (open symbols) (in the case of ArsR this refers to residues that were not selected, for inability to respond to inducer, following random mutagenesis), or untested (shaded symbols) for inducer recognition are marked. There is evidence to suggest that cysteine 14 of SmtB (hatched box) may be involved in Zn²⁺ sensing, although it is not essential. Differences in the arrangements of cysteine and histidine residues are proposed to account for differences in the metal specificities of these proteins, and cause a shift in the spectra of metals sensed.

motifs (Figure 14.4 and 14.7) (Bairoch 1993). These cysteine residues have been predicted to bind metal ions and, as a result, inhibit binding to DNA via the adjacent helix-turn-helix regions (Bairoch 1993; Shi et al. 1994). Consistent with this prediction, ArsR (of the *E. coli* plasmid R337) mutants in either of these cysteine residues (cysteine 32 to tyrosine, cysteine 32 to phenylalanine, and cysteine 34 to tyrosine) have been selected based upon their ability to bind to the *ars* promoter and inability to respond to metal ions (Shi et al. 1994). At least one, but generally two, histidine residues are also conserved within the second helix of predicted helix-turn-helix regions (Figure 14.7) and these may also be involved in metal-coordination (Bairoch 1993). However, mutation of such a histidine residue in ArsR resulted in constitutive expression from the *ars* operator-promoter, indicating that this mutation (within the DNA binding motif) results in loss of DNA binding (Shi et al. 1994).

Four SmtB mutants (T11S/C14S, C61S, C121S, and H105R/H106R; Figure 14.4) have been expressed in *smt* mutants of *Synechococcus* PCC 7942. In the absence of elevated concentrations of Zn^{2+}, these mutant proteins repressed transcription from the *smtA* operator-promoter, demonstrating that they have retained repressor function. Furthermore, electrophoretic mobility shift assays with extracts from these cells expressing the mutant SmtB proteins detected multiple SmtB-dependent complexes with the *smt* operator-promoter (equivalent to those observed with wild-type cells). The addition of Zn^{2+} alleviated repression *in vivo* by all of the mutants except H105R/H106R. Histidine 105 and/or histidine 106 are therefore essential for Zn^{2+} sensing by SmtB whereas, contrary to expectations, the cysteine residues are not. Cysteine 61 in SmtB corresponds to cysteine 32 in ArsR and is located at the amino-terminal end of the helix-turn-helix region (Figure 14.7). It is of significance that in SmtB this is a lone cysteine, whereas in other members of the group of related proteins it forms part of a cysteine-valine or alanine-cysteine motif (Figure 14.4). In common with SmtB, there is only a single cysteine residue at this site in MtnB (Figure 14.4).

Cells containing T11S/C14S mutated SmtB showed reduced expression from the *smtA* operator-promoter compared to cells containing wild-type SmtB, although expression was stimulated by Zn^{2+} (Turner et al. 1996). This suggests increased DNA binding associated with one or both of these mutations (conversion of threonine 11 to serine arose spontaneously during generation of a C14S mutated SmtB). It is therefore possible that cysteine 14 has some role in inducer recognition although it is clearly not required for Zn^{2+} sensing *in vivo*. Cysteine 14 is conserved in some members of the family of related proteins and is located in an amino-terminal domain that is present within the Zn^{2+}/Cd^{2+}-responsive members but absent from the ArsR proteins (Figure 14.7).

Different members of the SmtB-related group of metal-responsive transcriptional repressors have alternative metal-sensing mechanisms. This reveals the exciting possibility of a family of related bacterial repressors possessing distinct metal binding sites which give diversity in the spectra of metals sensed (Figure 14.5). As more of the residues that are (or are not) required for inducer recognition are identified within the different members of this family (Figure 14.7), the arrangements of cysteine and histidine residues may allow predictions to be made as to the "preferred" inducer of any newly identified members of this group.

9. SmtB-independent Complexes Also Form with the *smt* Operator-Promoter

In addition to the SmtB-DNA complex MAC1, two additional complexes (MAC2 and MAC3) form with the *smt* operator-promoter and extracts from *Synechococcus* PCC 7942 (Morby et al. 1993). Under some electrophoretic mobility shift assay conditions (Erbe et al. 1995) these SmtB-independent complexes are not detected (Turner et al. 1996). MAC2 and MAC3 were retained when extracts from *smt* mutants were used (Morby et al. 1993). MAC2 was shown to bind to sequences between the divergent *smtA* and *smtB* promoters which includes a perfect inverted 7-2-7 repeat (Morby et al. 1993) (Figure 14.8). It is noted that SmtB has also been shown to bind to this region *in vitro* at high protein concentrations (although it is not known whether SmtB binding to this site is physiologically relevant) (described above). If SmtB does interact with this second region *in vivo*, there is likely to be competition between SmtB and the protein component of MAC2 for binding (Figure 14.8). The identity of the MAC2 protein and function of MAC2 binding are not yet known.

The MAC3 complex forms on a region of the *smt* operator-promoter most distal to *smtA* (Figure 14.8). A 5-bp overlapping direct repeat (CCACC), immediately upstream of the MAC2 binding site, is a candidate binding site for MAC3 (Morby et al. 1993). In *smt* mutants, expression of *lacZ* driven by a truncated *smtA* operator-promoter was lower (ca. 2-fold) than that obtained for *lacZ* driven by the intact operator-promoter (Morby et al. 1993). The region lost in the truncated derivative contained the MAC3 binding site (Morby et al. 1993). MAC3 was therefore suggested to be an activator of *smtA* expression. There is also circumstantial data suggesting that the MAC3 activator is subject to feedback inhibition by SmtA (Turner et al., 1995). The proposed (negative, with respect to *smtA* expression) influence of SmtA on MAC3 may be indirect but it is not overcome by elevated concentrations of Zn^{2+}. Maximum expression from the *smtA* operator-promoter is not attained even in response to maximum permissive, or to inhibitory, metal concentrations for growth when this feedback inhibition is operational (in cells containing SmtA). This could reflect other functions, in addition to metal detoxification, for SmtA (Turner et al. 1995).

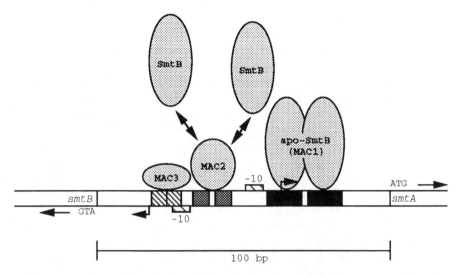

Figure 14.8 Protein interactions within the *smt* operator-promoter. Three protein-DNA complexes (MAC1, MAC2, and MAC3) form within the 100-bp *smt* operator-promoter region. The regions of DNA involved in these interactions have been identified approximately for MAC2 and MAC3 and more critically defined for SmtB (MAC1). Within these regions are repeat sequences that are candidate protein binding sites. An imperfect 12-2-12 hyphenated inverted repeat (solid), a 7-2-7 hyphenated inverted repeat (shaded), and an overlapping direct repeat (diagonal shading) are illustrated. At high concentrations *in vitro*, (apo-) SmtB interacts with nucleotides located between the *smtA* and *smtB*-10 consensus promoter sequences (small hatched boxes) in addition to interacting with nucleotides of the imperfect 12-2-12 inverted repeat. If SmtB binds to this second site *in vivo*, there is likely to be competition between apo-SmtB and the protein component of MAC2 for binding. The *smtA* and *smtB* transcription start sites (bent arrows) are marked.

10. A Putative Regulatory Role for SmtA

10.1 Proposed Regulatory Role for Animal Metallothioneins

In eukaryotes, metallothioneins sequester, and hence detoxify, excess amounts of some metal ions. Metallothionein expression increases in response to elevated concentrations of these ions [for a review, see Kägi (1991)]. However, animal metallothionein also shows (1) developmentally programmed changes in expression (Quaife et al. 1994) and (2) endogenously programmed changes in intracellular location (Tsujikawa et al. 1991), and additional functions have been proposed.

Zn^{2+} is, of course, a catalytic and/or structural component of many eukaryotic proteins (Vallee and Falchuk 1993). The physical and chemical properties of

Zn^{2+} make it ideally suited to certain types of catalysis (Vallee and Falchuk 1993). By contrast, it has been argued that the functional surfaces created by structures that contain Zn^{2+} could have evolved without a metal dependency (O'Halloran 1993). Many of these surfaces, for example in Zn^{2+}-finger proteins, allow interactions with other macromolecules, proteins, or nucleic acids. These interactions are thus Zn^{2+} dependent. Such a dependency could be a relic of a former regulatory role for Zn^{2+} or a clue to an existing role. It has been hypothesized that Zn^{2+} would have been relatively unavailable prior to the appearance of dioxygen in quantity, and is unlikely to have been "recruited" by proteins during early evolution (Fraústo da Silva and Williams 1993). The appearance of an atmosphere rich in dioxygen coincided with and was possibly caused by the evolution of oxygen-generating, photosynthetic cyanobacteria. Any hypothetical regulatory role for Zn^{2+} could have emerged initially in these bacteria.

Berg and Shi (1996) are among those that have surmised that Zn^{2+} levels may be used *in vivo* in an information-carrying role. Overall changes in the availability of Zn^{2+} (analogous to Ca^{2+} "gating") and/or Zn^{2+} exchange between specific donor and acceptor proteins (analogous to regulation by PO_4^{2-}) (O'Halloran 1993) could be mediated by metallothionein. Animal metallothioneins have a high affinity for Zn^{2+} (SmtA has an even higher Zn^{2+} affinity), but Zn^{2+} associated with eukaryotic metallothionein is also highly labile, allowing for rapid metal exchange with other ligands [cited in Kägi (1991)].

The synthesis of key informational macromolecules depends upon proteins that are themselves dependent upon Zn^{2+}. DNA primases (Griep and Lokey 1996), RNA polymerases (Vallee and Falchuk 1993), and tRNA synthetases (Landro and Schimmel 1994) are among such subsets of eukaryotic proteins involved in DNA, RNA, and protein synthesis, respectively. Thionein (apometallothionein) can readily inactivate the Zn^{2+}-requiring transcription factor Sp1 (human) and also acquires Zn^{2+} from *Xenopus laevis* transcription factor IIIA *in vitro* (Zeng et al. 1991a,b). Reversible Zn^{2+} exchange between metallothionein and the estrogen receptor Zn^{2+} finger has recently been demonstrated *in vitro* (Cano-Gauci and Sarkar 1996). Furthermore, in cultured animal cells metallothionein is exclusively located within the cytoplasm at G_0 and G_1, but is concentrated in nuclei at early S-phase, with maximal nuclear accumulation of metallothionein immediately preceding DNA synthesis (Tsujikawa et al. 1991).

10.2 Other Genes in the Cyanobacterial Metallothionein Gene Region

An ORF has recently been identified upstream of *smt* which encodes a product with significant similarity to known bacterial DNA primases (required for DNA synthesis) and has hence been designated *dnaG* (EMBL X94247; Figure 14.2). During attempts to delete sequences at the 3′ end of *dnaG*, homozygous mutants of *Synechococcus* PCC 7942 did not segregate, resulting in the production of merodiploids (Bird et al. unpublished), consistent with this region of DNA being essential for viability and the assignment of *dnaG* being correct. In other analyzed bacteria *dnaG* is located at the center of a conserved macromolecular synthesis operon (Versalovic et al. 1993). The other genes of this conserved operon are *rpsU* encoding ribosomal protein S21 and *rpoD* (*sigA*) encoding the primary σ subunit of RNA polymerase, required to initiate the synthesis of protein and (many species of) RNA respectively (Versalovic et al. 1993). The conserved macromolecular synthesis operon structure, even in distantly related bacterial species, is suggested to allow some coordination of the synthesis of key macromolecules (Versalovic et al. 1993). In contrast, in *Synechococcus* PCC 7942, *dnaG* is flanked by *pheT* (Cranenburgh and Robinson 1996), encoding the β subunit of phenylalanyl tRNA synthetase, and the *smt* divergon (Figure 14.1). There is cotranscription of *pheT-dnaG* (Bird et al. unpublished). It is noteworthy that the *pheT* gene product is (in common with *rpsU*) part of the translational machinery.

The *Synechococcus* PCC 7942 *dnaG* is the first cyanobacterial DNA primase gene to be reported. A gene *rpoD1*, thought to encode the primary σ factor, is located elsewhere (separate from *dnaG*) on the *Synechococcus* PCC 7942 chromosome (Tanaka et al. 1992). In another cyanobacterium, *Anabaena* PCC 7120, a monocistronic *sigA* that is not associated with *dnaG* has also been isolated (Brahamsha and Haselkorn 1991). A *dnaG* from *Synechocystis* PCC 6803 has recently been identified as part of the *Synechocystis* PCC 6803 genome project (available in Cyanobase) and is not associated with *rpsU* or *rpoD*. Cyanobacteria therefore appear to have an atypical arrangement of these genes involved in macromolecular synthesis, which could imply atypical coordination of the synthesis of key macromolecules. The regulation of DNA replication and cell division in some, but not all, cyanobacteria (including *Synechococcus* PCC 7942) has "unique features which have not to date been reconciled with the classical prokaryotic paradigm" (Binder and Chisholm 1990). In contrast to other bacteria, *Synechococcus* PCC 7942 shows asynchronous initiation of chromosome replication. It is hypothesized that atypical coordination of DNA synthesis could be due to atypical regulation of DNA primase.

The significance, if any, of the proximity of *smt* to the *pheT-dnaG* operon and amplification of *dnaG* in metal-resistant cyanobacteria (Figure 14.2) is unknown. It is common for functionally related genes to be clustered in cyanobacterial (and other prokaryotic) genomes (Grossman et al. 1994; Kaplan et al. 1994). At the amino-terminal end of *Synechococcus* PCC 7942 DNA primase, is a conserved Zn^{2+} binding region found in other bacterial primases (Stamford et al. 1992) (Figure 14.9), the purified DNA primase from *E. coli* having been shown to contain tightly bound Zn^{2+} (Stamford et al. 1992; Griep and Lokey 1996). Within the carboxy-terminal region of *Synechococcus* PCC 7942 DnaG is a predicted cysteine$_2$-histidine$_2$ type Zn^{2+}-finger motif (Figure 14.9) that is absent from the

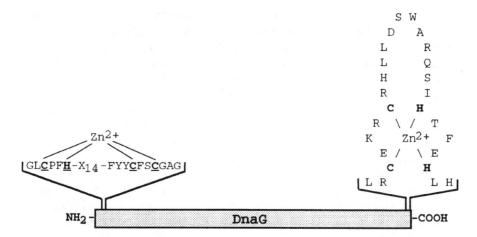

Figure 14.9 The Zn^{2+} binding regions of the deduced *Synechococcus* PCC 7942 DNA primase. At the amino-terminal of *Synechococcus* PCC 7942 DnaG is a predicted Zn^{2+} binding region that is conserved in all other known bacterial DNA primases (Stamford et al. 1992). Residues whose side chains are proposed to coordinate Zn^{2+} are indicated (bold/underlined). A diagrammatic representation of the predicted cysteine$_2$-histidine$_2$ type Zn^{2+}-finger motif located at the carboxy-terminal of *Synechococcus* PCC 7942 DnaG is also shown. This motif is not conserved in the other (so far) reported bacterial DNA primases.

other reported primases. Furthermore, Zn^{2+} binding *in vitro* was demonstrated following expression of the carboxy-terminal region of this DnaG in *E. coli* (Bird et al. unpublished). A putative Zn^{2+} binding site has been proposed for phenylalanyl-tRNA synthetase in *E. coli*, and other amino acyl-tRNA synthetases are known to require Zn^{2+} (Nureki et al. 1993). Furthermore, mutations that impair Zn^{2+} binding to isoleucyl-tRNA synthetase are known to confer Zn^{2+}-dependent cell growth in *E. coli* (Landro and Schimmel 1994). The location of *pheT-dnaG*, adjacent to *smt*, could be important due to a predicted requirement of their products for Zn^{2+}. It is formally possible that by regulating the availability of Zn^{2+}, SmtA may influence the activity of such Zn^{2+}-requiring cyanobacterial proteins with some analogy to suggested roles for animal metallothioneins. SmtA could act as an intracellular Zn^{2+} buffer or be part of a more dynamic mechanism donating and removing Zn^{2+} to and from Zn^{2+}-requiring proteins. It is tempting to speculate that some facet of the coordination of macromolecular synthesis, affected by macromolecular synthesis operons in other bacteria, is mediated by SmtA and Zn^{2+} in cyanobacteria and perhaps by other metallothioneins and Zn^{2+} in higher eukaryotes.

11. Prospectives and Perspectives

Several lines of evidence show that the cyanobacterial metallothionein divergon, *smt*, is involved in providing resistance to excess Zn^{2+}: SmtA recovered from cyanobacteria is associated with Zn^{2+} (Olafson et al. 1988); SmtA enhances Zn^{2+} accumulation in *E. coli* (Shi et al. 1992) and in *Synechococcus* PCC 7942 (Turner et al. 1995); recombinant SmtA has a high affinity for Zn^{2+} *in vitro* (Shi et al. 1992); and mutants deficient in *smtA* are hypersensitive to Zn^{2+} (Turner et al. 1993). In comparison with other metals, at maximum permissive concentrations, expression from the *smtA* operator-promoter is maximally induced by Zn^{2+} (Huckle et al. 1993).

Metal-induced transcription of *smtA* is mediated, at least in part, by SmtB, a metal-inducible *trans*-acting negative regulator (Huckle et al. 1993). At least some of the nucleotides and amino acid residues involved in DNA-SmtB interaction have been determined (Erbe et al. 1995) and inferred (Turner et al. 1996). SmtB directly binds Zn^{2+} (Erbe et al. 1995), and some progress has been made in defining the residues involved in metal perception (Turner et al. 1996). SmtB is a member of a family of prokaryotic metal-responsive repressor proteins (Figure 14.4), each of which may sense a different spectrum of metal ions (Figure 14.5). We foresee the intriguing prospect of the elucidation of structural features that determine the metal specificities of each individual protein (Figure 14.7). Although continued site-directed mutagenesis will establish which of the illustrated (Figure 14.7) histidine and cysteine residues are involved in metal sensing by each of these proteins, random mutagenesis followed by selection promises to reveal additional "unexpected" required residues.

In addition to understanding how SmtA is regulated by metals, future studies will address whether or not metal buffering by SmtA is, in turn, regulatory. Genes encoding Zn^{2+}-requiring proteins involved in macromolecular synthesis have now been identified in the metallothionein gene region (Figure 14.1). Does SmtA influence the metallation status of these proteins in a manner that is independent of exogenous changes in Zn^{2+} levels? The identification of other (than SmtB) factors that bind to the *smtA* operator-promoter is of importance because it is these factors that could influence SmtA production in response to an endogenous program (Figure 14.10). The hypothetical scheme shown in Figure 14.10 speculates that such changes in SmtA expression (perhaps at cell division, during DNA replication, or in stationary phase cultures) shift the endogenous Zn^{2+} levels (from n to $n - y$) and thereby influence the Zn^{2+} status of proteins such as DnaG. Future experiments should examine whether or not expression of *smtA* does change independently of exogenous metal levels (e.g., at different phases of culture growth). Furthermore, does the abundance of factors (other than SmtB) that bind to the *smtA* operator-promoter change during the culture

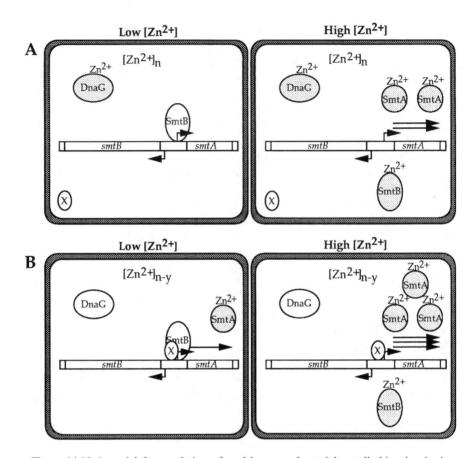

Figure 14.10 A model for regulation of, and by, cyanobacterial metallothionein. A. At low concentrations of Zn^{2+}, apo-SmtB (unshaded) binds to the *smtA* operator-promoter and represses transcription. At elevated concentrations of Zn^{2+}, Zn^{2+}-SmtB (shaded) dissociates from the *smtA* operator-promoter and *smtA* transcription proceeds. SmtB thus regulates the synthesis of SmtA in response to changes in available metal ion concentrations to maintain intracellular Zn^{2+} levels at $[Zn^{2+}]_n$. The Zn^{2+} status of Zn^{2+}-requiring proteins, such as DnaG, is thus preserved even when the exogenous $[Zn^{2+}]$ fluctuates. B. In addition to *smtA* transcription being regulated in response to metal ions (by SmtB) we propose that under certain conditions a second factor or factors (x) also regulate(s) *smtA*. These conditions are not yet determined but could, for example, relate to the cells' proliferative state. It is proposed that this modulation in the level at which SmtA "buffers" intracellular Zn^{2+} (to $[Zn^{2+}]_{n-y}$) would influence the Zn^{2+}-status of other proteins (represented by apo-DnaG). Within the proposed model, x is shown as an activator (because of circumstantial evidence that MAC3 is an activating complex), but it is possible that other forms of regulation could be involved.

cycle or at cell division? Can SmtA affect the Zn^{2+} status of Zn^{2+}-requiring proteins *in vitro*? Do changes in *smtA* expression coincide with changes in the synthesis of key macromolecules, and are changes in macromolecule composition detectable when cells deficient in *smtA* are grown under high- or low-Zn^{2+} regimes? Some cyanobacteria undergo differentiation events to generate specialized structures such as heterocysts [reviewed in Wolk et al. (1994)]. It would be of further interest to examine putative programmed changes in expression of any *smtA* homologs in such strains.

Acknowledgments

This work was supported by the Biotechnology and Biological Sciences Research Council (GR/J37126) and the Natural Environmental Research Council (GR3/8660). The authors thank James W. Huckle for previously unpublished observations.

References

Bairoch, A. 1993. A possible mechanism for metal-ion induced DNA-protein dissociation in a family of prokaryotic transcription factors. *Nucl. Acids Res.* 21:2515.

Barbosa, M. S., D. R. Lowry, and J. T. Schiller. 1989. Papillomavirus polypeptides E6 and E7 are zinc-binding proteins. *J. Virol.* 63:1404–1407.

Berg, J. M., and Y. Shi. 1996. The galvanization of biology: A growing appreciation for the roles of zinc. *Science* 271:1081–1085.

Binder, B. J., and S. W. Chisholm. 1990. Relationship between DNA-cycle and growth-rate in *Synechococcus* sp. strain PCC 6301. *J. Bacteriol.* 172:2313-2319.

Brahamsha, B., and R. Haselkorn. 1991. Isolation and characterization of the gene encoding the principle sigma factor of the vegetative cell RNA polymerase from the cyanobacterium *Anabaena* sp. strain PCC 7120. *J. Bacteriol.* 173:2442–2450.

Brinster, R. L., H. Y. Chen, R. Warren, A. Sarthy, and R. D. Palmiter. 1982. Regulation of metallothionein-thymidine kinase fusion plasmids injected into mouse eggs. *Nature* 296:39–42.

Brugnera, E., O. Georgiev, F. Radtke, R. Heuchel, E. Baker, G. R. Sutherland, and W. Schaffner. 1994. Cloning, chromosomal mapping and characterization of the human metal-regulatory transcription factor MTF-1. *Nucl. Acids Res.* 22:3167–3173.

Cano-Gauci, D. F., and B. Sarkar. 1996. Reversible zinc exchange between metallothionein and the estrogen receptor zinc finger. *FEBS Lett.* 386:1–4.

Cranenburgh, R. M., A. Gupta, J. S. Turner, and N. J. Robinson. 1997. HIP1 and gene rearrangement in cyanobacteria. Ph.D. thesis, University of Newcastle.

Cranenburgh, R. M., and N. J. Robinson. 1996. Phenylalanyl-tRNA synthetase gene, *pheT*, from *Synechococcus* PCC 7942. *J. Appl. Phycol.* 8:81–82.

Dameron, C. T., D. R. Winge, G. N. George, M. Sansones, S. Hu, and D. Hamer. 1991.

A copper-thiolate polynuclear cluster in the ACE1 transcription factor. *Proc. Natl. Acad. Sci. U.S.A.* 88:6127–6131.

Dodd, I. B., and J. B. Egan. 1990. Improved detection of helix-turn-helix DNA-binding motifs in protein sequences. *Nucl. Acids Res.* 18:5019–5027.

Endo, G., and S. Silver. 1995. CadC, the transcriptional regulatory protein of the cadmium resistance system of *Staphylococcus aureus* plasmid pI258. *J. Bacteriol.* 177:4437–4441.

Erbe, J. L., K. B. Taylor, and L. M. Hall. 1995. Metalloregulation of the cyanobacterial *smt* locus: Identification of SmtB binding sites and direct interaction with metals. *Nucl. Acids Res.* 23:2472–2478.

Fraústo da Silva, J.J.R., and R.J.P. Williams. 1993. Zinc: Lewis acid catalysis and regulation, In *The Biological Chemistry of the Elements*. Chapter 11, pp. 299–318. Clarendon Press, Oxford.

Fürst, P., S. Hu, R. Hackett, and D. H. Hamer. 1988. Copper activates metallothionein gene transcription by altering the conformation of a specific DNA binding protein. *Cell* 55:705–717.

Griep, M. A., and E. R. Lokey. 1996. The role of zinc and the reactivity of cysteines in *Escherichia coli* primase. *Biochem.* 35:8260–8267.

Grossman, A. R., M. R. Schaefer, G. G. Chiang, and J. O. Collier. 1994. The responses of cyanobacteria to environmental conditions: Light and nutrients, In *The Molecular Biology of Cyanobacteria*. ed. D. A. Bryant. pp. 641–675. Kluwer Academic Press, Dordrecht.

Gupta, A., A. P. Morby, J. S. Turner, B. A. Whitton, and N. J. Robinson. 1993. Deletion within the metallothionein locus of Cd-tolerant *Synechococcus* PCC 6301 involving a highly iterated palindrome (HIP1). *Mol. Microbiol.* 7:189–195.

Heuchel, R., F. Radtke, O. Georgiev, G. Stark, M. Aguet, and W. Schaffner. 1994. The transcription factor MTF-1 is essential for basal and heavy metal-induced metallothionein gene expression. *EMBO J.* 13:2870–2875.

Huckle, J. W., A. P. Morby, J. S. Turner, and N. J. Robinson. 1993. Isolation of a prokaryotic metallothionein locus and analysis of transcriptional control by trace metal ions. *Mol. Microbiol.* 7:177–187.

Ivey, D. M., A. A. Guffanti, Z. Shen, N. Kudyan, and T. A. Krulwich. 1992. The *cadC* gene product of alkiliphilic *Bacillus firmus* OF4 partially restores Na⁺ resistance to an *Escherichia coli* strain lacking an Na⁺/H⁺ antiporter (NhaA). *J. Bacteriol.* 174:4878–4884.

Ji, G., and S. Silver. 1992. Regulation and expression of the arsenic resistance operon from *Staphylococcus aureus* plasmid pI258. *J. Bacteriol.* 174:3684–3694.

Kägi, J. H. R. 1991. Overview of metallothionein. *Methods Enzymol.* 205:613–626.

Kaneko, T., A. Tanaka, S. Sato, H. Kotani, T. Sazuka, N. Miyajima, M. Sugiura, and S. Tabata. 1995. Sequence analysis of the genome of the unicellular cyanobacterium *Synechocystis* sp. strain PCC6803. I. Sequence features in the 1 Mb region from map positions 64% to 92% of the genome. *DNA Res.* 2:153–166.

Kaplan, A., R. Schwarz, J. Lieman-Hurwitz, M. Ronen-Tarazi, and L. Reinhold. 1994. Physiological and molecular studies on the response of cyanobacteria to changes in the ambient inorganic carbon concentration, In *The Molecular Biology of Cyanobacteria.* ed. D. A. Bryant. pp. 469–485. Kluwer Academic Press, Dordrecht.

Kojima, Y. 1991. Definitions and nomenclature of metallothioneins. *Methods Enzymol.* 205:8–10.

Kondorosi, E., M. Pierre, M. Cren, U. Haumann, M. Buire, B. Hoffmann, J. Schell, and A. Kondorosi. 1991. Identification of NoIR, a negative transacting factor controlling the *nod* regulon in *Rhizobium meliloti. J. Mol. Biol.* 222:885–896.

Labbé, S., J. Prevost, P. Remondelli, A. Leone, and C. Séguin. 1991. A nuclear factor binds to the metal regulatory elements of the mouse gene encoding metallothionen-1. *Nucl. Acids Res.* 19:4225–4231.

Landro, J. A., and P. Schimmel. 1994. Zinc-dependent cell growth conferred by mutant tRNA synthetase. *J. Biol. Chem.* 269:20217–20220.

Lebrun, M., A. Audurier, and P. Cossart. 1994. Plasmid-borne cadmium resistance genes in *Listeria monocytogenes* are similar to *cadA* and *cadC* of *Staphylococcus aureus* and are induced by cadmium. *J. Bacteriol.* 176:3040–3048.

McFall, E. 1986. *cis*-acting proteins. *J. Bacteriol.* 167:429–432.

Morby, A. P., J. S. Turner, J. W. Huckle, and N. J. Robinson. 1993. SmtB is a metal regulated repressor of the cyanobacterial metallothionein gene *smtA*: identification of a Zn inhibited DNA-protein complex. *Nucl. Acids Res.* 21:921–925.

Neyt, C., M. Iriarte, V. H. Thi, and G. R. Cornelis. 1997. Virulence and arsenic resistance in Yersiniae. *J. Bacteriol.* 179:612–619.

Nureki, O., T. Kohno, T. Miyazawa, and S. Yokoyama. 1993. Chemical modification and mutagenesis studies on zinc binding of aminoacyl-tRNA synthetases. *J. Biol. Chem.* 268:15368–15373.

O'Halloran, T. V. 1993. Transition metals in control of gene expression. *Science* 261:715–725.

Olafson, R. W., S. Loya, and R. G. Sim. 1980. Physiological parameters of prokaryotic metallothionein induction. *Biochem. Biophys. Res. Comm.* 95:1495–1503.

Olafson, R. W., W. D. McCubbin, and C. M. Kay. 1988. Primary- and secondary-structural analysis of a unique prokaryotic metallothionein from a *Synechococcus* sp. cyanobacterium. *Biochem. J.* 251:691–699.

Otsuka, F., A. Iwamatsu, K. Suzuki, M. Ohsawa, D. H. Hamer, and S. Koizumi. 1994. Purification and characterization of a protein that binds to the metal responsive elements of the human metallothionein IIA gene. *J. Biol. Chem.* 269:23700–23707.

Palmiter, R. D., R. L. Brinster, R. E. Hamer, M. E. Trumbauer, M. G. Rosenfeld, N. C. Birnberg, and R. M. Evans. 1982. Dramatic growth of mice that develop from eggs microinjected with metallothionein-growth hormone fusion genes. *Nature* 300:611–615.

Palmiter, R. D., G. Norstedt, R. E. Gelinas, R. E. Hamer, and R. L. Brinster. 1983. Metallothionein-GH fusion genes stimulate growth of mice. *Science* 222:809–814.

Quaife, C. J., S. D. Findley, J. C. Erickson, G. J. Froelick, E. J. Kelly, B. P. Zambrowiez, and R. Palmiter. 1994. Induction of a new metallothionein isoform (MT-IV) occurs during differentiation of stratified squamous epithelia. *Biochem.* 33:7250–7259.

Radtke, F., O. Georgiev, H. P. Müller, E. Brugnera, and W. Schaffner. 1995. Functional domains of the heavy metal-responsive transcription factor MTF-1. *Nucl. Acids Res.* 23:2277–2286.

Radtke, F., R. Heuchel, O. Georgiev, M. Hergersberg, M. Gariglio, Z. Dembic, and W. Schaffner. 1993. Cloned transcription factor MTF-1 activates the mouse metallothio-nein-1 promoter. *EMBO J.* 12:1355–1362.

Robinson, N. J., A. Gupta, A. P. Fordham-Skelton, R.R.D. Croy, B. A. Whitton, and J. W. Huckle. 1990. Prokaryotic metallothionein gene characterisation and expression: Chromosome crawling by ligation mediated PCR. *Proc. R. Soc. Lond.* B 242:241–247.

Robinson, N. J., P. J. Robinson, A. Gupta, A. J. Bleasby, B. A. Whitton, and A. P. Morby. 1995. Singular over-representation of an octameric palindrome, HIP1, in DNA from many cyanobacteria. *Nucl. Acids Res.* 23:729–735.

Rosenstein, R., K. Nikoleit, and F. Götz. 1994. Binding of ArsR, the repressor of the *Staphylococcus xylosus* (pSX267) arsenic resistance operon to a sequence with dyad symmetry within the *ars* promoter. *Mol. Gen. Genet.* 242:566–572.

San Fransisco, M.J.D., C. L. Hope, J. B. Owolabi, L. S. Tisa, and B. P. Rosen. 1990. Identification of the metalloregulatory element of the plasmid-encoded arsenical resis-tance operon. *Nucl. Acids Res.* 18:619–624.

Sedlmeier, R., and J. Altenbuchner. 1992. Cloning and DNA-sequence analysis of the mercury resistance genes of *Streptomyces lividans*. *Mol. Gen. Genet.* 236:76–85.

Shi, J., W. P. Lindsay, J. W. Huckle, A. P. Morby, and N. J. Robinson. 1992. Cyanobacterial metallothionein gene expressed in *Escherichia coli*-metal binding properties of the expressed protein. *FEBS Lett.* 303:159–163.

Shi, W., J. Wu, and B. P. Rosen. 1994. Identification of a putative metal binding site in a new family of metalloregulatory proteins. *J. Biol. Chem.* 269:19826–19829.

Shimizu, T. T., T. Hiyama, M. Ikeuchi, and Y. Inoue. 1992. Nucleotide sequence of a metallothionein gene of the thermophilic cyanobacterium *Synechococcus vulcanus*. *Plant Mol. Biol.* 20:565–567.

Stamford, N.P.J., P. E. Lilley, and N. E. Dixon. 1992. Enriched sources of *Escherichia coli* replication proteins. The *dnaG* primase is a zinc metalloprotein. *Biochim. Biophys. Acta* 1132:17–25.

Tanaka, K., S. Masuda, and H. Takahashi. 1992. The complete nucleotide sequence of the gene (*rpoD1*) encoding the principal σ factor of the RNA polymerase from the cyanobacterium *Synechococcus* sp. strain PCC 7942. *Biochim. Biophys. Acta* 1132:94–96.

Thiele, D. J. 1988. *ACE1* regulates expression of the *Saccharomyces cerevisiae* metallothio-nein gene. *Mol. Cell Biol.* 8:2745–2752.

Tsujikawa, K., T. Imai, M. Kakutani, Y. Kayamori, T. Mimura, N. Otaki, M. Kimura,

R. Fukuyama, and N. Shimizu. 1991. Localization of metallothionein in nuclei of growing primary cultured adult rat hepatocytes. *FEBS Lett.* 283:239–242.

Turner, J. S., P. D. Glands, A. C. R. Samson, and N. J. Robinson. 1996. Zn^{2+}-sensing by the cyanobacterial metallothionein repressor SmtB: Different motifs mediate metal-induced protein-DNA dissociation. *Nucl. Acids Res.* 24:3714–3721.

Turner, J. S., A. P. Morby, B. A. Whitton, A. Gupta, and N. J. Robinson. 1993. Construction and characterisation of Zn^{2+}/Cd^{2+} hypersensitive cyanobacterial mutants lacking a functional metallothionein locus. *J. Biol. Chem.* 268:4494–4498.

Turner, J. S., and N. J. Robinson. 1995. Cyanobacterial metallothioneins: Biochemistry and molecular genetics. *J. Ind. Microbiol.* 14:119–125.

Turner, J. S., N. J. Robinson, and A. Gupta. 1995. Construction of Zn^{2+}/Cd^{2+}-tolerant cyanobacteria with a modified metallothionein divergon: Further analysis of the function and regulation of *smt. J. Ind. Microbiol.* 14:259–264.

Vallee, B. L., and K. H. Falchuk. 1993. The biochemical basis of zinc physiology. *Physiol. Rev.* 73:79–118.

Versalovic, J., T. Koeuth, R. Britton, K. Geszvain, and J. R. Lupski. 1993. Conservation and evolution of the *rpsU-dnaG-rpoD* macromolecular synthesis operon in bacteria. *Mol. Microbiol.* 8:343–355.

Welch, J., S. Fogel, C. Buchman, and M. Karin. 1989. The *CUP2* gene product regulates the expression of the *CUP1* gene, coding for yeast metallothionein. *EMBO J.* 8:255–260.

Williams, S. G., S. R. Attridge, and P. A. Manning. 1993. The transcriptional activator HlyU of *Vibrio cholerae* nucleotide sequence and role in virulence gene expression. *Mol. Microbiol.* 9:751–760.

Wolk, C. P., A. Ernst, and J. Elhai. 1994. Heterocyst metabolism and development, In *The Molecular Biology of Cyanobacteria*, ed. D. A. Bryant. pp. 769–823. Kluwer Academic Press, Dordrecht.

Xu, C., W. Shi, and B. P. Rosen. 1996. The chromosomal *arsR* gene of *Escherichia coli* encodes a *trans*-acting metalloregulatory protein. *J. Biol. Chem.* 271:2427–2432.

Zeng, J., R. Heuchel, W. Schaffner, and J. H. R. Kägi. 1991a. Thionein (apometallothionein) can modulate DNA binding and transcription activation by zinc finger containing factor Sp1. *FEBS Lett.* 279:310–312.

Zeng, J., B. L. Vallee, and J. H. R. Kägi. 1991b. Zinc transfer from transcription factor IIIA fingers to thionein clusters. *Proc. Natl. Acad. Sci. U.S.A.* 88:9984–9988.

Zhou, P., and D. J. Thiele. 1993. Copper and gene regulation in yeast. *Biofactors* 4:105–115.

15

Metallothionein-like Genes and Phytochelatins in Higher Plants

Anthony P. Fordham Skelton, Nigel J. Robinson, and Peter B. Goldsbrough

1. Introduction

The mineral composition of soils is highly variable both spatially and temporally (for example, coincident with changes in soil moisture content). For plants to grow and reproduce they must adapt to at least some variation in this component of the environment. Adaptive mechanisms include regulating the uptake of nutrients from the soil, typically by transport activities in roots, and these may be complemented by other intracellular mechanisms that regulate availability and prevent toxicity of nutrients.

Plants contain phytochelatins (PCs) and metallothionein (MT)-like polypeptides. Although there is clear evidence of a role for PCs in the detoxification of excess Cd^{2+} in plants, the data supporting a role for MT proteins in the sequestration of metals of the copper and zinc triads in plants is more circumstantial.

The synthesis of PCs increases in response to elevated exogenous concentrations of Cd^{2+} and a number of other metal ions. Unlike the more familiar dependence on transcriptional regulation of new mRNA followed by protein synthesis, rapid metal-induced synthesis of PCs is mediated by the enhanced activity of preexisting enzymes. In addition, transcripts encoded by at least some representatives of plant MT-like gene families increase in abundance in response to changes in the exogenous concentration of a number of metal ions in some tissues and under some environmental conditions. Other endogenous and exogenous factors also influence MT-like transcript abundance in plants. Current information on the expression of such genes is largely an eclectic survey of different tissue types, environmental factors, species, developmental states, and selected representatives of gene families. In comparison to the understanding of metal-induced expression of MT genes in animals, yeasts, and some bacteria (refer to Chapters 9, 11, 12 and 14 of this volume), our understanding of the regulation of MT-like genes in plants is rudimentary. Putative *cis*-acting metal response elements (MREs) remain to be defined, no metal-responsive transcription factors are known, and there is little evidence to implicate any known signaling pathways.

2. Phytochelatins

2.1 Structure

On the basis of amino acid composition, Wagner (1984) was the first to propose that a low-M_r Cd^{2+} ligand (isolated from cabbage) was related to the tripeptide glutathione (γ-GluCysGly; GSH). Detailed structural analysis of equivalent ligands from *Rauvolfia serpentina* (Grill et al. 1985), tomato (Steffens et al. 1986), and *Datura innoxia* (Jackson et al. 1987) confirmed this hypothesis. The Cd^{2+} ligands isolated from cell suspension cultures of these species were comprised of a family of related peptides composed of only three amino acids: glutamate, cysteine, and glycine.

The general structure of these peptides is $(\gamma$-GluCys$)_n$-Gly, where n is >1 (for GSH, $n = 1$). They have been identified in a wide variety of plant species including dicots, monocots, gymnosperms and algae (Gekeler et al. 1989). Peptides with the same structure had been previously identified as the principle Cd^{2+} ligand in *Schizosaccharomyces pombe* (Kondo et al. 1983). Synthesis of PCs in fungi is not restricted to *S. pombe*, but has also been described in *Candida glabrata* (Mehra et al. 1988), *Saccharomyces cerevisiae*, and *Neurospora crassa* (Kneer et al. 1992). Many trivial names have been given to these peptides, including cadystins, γ-glutamyl peptides, and PCs. The latter term has gained wide acceptance and is used here. PCs are categorized as Class III MTs based on their structure and metal binding properties (Kägi and Schaffer 1988). Unlike Class I and II MTs, PCs are not gene products derived from the translation of mRNA by ribosomes but are instead synthesized enzymatically.

The structure of PCs indicates that they are related to GSH. In some plants, notably in the legume family, the major thiol compound is not GSH but a related tripeptide, γ-GluCys-β-Ala (Klapheck 1988). Upon exposure to Cd^{2+}, these plants synthesize polymers of γ-GluCys with a carboxy-terminal β-Ala, providing further evidence that PCs are related to GSH (Grill et al. 1986). The spectrum of compounds related to PCs has recently expanded with the addition of γ-(GluCys)$_n$-Ser in *Agrostis* and rice (Klapheck et al., 1994) and γ-(GluCys)$_n$-Glu in maize (Meuwly et al. 1995). These peptides are likely to be synthesized from the corresponding tripeptide related to GSH. Polymers of γ-GluCys have also been found in maize, *C. glabrata*, and *S. pombe*.

2.2 Synthesis of PCs

Many studies on the synthesis of PCs in plants have used cell suspension cultures as an easily manipulated system to examine responses to metals. PCs are present at low levels in cell cultures of many species but increase rapidly upon addition of a variety of metals. The cellular level of GSH declines when PC synthesis is induced, suggesting that GSH is involved in the synthesis of PCs (Grill et al. 1987; Scheller et al. 1987). Two other lines of evidence indicate that GSH is the substrate for PC synthesis. Buthionine sulfoximine, an inhibitor of GSH synthesis, blocks the metal-induced synthesis of PCs once the available pool of GSH is depleted (Grill et al. 1987; Scheller et al. 1987, Steffens et al. 1986; Reese and Wagner 1987). Second, addition of GSH to the medium restores synthesis of PCs in plant cells treated with buthionine sulfoximine and can stimulate PC synthesis under normal conditions (Mendum et al. 1990).

Grill et al. (1989) identified an enzyme activity in extracts from cells of *Silene cucubalus* that synthesized PCs from GSH by transferring γ-GluCys from a donor to an acceptor molecule (Figure 15.1). Both donor and acceptor could be either GSH or an extant PC molecule. This γ-GluCys dipeptidyl transpeptidase has been named PC synthase. Similar enzyme activities have been found in pea (Klapheck et al. 1995) and tomato (Chen et al. 1997). *In vitro* synthesis of PCs requires the presence of one of a number of metal ions, and synthesis continues until the metal ion is chelated, either by PCs or by addition of an exogenous chelator such as EDTA (Loffler et al. 1989). This provides one mechanism to autoregulate the accumulation of PCs in response to metal exposure. However, regulation of PC synthesis *in vivo* is likely to be more complex. The level of PCs in cells exposed to Cd^{2+} can be several fold higher than the measured cellular concentration of Cd^{2+} (Gupta and Goldsbrough 1990; Vogeli-Lange and Wagner,

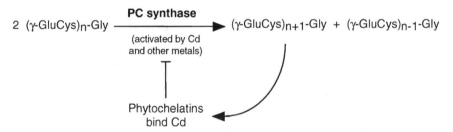

Figure 15.1 Synthesis of phytochelatins. Phytochelatin synthase catalyses the transfer of γ-GluCys (from GSH or a PC) to either GSH, to produce (γ-GluCys)$_2$-Gly, or a PC, to produce larger PCs. The enzyme requires Cd^{2+} or a number of other metal ions for activation. *In vitro*, when the available metal ion has been chelated by PCs or EDTA, PC synthesis ceases.

1996), perhaps reflecting the localization of Cd^{2+} and PCs in different metabolic compartments.

Synthesis of PCs in plant cells does not require *de novo* protein synthesis, and PC synthase is present in cells grown in standard cell culture media that contain relatively low levels of essential metals (Grill et al. 1989). The enzyme is also present in the roots and stems of tomato plants grown under standard greenhouse conditions, but the activity was not detected in leaves or fruits (Chen et al. 1997). Analysis of PC synthase activity from pea, using GSH, homo-GSH (γ-GluCys-β-Ala) and γ-GluCysSer as substrates, indicates that this enzyme has a γ-GluCys acceptor site that can utilize any of these substrates for polymerization of PCs, whereas the γ-GluCys donor site is more specific for GSH (Klapheck et al. 1995). It is not clear how this would function in species that produce little or no GSH, but instead have only homo-GSH. A second pathway for synthesis of PCs has been described in *S. pombe*, involving first the polymerization of γ-GluCys, followed by addition of Gly by GSH synthase (Hayashi et al. 1991). In *S. pombe*, this pathway does not require metal activation. A similar pathway for PC synthesis has not been described in plants.

2.3 Structure of PC-metal complexes

PCs readily chelate Cd^{2+} and other metals *in vitro*. However, the formation of complexes *in vivo* is more complicated, involving transport of both PCs and Cd^{2+} into the vacuole and the incorporation of cadmium sulfide. Cd^{2+}-PC complexes isolated from *S. pombe* could be resolved into two separate classes based on gel filtration chromatography (Hayashi et al. 1988). The low-molecular-weight (LMW) complexes appear first in response to Cd^{2+} and are comprised of only PCs and Cd^{2+}, with Cd^{2+} bound through the cysteine thiols. As discussed below, these LMW complexes, also known as Cd-BP2, are not sufficient to impart normal, wild-type levels of Cd^{2+} tolerance in *S. pombe*. The high-molecular-weight (HMW) complexes, also referred to as Cd-BP1, contain PCs, Cd^{2+} and CdS. These complexes are now believed to consist of a CdS crystallite core coated with PCs (Dameron et al. 1989). Incorporation of sulfide into the HMW complexes both increases the amount of Cd^{2+} bound per unit of PC and increases the stability of these complexes (Reese and Winge 1988). The HMW Cd^{2+}-PC complexes in both *C. glabrata* and *S. pombe* are localized in the vacuole (Mehra et al. 1994; Ortiz et al. 1995). The complexes have not been as well studied in plants, but there is now a growing body of evidence showing that Cd^{2+}-PC complexes formed in plants also contain sulfide (Reese et al. 1992; Speiser et al. 1992a) and are localized to the vacuole (Vogeli-Lange and Wagner 1990).

Although it is tempting to presume that PCs form complexes with all of the metals that induce PC synthesis, and mediate tolerance to these metals, the available evidence does not support this hypothesis. Maitani et al. (1996) have used inductively coupled plasma-atomic emission spectroscopy in combination with HPLC separation of native metal-PC complexes to demonstrate the *in vivo* formation of PC complexes with Cd^{2+}, Ag^+ and copper ions in root cultures of *Rubia tinctorum*. The PC complexes formed in response to Pb^{2+} and As^{3+} contained copper ions, but not the metal or metalloid used to induce synthesis of PCs. The only complexes identified *in vivo* are with Cd^{2+}, Ag^+ and copper ions, but complexes formed *in vitro* between PCs and Pb^{2+}, Hg^{2+} and Ag^+ have also been studied (Mehra et al. 1995, 1996a, b). As described in the following sections, there is good genetic evidence that PCs are not required for tolerance to all of the metals that induce PC synthesis. Tolerance of PC-deficient mutants of *Arabidopsis thaliana* to copper ions and Zn^{2+} is similar to that of the wild type (Howden and Cobbett 1992). It is possible that PCs serve as a primary tolerance system for a number of nonessential metal ions, whereas other systems, perhaps including MTs, predominate in the homeostasis of essential metal ions.

2.4 Genetic Analysis of PC Synthesis and Function

Greater understanding of the role of PCs in metal tolerance in both plants and fungi has come from analysis of mutants with reduced tolerance to Cd^{2+} which are altered in the synthesis of PCs and/or the formation of Cd^{2+}-PC complexes. Genetic analysis of Cd^{2+} sequestration by PCs was initiated in *S. pombe* and has now been extended to *Arabidopsis*. In both yeasts and plants, mutants with reduced Cd^{2+} tolerance were identified by their inability to grow when transferred to Cd^{2+}-containing media, and subsequent studies implicated modified PC metabolism. Results from studies of *Arabidopsis* mutants will be discussed first as they have altered synthesis of PCs and demonstrate conclusively the importance of PCs for Cd^{2+} (and Hg^{2+}) tolerance in plants. Most of the detailed analysis of *S. pombe* mutants centers on the assembly and compartmentation of Cd^{2+}-PC complexes.

ANALYSIS OF *ARABIDOPSIS* MUTANTS

Howden and Cobbett (1992) identified mutants of *Arabidopsis* with reduced tolerance to Cd^{2+} by screening for seedlings with metal-induced inhibition of seedling root growth. A modification of this method has been developed in which seedlings are transferred between different media *en masse* on a nylon mesh (Murphy and Taiz 1995a). This allows for rapid screening of larger populations and can be adapted to virtually any compound that can inhibit root growth. Cd^{2+}-sensitive mutants also accumulate a brown pigment in their roots after exposure to Cd^{2+}, and this phenotype has been used to identify additional Cd^{2+}-sensitive mutants (Howden et al. 1995a, b).

In *Arabidopsis*, two loci have been identified that modulate Cd^{2+} tolerance;

CAD1 and *CAD2*. It is noted that we have followed the most commonly used nomenclature for *Arabidopsis* genes: the wild-type allele is in capitals and italicized, e.g., *CAD1*; mutant alleles are in lowercase italics, e.g. *cad1*; different alleles of the same gene are distinguished numerically, e.g., *cad1-1*, *cad1-2*; and the protein product of a gene is indicated by plain capitals, e.g., CAD1.) Four *cad1* mutants (Howden and Cobbett 1992; Howden et al. 1995a) and a single *cad2* mutant (Howden et al. 1995b) have been characterized. All of the mutants accumulated less Cd^{2+} than wild-type plants, suggesting that these plants could be defective in Cd^{2+} sequestration. The mutants also had reduced accumulation of Cd^{2+}-binding complexes, measured by gel filtration chromatography of soluble plant extracts. The inability of both *cad1* and *cad2* mutants to form these complexes resulted from reduced synthesis of PCs.

In the four *cad1* mutants, the level of PCs synthesized correlated with the degree of Cd^{2+} tolerance (Howden et al. 1995a). Plants homozygous for the most severe allele, *cad1-3*, produced no detectable PCs when exposed to Cd^{2+} and were the most sensitive to Cd^{2+}. *cad1* mutants were also deficient in PC synthase activity, having less than 1% of the activity found in either wild-type or *cad2* plants. In response to Cd^{2+}, *cad1* mutants accumulated GSH to more than twice the level in wild-type plants, indicating that GSH synthesis is induced by Cd^{2+} exposure even in the absence of PC synthesis. Exposure to Cd^{2+} increases the activity of γ-GluCys synthetase in maize (Ruegsegger and Brunold 1992). The increase in cellular thiol content from GSH accumulation in *cad1* mutants was even greater than that from PC synthesis in wild-type plants. In the absence of PC synthesis, however, this response did not result in Cd^{2+} tolerance, indicating that GSH alone cannot provide cellular detoxification of Cd^{2+}. Treatment of the PC-deficient *cad1-3* mutant with buthionine sulfoximine did not increase significantly the Cd^{2+} sensitivity of these mutants, providing further evidence that GSH itself is not directly involved in Cd^{2+} detoxification in *Arabidopsis*.

Although it is possible that CAD1 may have a regulatory role, all of the mutants that lack PC synthase activity map to the *CAD1* locus. It is therefore likely that *CAD1* encodes either PC synthase or an essential component of this enzyme. The lack of any detectable PCs in the *cad1-3* mutant argues against the presence of a second pathway of PC synthesis in plants, similar to that described in *S. pombe*. If γ-GluCys polymerization followed by addition of a carboxy terminal Gly is a functional mechanism of PC synthesis in plants, then it must either involve the *CAD1* gene product or be of no physiological importance. Efforts are underway to isolate this gene, which lies on chromosome 5, using positional cloning techniques (C. Cobbett, personal communication).

The second locus identified as being involved in Cd^{2+} tolerance is *CAD2*, and only one *cad2* mutant allele has been identified so far (Howden et al. 1995b). The phenotype of plants homozygous for *cad2* is similar to that described for *cad1*: sensitivity to Cd^{2+}, reduced accumulation of Cd^{2+}, synthesis of a brown pigment in roots, and a reduced level of Cd^{2+}-binding complexes, especially the HMW complexes. Synthesis of PCs in *cad2* plants was reduced to 10% of that in the wild type, although these mutants had a level of PC synthase activity similar to that of wild-type plants. Under normal growth conditions, the cellular GSH concentration in the *cad2* mutant was only 20% of that in wild type plants. Although some PCs were produced in response to Cd^{2+}, lack of GSH prevented synthesis of sufficient PCs for effective sequestration of Cd^{2+} and expression of Cd^{2+} tolerance. A number of mechanisms might result in a lower level of GSH, including reduced synthesis or altered metabolism in one of the many pathways that involve GSH. However, the *cad2* mutation also cosegregates with a restriction fragment length polymorphism detected by a cDNA probe encoding γ-GluCys synthetase (May and Leaver 1994; C. Cobbett, personal communication), suggesting that *CAD2* encodes the first enzyme of GSH synthesis. This conclusion is supported by the observation that *cad2* plants do not accumulate the intermediate γ-GluCys. Figure 15.2 illustrates the synthesis of PCs, mechanisms of Cd^{2+} tolerance, and the likely functions of CAD1 and CAD2.

The Cd^{2+}-sensitive mutants of *Arabidopsis* have been crucial for establishing the role of PCs in metal tolerance in plants. PCs are essential for tolerance to both Cd^{2+} and Hg^{2+}, the *cad1* and *cad2* mutants being sensitive to both of these metals. The lack of any effect of buthionine sulfoximine on tolerance of tobacco cells to copper ions and Zn^{2+} indicated that PCs were not required for tolerance to these metals (Reese and Wagner, 1987). The genetic evidence that PCs do not play a major role in tolerance to excessive concentrations of copper ions and Zn^{2+} is now conclusive. The tolerance of *cad1* and *cad2* mutants to these essential metals is only slightly less than that of the wild type. This raises one of the continuing unanswered questions about the function of these peptides: What role, if any, do PCs play in response to these other metals? PC synthesis can be induced, both *in vivo* and *in vitro*, by a wide variety of metals, including copper ions and Zn^{2+}, and PCs can bind some of these metals. Activation of PC synthase by so many metals may reflect a fairly nonselective interaction at a metal binding site on the enzyme, or as a result of interaction between metals and the substrate GSH. If so, synthesis of PCs in response to some metals may be gratuitous. However, there may be other explanations for this phenomenon that could provide insight into the complex relationships between different metal ions in plants and other organisms.

ANALYSIS OF *S. POMBE* MUTANTS

Mutoh and Hayashi (1988) isolated Cd^{2+}-sensitive mutants of *S. pombe* which were analyzed for GSH content and accumulation of LMW and HMW Cd^{2+}-PC complexes in response to Cd^{2+}. The mutants were grouped into four classes with the following characteristics: (1) deficient in GSH; (2) unable to synthesize PCs

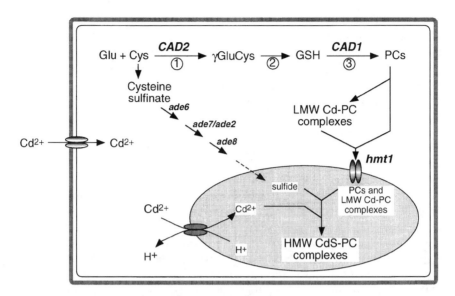

Figure 15.2 An integrated overview of the formation of Cd-PC complexes in plants and *S. pombe*. PC synthesis requires the synthesis of GSH, by γ-GluCys synthetase (1) and GSH synthetase (2), followed by PC synthase (3). PCs and LMW Cd-PC complexes are transported into the vacuole (shaded) by the HMT1 transporter. In the vacuole, PCs, Cd, and inorganic sulfide are utilized to assemble the HMW Cd-PC complexes required for expression of Cd tolerance. Transport of Cd into the vacuole can also occur via a Cd^{2+}/H^+ antiporter. Sulfide is generated by the action of several genes involved in adenine biosynthesis, probably using substrates derived from cysteine sulfinate, but there is no information on how sulfide enters the vacuole. Similarly, there is no information on how Cd enters the cell. Genes that are known to be essential for Cd tolerance in *Arabidopsis* (*CAD1* and *CAD2*) and in *S. pombe* (*ade2, ade6, ade7, ade8* and *hmt1*) are indicated.

but with normal levels of GSH; (3) deficient in HMW Cd^{2+}-PC complexes (Cd-BP1); (4) indistinguishable from wild type with respect to these parameters.

The GSH deficient mutants lacked either γ-GluCys synthetase (analogous to the *Arabidopsis cad2* mutant) or GSH synthetase. Using genetic complementation to restore Cd^{2+}-resistance to these mutants, genes encoding the enzymes of GSH synthesis have been cloned (Mutoh et al. 1991, 1995). The second class of mutants might be equivalent to the *Arabidopsis cad1* mutants and lack PC synthase. Complementation of these mutants would appear to be an appropriate method to clone the gene encoding this elusive enzyme, but, unfortunately, no further analysis of these mutants has been reported. Glaeser et al. (1991) also selected GSH-deficient mutants of *S. pombe* on the basis of resistance to N-methyl N′-nitro-N-nitrosoguanidine. This mutagen is activated intracellularly by thiol compounds such as GSH, and resistance might result from a deficiency in GSH. Because GSH-deficient mutants should be unable to produce PCs, sensitivity to Cd^{2+} was included as a second selection step. The GSH-deficient mutants had greatly reduced activity for either γ-GluCys synthetase or GSH synthetase. These mutants were also more sensitive to Ag^{2+}, Pb^{2+}, and Hg^{2+}, suggesting that PCs may be involved in reducing the cellular toxicity of these metals. However, tolerance to copper ions was not reduced in the GSH-deficient mutants, again indicating that PCs are not required for copper detoxification.

Detailed molecular genetic characterization of other *S. pombe* mutants that make PCs but do not form the HMW Cd^{2+}-PC complexes has been critical in establishing the importance of compartmentation and modification of Cd^{2+}-PC complexes to Cd^{2+} tolerance. Complementation of Cd^{2+}-sensitive mutants has identified a number of essential components of the PC-based Cd^{2+} tolerance mechanism. The first mutation characterized in this way was found to be in the *ade2* gene encoding an enzyme of adenine synthesis (Speiser et al. 1992b). Further analysis revealed that Cd^{2+} sensitivity required not only a lesion in *ade2* but also a mutation in either *ade6* or *ade7*. These genes are believed to influence Cd^{2+} tolerance by involvement in the synthesis of sulfide for incorporation into HMW complexes. ADE2 and ADE7 catalyze similar reactions, namely the addition of aspartate to an intermediate in adenine synthesis. Both enzymes can also utilize cysteine sulfinate, a sulfur-containing analog of aspartate and a likely metabolite of cysteine, as a substrate to generate 3-sulfinoacrylate (Juang et al. 1993). The sulfide incorporated into HMW Cd^{2+}-PC complexes may be derived from 3-sulfinoacrylate, although this has not yet been demonstrated. A single mutation in either *ade2* or *ade7* does not block synthesis of 3-sulfinoacrylate because of redundancy in the adenine biosynthesis pathway. Therefore, a double mutation (*ade2ade6* or *ade2ade7*) or a mutation in the gene for the next common step in the pathway, *ade8*, is required to give the Cd^{2+}-sensitive phenotype (Figure 15.2). This analysis has not only confirmed the importance of sulfide incorporation into HMW complexes as an essential component for Cd^{2+} tolerance but has indicated the likely source of sulfide to produce CdS. Although it is now clear that HMW Cd^{2+}-PC complexes containing CdS also accumulate in plants (Reese et al. 1992; Speiser et al. 1992a), there is as yet no data to indicate that the sulfide is generated in a manner similar to that proposed for *S. pombe*.

The second process required for Cd^{2+} tolerance highlighted by this mutational approach is the vacuolar accumulation of HMW Cd^{2+}-PC complexes. In *S. pombe* a mutation in the *hmt1* (heavy metal tolerance 1) locus reduces both Cd^{2+} tolerance and accumulation of HMW complexes (Ortiz et al. 1992). *hmt1* encodes a 90-kDa membrane transport protein that is localized to the vacuolar membrane. HMT1 contains a number of putative membrane-spanning domains, as well as two ATP binding cassette (ABC) domains. This structure is similar to that of other ABC-type transport proteins such as the mammalian multiple drug resistance proteins. However, most of these other transporters are involved in export of toxic compounds from the cell. In contrast, HMT1 mediates the intracellular compartmentation of Cd^{2+}-PC complexes within the vacuole (Figure 15.2). More detailed biochemical analysis of this protein has shown that HMT1 can transport both native PCs and LMW Cd^{2+}-PC complexes, but not the HMW complexes (Ortiz et al. 1995). This transport activity is ATP dependent and sensitive to vanadate, but does not require a pH gradient across the vacuolar membrane. A similar PC transport activity has been described recently in vacuole membrane preparations from oat (Salt and Rauser 1995). Both oats and *S. pombe* also show Cd^{2+}/H^+ antiport activity in the vacuole membrane, but this activity can be separated, both genetically and biochemically, from the PC transport activity (Salt and Wagner 1993; Ortiz et al. 1995).

The analysis of Cd^{2+}-sensitive mutants of *S. pombe* and *Arabidopsis* has revealed a number of critical steps in PC biosynthesis, localization, and modification that are essential for PC-mediated Cd^{2+} tolerance in both plants and yeasts. The biochemical studies of PC synthesis and vacuolar transport activities indicate that there are many similarities between plants and eukaryotic microbes in how PCs sequester and thereby reduce the toxicity of Cd^{2+}. There are likely to be many more processes that influence Cd^{2+} tolerance which have not yet been identified. Common to both plants and fungi might be genes involved in Cd^{2+} uptake and the assembly of Cd^{2+}-PC complexes. In plants, transport of Cd^{2+} from root to shoot may also be an important factor in metal tolerance at the whole plant level, and this is an area where our knowledge is at present limited.

2.5 PCs and Hypertolerance

In efforts to understand how plants adapt to toxic concentrations of specific metals, two approaches have been widely used. The first is to study plants growing in soils containing high concentrations of available metals, either from natural geologic conditions, or from anthropogenic pollution. Plants that grow in these environments have been studied as models for adaptation and selection. The second approach is to select plant cell cultures for growth in the presence of normally inhibitory concentrations of metal ions. Metal-resistant cell lines have been used to examine cellular mechanisms of metal detoxification. Those plants and cells can also be used to determine if PCs play a role in metal hypertolerance.

Ecotypes of *Silene vulgaris* with tolerance to Cd^{2+}, Zn^{2+} and copper ions were identified growing on mine tailings in Europe. When tested in hydroponic culture conditions, the metal-resistant ecotype produced less PCs than sensitive plants when exposed to the same external metal concentration (de Knecht et al. 1994). Other possible modifications that could contribute to tolerance, such as differences in PC chain length and sulfide incorporation into Cd^{2+}-PC complexes, were not significantly different between tolerant and unselected plants. These results indicate that PCs are not responsible for differential tolerance, at least in these eco-types.

A number of plant cell lines have been selected in culture for their ability to grow in the presence of normally toxic concentrations of Cd^{2+}. Where the concentration of PCs has been measured directly, Cd^{2+}-tolerant cells have been found to contain higher levels of PCs, and virtually all of the cellular Cd^{2+} is complexed with these PCs (Jackson et al. 1987; Steffens et al. 1986; Gupta and Goldsbrough 1991). Cd^{2+} resistance in *C. glabrata* was also accompanied by increased accumulation of HMW complexes (Mehra et al. 1994). Two independently selected Cd^{2+}-tolerant tomato cell lines have been shown to have increased activity of γ-GluCys synthetase (Chen and Goldsbrough, 1994; Steffens and Williams 1989). As the activity of this enzyme is likely to be the rate limiting step in GSH synthesis, increased activity of γ-GluCys synthetase could increase the supply of substrate for PC synthesis, especially when the pool of GSH is depleted upon exposure to Cd^{2+}. However, it remains to be proven that this adaptation in Cd^{2+}-tolerant tomato cells is responsible for Cd^{2+} tolerance. With the isolation of a cDNA encoding this enzyme from *Arabidopsis* (May and Leaver 1994), transgenic plants that overexpress this enzyme can be produced to test if this results in increased Cd^{2+} tolerance. In Cd^{2+}-resistant cells of *Datura innoxia*, the rate of formation of Cd^{2+}-PC complexes after exposure to Cd^{2+} was more rapid than in unselected cells (Delhaize et al. 1989). It is not known if this reflected a difference between cell lines specifically in the formation of HMW Cd^{2+}-PC complexes, but the initial rates of synthesis of (apo-) PCs did not differ between these lines. In *S. pombe*, overexpression of the HMT1 vacuolar transporter results in Cd^{2+} hypertolerance (Ortiz et al. 1992). These studies suggest a number of strategies to increase Cd^{2+} tolerance and Cd^{2+} accumulation, or to alter the tissue distribution of Cd^{2+}, in plants that are grown on Cd^{2+}-contaminated soils.

2.6 Other Functions for PCs

The observation that PCs are present in plants grown without deliberate addition of toxic metal ions suggests that PCs may be involved in other aspects of plant biochemistry. It has been postulated that PCs are involved in sulfur reduction and metabolism (Steffens 1990), and in metal ion homeostasis (Grill et al. 1988). GSH is also involved in a large number of stress response mechanisms, in the main providing protection against oxidative damage, and in detoxification of xenobiotics, including many herbicides (Rennenberg and Brunold 1994). Although the *cad1* and *cad2* mutants have no obvious phenotype in the absence of added Cd^{2+} or Hg^{2+}, studies of these plants under a more extensive range of environmental conditions may reveal other functions for PCs and GSH. The effects of specific mineral deficiencies as well as conditions that promote oxidative stress, such as high light intensity or low temperatures, might prove to be informative. There is also a need to examine the interactions between PCs and other metal ligand systems that plants possess, including organic acids such as malate and citrate, and MTs.

3. Plant Metallothionein-Like Genes

The identification of a large number of plant MT-like gene sequences in recent years has not been matched, in general, by either a corresponding increase in our understanding of the detailed mechanisms by which these genes are regulated or a definitive demonstration of the function(s) of the gene products. Plant genes encoding proteins with similarity to animal and microbial MTs were initially reported in 1990, and the literature was reviewed in 1993 by Robinson et al. This section therefore mainly concentrates on the relatively limited progress that has been made since 1993 in understanding both the regulation of expression and the function of plant MT-like genes.

3.1 Structure and Classification of Plant MT-Like Genes

The classification of plant MT-like genes is based upon the arrangement of Cys residues within their predicted products which are characteristic of metal binding motifs (Robinson et al. 1993). Figure 15.3 illustrates that predicted type 1 and type 2 MT-like proteins contain Cys-rich amino- and carboxy-terminal domains separated by a "spacer" region devoid of Cys residues. A conserved "GV" motif is present in this spacer within a region predicted to form a β-strand (Robinson et al. 1997). The functional significance of this conserved region is at present unknown. The E_c-type MTs (see Section 3.2) differ from the type 1 and type 2 categories in that they have Cys residues distributed throughout the length of the polypeptide.

The original classification was based upon a relatively small number of gene products, nine in total. As shown in Figure 15.3 there are now many more such genes. This increase in the number of examples of plant MT-like genes is due to two main factors. First, they have been commonly identified in differential screening strategies (Table 15.1). Second, the database of expressed sequence tags (ESTs) produced from random cDNA sequencing programs also contains a large number of sequences encoding MT-like gene products, examples of which are also illustrated in Figure 15.3.

Variants of note to the original classification include the products of the type 1 MT-like genes of the Brassicaceae (*Arabidopsis* and *Brassica napus*). Here the conserved "spacer" region between the amino- and carboxy-terminal Cys-rich domains is lacking, and there is an additional Cys residue within the carboxy-terminal domain. Whether or not these features are restricted to this family

```
TYPE   1
Pea           MSGCGCGSS  CNCGDSCKCN  KRSSGLSYSE  METTE...TV  ILGVGPAKIQ  FEGAEMSAA.  .SEDG.GCKC  GDNCTCDPCN  CK
Clover        MSGCNCGSS  CNCGDSCKCN  KRSSGLNYVE  AETTE...TV  ILGVGPAKIQ  FEDAEMGVA.  .AEDS.GCKC  GSSCTCDPCN  CK
Broad bean    MSGCNCGSS  CNCGDSCKCN  KRSSGLSYSE  VETKE...TV  ILGVGPAKIQ  FEGAEMSFA.  .SKEG.GCKC  GDNCTCDPCN  CK
Mimulus       MSGSCGSG   CKCGDNCSC.  SMYPD.....  METNTTV.TM  IEGVAPLKMY  SEGSEKSFG.  .AEGGNGCKC  GSNCKCDPCN  C
Maize         MSCSCGSS   CGGSSCKCG   KKYPDL.EET  STAAQ..PTV  VLGVAPEKKA  APFVEAAAE   SGEAAHGCSC  GSGCKCDPCN  C
Barley        MSCSCGSS   CGCGSNCNCG  KMYPDL.EEK  SGATMQVTVI  VLGVGSAKV.  ...QFEEAAE  FGEAAHGCSC  GANCKCNPCN  C
Wheat         MSCNCGSG   CSCGSDCKCG  KMYPDLTEQG  SAAAQVAAVV  VLGVAPENKA  G..QFEVAA.  .GQSGEGCSC  GDNCKCNPCN  C
Rice          MSCSCGSS   CSCGSNCSCG  KKYPDL.EEK  SSSTK..ATV  VLGVAPEKKA  Q..QFEAAAE  SGETAHGCSC  GSSCRCNPCN  C
Consensus     C CGS      C  C                               GV                      GC C G      C C PCN     C

A.thaliana a  MADSNCGCGS  SCKCGDSCSC  EKNYNKECDN  CSCGSNCSCG  SNCNC
A.thaliana c  MAGSNCGCGS  GCKCGDSCSC  EKNYNTECDS  CSCGSNCSCG  DSCSC
B.napus       MADSNCGCGS  SCKCGDSCSC  EKNYNKECDN  CSCGSNCSCG  SNCNC
Consensus     MA SNCGCGS  CKCGDSCSC   EKNYN.ECD   CSCGSNCSCG  C C

TYPE   2
Clover        MSCCGGNCGC  GSACKCGNGC  GGCKMNADLS  YT.ESTTTET  IVMGVGSAKA  QFEGAEMG..  ..AESGGCK   CGANCTCDPC  TCK
Broad bean    MSCCGGNCGC  GSSCKCGSGC  GGCKMYADLS  YT.ESTTSET  LIMGVGSEKA  QYESAEMG..  ..AENDGCK   CGANCTCNPC  TCK
Soybean       MSCCGGNCGC  GSSCKCGNGC  GGCKMYPDLS  YT.ESTTTET  LVMGVAPVKA  QYESAEMG..  ..AENDGCK   CGANCTCNPC  TCK
Castor bean   MSCCGGNCGC  GSGCKCGNGC  GGCKMYPDMS  FS.EKTTTET  LVLGVGAEKA  HFEGGEMGVV  ..GAEEGGCK  CGDNCTCNPC  TCK
Kiwifruit     MSCCGGKCGC  GSSCSCGSGC  GGCGMYPDLS  YS.EMTTTET  LIVGVAPQKT  YFEGSEMGV.  ..AAENGCK   CGSDCKCDPC  TCK
A.thaliana a  MSCCGGNCGC  GSGCKCGNGC  GGCKMYPDLG  FSGETTTTET  FVLGVAPAMK  NQYEASGESN  ..NAESDACK  CGSDCKCDPC  TCK
A.thaliana b  MSCCGGSGC   GSACKCGNGC  GGCKRYPDL.  ...ENTATET  LVLGVAPAMN  SQYEASGETF  ..VAENDACK  CGSDCKCNPC  TCK
B.campestris  MSCCGGNCGC  GSGCKCGNGC  GGCKMYPDLG  FSGESTTTET  FVFGVAPAMK  NQYEASGEGV  ..AENDRCK   CGSDCKCDPC  TCK
Coffee        MSCCGGNCGC  GAGCKCSGGC  GGCKMYPELS  YTENTAAETL  .ILGVAPPKT  TYLEGAGEEA  ..AAENGGCK  CGPDCKCNPC  NCK
Elder         MSCCGGKCGC  GSGCSCCTGC  NCGGMYPDI.  .ENTTAATII  ..EGVAPTKM  YAEGSEMSFM  ..AEGHACK   CGPNCTCDPC  NC
Tomato        MSCCGGSCGC  GSGCKCGNGC  GGCGMYPDME  KSATFSIVE.  ...GVAPVHN  YGRVEEKAAG  .....EGCK   CGSNCTCDPC  NC
Tomato        MSCCGGNCGC  GSSCKCDNGC  GGCGMYPDLE  STTTFTIIE.  ...GVAPMKN  YGVAEKATEG  ...GNGCK    CGSNCTCDPC  NC
Tomato        MSCCGGNCGC  GSSCKMYDMS  GGCKMYPDMS  YTESSTTTET  LVLGVAPEKT  SFGAMEMGES  ..PVAENGCK  CGSDCKCNPC  TCSK
Rice*         MSCCGGNCGC  GSGCQQGSGC  GGCKMYPEMA  ..EEVTTTQT  VIMGVAPSKG  HAEGLEAGAA  AXAGAENGCK  CGDNCTCNPC  NCGK
Consensus     MSCCG CGC   G  C GC GC                          GV                      CK          CG C C PC   C
```

```
Ec TYPE
Wheat        MGCDDKCGCA VPCPGGTGCR CTSARSGAAA G.EHTTCGCG EHCGCNPCAC GREGTPSGRA NRRANCSCG AACNCASCGS ATA
Maize*       MGCDDKCGCA VPCPGGKDCR CTSGSGG...  QREHTTCGCG EHCECSPCTC GRATMPSGRE NRRANCSCG ASCNCASCAS A
A.thaliana*  AGCNDSCGCP VPCPGGNSCR CRMREASAGE Q.GHMVCPCG EHCGCNPCNC PKTQTQTS.. ..DKGCTCG EGCTCASCDT
A.thaliana*  ASCNDRCGCP SPCPGGESCR CKMMSEASGG DQEHNTCPCG EHCGCNPCNC PKTQTQTS.. ..AKGCTCG EGCTCATCAA
consensus    C D CGC     PCPGG CR  C                      H TC CG EHCGC PC C                  C.CG   C CA C


Other metallothionein-like gene products

Tomato       MSGCGGSCNC GSSCSGKGG  GC.NMYPDLE KSTTLTIIE. ....GVAPMN NKGMVEGSIE KATEGN...  GCKCGGSCKC DPCNCCSASTIWT
Rice*        MSCGGSCNC  G.SCGC..GG GCGKMYPDLA EKINTTITTA TTVLGVAPEK GHFEVMVG.. KAGESGEAAH GSCGGSSCKC NPCNC
Consensus    CCGSCNC    G.SC C  GG GC  MYPDLA  TI         GVAP                  KA  G     GC CGSSCKC  PCNC

Kiwifruit    MSDKGCNCDC ADSSQCVKKG NS..IDIVET DKSYIEDVVM GVPAAESGG. .KCKCGTSCP CVNCTCD
Rice*        MSDKGCNCDC ADKSQCVKKG TSYGVVIVEA EKSHFEEV.. .AAGXENGG. .CKCGTSCS CTDKCGK
A.thaliana*  MSSNCGSCDC ADKTQCVKKG TSYTFDIVET QESYKEAMM  DVGAEENNAN CKCKCGSSCS CVNCTCCPN
Consensus    MSDKGCNCDC AD SQCVKKG  S         IVE        E       E   CKC   SC   C C C

Barley       MSCCGGKCGC GAGCQCGTGC GGCKMFPDVE ATAGAAAMVM PTASHKGSSG GFEMAGGETG GCDCATCKCG TRAAAPAAAA
             SEPAPGRPAG RGEHEDERRT SNTNQAPSPS PSYHQ
```

Figure 15.3 Comparison of the amino acid sequences of type 1, 2, E$_c$ and other related plant metallothionein-like gene products. The derived amino acid sequences of MT-like gene products are shown, conserved Cys residues and a "GV" motif are in boldcase, and invariant residues are also indicated in each consensus sequence. The type 1 sequences from *Arabidopsis thaliana* and *Brassica napus* do not contain the central 'spacer' region and an extra Cys residue is present in the carboxy-terminal domain. Where further studies have been performed to examine expression patterns refer to Table 15.1 for the relevant reference. Unpublished sequences present in the EMBL database were retrieved using text-based searches. An * next to a database accession indicates that the sequence is a translation of an expressed sequence tag (EST) derived from the random cDNA sequencing programs. Database accessions are as follows. **Type 1:** pea Z23097; clover Z26493; broad bean X91078; *Mimulus guttatus* X51993; maize S57628; barley X58540; wheat L11879; rice U18404; *Arabidopsis thaliana a* U11253;; *Arabidopsis thaliana c* U11255 *Brassica napus* U20236. **Type 2:** clover Z26492; broad bean X77254; soybean not available; castor bean L02306; kiwifruit L27813; *Arabidopsis thaliana a* D11394; *Arabidopsis thaliana b* U11256; *Brassica campestris* L31940; coffee U11423; elder X83439; tomato Z68309, Z68310, Z68138; rice* D15602. **E$_c$ type:** wheat X68288-X68291; maize U10696; *Arabidopsis thaliana* Z32602, Z27049. **Other metallothionein-like gene products:** tomato Z68185; rice* D25070; kiwifruit L27811; rice* D21979; *Arabidopsis thaliana* T43340; barley S53707.

Table 15.1 Expression patterns of plant MT-like genes.

Species	MT-like genes characterized	Patterns of Expression	References
Pea	Type 1, cDNA and genomic.	*PsMT* genes comprise multigene family. Transcript abundance increases in iron-deficient roots. Under iron-sufficient conditions transcripts increase in response to added copper. Significantly lower levels of expression observed in leaves and developing cotyledons.	Evans et al. (1990) Robinson et al. (1992, 1993)
Mimulus guttatus	Type 1, cDNA.	Multigene family. Expression greater in roots than leaves, transcript levels repressed by copper shock (and zinc and cadmium treatments).	de Miranda et al. (1990)
Arabidopsis thaliana	Type 1 and 2, cDNA, and genomic.	Multigene family. *MT1* transcripts predominate in roots; *MT2* transcripts expressed at higher levels in leaves. *MT1a*, greater expression in roots, induced by copper in leaves; *MT1b*, probable pseudogene; *MT1c*, expressed in roots and leaves; *MT2a* induced by copper in seedlings. *MT2* transcripts are also induced in seedlings treated with Ag^+, Cd^{2+}, Zn^{2+}, Ni^{2+}, or heat shock but not salicylic acid and are elevated in the roots of the *Arabidopsis* copper-sensitive mutant *cup1-1*.	Zhou and Goldsbrough (1994, 1995) Murphy and Taiz (1995b) van Vliet et al. (1995)
	E_c type, cDNA.	Expressed sequence tags encoding E_c products sequenced from cDNA libraries constructed from seed material.	
Brassica napus	Type 1, cDNA.	Expressed during leaf senescence and in flowers (may also be senescence associated).	Buchanan-Wollaston (1994)
Barley	Type 1, cDNA.	Induced by iron deficiency in roots, transcripts not detected in leaves.	Okumura et al. (1991, 1992)
Wheat	Type 1, cDNA.	Aluminum induced in roots, not by heat shock, expressed in leaves. Expression in roots also induced by other metals, including copper, low calcium treatment, and wounding of roots (and leaves).	Snowden and Gardner (1993) Snowden et al. (1995)

Continued

Table 15.1 Continued

Species	MT-like genes characterized	Patterns of Expression	References
Wheat (continued)	E_c type, cDNA.	Single-copy gene within haploid genome. Highly expressed during early stages of embyogenesis. Expression increases in response to abscisic acid but not Zn^{2+}. Promoter region contains putative *cis*-acting abscisic acid-responsive elements.	Kawashima et al. (1992)
Maize	Type 1, cDNA, and genomic. E_c type, cDNA.	Highly expressed in roots, less in leaves, pith, and kernel. In excised root tips, expression is induced by glucose starvation and heavy metal stress but not heat shock. Probable single-copy gene in haploid genome. Transcripts are embryo specific and their abundance is maximal in mid-maturation phase. Abscisic acid-inducible requiring the product of the *viviparous-1* gene (regulates expression of genes with *cis*-acting abscisic acid-response elements).	de Framond (1991), Chevalier et al. (1995) White and Rivlin (1995)
Rice	Type 1, cDNA, and genomic.	In cell cultures, expression induced by copper ions, dimethylsulfoxide, sucrose starvation, and heat shock but is repressed by abscisic acid. Expression greater in roots than sheaths or leaves but increase during leaf senescence	Hsieh et al. (1995)
Soybean	Type 2, cDNA.	Greater expression in leaves compared with roots, not induced by copper treatment roots.	Kawashima et al. (1991)
Broad bean	Type 2, cDNA.	Expressed in trichome-containing leaf tissue, reduced root expression, also expressed in flowers and stems. Not induced by metal, UV light, or salicylic acid treatments.	Foley and Singh (1994)
Kiwifruit	Type 2 and atypical, cDNA.	Type 2 sequence highly expressed 8–10 days after anthesis, later during fruit ripening and in roots. Gene encoding atypical product expressed during late fruit development and subsequent ripening.	Ledger and Gardner (1994)
Elder	Type 2, cDNA.	Multigene family. Expressed in leaflet abscission zones, ethylene increases transcript abundance. Cognate transcripts expressed in senescent leaves.	Coupe et al. (1995)

remains to be established as similar gene products have not been identified in species outside the Brassicaceae. Conversely, it remains to be determined if type 1 gene products containing the central "spacer" region are present in *Arabidopsis* or related species. In addition, there are now several examples of plant MT-like sequences in which the Cys motifs within the predicted products do not conform to either the type 1 or type 2 categories. For example, the rice sequence (accession D25070, Figure 15.3) has atypical Cys motifs within its amino-terminal domain but appears to be similar to the type 1 MT-like genes within the carboxy-terminal region. Given the overall success with which most of the new genes are accommodated within this classification, we have not adopted separate categories for these new variants.

The presence of representatives of each MT-like gene category within a single species (*Arabidopsis*) or within closely related species (wheat, barley, and rice) illustrates that plants contain a complement of different MT-like gene types, and that each type may also be encoded by a multigene family (see Figure 15.3 and Table 15.1). This is significant, as it suggests that the predicted products of plant MT-like genes could have distinct but possibly overlapping functions, for example, differing affinities for metal ions.

3.2   The Zn^{2+}-Binding E_c Protein of Wheat

The wheat E_c protein is, to date, the only plant protein that can be unequivocally designated an MT as the isolated gene product is associated with Zn^{2+} (Hofmann et al. 1984; Lane et al. 1987). Transcripts encoding the E_c protein are present at high levels during embryogenesis and expression appears to be embryo-specific (Kawashima et al. 1992). A cDNA encoding the E_c-type MT has also been cloned in maize (White and Rivin 1995) but E_c-type genes are not restricted to monocots as *Arabidopsis* ESTs encoding E_c homologs have been identified in cDNA libraries constructed from seed tissue (Figure 15.3).

One possible function for the E_c protein is in Zn^{2+} homeostasis and, by analogy to animal MTs, it could play a role in either the donation or removal of Zn^{2+} from metalloproteins such as Zn^{2+}-requiring transcription factors (Zeng et al. 1991a, b; Cano-Gauci and Sarkar 1996). The E_c protein binds only about 5% of the total Zn^{2+} in mature embryos, suggesting that it is not the main Zn^{2+} store. In germinating embryos, E_c gene expression is not increased by the addition of Zn^{2+} but does respond to the plant hormone abscisic acid (ABA), consistent with the presence of ABA-responsive *cis*-acting regulatory elements within the promoter region (Kawashima et al. 1992). In maize, E_c gene expression is also modulated by ABA and, furthermore, this requires the product of the *viviparous-1* gene (which regulates the expression of promoters containing ABA-responsive elements). In *viviparous-1* mutants, E_c genes are not expressed (White and Rivin 1995). These observations again support a role for E_c in Zn^{2+} homeostasis as opposed to detoxification. Whether the corresponding E_c genes of *Arabidopsis* are also subject to control by ABA remains to be determined.

3.3 Plant MT-like Genes Expressed in Vegetative Tissues

Identification of the first plant MT-like genes expressed in vegetative tissues came from studies designed to isolate genes expressed either in specific organs or in response to fluctuations in environmental conditions. In the case of the type 1 MT-like genes of pea, *PsMT*, cDNAs encoding *PsMT* transcripts were isolated in the process of identifying genes that exhibited either root-specific or root-enhanced expression, and a genomic clone, encoding *PsMT*$_A$, was subsequently characterized (Evans et al. 1990). By contrast, cDNAs encoding the type 1 MT-like proteins of *Mimulus guttatus* were isolated from a root cDNA library on the basis of reduced RNA expression in response to copper shock (de Miranda et al. 1990). Subsequent to these reports MT-like genes have been identified in a wide range of plant species. Table 15.1 summarizes the available data on the expression patterns of plant MT-like genes.

3.4 Evidence That Plant MT-Like Gene Products May Be Involved in Trace Metal Ion Metabolism

Although the wheat E_c protein had been purified from wheat embryos (Lane et al. 1987), efforts to identify MT proteins in vegetative tissues of plants had been unsuccessful. However, evidence has been obtained recently that at least some of the *Arabidopsis* MT genes are indeed translated into proteins (Murphy et al. 1997). In efforts to provide indirect evidence that the products of these genes do play a role in trace metal ion metabolism, plant MT-like coding sequences have been expressed in heterologous hosts. These studies have utilized the products of the pea MT-like gene, *PsMT*$_A$, and the *Arabidopsis* MT-like genes, *MT1* and *MT2*.

EXPRESSION OF PSMT$_A$ IN *E. COLI* AND *ARABIDOPSIS*:
METAL BINDING PROPERTIES

When compared with cells expressing equivalent control constructs, expression of the *PsMT*$_A$ coding sequence in *E. coli* was found to result in increased accumulation of Cd^{2+} and copper ions in cells grown in media supplemented with either of the respective metals (Kille et al. 1991; Evans et al. 1992). The recombinant protein associated with Cd^{2+} with an estimated stoichiometry (Cd^{2+}/PsMT$_A$) of between 5.6 and 6.1 g-atoms of Cd^{2+} per mole of protein (Kille et al. 1991). The pH of half-dissociation of Zn^{2+}, Cd^{2+} and copper ions suggested that the recombinant protein had slightly lower affinities for Cd^{2+} and Zn^{2+} compared to equine renal MT, but comparable affinities for copper ions (Tommey et al. 1991). It is possible that the products of other types of plant MT-like genes may have differing metal binding specificities when compared with $PsMT_A$, because of differences in the spacing and number of Cys residues (see the following section).

A construct containing the $PsMT_A$ coding sequence under the control of the constitutive cauliflower mosaic virus 35S promoter was introduced into *Arabidopsis* (Evans et al. 1992). Approximately 75% of the segregating progeny, derived from a selfed parent expressing this construct, accumulated more copper than the highest-accumulating seedling from the progeny of a control plant. This is consistent with the $PsMT_A$ protein binding copper ions *in planta*.

EXPRESSION OF *ARABIDOPSIS* MT-LIKE GENES IN MICROBIAL HOSTS

Complementary approaches to determining the contribution of plant MT-like gene products to trace metal metabolism/detoxification have also been undertaken using metal-sensitive microbial mutants unable to express their endogenous MT genes.

Using a yeast (*S. cerevisiae*) strain in which the endogenous MT gene, *CUP1*, had been deleted, Zhou and Goldsbrough (1994) demonstrated that constitutive expression of either the *Arabidopsis MT1a* or *MT2a* coding sequences in this host increased metal tolerance. Transformants expressing either of these genes were able to grow on media containing concentrations of Cd^{2+} or copper ions that inhibited growth of the *cup1* mutant. Over the range of metal levels examined, cells expressing either *MT1a* or *MT2a* were viable on media containing up to 3 mM $CuSO_4$. In contrast, the degree of restored tolerance to Cd^{2+} differed between *MT1a* or *MT2a*; cells expressing *MT1a* were viable only on media containing up to 10 μM $CdSO_4$, whereas those expressing *MT2a* grew on media containing 100 μM $CdSO_4$. Again, this difference may reflect different metal affinities between the two types of MT-like gene products due to differences in Cys spacing and arrangement, or, alternatively, the presence of the spacer region in the MT2a protein may effect Cd^{2+} binding.

The *Arabidopsis MT2a* coding sequence has also been expressed in a Zn^{2+}-sensitive cyanobacterial mutant (Robinson et al. 1996). Deletion of the *smtA* MT gene of *Synechococcus PCC 7942* results in hypersensitivity to Zn^{2+} but not to copper ions (refer to Robinson et al. Chapter 15, this volume). Expression of *MT2a* in this mutant, under the control of the metal-responsive *smtA* promoter region, results in the partial restoration of tolerance to Zn^{2+}.

The inability of *MT2a* expression to restore tolerance to Zn^{2+} to the level observed in wild type cells expressing *smtA* may be due to a difference in the relative affinities of MT2a and SmtA for Zn^{2+}. SmtA has the highest affinity for Zn^{2+} of any known MT, with a pH of half dissociation of 4.1 determined for the recombinant protein (Shi et al. 1992). The recombinant MT2a protein binds Zn^{2+}

with an affinity comparable to that of recombinant $PsMT_A$ but less than that of SmtA. Mean values of 5.05 and 5.30 for MT2a and $PsMT_A$, respectively, have been obtained (Robinson et al. 1996). Whether the slight (0.25 pH unit) difference in the affinity for Zn^{2+} between MT2a (type 2) and $PsMT_A$ (type 1) reflects a difference in their respective affinities for trace metals *in vivo* remains to be established.

3.5 Metals and Plant MT-Like Gene Expression

The expression of some plant MT-like genes did not increase in response to elevated levels of trace metals (de Miranda et al. 1990; Kawashima et al. 1991; Foley and Singh 1994). However, as discussed in the following sections, it is possible that plant growth conditions and/or the composition of nutrient media may have masked any effect of added metal ions on MT-like gene expression.

REGULATION OF EXPRESSION OF THE PEA MT-LIKE GENES, *PSMT*

PsMT transcripts are present at high levels in the roots of hydroponically grown peas but are much reduced in leaves and developing cotyledons (Evans et al. 1990; Robinson et al. 1992). The expression in roots increases over a 15-day period starting from germination. Initially it appeared that expression in roots was constitutive as the hydroponic media did not contain supraoptimal levels of trace metal ions. However, these experiments were performed with no attempt to maintain the plants under conditions of iron sufficiency. Subsequent to the identification of the *PsMT* genes, a cDNA, *ids1*, encoding a type 1 MT-like protein from barley was reported to be expressed in the roots of plants subjected to conditions of iron deficiency but not in roots supplemented with highly available sources of iron (Okumura et al. 1991, 1992). Similar observations have subsequently been reported for *PsMT* transcripts in pea (Robinson et al. 1993).

Under conditions of iron deficiency there is increased accumulation of copper in the roots and shoots of pea plants (Robinson et al. 1993; Welch et al. 1993). The finding that iron and copper status affects the levels of expression of *PsMT* genes in roots may explain why many other type 1 MT-like sequences have been found to be expressed at high levels in roots. The stringent growth conditions required to maintain an adequate supply of iron were not employed in these studies, and the resulting plant material may have been iron deficient, with a high level of MT-like gene expression that was refractory to further induction by other metals.

SEPARATE PROMOTER REGIONS MEDIATE THE EXPRESSION OF THE *PSMT_A*
GENE IN ROOTS AND AERIAL TISSUES

As a first step in defining the *cis*-acting regulatory sequences of a plant MT-like gene, the 5′ flanking sequences of the $PsMT_A$ gene from the pea were fused to the reporter gene encoding β-glucuronidase (*GUS*) to create a translational fusion including the initiation codon and first six codons of the *PsMT*$_A$ coding sequence. A series of promoter deletions have also been made and all of these constructs introduced into *Arabidopsis* (Figure 15.4). GUS activity was observed in roots, in the leaves of seedlings grown in culture and in a variety of senescent aerial tissues (including leaves, cotyledons, floral organs, and siliques). All of the deletion constructs exhibited similar patterns of *GUS* expression with the exception of the shortest construct (deletion 3), which retained expression in the

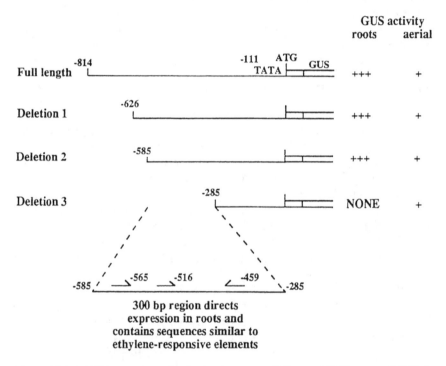

Figure 15.4 A 300-base-pair region in the promoter of the pea MT-like gene *PsMT*$_A$ is required for expression in roots. Constructs containing the promoter region of the *PsMT*$_A$ gene and deletions derived from it were fused to the reporter gene *GUS* and introduced into *Arabidopsis*. Several independent transgenic lines containing each construct were analyzed for GUS activity by histochemical staining of seedings from T$_1$ generations grown on tissue culture medium. GUS activity in roots was detected in seedlings containing the full-length construct or deletion constructs 1 and 2 but was dramatically reduced in lines containing the deletion 3 construct. In aerial tissues GUS activity was observed for all the constructs. The positions of a putative TATA box and sequences with similarity to ethylene-responsive elements are indicated. Nucleotide numbering is taken from the first base of the methionine initiation codon.

aerial tissues of seedlings grown in tissue culture but, most significantly, was deficient in expression within the roots and within senescent aerial tissues of soil-grown plants (Figure 15.4) (Fordham-Skelton et al. 1997). The 300-base-pair region between deletions 2 and 3 contains sequences similar to elements that bind nuclear proteins found in the promoter regions of a subset of ethylene-responsive genes (Itzhaki et al. 1994). At present it is unclear whether this region of the promoter is also responsible for the regulation of *PsMT* expression in response to changes in iron and copper status. It is tempting to speculate that the expression of $PsMT_A$ in roots and senescent aerial tissues, but not in the aerial parts of seedlings grown in tissue culture, may be ethylene mediated. There is evidence that root surface iron-chelate reductase activity is controlled by ethylene in response to iron deficiency (Romera et al. 1996a, b). This would explain the link between the expression of MT-like genes in roots and the activity of iron-chelate reductases, as has previously been observed (Robinson et al. 1993). It is formally possible that ethylene, or at least some facet of the ethylene-signaling pathway, may have a general role in signaling associated with the perception of metals by plant roots.

THE MT-LIKE GENES OF *ARABIDOPSIS*

Arabidopsis is the only plant species in which type 1 and type 2 MT-like genes and the E_c encoding genes have been identified (Zhou and Goldsbrough 1995). The *MT1a* and *MT1c* genes are linked as an inverted duplication, and the other genes are present at different positions in the genome. The type 1 and type 2 gene families are each comprised of two expressed members. A third gene, *MT1b*, is most probably a pseudogene as it encodes a predicted protein that, when compared with other type 1 gene products, contains an atypical amino-terminal region and is truncated within the carboxy-terminal region. Additionally, transcripts were not detectable using reverse transcriptase PCR. ESTs indicate the presence of at least one more expressed MT-like gene in the *Arabidopsis* genome (see Figure 15.3).

RNA gel blot analyses indicate that individual members of the family are differentially expressed (Table 15.1). In general, expression of the *MT1* genes predominates in roots and etiolated seedlings whereas the *MT2* genes are expressed at higher levels in leaves. In leaves exposed to copper ions, *MT1* transcript abundance increases and there is also, to a lesser degree, induction by either Zn^{2+} or Cd^{2+} (Zhou and Goldsbrough 1994). *MT2a* and *MT2b* genes are differentially responsive to copper ions. In seedlings exposed to copper ions, *MT2a* transcript levels increase, whereas those of *MT2b* are elevated even in the absence of added copper (Zhou and Goldsbrough 1995). Murphy and Taiz (1995b) demonstrated that *MT2* transcript abundance also increased in seedlings treated with Ag^+, Cd^{2+}, Zn^{2+}, Ni^{2+}, or heat shock but were unaffected by salicylic acid.

Other evidence for a linkage between metal homeostasis/detoxification and *Arabidopsis* MT-like gene expression has come from the studies of van Vliet et al. (1995). Expression of the *MT2a* gene is elevated in roots of the copper-sensitive *Arabidopsis* mutant, *cup1-1*, and induction of *MT2a* transcripts in roots by copper ions is seen at lower concentrations than in wild-type plants. The *cup1-1* mutant is not deficient in PC biosynthesis and is therefore distinct from the *cad1* and *cad2* mutants described above. Increased levels of copper ions and Cd^{2+} are found in the *cup1-1* mutant when compared with wild-type plants and this correlates with increased *MT2a* expression. The nature of the *CUP1* gene product is not yet known.

METALLOREGULATION OF OTHER MT-LIKE GENES

In addition to pea and *Arabidopsis*, there are some reports of the metalloregulation of MT-like genes in other species. In rice suspension cultures, transcripts encoding a type 1 MT-like product, *rgMT*, increase after treatment with copper ions but not with other metals (Hsieh et al. 1995). Aluminum stress increases the abundance of transcripts encoding a type 1 MT-like product, *wali1*, in wheat roots (Snowden and Gardner 1993). Elevated levels of copper, cadmium, and other metals (to a lesser extent) and depleted calcium levels also caused an increase in transcript abundance in roots, but in all of these treatments root growth was inhibited when compared with control material (Snowden et al. 1995). Ledger and Gardner (1994) noted that expression of an MT-like gene during the early development of kiwifruit coincided with maximal accumulation of copper and other metals. However, expression subsequently declined and then increased in the later stages of development, but at these stages there was not a correlation with metal content.

It is noted that both the *rgMT* and *wali1* transcripts also accumulate in response to other treatments, as discussed below. In cell cultures *rgMT* transcript abundance increased in response to heat shock, exposure to dimethylsulfoxide, and sucrose starvation but was repressed by ABA (Hsieh et al. 1995). Whether the *rgMT* gene is responsive in a similar manner *in planta* remains to be determined. Wounding of roots and leaves increased *wali1* transcript levels, but heat shock had no effect (Snowden et al. 1995).

3.6 Developmental and Environmental Regulation of MT-Like Genes

In addition to elevated levels of trace metals, MT gene expression in animals is also responsive to a wide range of factors including hormones, growth factors, and stresses such as starvation and ultraviolet irradiation [see the table presented by Kägi (1991)]. It is now clear that transcripts encoding MT-like gene products also change greatly in plant tissues in response to a diverse range of developmental cues and environmental factors including senescence, abscission, ripening, heat shock, and sugar deficiency (Table 15.1).

It is formally possible that (at least some) changes in expression mediated by nonmetal stimuli are due to pleiotropic effects on metal ion concentrations. For example, studies on the yeast *S. cerevisiae* have revealed interactions between oxygen availability, intracellular copper levels and MT gene expression. The two copper MT genes *CUP1* and *CRS5* have high levels of basal activity in yeast cultures grown anaerobically but are repressed in cultures exposed to air. Strain and Culotta (1996) demonstrated that yeast cells under anaerobic conditions accumulate more copper than those grown under aerobic conditions, and the intracellular copper is rapidly released when cultures are exposed to air. Thus the effect of oxygen on MT expression is due to coincident changes in copper levels which directly control MT expression via activation of the copper-containing ACE1 transcription factor. Additionally, there is some evidence to suggest that changes in sugar levels, in combination with oxygen availability, also affect copper levels and *CUP1* gene expression (Galiazzo et al. 1991).

By analogy to yeast, the induction of plant MT-like transcripts by sugar deficiency (Chevalier et al. 1995; Hsieh et al. 1995) may be due to changes in the levels of intracellular metal ions which occurs as a consequence of these treatments. Such coincident changes in metal levels could also be responsible for the increased expression of MT-like genes during the senescence of plant tissues (Buchanan-Wollaston 1994; Coupe et al. 1995; Hsieh et al. 1995) as metals are released from ligands for subsequent mobilization to other parts of the plant. An alternative hypothesis is that the products of these genes play some other role, such as acting as free radical scavengers (Buchanan-Wollaston 1994), as senescence results in the enhanced production of reactive oxygen species (Thompson et al. 1987).

3.7 Metalloregulation of Plant MT-Like Genes; Future Prospects

How expression of some plant MT-like genes is altered in response to metals remains an open question. The characterization of genomic clones encoding MT-like genes has naturally led to a search for known MREs within their promoter regions. However, sequences exactly matching the consensus sequence for either the core MREs of animal MT genes or the upstream activation sequences (UASs) of *CUP1* are not present within the MT-like genes of pea (Evans et al. 1990), maize (de Framond 1991), or *Arabidopsis* (Zhou and Goldsbrough 1995). In addition, experimental evidence argues against plants having a regulatory mechanism similar to that of *CUP1*. The UAS of *CUP1* has been placed upstream of a reporter gene and introduced into tobacco (Mett et al. 1993). Transcription of this gene was dependent upon the introduction of ACE1 which binds to this UASs. This demonstrates that tobacco lacks proteins capable of stimulating transcription from a fungal UASs. MREs distinct from those of animals and fungi have yet to be described within plant MT-like genes.

The availability of a number of MT-like gene promoter regions from a single plant (*Arabidopsis*) should facilitate a systematic analysis of the *cis*-acting sequences and *trans*-acting factors responsible for their different patterns of expression and, at least for some of the genes, metal-induced gene expression. *Arabidopsis* provides a relatively tractable system in which to isolate mutants in trace metal homeostasis/detoxification (for example, the cadmium- and copper-sensitive mutants described here). An array of such mutants will prove to be valuable in future research directed toward a greater understanding of the regulation and function of plant MT-like genes.

4. Perspective

A rapid increase in the number of characterized plant genes whose products are involved in metal perception, uptake, sequestration and distribution is imminent. This will result from (a) expansion of sequence information derived from both cDNA and genomic sequencing programs, especially in *Arabidopsis*, (b) characterization of numerous mutants impaired in trace metal ion homeostasis, and (c) widespread use of complementation and "gain-of-function" cloning of plant genes in microbial mutants. Reports of genes identified by such approaches are beginning to appear. For example, an *Arabidopsis* mutant has been described which exhibits increased accumulation of manganese (and other metals) and also has enhanced root surface ferric chelate reductase activity (Delhaize 1996); conversely, other mutants are unable to induce this reductase activity in response to iron deficiency (Yi and Guerinot 1996). Complementation of yeast mutants with an *Arabidopsis* expression library has identified a copper transporter (Kampfenkel et al. 1995) and a similar approach resulted in the isolation of a gene encoding a putative Fe^{2+} transporter whose expression is induced in iron-deficient roots (Eide et al. 1996). It is also noted that the plant EST databases contain numerous predicted gene products similar to known proteins involved in metal ion homeostasis. Examples include EST homologs of heavy metal transporting 'CPX-type' ATPase and putative zinc transporters.

Challenges for the future include determining the extent to which the products of genes homologous to, or functionally complementary to, those of microbes (or animals) perform equivalent roles in metal ion homeostasis within plants. Of course plants, compared with microorganisms, possess additional features such as the translocation of metals from roots and their subsequent distribution to organs throughout the life cycle. The results of such studies will have obvious implications for improving the mineral efficiency of crops and indeed their nutritional "quality." Sessile plants must respond to temporal and spatial changes in the availability of metals in soils. There is likely to be a growing interest in understanding how metals are perceived by plants as the catalog of known metal-responsive genes expands.

Acknowledgements

AFS and NJR acknowledge the support of research grants PO802 and 13/C01067 from the UK Biotechnology and Biological Sciences Research Council. AFS is a University of Newcastle Research Fellow. Research in PBG's laboratory has been supported primarily by grants from USDA-NRI.

References

Buchanan-Wollaston, V. 1994. Isolation of cDNA clones for genes that are expressed during leaf senescence in *Brassica napus*—Identification of a gene encoding a senescence-specific metallothionein-like protein. *Plant Physiol.* 105:839–846.

Cano-Gauci, D. F., and B. Sarkar. 1996. Reversible zinc exchange between metallothionein and the estrogen receptor zinc finger. *FEBS Lett.* 386:1–4.

Chen, J., and P. B. Goldsbrough. 1994. Increased activity of γ-glutamylcysteine synthetase in tomato cells selected for cadmium tolerance. *Plant Physiol.* 106:233–239.

Chen, J., J. Zhou, and P. B. Goldsbrough. 1997. Characterization of phytochelatin synthase from tomato. *Physiol. Plant.* (in press).

Chevalier, C., E. Bourgeois, A. Pradet, and P. Raymond. 1995. Molecular cloning and characterization of six cDNAs expressed during glucose starvation in excised maize (*Zea mays* L.) root tips. *Plant Mol. Biol.* 28:473–485.

Coupe, S. A., J. E. Taylor, and J. A. Roberts. 1995. Characterisation of an mRNA encoding a metallothionein-like protein that accumulates during ethylene-promoted abscission of *Sambucus nigra* L. leaflets. *Planta* 197:442–447.

Dameron, C. T., R. N. Reese, R. K. Mehra, A. R. Kortan, P. J. Carroll, M. L. Steigerwald, L. E. Brus, and D. R. Winge. 1989. Biosynthesis of cadmium sulfide quantum semiconductor crystallites. *Nature* 338:596–597.

de Framond, A. J. 1991. A metallothionein-like gene from maize (*Zea mays*). *FEBS Lett.* 290:103–106.

de Knecht, J. A., M. van Dillen, P. L. M. Koevoets, H. Schat, J.A.C. Verkleij, and W.H.O. Ernst. 1994. Phytochelatins in cadmium-sensitive and cadmium-tolerant *Silene vulgaris*: Chain length distribution and sulfide incorporation. *Plant Physiol.* 104:255–261.

Delhaize, E., P. J. Jackson, L. D. Lujan, and N. J. Robinson. 1989. Poly(γ-glutamylcysteinyl)glycine synthesis in *Datura innoxia* and binding with cadmium. *Plant Physiol.* 89:700–706.

Delhaize, M. 1996. A metal-accumulator mutant of *Arabidopsis thaliana*. *Plant Physiol.* 111:849–855.

de Miranda, J. R., M. A. Thomas, D. A. Thurman, and A. B. Tomsett. 1990. Metallothionein genes from the flowering plant *Mimulus guttatus*. *FEBS Lett.* 260:277–280.

Eide, D., M. Broderius, J. Fett, and M. L. Guerinot. 1996. A novel iron-regulated metal transporter from plants identified by functional complementation in yeast. *Proc. Natl. Acad. Sci U.S.A.* 93:5624–5628.

Evans, I. M., L. N. Gatehouse, J. A. Gatehouse, N. J. Robinson, and R.R.D. Croy. 1990. A gene from pea (*Pisum sativum* L.) with homology to metallothionein genes. *FEBS Lett.* 262:29–32.

Evans, K. M., J. A. Gatehouse, W. P. Lindsay, J. Shi, A. M. Tommey, and N. J. Robinson. 1992. Expression of the pea metallothionein-like gene *PsMT*$_A$ in *Escherichia coli* and *Arabidopsis thaliana* and analysis of trace metal ion accumulation: Implications for *PsMT*$_A$ function. *Plant Mol. Biol.* 20:1019–1028.

Foley, R. C., and K. B. Singh. 1994. Isolation of a *Vicia faba* metallothionein-like gene: Expression in foliar trichomes. *Plant Mol. Biol.* 26:435–444.

Fordham-Skelton, A. P., C. Lilley, P. E. Urwin, and N. J. Robinson. 1997. GUS expression in Arabidopsis directed by 5′ regions of a pea metallothionein-like gene, *PsMT*$_A$. *Plant Mol. Biol.* 34:659–668.

Galiazzo, F., M. R. Ciriolo, M. T. Carri, P. Civitareale, L. Marcocci, F. Marmocchi, and G. Rotilio. 1991. Activation and induction by copper of Cu/Zn superoxide dismutase in *Saccharomyces cerevisiae*. *Eur. J. Biochem.* 196:545–549.

Gekeler, W., E. Grill, E.-L. Winnacker, and M. H. Zenk. 1989. Survey of the plant kingdom for the ability to bind heavy metals through phytochelatins. *Zeit. Naturfor. Sec. C Biosci.* 44:361–369.

Glaeser, H., A. Coblenz, R. Kruczek, I. Ruttke, A. Ebert-Jung, and K. Wolf. 1991. Glutathione metabolism and heavy metal detoxification in *Schizosaccharomyces pombe*. *Current Genet.* 19:207–213.

Grill, E., S. Loffler, E.-L. Winnacker, and M. H. Zenk. 1989. Phytochelatins, the heavy-metal-binding peptides of plants, are synthesized from glutathione by a specific γ-glutamylcysteine dipeptidyl transpeptidase (phytochelatin synthase). *Proc. Natl. Acad. Sci. USA* 86:6838–6842.

Grill, E., J. Thumann, E.-L. Winnacker, and M. H. Zenk. 1988. Induction of heavy-metal binding phytochelatins by innoculation of cell cultures in standard media. *Plant Cell Rep.* 7:375–378.

Grill, E., E.-L. Winnacker, and M. H. Zenk. 1985. Phytochelatins, the principal heavy-metal complexing peptides of higher plants. *Science* 230:674–676.

Grill, E., E.-L. Winnacker, and M. H. Zenk. 1986. Homo-phytochelatins are heavy-metal-binding peptides of homo-glutathione containing Fabales. *FEBS Lett.* 205:47–50.

Grill, E., E.-L. Winnacker, and M. H. Zenk. 1987. Phytochelatins, a class of heavy-metal-binding peptides from plants are functionally analogous to metallothioneins. *Proc. Natl. Acad. Sci. USA* 84:439–443.

Gupta, S. C., and P. B. Goldsbrough. 1990. Phytochelatin accumulation and stress tolerance in tomato cells exposed to cadmium. *Plant Cell Rep.* 9:466–469.

Gupta, S. C., and P. B. Goldsbrough. 1991. Phytochelatin accumulation and tolerance in selected tomato cell lines. *Plant Physiol.* 97:3306–3312.

Hayashi, Y., C. W. Nakagawa, N. Mutoh, M. Isobe, and T. Goto. 1991. Two pathways in the biosynthesis of cadystins (γ-EC)$_n$G in the cell-free system of the fission yeast. *Biochem. Cell Biol.* 69:115–121.

Hayashi, Y., C. W. Nakagawa, D. Uyakul, K. Imai, M. Isobe, and T. Goto. 1988. The change of cadystin components in Cd-binding peptides from the fission yeast during their induction by cadmium. *Biochem. Cell Biol.* 66:288–295.

Hofmann, T., D.I.C. Kells, and B. G. Lane. 1984. Partial amino acid sequence of the wheat germ E$_c$ protein. Comparison with another protein very rich in half-cystine and glycine: wheat germ agglutinin. *Can. J. Biochem. Cell Biol.* 62:908–913.

Howden, R., C. R. Andersen, P. B. Goldsbrough, and C. S. Cobbett. 1995b. A cadmium-sensitive, glutathione-deficient mutant of *Arabidopsis thaliana. Plant Physiol.* 107:1067–1073.

Howden, R., and C. S. Cobbett. 1992. Cadmium-sensitive mutants of *Arabidopsis thaliana. Plant Physiol.* 100:100–107.

Howden, R., P. B. Goldsbrough, C. R. Andersen, and C. S. Cobbett. 1995a. Cadmium-sensitive, *cad1*, mutants of *Arabidopsis thaliana* are phytochelatin deficient. *Plant Physiol.* 107:1059–1066.

Hsieh, H.-M., W.-K. Liu, and P. C. Huang. 1995. A novel stress-inducible metallothionein-like gene from rice. *Plant Mol. Biol.* 28:381–389.

Itzhaki, H., J. M. Maxson, and W. R. Woodson. 1994. An ethylene-responsive enhancer element is involved in the senescence-related expression of the carnation glutathione-S-transferase (*GST1*) gene. *Proc. Natl. Acad. Sci. U.S.A.* 91:8925–8929.

Jackson, P. J., C. J. Unkefer, J. A. Doolen, K. Watt, and N. J. Robinson. 1987. Poly(γ-glutamylcysteinyl)glycine: Its role in cadmium resistance in plant cells. *Proc. Natl. Acad. Sci. U.S.A.* 84:6619–6623.

Juang, R.-H., K. F. MacCue, and D. W. Ow. 1993. Two purine biosynthetic enzymes that are required for cadmium tolerance in *Schizosaccharomyces pombe* utilize cysteine sulfinate in vitro. *Arch. Biochem. Biophys.* 304:392–401.

Kägi, J.H.R. 1991. Overview of metallothionein. *Methods Enzymol.* 205:613–626.

Kägi, J.H.R., and A. Schaffer. 1988. Biochemistry of metallothionein. *Biochem.* 27:8510–8515.

Kampfenkel, K., S. Kushnir, E. Babiychuk, D. Inze, and M. van Montagu. 1995. Molecular characterization of a putative *Arabidopsis thaliana* copper transporter and its yeast homologue. *J. Biol. Chem.* 270:28479–28486.

Kawashima, I., Y. Inokuchi, M. Chino, M. Kimura, and N. Shimizu. 1991. Isolation of a gene for a metallothionein-like protein from soybean. *Plant Cell Physiol.* 32:913–916.

Kawashima, I., T. D. Kennedy, M. Chino, and B. G. Lane. 1992. Wheat E$_c$ metallothionein genes. *Eur. J. Biochem.* 209:971–976.

Kille, P., D. R. Winge, J. L. Harwood, and J. Kay. 1991. A plant metallothionein produced in *E. coli. FEBS Lett.* 295:171–175.

Klapheck, S. 1988. Homoglutathione: Isolation, quantification and occurrence in legumes. *Physiol. Plant.* 74:727–732.

Klapheck, S., W. Fliegner, and I. Zimmer. 1994. Hydroxymethyl-phytochelatins [(γ-glutamylcysteine)$_n$-serine] are metal induced peptides of the Poaceae. *Plant Physiol.* 104:1325–1332.

Klapheck, S., S. Schlunz, and L. Bergmann. 1995. Synthesis of phytochelatins and homo-phytochelatins in *Pisum sativum* L. *Plant Physiol.* 107:515–521.

Kneer, R., T. M. Kutchan, A. Hochberger, and M. H. Zenk. 1992. *Saccharomyces cerevisiae* and *Neurospora crassa* contain heavy metal sequestering phytochelatin. *Arch. Microbiol.* 157:305–310.

Kondo, N., M. Isobe, K. Imai, and T. Goto. 1983. Structure of cadystin, the unit peptide of cadmium-binding peptides induced in the fission yeast *Schizosaccharomyces pombe. Tetrahed. Lett.* 24:925–928.

Lane, B., R. Kajioka, and T. Kennedy. 1987. The wheat-germ E_c protein is a zinc-containing metallothionein. *Biochem. Cell Biol.* 65:1001–1005.

Ledger, S. E., and R. C. Gardner. 1994. Cloning and expression of five cDNAs for genes differentially expressed during fruit development of kiwifruit (*Actinidia deliciosa* var. deliciosa). *Plant Mol. Biol.* 25:877–886.

Loffler, S., A. Hochberger, E. Grill, E.-L. Winnacker, and M. H. Zenk. 1989. Termination of the phytochelatin synthase reaction through sequestration of heavy metals by the reaction product. *FEBS Lett.* 258:42–46.

Maitani, T., H. Kubota, K. Sato, and T. Yamada. 1996. The composition of metals bound to class III metallothionein (phytochelatin and its desglycyl peptide) induced by various metals in root cultures of *Rubia tinctorum. Plant Physiol.* 110:1145–1150.

May, M. J., and C. J. Leaver. 1994. *Arabidopsis thaliana* γ-glutamylcysteine synthetase is structurally unrelated to mammalian, yeast, and *Escherichia coli* homologs. *Proc. Natl. Acad. Sci. U.S.A.* 91:10059–10063.

Mehra, R. K., V. R. Kodati, and R. Abdullah. 1995. Chain-length-dependent Pb(II)-coordination in phytochelatins. *Biochem. Biophys. Res. Comm.* 215:730–736.

Mehra, R. K., J. Miclat, V. R. Kodati, R. Abdullah, T. C. Hunter, and P. Mulchandani. 1996a. Optical spectroscopic and reverse phase HPLC analyses of Hg(II) binding to phytochelatins. *Biochem. J.* 314:73–82.

Mehra, R. K., P. Mulchandani, and T. C. Hunter. 1994. Role of CdS quantum crystallites in cadmium resistance in *Candida glabrata. Biochem. Biophys. Res. Comm.* 200:1193–1200.

Mehra, R. K., E. B. Tarbet, W. R. Gray, and D. R. Winge. 1988. Metal-specific synthesis of two metallothioneins and γ-glutamyl peptides in *Candida glabrata. Proc. Natl. Acad. Sci. U.S.A.* 85:8815–8819.

Mehra, R. K., K. Tran, G. W. Scott, P. Mulchandani, and S. S. Sani. 1996b. Ag(I)-binding to phytochelatins. *J. Inorg. Biochem.* 61:125–142.

Mendum, M. L., S. C. Gupta, and P. B. Goldsbrough. 1990. Effect of glutathione on phytochelatin synthesis in tomato cells. *Plant Physiol.* 93:484–488.

Mett, V. L., L. P. Lochhead, and H. S. Reynolds. 1993. Copper-controllable gene expression system for whole plants. *Proc. Natl. Acad. Sci. U.S.A.* 90:4567–4571.

Meuwly, P., P. Thibault, A. L. Schwan, and W. E. Rauser. 1995. Three families of thiol peptides are induced by cadmium in maize. *Plant J.* 7:391–400.

Murphy, A., and L. Taiz. 1995a. A new vertical mesh transfer technique for metal tolerance studies in *Arabidopsis. Plant Physiol.* 108:29–38.

Murphy, A., and L. Taiz. 1995b. Comparison of metallothionein gene expression and nonprotein thiols in ten *Arabidopsis* ecotypes. *Plant Physiol.* 109:945–954.

Murphy, A., J. Zhou, P. B. Goldsbrough, and L. Taiz. 1997. Purification and immunological identification of metallothioneins 1 and 2 from *Arabidopsis thaliana. Plant Physiol.* 113:1293–1301.

Mutoh, N., and Y. Hayashi. 1988. Isolation of mutants of *Schizosaccharomyces pombe* unable to synthesize cadystin, small cadmium-binding peptides. *Biochem. Biophys. Res. Comm.* 151:32–39.

Mutoh, N., C. W. Nakagawa, S. Ando, K. Tanabe, and Y. Hayashi. 1991. Cloning and sequencing of the gene encoding the large subunit of glutathione synthetase of *Schizosaccharomyces pombe. Biochem. Biophys. Res. Comm.* 181:430–436.

Mutoh, N., C. W. Nakagawa, and Y. Hayashi. 1995. Molecular cloning and nucleotide sequencing of the γ-glutamylcysteine synthetase gene of the fission yeast *Schizosaccharomyces pombe. J. Biochem.* 117:283–288.

Okumura, N., N.-K. Nishizawa, Y. Umehara, and S. Mori. 1991. An iron deficiency-specific cDNA from barley roots having two homologous cysteine-rich MT domains. *Plant Mol. Biol.* 17:531–533.

Okumura, N., N.-K. Nishizawa, Y. Umehara, T. Ohata, and S. Mori. 1992. Iron deficiency specific cDNA (*Ids1*) with two homologous cysteine rich MT domains from the roots of barley. *J. Plant Nutr.* 15:2157–2172.

Ortiz, D. F., L. Kreppel, D. M. Speiser, G. Scheel, G. McDonald, and D. W. Ow. 1992. Heavy-metal tolerance in the fission yeast requires an ATP-binding cassette-type vacuolar membrane transporter. *EMBO J.* 11:3491–3499.

Ortiz, D. F., T. Ruscitti, K. F. McCue, and D. W. Ow. 1995. Transport of metal-binding peptides by HMT1, a fission yeast ABC-type vacuolar membrane protein. *J. Biol. Chem.* 270:4721–4728.

Reese, R. N., and G. J. Wagner. 1987. Effects of buthionine sulfoximine on Cd-binding peptide levels in suspension-cultured tobacco cells treated with Cd, Zn, or Cu. *Plant Physiol.* 84:574–577.

Reese, R. N., C. A. White, and D. R. Winge. 1992. Cadmium sulfide crystallites in Cd-(γ-EC)$_n$G peptide complexes from tomato. *Plant Physiol.* 98:225–229.

Reese, R. N., and D. R. Winge. 1988. Sulfide stabilization of the cadmium-γ-glutamyl peptide complex of *Schizosaccharomyces pombe*. *J. Biol. Chem.* 263:12832—12835.

Rennenberg, H., and C. Brunhold. 1994. Significance of glutathione metabolism in plants under stress. *Prog. Bot.* 55:142–156.

Robinson, N. J., I. M. Evans, J. Mulcrone, J. Bryden, and A. M. Tommey. 1992. Genes with similarity to metallothionein genes and copper, zinc ligands in *Pisum sativum* L. *Plant Soil* 146:291–298.

Robinson, N. J., A. M. Tommey, C. Kuske, and P. J. Jackson. 1993. Plant metallothioneins. *Biochem. J.* 295:1–10.

Robinson, N. J., J. R. Wilson, J. S. Turner, A. P. Fordham-Skelton, and Q. J. Groom. 1997. Metal gene interaction in roots: Metallothionein-like genes and iron reductases. In *Plant Roots—From Cells to Systems,* eds. H. M. Anderson et al., pp. 117–130. Kluwer Academic Publishers, The Netherlands.

Robinson, N. J., J. R. Wilson, and J. S. Turner. 1996. Expression of the type 2 metallothionein-like gene *MT2* from *Arabidopsis thaliana* in Zn^{2+}-metallothionein deficient *Synechococcus* PCC 7942: Putative role for MT2 in Zn^{2+}-metabolism. *Plant Mol. Biol.* 30:1169–1179.

Romera, F. J., R. M. Welch, W. A. Norvell, and S. C. Schaefer. 1996a. Iron requirement for and effects of promoters and inhibitors of ethylene action on stimulation of Fe(III)-chelate reductase in roots of strategy I species. *Bio Metals* 9:45–50.

Romera, F. J., R. M. Welch, W. A. Norvell, S. C. Schaefer and L. V. Kochian. 1996b. Ethylene involvement in the over-expression of Fe(III)-chelate reductase by roots of *E107* pea [*Pisum sativum* L. (*brz, brz*)] and *chloronerva* tomato (*Lycopersicon esculentum* L.) mutant genotypes. *Bio Metals* 9:38–44.

Ruegsegger, A., and C. Brunold. 1992. Effect of cadmium on γ-glutamylcysteine synthesis in maize seedlings. *Plant Physiol.* 99:428–433.

Salt, D. E., and W. E. Rauser. 1995. MgATP-dependent transport of phytochelatins across the tonoplast of oat roots. *Plant Physiol.* 107:1293–1301.

Salt, D. E., and G. J. Wagner. 1993. Cadmium transport across tonoplast of vesicles from oat roots: Evidence for a $Cd^{2(+)}/H^{(+)}$ antiport activity. *J. Biol. Chem.* 268:12297–12302.

Scheller, H. V., B. Huang, E. Hatch, and P. B. Goldsbrough. 1987. Phytochelatin synthesis and glutathione levels in response to heavy metals in tomato cells. *Plant Physiol.* 85:1031–1035.

Shi, J., W. P. Lindsay, J. W. Huckle, A. P. Morby, and N. J. Robinson. 1992. Cyanobacterial metallothionein expressed in *Escherichia coli. FEBS Lett.* 303:159–163.

Snowden, K. C., and R. C. Gardner. 1993. Five genes induced by aluminum in wheat (*Triticum aestivum* L.) roots. *Plant Physiol.* 103:855–861.

Snowden, K. C., K. D. Richards, and R. C. Gardner. 1995. Aluminum-induced genes: Induction by toxic metals, low calcium, and wounding and pattern of expression in root tips. *Plant Physiol.* 107:341–348.

Speiser, D. M., S. L. Abrahamson, G. Banuelos, and D. W. Ow. 1992a. *Brassica juncea* produces a phytochelatin-cadmium-sulfide complex. *Plant Physiol.* 99:817–821.

Speiser, D. M., D. F. Ortiz, L. Kreppel, and D. W. Ow. 1992b. Purine biosynthetic genes are required for cadmium tolerance in *Schizosaccharomyces pombe*. *Mol. Cell. Biol.* 12:5301–5310.

Steffens, J. C. 1990. The heavy metal-binding peptides of plants. *Ann. Rev. Plant Physiol. Plant Mol. Biol.* 41:533–575.

Steffens, J. C., D. F. Hunt, and B. G. Williams. 1986. Accumulation of non-protein metal-binding polypeptides(γ-glutamyl-cysteinyl)$_n$-glycine in selected cadmium-resistant tomato cells. *J. Biol. Chem.* 261:13879–13882.

Steffens, J. C., and B. G. Williams. 1989. Increased activity of γ-glutamylcysteine synthetase in DMSO-permeabilized cadmium-resistant plant cells. In *Metal Ion Homeostasis: Molecular Biology and Biochemistry*, ed. D. H. Hamer and D. R. Winge, pp. 359–366. Alan R. Liss, Inc, New York.

Strain, J., and V. C. Culotta. 1996. Copper ions and the regulation of *Saccharomyces cerevisiae* metallothionein genes under aerobic and anaerobic conditions. *Mol. Gen. Genet.* 251:139–145.

Thompson, J. E., R. L. Legge, and R. F. Barber. 1987. The role of free radicals in wounding and senescence. *New Phytol.* 105:317–344.

Tommey, A. M., J. Shi, W. P. Lindsay, P. E. Urwin, and N. J. Robinson. 1991. Expression of the pea gene *PsMT$_A$* in *E. coli. FEBS Lett.* 292:48–52.

van Vliet, C., C. R. Andersen, and C. S. Cobbett. 1995. Copper-sensitive mutant of *Arabidopsis thaliana. Plant Physiol.* 109:871–878.

Vogeli-Lange, R., and G. J. Wagner. 1990. Subcellular localization of cadmium and cadmium-binding peptides in tobacco leaves. Implication of a transport function for cadmium-binding peptides. *Plant Physiol.* 92:1086–1093.

Vogeli-Lange, R., and G. J. Wagner. 1996. Relationship between cadmium, glutathione and cadmium-binding peptides (phytochelatins) in leaves of intact tobacco seedlings. *Plant Sci.* 114:11–18.

Wagner, G. J. 1984. Characterization of a cadmium-binding complex of cabbage leaves. *Plant Physiol.* 76:797–805.

Welch, R. M., W. A. Norvell, S. C. Schaefer, J. E. Shaff, and L. V. Kochian. 1993. Induction of iron (III) and copper (II) reduction in pea (*Pisum sativum* L.) roots by Fe and Cu status: Does the root-cell plasmalemma Fe(III)-chelate reductase perform a general role in regulating cation uptake? *Planta* 190:555–561.

White, C. N., and C. J. Rivin. 1995. Characterisation and expression of a cDNA encoding a seed-specific metallothionein in maize. *Plant Physiol.* 108:831–832.

Yi, Y., and M. L. Guerinot. 1996. Genetic evidence that induction of root Fe(III) chelate reductase activity is necessary for iron uptake under iron deficiency. *Plant J.* 10:835–844.

Zeng, J., R. Heuchel, W. Schaffner, and J.H.R. Kägi. 1991a. Thionein (apometallothionein) can modulate DNA binding and transcription activation by zinc finger containing factor Sp1. *FEBS Lett.* 279:310–312.

Zeng, J., B. L. Vallee, and J.H.R. Kägi. 1991b. Zinc transfer from transcription factor-IIIA to thionein clusters. *Proc. Natl. Acad. Sci. U.S.A.* 88:9984–9988.

Zhou, J., and P. B. Goldsbrough. 1994. Functional homologs of fungal metallothionein genes from *Arabidopsis. Plant Cell* 6:875–884.

Zhou, J., and P. B. Goldsbrough. 1995. Structure, organization and expression of the metallothionein gene family in *Arabidopsis. Mol. Gen. Genet.* 248:318–328.

16

Regulation of Ferritin Synthesis and Degradation in Plants

Jean-François Briat and Stéphane Lobréaux

1. Introduction

Iron is essential for plants, and either starvation or excess of this element can be responsible for severe nutritional disorders deeply affecting the physiology of plants. For example, iron is known to play an important role in photosynthesis, which takes place in the chloroplasts in plants. Iron starvation is responsible for the syndrome of chlorosis, which leads to a yellowing of leaves and to a disorganization of the photosynthetic apparatus. Treatment of roots and leaves with iron salts can relieve chlorosis (Gris 1844; Briat et al. 1995a, b). Besides photosynthesis, iron is also important in plant physiology for nitrogen fixation, which involves iron proteins such as leghemoglobin and nitrogenase (Nash and Shulman 1976; Appleby et al. 1969). It is also associated with key enzymes such as lipoxygenases and 1-aminocyclopropane-1-carboxylic acid (ACC) oxidase involved in plant hormone synthesis (Siedow 1991; Bouzayen et al. 1991). Although iron is essential for plants, in excess it can be highly damaging, leading to a variety of morphological symptoms. An excess of iron can occur under anaerobic and low pH conditions. This environment favors stability of ferrous iron ions, which are readily taken up by the roots of plants; this can be a natural environmental problem in acidic flooded soils. Among the morphological symptoms due to iron excess, the one best characterized is leaf bronzing in rice (Ponnamperuma et al. 1955).

Therefore, utilization of iron by plants, as for animals and bacteria, results from a fine balance requiring the integration of assimilation and storage of this element to avoid insolubility and toxicity problems (Halliwell and Gutteridge 1986). In other words, genetic systems have evolved in plants in order to maintain iron in appropriate quantities and forms, whatever the environmental conditions. Iron uptake is an obvious step for regulation in response to variations in environmental iron concentration. In plants, this has been studied essentially at the physiological level (Marshner et al. 1986). Once inside the cells, iron is kept in a soluble, nontoxic, bioavailable form by various molecules, including the ferritins. The steady-state level of this class of proteins is developmentally and environmentally regulated in plants. The mechanisms involved in these controls are the subject of this chapter.

2. Structure and Cellular Localization of Plant Ferritin

It was suggested 40 years ago that plants may contain ferritin (Mikhlin and Pshenova 1953), and the first characterization of this protein in plants was reported 10 years later (Hyde et al. 1963). Ferritin has been described in almost all plant species and has been detected in many cells by classical electron microscopy methods. Inside cells, ferritin is always found in the stroma of various plastids; although described in chloroplasts, ferritin has been mainly observed in nonphotosynthetic plastids such as proplastids, etioplasts, chromoplasts, and amyloplasts. Accumulation of this protein occurs in the plastids of storage tissues like seeds or shoot and root apices. In addition, iron overload of iron-starved plants, chemical treatments with nicotine, ozone, streptomycin, or ethylene, and conditions of impaired photosynthetic activity also favor ferritin accumulation in plant plastids (Seckback 1982). Biochemical confirmation of the plastid localization of plant ferritin was obtained by showing that the subunit of the protein synthesized in response to iron treatment of bean leaves was synthesized as a precursor with a molecular weight slightly higher than that of the mature subunit found inside the chloroplasts after its translocation; this ferritin precursor was taken up *in vitro* by purified chloroplasts (Van der Mark et al. 1983a). The same observation was reported more recently for ferritin of cultured soybean cells overloaded with iron (Proudhon et al. 1989; Lescure et al. 1991). Demonstration of the targeting of the ferritin precursor to the plastids in this cell system was obtained by direct observation of the protein by immunoelectron microscopy methods (Lescure et al. 1991). Plastid targeting of plant protein precursors requires an amino-terminus transit peptide, which is cleaved during the cross through the plastid envelope (Von Heijne et al. 1989). Evidence of such a transit peptide responsible for plastid targeting of plant ferritins was obtained by comparing the NH_2 terminal amino acid sequence determined by microsequencing of purified mature subunits of various ferritins (Lescure et al. 1990; Ragland et al. 1990; Lobréaux et al. 1992a) with the amino acid sequences deduced from the nucleotide sequences of various plant ferritin cDNAs that have recently been cloned (Lescure et al. 1990; Spence et al. 1991; Lobréaux et al. 1992a, b; Gaymard et al. 1996).

Comparison of the amino acid sequences of ferritin from various plant species with their animal counterparts revealed an identity ranging from 39 to 49% (Andrews et al. 1992). Therefore, although their cytological localization is different (plastids vs cytoplasm), plant and animal ferritins arise from a common ancestor. The amino acids involved in the ferroxidation process of animal H-type ferritins, which allows a rapid uptake of iron inside the mineral core (Bauminger et al. 1991; Lawson et al. 1991), are conserved in the plant ferritin subunit. However, the surface of the plant ferritin cavity is enriched in carboxylic residues characteristic of animal L-type subunits, which is known to be responsible for efficient iron nucleation and increased stability of the mineral core. Therefore, it can be concluded that although only one type of plant ferritin subunit has been described so far, it has the characteristics of both H and L animal ferritin subunits; however, microheterogeneity has been reported in the amino acid sequences of the maize and pea ferritin subunit (Lobréaux et al. 1992a; Van Wuytswinkel et al. 1995). A very high level of secondary structural similarity between pea seed ferritin and mammalian ferritin has been reported (Lobréaux et al. 1992b). In addition, a model of the three-dimensional structure of the pea seed ferritin subunit has been proposed, highlighting the remarkable conservation of the 3-D core structure of plants as modeled to human ferritins, and the predicted regions of change (Lobréaux et al. 1992b). Among these regions of change, two could be of importance in some of the structure/function relationships of this protein. First, the channels found at the 4-fold symmetry axes of the molecule are hydrophobic in animal ferritin, but hydrophilic in plant ferritin. Second, a conserved additional sequence of 24 amino acids is found at the amino-terminus of plant ferritin and could play an important role in the control of plant ferritin degradation (see Section 3).

3. Transcriptional Regulation of Plant Ferritin Synthesis in Response to Iron

The major control of ferritin synthesis in animal systems is known to occur at the translational level (Zahringer et al. 1976; Shull and Theil 1982). The first evidence for such a control came from the observation that actinomycin D, at concentrations that inhibited mRNA synthesis, had no effect on the iron-induced mobilization of ferritin mRNA in the polysomal fractions or on ferritin synthesis in rat liver (Zahringer et al. 1976). Furthermore, in the rat system, the steady-state level of liver ferritin mRNA was unaffected by iron treatment, whereas ferritin protein accumulated several-fold (Aziz and Munro 1986). At the molecular level, translational control of animal ferritin synthesis is mediated through translational repressors, the iron responsive proteins (IRPs) which bind a specific stem-loop structure, the iron responsive element (IRE), found in the 5′ untranslated region of ferritin mRNA (Klausner et al. 1993; Theil 1994).

In plants, 5′ untranslated regions of ferritin mRNA do not contain IRE (Lescure et al. 1991; Spence et al. 1991; Lobréaux et al. 1992a, b; Gaymard et al. 1996), nor it is found in known soybean and maize ferritin genes (Fobis-Loisy et al. 1995; Proudhon et al. 1996). In addition, no IRP activity, as examined by RNA gel shift mobility assays using an animal IRE as a probe, has been detected in yeast and plant extracts (Rothenberger et al. 1990). Consistent with this observation, animal ferritin mRNAs are perfectly well translated in an *in vitro* wheat germ system, whereas they are repressed in a rabbit reticulocyte system that contains endogenous IRP; this translational repression can be conferred to the plant translation system by biochemical complementation with purified IRP1 from rabbit liver (Brown et al. 1989). Moreover, fusion of an animal IRE to soybean ferritin cDNA resulted in no specific translational control of soybean ferritin mRNA (Dix et al. 1993; Kimata and Theil 1994), although IRE fused to reporter genes such as chloramphenicol acetyl transferase or human growth hormone led to translational regulation of the mRNA (Aziz and Munro 1987; Hentze et al. 1987). Computer modeling has predicted that the formation of the typical hairpin loop structure of the IRE is prevented in a frog IRE-soybean ferritin mRNA chimera (Kimata and Theil 1994). Alternative hairpins between the crucial CAGUG loop sequence of the IRE stem-loop (Dix et al. 1993) and the transit peptide sequences of soybean ferritin mRNA can form in the IRE-soybean ferritin mRNA chimera, preventing IRE function (Kimata and Theil 1994). Finally, although one IRP (IRP1) is a member of an ancient protein family, the aconitases (Kennedy et al. 1992), common to plants and animals, it is important to keep in mind that the iron-sulfur cluster of mitochondrial aconitase from potato tubers is very different from its animal counterpart; this novel plant iron center structure involves a distorted 3Fe-type cluster with a nearby Fe atom at a distance of 4.9Å (Jordanov et al. 1992). Although an aconitase plant ferritin cDNA has been recently cloned (Peyret et al. 1995) the structure of the Fe-S cluster of the plant cytosolic aconitase is at present unknown, and the involvement of this enzyme as a regulator of plant ferritin expression switched through the disassembly of its iron-sulfur cluster has yet to be demonstrated.

For the reasons detailed above, it was therefore very unlikely that the translational control that operates in animals through the IRE/IRP system would have been conserved in plants. In agreement with this statement, plant ferritin synthesis in response to iron could have been mediated at the transcriptional level, the mRNA stability level, or both. Van der Mark et al. (1983a) have shown that iron loading of bean leaves results in an increase in the concentration of translatable ferritin mRNA. This observation was confirmed by using cultured soybean cells (Lescure et al. 1991). In addition, when actinomycin D was added to these soybean cells, at concentrations that inhibited mRNA synthesis but still allowed protein synthesis at the same rate as in the control, the iron induction of ferritin protein accumulation was strongly inhibited (Proudhon et al. 1989). Also, the observed increase of ferritin protein in response to increased iron concentration is accompanied by an equivalent accumulation of hybridizable ferritin mRNA in cultured soybean cells (Lescure et al. 1991), in leaves and roots of maize plantlets (Lobréaux et al. 1992a), and in *Arabidopsis thaliana* seedlings (Gaymard et al. 1996). Results of nuclear run-on experiments using the soybean cell system clearly demonstrate that in this case the increase in ferritin mRNA can be accounted for by an increase in transcription of ferritin genes (Lescure et al. 1991).

4. Iron Sensing in Plants: A Link Between Iron-dependent Metabolic Pathways and the Control of Plant Ferritin Synthesis

4.1 Is There Evidence for Trans-acting Factors as Iron-sensing Proteins in Plants?

In recent years, in animal and bacterial systems, iron has been found associated with *trans*-acting factors, the presence, or the state of oxidation, of the metal being important in controlling the activity of these regulators (Hantke 1981; De Lorenzo et al. 1987; Bagg and Neilands 1987; Hidalgo and Demple 1994; Klausner et al. 1993; Haile et al. 1991; Rouault and Klausner 1996; Paraskeva and Hentze 1996). These iron-proteins are supposed to sense the iron status of the cells and to interact with DNA or RNA specific sequences in order to regulate gene expression, either by repression or activation.

As we have already mentioned, no IRE has been described in plant ferritin mRNA. Furthermore, no IRP activity has been detected in plant extracts when using an animal IRE as a probe (Rothenberger et al. 1990). Because plant ferritin, in contrast with animal ferritin, appears to be regulated at the transcriptional level in response to iron, it is likely that *cis*-regulatory elements responsible for this control will be located in the promoter region of plant ferritin genes. In order to address this point, various ferritin genes from soybean and maize have been isolated and characterized (Fobis-Loisy et al. 1995; Proudhon et al. 1996). Although consensus metal regulatory elements (MREs), as well as as other *cis*-regulatory elements, have been noted within 1 (kbp) upstream of the transcription initiation sites of these genes, no evidence for the functionality of these *cis*-elements has been reported so far. Classical promoter fusions to a reporter gene are currently being performed both with maize and soybean promoters to characterize important DNA regions responsible for the iron response in plants.

4.2 The Iron Status of Plants Controls the Activity of Key Enzymes Involved in the Synthesis of Stress Hormones

An alternative to control of gene expression through direct interaction of iron with a *trans*-acting factor could be that environmental iron concentration variations modulate expression through complex transduction pathways involving sensor, transducer, and regulator components. Control of expression of plant genes in response to a variety of environmental stresses is often mediated by transduction pathways that involve the transient accumulation of the plant hormones abscisic acid (Giraudat et al. 1994; Giraudat 1995) and/or ethylene (Yang and Hoffman 1984; Kende 1993; Chang 1996). Abscisic acid (ABA) is a 15-atom sesquiterpene (Figure 16.1). Biochemical and genetic approaches have demonstrated that in higher plants ABA is synthesized by cleavage of epoxy-carotenoids (e.g., violaxanthin, zeaxanthin) to xanthoxin, which is converted to ABA-aldehyde by xanthoxin oxidase. ABA-aldehyde then yields ABA through the action of an aldehyde oxidase (Taylor 1992; Zeevaart et al. 1992). Work has been carried out to see if carotenoid biosynthesis (i.e., the pathway for the synthesis of ABA precursors) is dependent on the iron status of plant cells. Using

Abscisic Acid

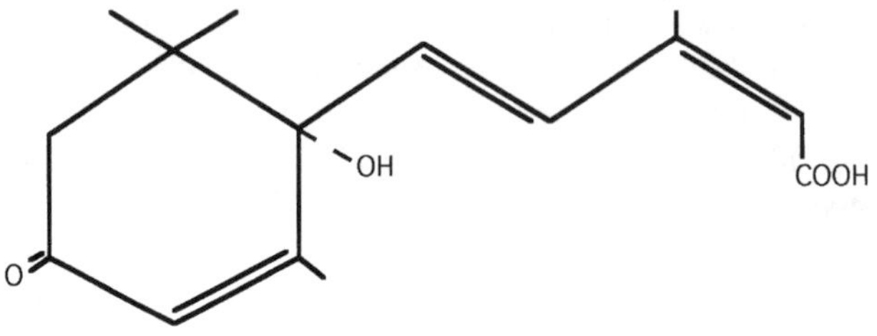

Ethylene

$$CH_2 = CH_2$$

Figure 16.1 Chemical structure of the hormones abscisic acid (ABA) and ethylene involved in plant responses to stress.

cultured sycamore cells it was shown that iron starvation strongly decreased the epoxy-carotenoid concentration, and consequently ABA was undetectable (Pascal et al. 1995). This decrease in the concentration of ABA precursors in response to iron starvation was explained by an inhibition of the activity of the phytoene desaturase, which catalyzes the first desaturation step in the carotenogenesis pathway; as a consequence an accumulation of phytoene and phytofluene was observed. The link between iron status and ABA synthesis was also documented by physiological data as the suberization process in plants, which is also under the control of ABA (Roberts and Kollatukudy 1989), is repressed under iron starvation conditions (Sijmons et al. 1985).

The second plant hormone involved in signal/integrated transduction is the gas ethylene (Figure 16.1), which is synthesized in response to a variety of environmental stresses (Yang and Hoffman 1984). The ethylene biosynthetic pathway has been deduced this past decade (Kende 1993): S-adenosyl methionine is converted into 1-aminocyclopropane-1-carboxylic acid (ACC) by ACC synthase; ethylene is then produced through the action of ACC oxidase. This latter enzyme, which requires Fe^{2+} and ascorbate to function, is inhibited when ortho-phenanthroline is added to cultured tomato cells (Bouzayen et al. 1991). This reversible inhibition is also obtained when cells are cultured under iron starvation conditions; under these conditions almost no ethylene is synthesized. At the other extreme of iron starvation is the effect of iron toxicity, which is responsible for a nutritional disorder of rice plants widely distributed in tropical lowlands and called bronzing (Ponnamperuma et al. 1955). When Peng and Yamauchi (1993) experimentally induced bronzing by dipping the cut end of rice leaves into ferrous sulfate at pH 3.5, they observed a 20-fold increase in ethylene production with a maximum reached 24 hours after the beginning of the treatment. Aminoethoxy-vinylglycine and Co^{2+}, the inhibitors of ACC synthase and of ACC oxidase, respectively (Kende 1993), decreased strongly the ethylene production in response to this iron overload. It should, however, be noted that free radicals can also convert ACC to ethylene. This is particularly relevant due to the report that this reaction can be initiated by Fenton chemistry, ACC being converted to ethylene by hydroxyl radicals produced by the reactivity of ferrous iron with H_2O_2 (Legge et al. 1982).

It can be concluded, therefore, that iron is an important element for the production of abscisic acid and ethylene in plants, through its effects on the activities of key enzymes such as phytoene desaturase and ACC oxidase (Figure 16.2). However, it cannot be ruled out that other steps in the pathways responsible for the synthesis and/or for the responses to these plant hormones could be affected by the iron status of the plants. The question to address now concerns the links that may exist between the control of the synthesis of abscisic acid and ethylene through the iron status of plants, and the control of ferritin synthesis.

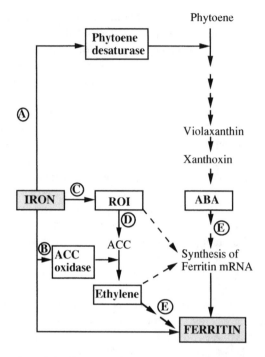

Figure 16.2 Illustration of the links between iron and ferritin synthesis in plants through stress hormone synthesis. Iron is required for the synthesis of abscisic acid (ABA) and ethylene, which are plant hormones involved in the control of ferritin synthesis. Iron is also required for full accumulation of the protein. Iron availability in the plant controls the activities of phytoene desaturase and ACC oxidase, which are key enzymes for the synthesis of these hormones. Iron toxicity results from its reactivity with reduced forms of oxygen to produce reactive oxygen intermediates (ROI), known to be able to convert 1-aminocyclopropane-1-carboxylic acid (ACC) into ethylene. Dashed arrows indicate possible effects that have not been demonstrated experimentally.

4.3 Hormonal Regulation of Plant Ferritin Synthesis in Response to Iron

Ferritin mRNA and protein are almost undetectable in the leaves and roots of plantlets of maize starved of iron for one week (Lobréaux et al 1992a). After iron is resupplied to these plants, the concentration of abscisic acid (ABA) in roots and leaves transiently increases to a maximum of 5-fold within 3 to 6 hours; as a consequence of this increase in hormone concentration, mRNAs known to be induced by abscisic acid accumulate (Lobréaux et al. 1993). Furthermore, treatment of plantlets with exogenous ABA induces ferritin mRNA accumulation, although ferritin protein accumulation is less than that induced by the iron treatment. Proof that iron resupply to iron-starved plantlets induces ferritin synthesis, through a transduction pathway involving the plant hormone abscisic acid, came from an experiment using an ABA-deficient mutant. This deficiency of the *vp2* maize mutant, which is due to a single mutation in one enzyme of the carotenoid biosynthetic pathway, also abolishes ferritin mRNA accumulation in response to iron treatment; ferritin mRNA accumulation can, however, be rescued by exogenous ABA treatment of maize plantlets (Lobréaux et al. 1993). Interestingly, simultaneous addition of saturating concentrations of exogenous ABA and iron does not result in more ferritin mRNA accumulation than iron treatment alone; the two responses are not additive. Additionally, ferritin mRNA accumulation in response to exogenous ABA is 30 to 40% lower than that in response to iron treatment (Lobréaux et al. 1993). It can be concluded, therefore, that ABA is involved in the iron response in plants leading to ferritin synthesis, but it does not account for the induction of all this response. Using de-rooted maize plantlets, where iron overloading of leaves can be easily carried out, a very rapid accumulation of ferritin mRNA and protein was observed (Lobréaux et al. 1995). This accumulation was shown to be independent of the ABA pathway. Instead, reactive oxygen intermediates (ROIs) were demonstrated to be involved in this pathway, as pretreatment of the de-rooted plantlets with antioxidants such as *N*-acetyl cysteine or glutathione completely antagonized the response, whereas a pro-oxidant treatment (H_2O_2) in presence of a low concentration of iron promoted ferritin mRNA accumulation (Lobréaux et al. 1995). Indeed, using specific probes from the 3′ untranslated regions of the two maize ferritin genes so far characterized, it has been recently possible to show that maize ferritin synthesis in response to iron is the sum of a rapid ABA-independent induction of the *ZmFer1* gene expression, mediated by an oxidative pathway, and of a general stress response involving ABA activation of the *ZmFer2* gene expression (Fobis-Loisy et al. 1995; Savino, Briat and Lobréaux submitted for publication). Activation by iron through an oxidative pathway independent of ABA of a ferritin gene from *Arabidopsis thaliana* has also been reported very recently (Gaymard et al. 1996).

Evidence for a link between ethylene and ferritin synthesis comes from experiments where exogenous ethylene has been applied to morning glory plants, resulting in ferritin protein accumulation in the chloroplasts of their leaves (Toyama 1980). Interestingly, as mentioned above, iron overload of de-rooted rice plantlets to induce experimental "bronzing" also induced ethylene synthesis (Peng and Yamauchi 1993). The conditions used to induce maize ferritin synthesis by iron excess in de-rooted maize plantlets are very similar to the one used in the rice "bronzing" experiment. However, it is unknown if ethylene participates as a cellular relay to induce ferritin synthesis in this maize system in response to an iron-mediated oxidative stress.

4.4 Integration of Iron-dependent Metabolic Pathways and the Control of Plant Ferritin Synthesis in Response to Iron: A Hypothetical Scheme

A general scheme that integrates the different known and proposed control pathways for the synthesis of plant ferritin in response to iron can be drawn (Figure 16.2).

As detailed in Section 4.2, the ability of iron to regulate both abscisic acid and ethylene synthesis has been demonstrated through its control of phytoene desaturase (A in Figure 16.2) and ACC oxidase (B in Figure 16.2) activities, respectively. Additionally, ethylene production may be induced via oxidative stress (C in Figure 16.2), as a result of iron overloading, and from the conversion of ACC to ethylene by the action of hydroxyl radicals through the Fenton reaction (D in Figure 16.2). Also as mentioned in Section 4.3, both ABA and ethylene have been shown to induce ferritin synthesis (E in Figure 16.2).

In the case of ABA, the link between iron resupply, ABA synthesis, and maize ferritin mRNA accumulation has been clearly established, although in order to obtain maximal protein accumulation there is a requirement for iron (Lobréaux et al. 1993). This pathway could require the reactivation of phytoene desaturase in iron-starved plants, which would then allow the synthesis of epoxy carotenoids, *i.e.,* the ABA precursors, from the phytoene pool that accumulated during the iron starvation period (Pascal et al. 1995). It is possible, however, that other levels of control are involved in this pathway of iron response through ABA. Additional evidence for ABA control of ferritin mRNA synthesis is the presence of *cis* elements, known to participate in the ABA response in plants, in the promoter region sequence found upstream of the *ZmFer2* maize ferritin gene (Fobis-Loisy et al. 1995).

Ethylene has been shown to be able to induce ferritin protein accumulation, and iron has been found to be required for its formation. However, the link between ethylene biosynthesis in response to iron resupply or overload and the synthesis of ferritin has not been directly established as in the case of ABA. Furthermore, although ethylene is known to control gene expression at the transcriptional level in other systems (Montgomery et al. 1993), it remains to be demonstrated whether treatment by exogenous ethylene induces ferritin protein accumulation through an activation of ferritin mRNA accumulation.

It has been observed that induction of ferritin mRNA and protein accumulation via an ABA-independent pathway in derooted plantlets of maize, and in *Arabidopsis thaliana* seedlings, can be antagonized by pretreatment with antioxidants (Lobréaux et al. 1995; Gaymard et al. 1996). This may be because reactive oxygen intermediates, formed through iron catalysis, are able to act as secondary messengers and activate ferritin gene transcription. Activation of transcription factors by free radicals has been documented for animal systems, as, for example, in the case of NF-κB (Schreck et al. 1991). Determining whether plant ferritin synthesis is induced in this way may reveal another iron response pathway. Alternatively, unknown pathway(s) involving iron protein(s) as *trans* regulator(s), as in the case of animal and bacterial systems, could be revealed by the functional analysis of plant ferritin promoters.

5. Regulation of Plant Ferritin Synthesis and Degradation During Development

5.1 Examples of Developmental Regulation of Ferritin Synthesis and Evidence for Posttranscriptional Control

Besides the environmental control of ferritin synthesis during iron resupply to iron-starved plants or during iron overload, plant ferritin is also developmentally regulated. This is demonstrated by a few examples in the literature which are documented at the biochemical and molecular levels. First, during the life cycle of a plant, ferritin protein is almost undetectable in vegetative organs such as roots and leaves (Lobréaux and Briat 1991), as determined by immunological methods. Only degraded forms of the protein can be detected in the early days following germination. In contrast, the concentration of ferritin mRNA was reported to be higher in mature soybean leaves than in young leaves (Ragland et al. 1990), and this was the first indication that a posttranscriptional event could regulate ferritin synthesis during development. Using maize leaf as a model, an uncoordinated change in ferritin mRNA and ferritin protein concentrations was also reported, suggesting again a posttranscriptional control of ferritin synthesis during plant leaf development (Theil and Hase 1993). Second, developmental control of ferritin synthesis has also been documented by showing that the protein accumulated both in the cotyledon and embryo axis during pea seed maturation (Lobréaux and Briat 1991). However, ferritin mRNA concentrations and the rate of its synthesis were not measured during the course of this process, and it is therefore unknown at which level this control occurs. It is important to remember that a large number of genes that are expressed during seed formation are under the control of the plant hormone abscisic acid (Skriver and Mundy 1990). Because induction of ferritin synthesis in plants under iron stress conditions is controlled, at least in part, by ABA (Lobréaux et al. 1993), ferritin accumulation during seed formation could also be under the hormonal control of ABA. This is the situation for other plant genes regulated by ABA during embryogenesis and in response to salt and/or water stresses (Skriver and Mundy 1990) that are induced in vegetative tissues. There are also, however, examples of genes that are under ABA control during stress responses and yet are not dependent on this hormone for their expression during embryogenesis (Pena-Cortez et al. 1991). Therefore, the pathway responsible for the induction of ferritin synthesis and accumulation during seed formation remains to be characterized. During the early steps of germination, the disappearance of ferritin from the embryo axis (Lobréaux and Briat 1991) is regulated posttranslationally as explained in Section 5.2. Third, during the development of soybean nodules, leading to symbiotic nitrogen fixation, ferritin mRNA concentrations remain elevated, while the protein concentration decreases 4- to 5-fold (Ragland and Theil 1993), indicating posttranscriptional control of ferritin abundance during this plant specific developmental program.

5.2 Molecular Basis of Posttranscriptional Control of Ferritin Synthesis and Degradation During Development

These is no actual evidence that stability and/or translation of plant ferritin mRNA could be the targets for controlling ferritin synthesis. It should, however, be noted that three overlapping 5'AUUUA3' repeats are present in the 3' untranslated region of one of the three maize ferritin cDNAs sequenced (Briat et al. 1995b). Such repeats of the AUUUA sequence were first characterized in animal systems because of their role in the control of mRNA stability (Shaw and Kamen 1986). Generally, they are found in the 3' untranslated regions of short half-life mRNAs such as lymphokine, cytokine, and protooncogenes mRNAs, and their activity is likely to be regulated (Schuler and Cole 1988). In plants, 5'AUUUA3' repeats have already been suggested as mRNA stability determinants (Takahashi et al. 1989), and a synthetic 60-base sequence containing 11 copies of the AUUUA motif has been shown to be responsible for the destabilization of reporter genes in transgenic tobacco cell lines and tobacco plants, when inserted within their 3' untranslated regions (Ohme-Takagi et al. 1993). The sequence per se rather than the A+U content was shown to be responsible. As well as in maize ferritin mRNA this motif has also been found in the 3' untranslated region of a soybean ferritin mRNA (Lescure et al. 1991), although not in the same context as that of the maize ferritin. Whether this destabilizing sequence in the 3' untranslated region of some, but not all, plant ferritin transcripts represents a functional determinant for the control of plant ferritin synthesis remains to be determined. Other plant-specific sequences, the DST sequences found in the 3' untranslated region of a set of unstable soybean transcripts known as the small auxin-up RNAs (SAURs), are also involved in the control of plant mRNA stability (Newman et al. 1993). However such determinants, in contrast with 5'AUUUA3' motifs, are not observed in the 3' untranslated region of plant ferritin cDNA so far sequenced.

Posttranscriptional regulation of ferritin synthesis has been reported to occur during soybean nodule development (Ragland and Theil 1993), and it has been clearly demonstrated that this not due to translational control (Kimata and Theil 1994). It was therefore concluded that regulation of ferritin synthesis during nodule development was a posttranslational event that could be due to increased ferritin turnover or autocatalytic degradation in mature nodules, as already observed in germinating pea seeds (Lobréaux and Briat 1991).

The disappearance of ferritin protein from pea seed embryo axes during the first days following germination is accompanied by the appearance of two polypeptides of lower molecular weight than the basic ferritin subunit (Lobréaux and Briat 1991). These polypeptides, of molecular masses 26.5 and 25 kDa, respectively, react with antibodies raised against pea seed ferritin and have also been observed in samples of ferritin purified from pea seed soaked for 48 hours in aerated water (Laulhère et al. 1989). Also during soybean nodule development, before ferritin disappears (Ragland and Theil 1993), ferritin breakdown products have been observed. It is therefore likely that ferritin degradation requires processing of the ferritin subunit first. A mechanism responsible for this specific processing has been proposed based on iron release experiments performed *in vitro* (Laulhère et al. 1989, 1990). This mechanism was implicated by the finding that release of ferritin iron, when reduced by ascorbate or light, induces conversion of the ferritin subunit from 28 kDa to 26.5 and 25 kDa, and as a consequence pea seed ferritin then has a tendency to aggregate and to become insoluble (Laulhère et al. 1989). Fenton chemistry has been shown to be involved, as the processing is inhibited by free radical scavengers as well as desferrioxamine and ortho-phenanthroline, which chelate ferric and ferrous iron ions, respectively (Laulhère et al. 1989, 1990). Microsequencing of the mature 28-kDa subunit and of the processing product of 26.5 kDa has indicated that these free radical cleavages occurred within the plant-specific amino-terminus of the pea seed ferritin subunit (Laulhère et al. 1989). These observations have recently been confirmed by the demonstration that recombinant pea seed ferritin, deleted of its amino-terminus, is less soluble than recombinant wild-type pea seed ferritin (Van Wuytswinkel et al. 1995).

A model explaining the mechanism of ferritin degradation during germination has been proposed (Lobréaux and Briat 1991). Based on data reported on ferritin regulation during nodulation (Kimata and Theil 1994), this model could also apply to other plant-specific developmental processes. It can be assumed that an increase in the reducing power of plastids, where ferritin is located in germinating seeds and in aging nodules, induces ferritin iron release for metabolic needs. As a consequence, ferritin is processed at its amino-terminus by free radicals generated during the iron release. Free radical processed proteins are known to be more sensitive to protease attacks (Davies 1987), and it is well known that proteolytic activities are enhanced during germination (Vierstra 1993) and aging of nodules (Pfeiffer et al. 1983); therefore processed ferritin is likely to be actively degraded by proteases during the course of these developmental processes. It is still unknown whether iron, which is released inside the plastids, serves only for plastid functions, or if it can also be exported out of the plastids to be used for other cellular functions.

6. Conclusion and Perspectives

Developmental and environmental regulation of plant ferritin synthesis appears to be a complex interplay of transcriptional and posttranscriptional mechanisms involving cellular relays such as plant hormones. In the future it will be important to understand the molecular and cellular events controlling these different pathways. Of particular interest will be the characterization of (1) the mechanisms responsible for ferritin accumulation during developmental processes such as nodulation and seed formation, and (2) the characterization of the transduction pathways involved in ferritin synthesis in response to iron excess. A major challenge will also be to characterize molecular targets induced by iron starvation, in order to study their regulation and to understand integrated regulatory circuits.

References

Andrews, S. C., P. Arosio, W. Bottke, J. F. Briat, M. von Darl, P. M. Harrison, J. P. Laulhère, S. Levi, S. Lobréaux, and S. Yewdall. 1992. Structure, function and evolution of ferritins. *J. Inorg. Biochem.* 47:161–174.

Appleby, C. A. 1969. Properties of leghaemoglobin *in vivo* and its isolation as ferrous oxyleghaemoglobin. *Biochem. Biophys. Acta* 188:222–229.

Aziz, N., and H. N. Munro. 1986. Both subunits of rat liver ferritin are regulated at a translational level by iron induction. *Nucl. Acids Res.* 14:915–927.

Aziz, N., and H. N. Munro. 1987. Iron regulates ferritin mRNA translation through a segment of its 5' untranslated region. *Proc. Natl. Acad. Sci. U.S.A.* 84:8478–8482.

Bagg, A., and J. B. Neilands. 1987. Ferric uptake regulation protein acts as a repressor, employing iron(II) as a cofactor to bind the operator of an iron transport operon in *Escherichia coli. Biochem.* 26:5471–5477.

Bauminger, E. R., P. M. Harrison, D. Hechel, I. Nowik, and A. Treffry. 1991. Mössbauer spectroscopic investigation of structure-function relations in ferritins. *Biochem. Biophys. Acta* 1118:48–58.

Bouzayen, M., G. Felix, A. Latché, J. C. Pech, and T. Boller. 1991. Iron: An essential cofactor for the conversion of 1-aminocyclopropane-1-carboxylic acid to ethylene. *Planta* 184:244–247.

Briat, J. F., I. Fobis-Loisy, N. Grignon, S. Lobréaux, N. Pascal, G. Savino, S. Thoiron, N. von Wirèn, and O. Van Wuytswinkel. 1995a. Cellular and molecular aspects of iron metabolism in plants. *Biol. Cell.* 84:69–81.

Briat, J. F., A. M. Labouré , J. P. Laulhère, A. M. Lescure, S. Lobréaux, D. Proudhon, and O. Van Wuytswinkel. 1995b. Molecular and cellular biology of plant ferritins. In *Iron in Plants and Soils*, ed. J. Abadia pp. 265–276. Developments in Plant and Soil Sciences series, Kluwer Academic Publishers, The Netherlands.

Brown, H. P., S. D. McQueen, W. E. Walden, M. M. Patino, L. Gaffield, D. Bielser, and R. E. Thach. 1989. Requirements for the translational repression of ferritin transcripts in wheat germ extracts by 90 kDa protein from rabbit liver. *J. Biol. Chem.* 264:13383–13386.

Chang, C. 1996. The ethylene transduction pathway in *Arabidopsis*: An emerging paradigm? *Trends in Biochem. Sci.* 21:129–133.

Davies, K. J. A. 1987. Protein damage and degradation by oxygen radicals. *J. Biol. Chem.* 262:9895–9902.

De Lorenzo, V., S. Wee, M. Herrero, and J. B. Neilands. 1987. Operator sequences of the aerobactin operon of plasmid ColV-K30 binding the ferric uptake regulation (*fur*) repressor. *J. Bacteriol.* 169:2624–2630.

Dix, D. J., P. N. Lin, A. R. Mckenzie, W. E. Walden, and E. C. Theil. 1993. The influence of the base-paired flanking region on structure and function of the ferritin messenger RNA iron regulatory element. *J. Mol. Biol.* 231:230–240.

Fobis-Loisy, I., K. Loridon, S. Lobreaux, M. Lebrun, and J. F. Briat. 1995. Structure and differential expression of two maize ferritin genes in response to iron and abscisic acid. *Eur. J. Biochem.* 231:609–619.

Gaymard, F., J. Boucherez, and J. F. Briat. 1996. Characterization of a ferritin mRNA from *Arabidopsis thaliana* accumulated in response to iron through an oxidative pathway independent of abscisic acid. *Biochem. J.* (in press).

Giraudat, J. 1995. Abscisic acid signaling. *Curr. Opin. Cell Biol.* 7:232–238.

Giraudat, J., F. Parcy, N. Bertauche, F. Gosti, J. Leung, P. C. Morris, M. Bouvier-Durand, and N. Vartanian. 1994. Current advances in abscisic acid action and signalling. *Plant Mol. Biol.* 26:1557–1577.

Gris, E. 1844. Nouvelles expériences sur l'action de composés ferrugineux solubles appliqués à la végétation et spécialement en traitement de la chlorose et de la débilité des plantes. *C. R. Acad. Sci. Paris.* 19:1118–1119.

Haile, D. J., T. A. Rouault, J. B. Harford, M. C. Kennedy, G. A. Blondin, H. Beinert, and R. D. Klausner. 1991. Cellular regulation of the iron responsive element binding protein: Disassembly of the cubane iron-sulfur cluster results in high affinity RNA binding. *Proc. Natl. Acad. Sci. U.S.A.* 89:11735–11739.

Halliwell, B., and J. M. C. Gutteridge. 1986. Oxygen free-radicals and iron in relation to biology and medicine. Some problems and concepts. *Arch. Biochem. Biophys.* 246:501–508.

Hantke, K. 1981. Regulation of ferric iron transport in *E. coli*: Isolation of a constitutive mutant. *Mol. Gen. Genet.* 182:288–292.

Hentze, M. W., S. W. Caughman, T. A. Rouault, J. G. Barriocanal, A. Dancis, J. B. Harford, and R. D. Klausner. 1987. Identification of the iron-responsive element for the translational regulation of human ferritin mRNA. *Science* 238:1570–1572.

Hidalgo, E., and B. Demple. 1994. An iron-sulfur center essential for transcriptional activation by the redox-sensing SoxR protein. *EMBO J.* 13:138–146.

Hyde, B. B., A. J. Hodge, A. Kahn, and M. L. Birnstiel. 1963. Studies of phytoferritin. I. Identification and localization. *J. Ultrastruct. Res.* 9:248–258.

Jordanov, J., F. Courtois-Verniquet, M. Neubereger, and R. Douce. 1992. Structural investigations by extended X-ray absorption fine structure spectroscopy of the iron center of mitochondrial aconitase in higher plant cells. *J. Biol. Chem.* 267:16775–16778.

Kende, H. 1993. Ethylene biosynthesis. *Ann. Rev. Plant Physiol. Plant Mol. Biol.* 44:283–307.

Kennedy, M. C., L. Mende-Mueller, G. A. Blondin, and H. Beinert. 1992. Purification and characterization of cytosolic aconitase from beef liver and its relationship to the iron responsive element binding protein. *Proc. Natl. Acad. Sci. U.S.A.* 89:11730–11734.

Kimata, Y., and E. C. Theil. 1994. Posttranscriptional regulation of ferritin during nodule development in soybean. *Plant Physiol.* 104:263–270.

Klausner, R D, T. A. Rouault, and J. B. Harford. 1993. Regulating the fate of mRNA: The control of cellular iron metabolism. *Cell* 72:19–28.

Laulhère, J. P., A. M. Labouré, and J. F. Briat. 1989. Mechanism of the transition from plant ferritin to phytosiderin. *J. Biol. Chem.* 264:3629–3635.

Laulhère, J. P., A. M. Labouré, and J. F. Briat. 1990. Photoreduction and incorporation of iron into ferritins. *Biochem J.* 269:79–84.

Lawson, D. M., P. J. Artymiuk, S. J. Yewdall, J. M. A. Smith, J. C. Livingston, A. Treffry, A. Luzzago, S. Levi, P. Arosio, C. Cesareni, C. D. Thomas, W. V. Shaw, and P. M. Harrison. 1991. Solving the structure of human H ferritin by genetically engineering intermolecular crystal contacts. *Nature* 349:541–544.

Legge, R. L., J. E. Thompson, and J. E. Baker. 1982. Free radical mediated formation of ethylene from 1-aminocyclopropane-1-carboxylic acid: A spin-trap study. *Plant Cell Physiol.* 23:171–177.

Lescure, A. M., O. Massenet, and J. F. Briat. 1990. Purification and characterization of an iron-induced ferritin from soybean (*Glycine max*) cell suspensions. *Biochem. J.* 272:147–150.

Lescure, A. M., D. Proudhon, H. Pesey, M. Ragland, E. C. Theil, and J. F. Briat. 1991. Ferritin gene transcription is regulated by iron in soybean cell cultures. *Proc. Natl. Acad. Sci. U.S.A.* 88:8222–8226.

Lobréaux, S., and J. F. Briat. 1991. Ferritin accumulation and degradation in different organs of pea (*Pisum sativum*) during development. *Biochem. J.* 274:601–606.

Lobréaux, S., T. Hardy, and J. F. Briat. 1993. Abscisic acid is involved in the iron-induced synthesis of maize ferritin. *EMBO J.* 12:651–657.

Lobréaux, S., O. Massenet, and J. F. Briat. 1992a. Iron induces ferritin synthesis in maize plantlets. *Plant Mol. Biol.* 19:563–575.

Lobréaux, S., S. Yewdall, J. F. Briat, and P. M. Harrison. 1992b. Amino-acid sequence and predicted three-dimensional structure of pea seed (*Pisum sativum*) ferritin. *Biochem J.* 288:931–939.

Lobréaux, S., S. Thoiron, and J. F. Briat. 1995. Induction of ferritin synthesis in maize leaves by an iron-mediated oxidative stress. *Plant J.* 8:443–449.

Marshner, H., V. Römheld, and M. Kissel. 1986. Different strategies in higher plants in mobilization and uptake of iron. *J. Plant Nutr.* 86:695–713.

Mikhlin, D. M., and K. V. Pshenova. 1953. Compounds of copper and iron in plants. *Daukl. Akad. Nauk. SSSR Moscow* 90:433–435.

Montgomery, J., S. Goldman, J. Deikman, L. Margossian, and R. L. Fisher. 1993. Identification of an ethylene-responsive region in the promoter of a fruit ripening gene. *Proc. Natl. Acad. Sci. U.S.A.* 90:5939–5943.

Nash, D. T., and H. M. Shulman. 1976. Leghemoglobins and nitrogenase activity during soybean nodule development. *Can. J. Bot.* 54:2790–2797.

Newman, T. C., M. Ohme-Takagi, C. B. Taylor, and P. J. Green. 1993. DST sequences, highly conserved among plant *SAUR* genes, target reporter transcripts for rapid decay in tobacco. *Plant Cell* 5:701–714.

Ohme-Takagi, M., C. B. Taylor, T. C. Newman, and P. J. Green. 1993. The effect of sequences with high AU content on mRNA stability in tobacco. *Proc. Natl. Acad. Sci. U.S.A.* 90:11811–11815.

Paraskeva, E., and M. W. Hentze. 1996. Iron sulfur clusters as genetic regulatory switches: The bifunctional iron regulatory protein-1. *FEBS Lett.* 389:40–43.

Pascal, N., M. A. Block, K. E. Pallett, J. Joyard, and R. Douce. 1995. Inhibition of carotenoid biosynthesis in sycamore cells deprived of iron. *Plant Physiol. Biochem.* 33:97–104.

Pena-Cortez, H., L. Willmitzer, and J. J. Sanchez-Serrano. 1991. Abscisic acid mediates wound induction but not developmental-specific expression of the proteinase inhibitor II gene family. *Plant Cell* 3:963–972.

Peng, X. X., and M. Yamauchi. 1993. Ethylene production in rice bronzing leaves induced by ferrous iron. *Plant Soil* 149:227–234.

Peyret, P., P. Perez, and M. Alric. 1995. Structure, genomic organization and expression of the *Arabidopsis thaliana* aconitase gene. Plant aconitase shows significant homology with mammalian iron-responsive element-binding protein. *J. Biol. Chem.* 270:8131–8137.

Pfeiffer, N. E., C. M. Torre, and F. W. Wagner. 1983. Proteolytic activity in soybean root nodules. *Plant Physiol.* 71:797–802.

Ponnamperuma, F. N., R. Bradfield, and M. Peech. 1955. Physiological disease of rice attributable to iron toxicity. *Nature* 175:275.

Proudhon, D., J. F. Briat, and A. M. Lescure. 1989. Iron induction of ferritin synthesis in soybean cell suspensions. *Plant Physiol.* 90:586–590.

Proudhon, D., J. Wei, J. F. Briat, and E. C. Theil. 1996. Ferritin gene organization: Differences between plants and animals suggest possible kingdom-specific selective constraints. *J. Mol. Evol.* 42:325–336.

Ragland, M., J. F. Briat, J. Gagnon, J. P. Laulhère, O. Massenet, and E. C. Theil. 1990. Evidence for a conservation of ferritin sequences among plants and animals and for a transit peptide in soybean. *J. Biol. Chem.* 265:18339–18344.

Ragland, M., and E. C. Theil. 1993. Ferritin (mRNA, protein) and iron concentrations during soybean nodule development. *Plant Mol. Biol.* 21:555–560.

Roberts E., and P. E. Kollatukudy. 1989. Molecular cloning, nucleotide sequence, and abscisic acid induction of a suberization-associated highly anionic peroxidase. *Mol. Gen. Genet.* 217:223–232.

Rothenberger, S., E. W. Müllner, and L. C. Kühn. 1990. The mRNA-binding protein which controls ferritin and transferrin receptor expression is conserved during evolution. *Nucl. Acids Res.* 18:1175–1179.

Rouault, T. A., and R. D. Klausner. 1996. Iron-sulfur clusters as biosensors of oxidants and iron. *Trends in Biochem. Sci.* 21:174–177.

Schreck, R., P. Rieber, and P. A. Bauerle. 1991. Reactive oxygen intermediates as apparently widely used messengers in the activation of the NF-κB transcription factor and HIV-1. *EMBO J.* 10:2247–2258.

Schuler, G. D., and M. D. Cole. 1988. GM-CSF and oncogene mRNA stabilities are independently regulated in *trans* in a mouse monocytic tumor. *Cell* 55:1115–1122.

Seckback, J. J. 1982. Ferreting out the secret of plant ferritin—A review. *J. Plant Nutr.* 5:369–394.

Shaw, G., and R. Kamen. 1986. A conserved AU sequence from the 3′ untranslated region of GM-CSF mRNA mediates selective mRNA degradation. *Cell* 46:659–667.

Shull, G. E., and E. C. Theil. 1982. Translational control of ferritin synthesis by iron in embryonic reticulocytes of the bullfrog. *J. Biol. Chem* 257:14187–14191.

Siedow, J. N. 1991. Plant lipoxygenase: Structure and function. *Ann. Rev. Plant Physiol. Plant Mol. Biol.* 42:145–188.

Sijmons, P. C., P. E. Kollatukudy, and H. F. Bienfait. 1985. Iron deficiency decreases suberization in bean roots through a decrease in suberin-specific peroxidase activity. *Plant Physiol.* 78:115–120.

Skriver, K., and J. Mundy. 1990. Gene expression in response to abscisic acid. *Plant Cell* 2:503–512.

Spence, M. J., M. T. Henzl, and P. J. Lammers. 1991. The structure of a *Phaseolus vulgaris* cDNA encoding the iron storage protein ferritin. *Plant Mol. Biol.* 17:499–504.

Takahashi, Y., H. Kuroda, T. Tanaka, Y. Machida, I. Takebe, and T. Nagata. 1989. Isolation of an auxin-regulated gene cDNA expressed during the transition from G0 to S phase in tobacco mesophyll protoplasts. *Proc. Natl. Acad. Sci. U.S.A.* 86:9279–9283.

Taylor, I. B. 1992. Genetics of ABA synthesis. In *Abscisic Acid: Physiology and Biochemistry*, eds. W. J. Davies and H. G. Jones. pp. 23–38. βios Scientific Publishers, London, UK.

Theil, E. C. 1994. Iron regulatory elements (IREs): A family of mRNA non-coding sequences. *Biochem. J.* 304:1–11.

Theil, E. C., and T. Hase. 1993. Plant and microbial ferritins. In *Iron Chelation in Plants and Soil Microorganisms.* pp. 133–156. Academic Press, New York and San Diego.

Toyama, S. 1980. Electron microscope studies on the morphogenesis of plastids. X.

Ultrastructural changes of chloroplasts in morning glory leaves exposed to ethylene. *Amer. J. Bot.* 67:625–635.

Van der Mark, F., H. F. Bienfait, and H. Van der Ende. 1983a. Variable amounts of translatable ferritin mRNA in bean leaves with various iron contents. *Biochem. Biophys. Res. Comm.* 115:463–469.

Van der Mark, F., W. Van der Briel, and H. G. Huisman. 1983b. Phytoferritin is synthesized *in vitro* as a high-molecular-weight precursor. *Biochem. J.* 214:943–950.

Van Wuytswinkel, O., G. Savino, and J. F. Briat. 1995. Purification and characterization of recombinant pea seed ferritins expressed in *Escherichia coli*: Influence of amino terminus deletions on protein solubility and *in vitro* core formation. *Biochem. J.* 305:253–261.

Vierstra, R. D. 1993. Protein degradation in plants. *Ann. Rev. Plant Physiol. Plant Mol. Biol.* 44:385–410.

Von Heijne, G. V., H. Steppuhn, and R. G. Herrmann. 1989. Domain structure of mitochondrial and chloroplast targeting peptides. *Eur. J. Biochem.* 180:535–545.

Yang, S. F., and N. E. Hoffman. 1984. Ethylene biosynthesis and its regulation in higher plants. *Ann. Rev. Plant Physiol.* 35:155–189.

Zahringer, J. B., S. Baliga, and H. N. Munro. 1976. Novel mechanism for translational control in regulation of ferritin synthesis. *Proc. Natl. Acad. Sci. U.S.A.* 73:857–861.

Zeevaart, J. A. D., C. D. Rock, F. Fantauzzo, T. G. Heath, and D. A. Cage. 1992. Metabolism of ABA and its physiological implications. In *Abscisic Acid: Physiology and Biochemistry*, eds. W. J. Davies and H. G. Jones. pp. 39–52. βios Scientific Publishers, London, UK.

17

Reciprocal, Copper-Responsive Accumulation of Plastocyanin and Cytochrome c_6 in Algae and Cyanobacteria: A Model for Metalloregulation of Metalloprotein Synthesis

Sabeeha Merchant

1. Introduction

On the basis of nutritional value and cytotoxicity, three broad categories of metals have been distinguished: metals that function as cofactors in enzymes and are toxic only at extremely high concentrations, metals that have no known metabolic function and are highly toxic, and metals that are essential for life but toxic at higher than threshold concentrations. Each of the three classes of metals presents a different biological problem with its own distinctive solution. The third situation, exemplified by copper and iron, is perhaps the most interesting to consider from the perspective of regulation, because the organism has to maintain metalloprotein levels and a relatively constant intracellular utilization pool, despite variation in the supply of metal nutrients and the potential for metal toxicity. The essence of the problem lies in the need for tight coordination and complementary regulation of uptake, utilization, and chelation. Elegant biological systems designed to address some aspects of this problem in eukaryotic cells, specifically the regulation of uptake and chelation, have been described in previous chapters. Another important topic is metal ion control of metal-utilizing pathways (exemplified by the biosynthesis of catalytic metalloproteins such as redox enzymes or electron transfer proteins). Recent experimental efforts in this area have emphasized metalloregulation of gene expression in microorganisms.

Microorganisms are excellent experimental systems for the study of metal ion control of metalloprotein biosynthesis, as metal availability can be controlled simply by growing the experimental organism in defined media. Some examples of metal-regulated genes in microorganisms include (1) the reciprocal, copper-responsive regulation of (copper-containing) plastocyanin and (heme-containing) cytochrome c_6 synthesis in green algae and cyanobacteria, (2) the induction of alternate (molybdenum-, vanadium-, or iron-containing) forms of component I of dinitrogenase in some nitrogen-fixing bacteria in response to molybdenum and vanadium availability, and (3) the reciprocal, iron-responsive synthesis of (iron-containing) ferredoxins and (iron-free) flavodoxins in some cyanobacteria. In the case of plastocyanin and cytochrome c_6 synthesis, plastocyanin is the "preferred" catalyst for photosynthetic electron transfer. However, if copper is not available to support plastocyanin synthesis at the stoichiometry required for optimal operation of the pathway, plastocyanin accumulation is reduced and cytochrome c_6 synthesis is induced to compensate for the plastocyanin deficiency (Wood 1978; Sandmann et al. 1983). Likewise, in the nitrogen-fixing bacteria, component I of dinitrogenase contains an Fe-Mo cofactor if molybdenum is available in the growth environment, but under molybdenum-deficient growth conditions a V-Fe-containing protein encoded by a *different* gene substitutes. A third type of enzyme, free of either molybdenum or vanadium, is synthesized if neither molybdenum nor vanadium is available (Bishop et al. 1980; reviewed by Pau 1989). In cyanobacteria, ferredoxin is an abundant iron-containing redox protein that functions in numerous reductive pathways. In iron-supplemented media, ferredoxin is abundant, whereas in iron-deficient media, ferredoxin-encoding mRNAs are degraded and flavodoxins are induced as substitute catalysts in those pathways (Laudenbach et al. 1988; Sandmann et al. 1990; Bovy et al. 1993; Morand et al. 1994).

More recently, with the availability of molecular tools to study specific gene expression, additional examples of trace element-dependent synthesis of metalloproteins have been described in various microorganisms. These include the nickel-dependent synthesis of nickel-containing hydrogenase in *Bradyrhizobium japonicum*, copper-dependent synthesis of yeast copper- and zinc-containing superoxide dismutase, iron-dependent repression of *E. coli* manganese-containing superoxide dismutase, and selenium-dependent repression of NiFe hydrogenases in *Methanococcus voltae* (Kim and Maier 1990; Gralla et al. 1991; Privalle and Fridovich 1993; Berghofer et al. 1994). Cofactor-dependent regulation of metalloprotein-encoding genes thus appears to be widespread in nature.

This chapter reviews our understanding of the physiology and biochemistry of the copper-dependent reciprocal accumulation of plastocyanin and cytochrome c_6 in green algae and cyanobacteria as a well-studied representative example of metalloregulation of metalloenzyme synthesis.

2. Plastocyanin and Cytochrome c_6

2.1 Properties and Function

Plastocyanin is a small (97–104 amino acids), thylakoid lumen-localized, copper binding protein that functions in photosynthesis to catalyze electron transfer from cytochrome f of the membrane-associated cytochrome b_6f complex to P700$^+$ in photosystem I, and in respiration (in cyanobacteria) to catalyze electron transfer from the cytochrome b_6f complex to the terminal oxidase [reviewed in Morand et al. (1994); Redinbo et al. (1994); Table 17.1]. Plastocyanin contains a single redox active copper ($E_m \sim 370$ mV) and is referred to as a "blue" copper protein because of the absorption properties of the oxidized form of the protein ($\Delta\varepsilon \sim 4.7 \times 10^3$ M^{-1} cm^{-1} at 600 nM). In photosynthetic tissue (e.g., leaf tissue), plastocyanin can account for up to 80% of the intracellular copper. In green algae, which generally do not contain other abundant copper enzymes (like Cu, Zn-superoxide dismutase), plastocyanin must account for an even higher proportion of the total copper.

Cytochrome c_6 is one of the ubiquitously distributed heme proteins that function in the energy-transducing electron transfer chains of respiration and photosynthesis [reviewed in Moore and Pettigrew (1990); Morand et al. (1994); Table 17.1]. The heme group ($E_m \sim 370$ mV) is covalently attached to the polypeptide and is responsible for the characteristic visible absorption spectrum of reduced cytochromes ($\Delta\varepsilon \sim 21 \times 10^3$ M^{-1} cm^{-1} at 552 nm for cytochrome c_6). Cytochromes c_6, like plastocyanins, are small, soluble proteins (size range 80–90 amino acids) located in the thylakoid lumen. The biochemical function of cytochrome c_6 in

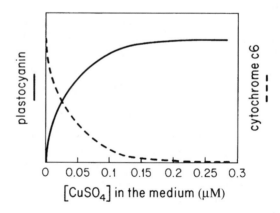

Figure 17.1 Reciprocal, copper-responsive accumulation of plastocyanin and cytochrome c_6. The plastocyanin and cytochrome c_6 contents of cultures of various algae and cyanobacteria were estimated as a function of the copper content (in μM) of the growth medium (Bohner et al. 1980b; Sandmann et al. 1983; Sandmann 1986; Hill and Merchant 1992; Bovy et al. 1992; Nakamura et al. 1992). This figure represents a generalized view of the results. Note that plastocyanin synthesis in response to copper availability is saturable. That is, plastocyanin abundance does not increase beyond a preset stoichiometry per cell (~ 2 per reaction center) when excess copper is provided.

Table 17.1 Plastocyanin and cytochrome c_6

Protein	Cofactor	Occurrence	Gene
Plastocyanin	Copper	Plants	*PetE*
		Algae	*Pcy1*
		Cyanobacteria	*petE*
Cytochrome c_6	Heme	Algae	*Cyc6*
		Cyanobacteria	*petJ*

green algae and cyanobacteria is identical to that of plastocyanin, viz. reduction of the photooxidized photosystem I reaction center.

2.2 Reciprocal, Copper-responsive Accumulation of Plastocyanin and Cytochrome c_6

Some algae and cyanobacteria have the capacity to synthesize either plastocyanin or cytochrome c_6 depending on the availability of copper as a micronutrient in the growth environment (Wood 1978; Sandmann et al. 1983; Ho and Krogmann 1984; Figure 17.1). If copper is available in amounts sufficient to satisfy the plastocyanin biosynthetic pathway, the organism accumulates this copper protein and uses it for the electron transfer pathway. Generally, cytochrome c_6 is not detected in such cultures. However, under conditions of copper deficiency (which would limit or prevent the synthesis of a functional form of plastocyanin), these algae and cyanobacteria remain photosynthetically competent by inducing the accumulation of heme-containing cytochrome c_6. At the level of primary sequence, there are no similarities between cytochromes c_6 and plastocyanins. However, Ho and Krogmann, who isolated both proteins from a wide variety of cyanobacterial species, noted that their pIs (which ranged from ~ 4 to ~ 9 depending on the species) covaried. This suggests that these two evolutionarily unrelated proteins may have coevolved within a particular species in response to alternations/mutations in shared reaction partners (Ho and Krogmann 1984). Their model implies that the genes for both proteins are expressed in these organisms in their natural environment. The wide distribution of cyanobacterial and algal species containing *both* plastocyanin and cytochrome c_6 indicates that adaptation to copper deficiency might be a common phenomenon in nature. Nevertheless, it should be pointed out that copper-responsive regulation of plastocyanin and cytochrome c_6 abundance occurs only in species that have the capacity to synthesize both proteins (Wood 1978; Sandmann et al. 1983; Sandmann 1986; van der Plas et al. 1989; Laudenbach et al. 1990; Bovy 1993). For instance, in species that can synthesize cytochrome c_6 but not plastocyanin, cytochrome c_6 accumulation is not regulated by copper.

2.3 Biosynthetic Pathway

In eukaryotes, plastocyanin and cytochrome c_6 are nucleus encoded (Merchant and Bogorad 1986a; Li and Merchant 1992; Nakamura et al. 1992; Table 17.1). The messenger RNAs encode higher-molecular-weight precursors that are targeted posttranslationally to the chloroplast (Figure 17.2). The pre-proteins are translocated across the envelope and the thylakoid membranes, processed to their mature forms by two proteolytic cleavage events, and assembled with their respective cofactors in the thylakoid lumen (Li et al. 1990; Howe and Merchant 1993, 1994). The pathway in cyanobacteria is not as complex owing to the reduced degree of compartmentation; nevertheless, the overall pathway is, in principle, similar. Regulation could occur at any one or more of these biosynthetic steps.

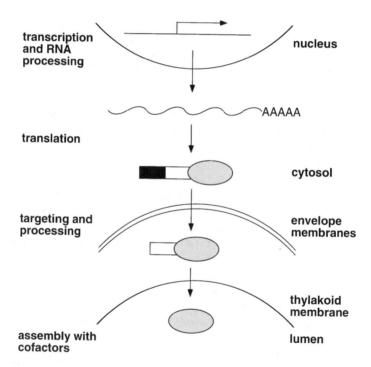

Figure 17.2 Pathway of plastocyanin and cytochrome c_6 synthesis in eukaryotes. Plastocyanin and cytochrome c_6 are encoded in the nucleus. The primary product of translation is a precursor protein with a two-domain transit sequence whose function is to target the mature protein to the thylakoid lumen. The transit sequence is removed sequentially upon translocation of the precursor across the envelope and thylakoid membranes.

3. Regulation of Plastocyanin vs. Cytochrome c_6 Accumulation

3.1 Characterization of the Signal

What is the signal? An important question raised by the preceding observations concerns the nature of the "signal" perceived by the organism so that it can respond to copper availability by synthesizing either plastocyanin or cytochrome c_6, as appropriate. Does the cell measure copper availability directly? Or is it responding to the holoplastocyanin content or the redox status of the electron transfer chain? The evidence argues most strongly for direct measurement of copper in both prokaryotes (cyanobacteria) and eukaryotes (green algae) (Merchant and Bogorad 1987a; Zhang et al. 1994). Specifically, synthesis of cytochrome c_6 continues to be copper regulated in plastocyanin-minus mutant variants of responsive strains. The absence of plastocyanin molecules or the loss of plastocyanin function does not induce cytochrome c_6 synthesis. This suggests that the primary factor involved in induction of cytochrome c_6 accumulation is copper ion availability.

SENSITIVITY AND SELECTIVITY

In most responsive organisms, differential synthesis of plastocyanin vs cytochrome c_6 is observed when the concentration of copper in the medium is varied between low nM to sub-µM (Bohner et al. 1980; Sandmann 1986; Bovy et al. 1992; Nakamura et al. 1992; Zhang et al. 1992; Ghassemian et al. 1994). This concentration range is consistent with a model in which the regulatory pathway represents an adaptation to copper deficiency. The absolute concentration of medium copper appears to be less meaningful than copper availability measured on a per-cell basis (Merchant et al. 1991). In the case of *Chlamydomonas*, maintenance of plastocyanin levels (at a calculated stoichiometry of $\sim 8 \times 10^6$ molecules/cell) requires about 9×10^6 Cu ions/cell. If the cell density is such that copper concentration falls below $\sim 9 \times 10^6$/cell, cytochrome c_6 synthesis is induced to compensate for the plastocyanin deficiency (Hill and Merchant 1992). The absolute concentration of copper in the medium is still relevant, as the existence of high-affinity (perhaps inducible) uptake systems for copper can be inferred from the ability of the organism to respond to low concentrations of copper in the medium. For instance, in *Chlamydomonas* and *Pediastrum boryanum* a response (alteration in gene expression) was detected when as little as 2–5 nM copper was added to copper-deficient medium (Merchant et al. 1991; Nakamura et al. 1992). Likewise, in cyanobacteria (and also *Scenedesmus*), significant alterations in the expression of the *petE* and *petJ* genes are observed at copper ion concentrations below 50 nM (Bohner et al. 1980, 1981; Bovy et al. 1992; Zhang et al. 1992; see Table 17.1 for definitions).

To assess the selectivity of this metal-responsive pathway, a variety of metals (including silver and mercury ions) have been tested for their ability to serve as coregulators of *Pcy1* and *Cyc6* expression in *Chlamydomonas* and *Scenedesmus* (Merchant et al. 1991; Hill and Merchant 1992; Li and Merchant 1992). In *Chlamydomonas*, the response of the *Cyc6* gene exhibits a 20-fold greater specificity for copper vs. mercury as the metal ion regulator, whereas silver ions are completely ineffective. The copper-dependent stimulation of plastocyanin accumulation, which in *Chlamydomonas* occurs at the level of holoprotein formation and stabilization (see Section 3.2), exhibits even greater specificity for copper. Neither silver nor mercury ions support plastocyanin accumulation in *Chlamydomonas* (Hill et al. 1991). In *Scenedesmus*, the plastocyanin biosynthetic pathway appears to be metal responsive at two levels (see Section 3.2). Metal insertion into the apoprotein displays high metal selectivity *in vivo*; however, the accumulation of apoplastocyanin may be slightly stimulated by the addition of silver and mercuric ions in the medium (Bohner et al. 1981; Li and Merchant 1992). The metal specificity of holoprotein formation is unlikely to result from the structure of the polypeptide, as *in vitro* metal-substitution experiments demonstrated that the metal binding site of plastocyanin can accommodate various other metals (Tamilasaran and McMillin 1986).

CONCLUDING REMARKS

Whether the metal selectivity and sensitivity of this regulatory response is mediated at the level of the copper sensor or through some other components (e.g., a copper uptake or delivery pathway) is not known at present. Also not known is whether the effects of silver and mercury ions occur via substitution at the metal binding site of the copper sensor or whether they represent nonspecific effects resulting from inactivation of critical thiols in some other regulatory components. Nevertheless, the metal selectivity and sensitivity of this metalloregulatory system distinguish it from those described in previous chapters.

3.2 Plastocyanin

LEVEL OF REGULATION

Regulation of plastocyanin accumulation has been examined in various species including *Chlamydomonas reinhardtii*, *Scenedesmus obliquus*, and *Pediastrum boryanum* among the green algae, and *Anabaena* and *Synechocystis* spp. among the cyanobacteria (Merchant and Bogorad 1986a,b; van der Plas et al. 1989; Briggs et al. 1990; Bovy et al. 1992; Li and Merchant 1992; Nakamura et al. 1992; Zhang et al. 1992; Ghassemian et al. 1994). Examination of mRNA and protein abundance as a function of copper concentration provides evidence for copper-responsive regulation at two stages in plastocyanin biosynthesis: (1) at the level of template accumulation (mRNA abundance), exemplified by *Scenedesmus obliquus* in Figure 17.3, and (2) at the level of accumulation of mature protein, exemplified by *Chlamydomonas reinhardtii* in Figure 17.3. In some cases both modes of regulation may operate.

This is best demonstrated in *Anabaena* where higher concentrations of copper are required to support plastocyanin accumulation as compared to the accumulation of plastocyanin-encoding mRNA (Bovy et al. 1992). For instance, as medium copper concentration is increased from 0 to 0.05 μM, plastocyanin-encoding messages increase to about 40% of maximum abundance, whereas the increase in protein abundance in the same culture is barely apparent until medium copper concentration reaches 0.1 μM. Again, the response with respect to mRNA accumulation is almost saturated by the addition of 0.2 μM $CuSO_4$ to the medium, whereas concentrations greater than 0.7 μM $CuSO_4$ are required before protein accumulation reaches a plateau. The same phenomenon was noted in *Synechocystis* and *Scenedesmus* (Briggs et al. 1990; Zhang et al. 1992; Quinn and Merchant, unpublished results). In *Synechocystis*, 0.3 μM copper is sufficient to saturate the response with respect to RNA accumulation but not with respect to plastocyanin accumulation. To assess more thoroughly the contribution of processes controlling mRNA vs protein abundance to the regulation of plastocyanin accumulation in various organisms, experiments in which the copper concentration is varied on a per-cell basis and where RNA and protein are prepared from the same batch of cells should be undertaken for each of the above experimental organisms.

TRANSCRIPTIONAL REGULATION

In *Anabaena* sp., the copper-dependent differential in the accumulation of plastocyanin-encoding mRNA transcripts is attributed to control at the level of transcription initiation, because the half-life of *petE* transcripts (measured in the presence of an inhibitor of bacterial transcription) was found to be unchanged in copper-deficient cells relative to copper-supplemented cells (Bovy et al. 1992). The induction of *petE* transcription exhibits a lag of approximately 20 minutes and appears to require *de novo* synthesis of an as yet unidentified protein. The lag time could be attributed to the time required for copper uptake and/or the synthesis or modification of regulatory factors and/or the synthesis of the transcript. A first step in the elucidation of the mechanism of regulation might be the identification of copper-responsive sequences associated with the *petE* gene, and this remains the subject of investigation in many laboratories (Bovy 1993; Nakamura et al. 1994; J. Quinn, unpublished work).

POSTTRANSLATIONAL REGULATION

Posttranslational processes affecting plastocyanin abundance have been studied primarily with *Chlamydomonas*, where differential accumulation is attributed

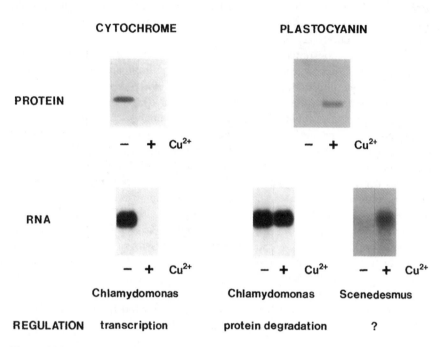

Figure 17.3 Examples of regulation of plastocyanin and cytochrome c_6 accumulation. Typical results of immunoblot (top panels) and RNA hybridization (bottom panels) analyses of proteins and RNAs isolated from copper-supplemented or copper-deficient algal or cyanobacterial cells are shown. The correspondence of the Northern and Western blots indicates that cytochrome c_6 synthesis generally appears to be controlled at the level of mRNA accumulation (left-hand side), probably via regulation of transcription initiation. Control at the transcriptional level has been demonstrated in only two cases; *Chlamydomonas reinhardtii* and *Anabaena* [Merchant et al. 1991; Bovy et al. 1992]. For plastocyanin (right-hand side), copper-responsive accumulation is mediated by control of mRNA accumulation in some organisms (exemplified by *Scenedesmus*) or by control of protein accumulation in other organisms (exemplified by *Chlamydomonas*). Where tested, lower concentrations of copper are required for saturating the former response, whereas higher concentrations are required for the latter (presumably owing to a stoichiometric demand for copper for holoprotein formation). See text for discussion of the work of Bovy et al. (1992).

entirely to rapid degradation of plastocyanin in copper-deficient cells vs. copper-supplemented cells ($t\frac{1}{2}$ < 20 minutes vs > several hours). "Pulse"-radiolabeling experiments indicate that translation of the message as well as import and processing of the translation product occur independently of copper availability (Merchant and Bogorad 1986b). In copper-deficient cells, mature plastocyanin is degraded after import and processing of the pre-protein, whereas in copper-supplemented cells, the mature protein is extremely stable. The active site of the protease responsible for the final processing step faces the thylakoid lumen; this observation suggests that the plastocyanin-degrading activity is located in the thylakoid lumen or can access substrates in the thylakoid lumen. Two models can be proposed to account for differential degradation of plastocyanin. In one model, a constitutively active protease, located in the thylakoid lumen, would recognize the unstable apoform but not the metal-stabilized holoform as a target for proteolysis. In such a situation, plastocyanin molecules that are destabilized by other means (e.g., by mutation of the copper binding site) should be degraded regardless of the availability of copper in the medium. Alternatively, a lumenal protease, specific for plastocyanin, would be induced or activated only in copper-deficient cells. According to this model, apoplastocyanins produced by mutation might accumulate in copper-supplemented cells despite their inability to bind copper, but would be rapidly degraded in copper-deficient cells. The rationale for copper-responsive regulation of plastocyanin degradation is self-evident if it is noted that apoplastocyanin is a branch-point intermediate in two pathways; in copper-supplemented cells apoplastocyanin is a substrate for a biosynthetic pathway (holoprotein formation), whereas in copper-deficient cells it is a substrate for a degradative pathway (Li and Merchant 1992). The physiological function of the degradative pathway might be to favor distribution of copper to other essential enzymes (e.g., cytochrome oxidase in respiration) under conditions of copper deficiency.

The mechanism of plastocyanin degradation is not presently known; preliminary information provides evidence for the operation of both models [Li and Merchant 1995]. Specifically, apoplastocyanin is a better substrate for proteolysis compared to holoplastocyanin, but the responsible protease may be active only in copper-deficient cells. Purified apoplastocyanin is highly susceptible to cleavage *in vitro* by chymotrypsin, whereas holoplastocyanin is quite resistant under the same conditions. This difference is probably attributable to the reduced thermodynamic stability of the folded apoform vs the holoform [Koide et al. 1993], and may be sufficient to account for differential accumulation of plastocyanin in copper-supplemented vs copper-deficient cells, provided a suitable protease is localized to the thylakoid lumen. However, during the characterization of a mutant strain that accumulated apoplastocyanin, Li and Merchant (1995) noted that this mutant strain accumulated apoplastocyanin in copper-supplemented cells but not in copper-deficient cells. This suggests that the apoplastocyanin-degrading activity is not expressed in copper-supplemented cells. The determination of whether the putative protease is allosterically regulated by copper, or whether its expression is subject to regulation, awaits its isolation and purification.

SUMMARY OF COPPER-RESPONSIVE PLASTOCYANIN BIOSYNTHESIS

Some organisms have evolved to regulate expression of the plastocyanin gene so that plastocyanin precursors are not synthesized, or are synthesized at only low levels, unless copper is available. Other organisms continue to expend energy by synthesizing and then degrading a nonfunctional apoprotein in copper-deficient medium. The trade-off appears to be between conservative utilization of resources vs the ability to respond rapidly to any available copper. By displaying both levels of regulation at different sensory ranges, *Anabaena* and *Synechocystis* spp. exemplify organisms that have settled for a compromise position. In these organisms, plastocyanin precursors are not synthesized unless a threshold amount of copper is available. This threshold amount is less than that required for saturation of holoprotein formation. The identity of the copper sensor and the path of signal transduction remain unknown.

3.3 Cytochrome c_6

LEVEL OF REGULATION

For cytochrome c_6, copper-responsive regulation of its biosynthesis appears to be mediated at the level of mRNA accumulation in all organisms where it has been studied (Merchant and Bogorad 1986a; 1987b; Bovy et al. 1992; Nakamura et al. 1992; Zhang et al. 1992; Ghassemian et al. 1994). In *Chlamydomonas* and *Anabaena*, further investigation indicates that differential accumulation is attributable largely to copper-responsive regulation at the level of transcription, with little or no contribution from posttranscriptional processes affecting mRNA stability. In *Chlamydomonas*, the half-life of the Cyc_6 message is the same in copper-supplemented vs copper-depleted cells (~ 45–60 minutes) and in *Anabaena*, the half-life of *petJ* transcripts in copper-replete cells (~ 10 minutes) is only slightly shorter relative to copper-deficient cells (~ 15 minutes) (Hill et al. 1991; Merchant et al. 1991; Bovy et al. 1992). Similarly, the protein, once synthesized, is stable; its degradation does not appear to be accelerated in copper-supplemented cells. The loss of immunoreactive protein in copper-supplemented cells results from reduced synthesis (owing to decreased template abundance) and dilution by cell growth and division. With respect to the sensitivity of the transcriptional response to copper, saturation of the response of the Cyc_6 gene does not occur until there is sufficient copper to satisfy the demands of the plastocyanin biosynthetic pathway (Merchant et al. 1991; Bovy et al. 1992; Zhang et al. 1992). This is not unexpected if induction of cytochrome c_6 synthesis is designed as an adaptation to inadequate plastocyanin levels resulting from copper deficiency.

TRANSCRIPTIONAL REGULATION

In *Chlamydomonas*, cytochrome c_6 is encoded by a single nuclear gene (*Cyc6*). Nuclei prepared from copper-deficient cells elongate *Cyc6* transcripts *in vitro* whereas nuclei isolated from copper-sufficient cells (or cells supplemented with copper salts for as little as 30 minutes) do not (Merchant et al. 1991; Howe et al. 1992). The regulatory molecules (e.g., transcription effectors) responsible for mediating this effect appear to be preexisting in copper-deficient cells, because the inhibition of *Cyc6* transcription in response to added copper occurs rapidly ($t\frac{1}{2} \sim 10$ minutes) without a requirement for protein synthesis. Quantitative Northern analyses indicate that the range of regulation of *Cyc6* message levels by copper is greater than 10^3-fold (Hill and Merchant 1992). Severely copper-deficient cells (< 3 nM copper salts in the medium) contain several hundred transcripts per cell, whereas fully copper-supplemented cells (> 1 μM copper salts in the medium) average less than 0.1 transcript per cell (Hill and Merchant 1992). In *Anabaena*, which also contains a single gene for cytochrome c_6 (*petJ*), the extent of regulation is not as dramatic; the abundance of *petJ* transcripts is decreased between 7–20-fold upon the addition of sufficient amounts of copper (> 1 μM) to the medium (Bovy et al. 1992). Unlike *Chlamydomonas*, *Anabaena* cells seem to contain cytochrome c_6-encoding transcripts at a low basal level even when copper availability is in excess of the amount required to saturate the plastocyanin biosynthetic pathway. The difference may be related to the dual role of plastocyanin and cytochrome c_6 in cyanobacteria, where the proteins function in photosynthesis as well as respiration (Lockau 1981).

In *Anabaena*, the lag time for the response of the *petJ* gene to the presence of copper in the medium is similar to that of the *petE* gene in this organism, viz., ~ 15 minutes, and both responses are blocked if translation is inhibited with chloramphenicol (Bovy et al. 1992). Specifically, the level of *petE* transcripts does not increase and the level of *petJ* transcripts does not decrease upon copper supplementation of chloramphenicol-treated *Anabaena* cells (Bovy et al. 1992). Examination of the nucleotide sequences flanking the start sites of transcription reveals two regions of conservation between the *petE* and *petJ* genes (Bovy 1993). One region, comprising about 20 bp, exhibits about 80% identity between the two genes. The other region consists of the sequences 5'–T_5C_3TA–3', which is found in both genes. Neither region has been subjected to functional analysis (see below). These preliminary observations are consistent with the operation of a common signal transduction pathway; however, there are no additional genetic or other substantive data at present to support this interesting model. The transcriptional responses of the *petE* and *petJ* genes exhibit somewhat different concentration dependencies with respect to copper (saturation of the response of the *petJ* gene requires more copper than for saturation of the transcriptional response of the *petE* gene), so the molecular details of the copper-responsive signal transduction pathway in *Anabaena* should provide some novel insights with respect to the interaction of metal binding sites in metal-dependent regulators.

The copper responsive elements (CuREs) associated with the *petE* and *petJ* genes of *Anabaena* have not yet been identified because the reporter gene fusion constructs designed for this purpose did not exhibit copper-responsive expression in the test organism (Bovy 1993). The finding that the *Anabaena petE* and *petJ* genes are not copper responsive in the test organism, *Synechococcus*, is not surprising, as *Synechococcus* has no endogenous *petE* gene and accordingly does not display copper-responsive expression of its *petJ* gene (Laudenbach et al. 1990). One or more components of the copper-responsive regulatory circuit may be missing in such a situation. Interestingly, the *petE*-driven reporter gene exhibited low constitutive expression, whereas the reporter gene driven by putative *petJ* regulatory sequences exhibited high constitutive expression (i.e., copper-independent) in the same genomic location. This suggests that, in *Anabaena*, copper-responsive expression of plastocyanin requires *activation* of the *petE* gene in copper-supplemented cells, whereas copper-responsive expression of cytochrome c_6 depends on *repression* in the presence of copper (Bovy 1993). The inhibitory effect of chloramphenicol (see preceding discussion) is not inconsistent with this model.

In contrast to the preceding model for *Anabaena*, the copper-responsive accumulation of *Cyc6* transcripts in *Chlamydomonas reinhardtii* is attributed to CuREs that function primarily as *activators* of transcription under conditions of copper deficiency. Analysis of the coding and flanking sequences of the *C. reinhardtii* *Cyc6* gene indicates that CuREs are located between positions -127 and -56 relative to the start site of transcription. Specifically, this fragment can confer copper-responsive expression to a promoter-less reporter gene or to a reporter gene driven by a minimal β-tubulin promoter (Quinn and Merchant 1995). The -127 to -56 fragment can be further divided into two distinct regions of CuRE activity, from -127 to -110 and from -110 to -56; each can function independently to confer copper responsiveness to the β-tubulin promoter. In the absence of CuREs, the β-tubulin promoter-driven constructs display low, copper-independent expression. When associated with CuREs, expression from the β-tubulin promoter increases several-fold in copper-deficient medium but does not change significantly in copper-supplemented medium. This behavior suggests that the CuREs function as binding sites for a transcriptional activator in copper-deficient cells. Note that this model is the opposite of the one described above for *Anabaena* where copper-responsive expression of its *petJ* gene is proposed to occur by repression in the presence of copper. The signal transduction pathway may well be quite different in the prokaryotic *Anabaena* system compared to the eukaryotic *Chlamydomonas* system.

SUMMARY OF CYTOCHROME C$_6$ REGULATION

Organisms that contain genetic information for both plastocyanin and cytochrome c_6 appear, quite uniformly, to regulate the expression of the latter via copper-responsive accumulation of mRNA. Transcriptional control seems to be the major contributor to differential mRNA accumulation in both eukaryotes (*Chlamydomonas*) and prokaryotes (*Anabaena*). It would not be unreasonable to assume that transcriptional regulation plays an important role in other species as well. The CuREs associated with the *Chlamydomonas Cyc6* gene have not yet been delineated by mutagenesis and functional assays; the CuRE binding proteins of *Chlamydomonas* likewise remain to be isolated.

It is not necessary to assume (nor is it likely) that the regulatory schemes of eukaryotes and prokaryotes employ related components. For instance, the putative CuRE-containing regions associated with the *Anabaena* and *Chlamydomonas* cytochrome c_6-encoding gene share little or no sequence similarity. The eukaryotic and prokaryotic organisms may well employ different signal transduction pathways. It is possible that the prokaryotic regulatory schemes are modeled after two-component regulatory systems (see Chapter 5), whereas the eukaryotic copper-responsive regulators are modeled after the yeast ACE1/AMT1-like transcription factors (see Chapters 11 and 12, also Jungmann et al. 1993).

4. Other Responses to Copper Deficiency

In addition to plastocyanin and cytochrome c_6, there may be other proteins whose expression is regulated by copper ion availability. In *Chlamydomonas*, one of these, originally identified as a soluble, 35-kDa protein that is coordinately expressed with cytochrome c_6, is the enzyme coproporphyrinogen oxidase (Merchant and Bogorad 1986b, 1987a; Hill and Merchant 1995). The expression of the *Cpx1* gene, encoding coproporphyrinogen oxidase, displays the same sensitivity and selectivity for regulation by copper ion as does the *Cyc6* gene. The *Cpx1* and *Cyc6* genes are therefore assumed to be downstream response targets of a common copper sensor. The induction of coprogen oxidase activity in copper-deficient cells is an interesting observation, because it suggests that the organism is compensating for the loss of one redox cofactor (copper) by increasing the intracellular availability of another (heme). This phenomenon provides yet another example of the tight relationship between copper and iron metabolism.

In other work, copper-deficient *Chlamydomonas* cells were noted to display greater sensitivity to silver toxicity than copper-supplemented cells (Howe and

Merchant 1992); this is attributed to increased expression of a copper uptake pathway which might include a transport component and a cell surface cupric reductase (Hill et al. 1996). Recently, genetic evidence suggests that a putative plastocyanin-degrading activity in *Chlamydomonas* may be responsive to regulation by copper (Li and Merchant 1995; discussed above). These findings suggest that control of plastocyanin and cytochrome c_6 expression may be only one of several adaptations of an organism to copper deficiency.

5. Future Directions

The emphasis in ongoing work in several laboratories is on the identification of the components of the regulatory pathways, specifically the CuREs associated with plastocyanin and cytochrome c_6-encoding genes of prokaryotes and eukaryotes, the CuRE-binding protein(s), the copper sensor, and the components of the signal transduction pathways. One approach towards this objective might be to isolate copper-insensitive regulatory mutants. The characterization of these mutants would provide insight into the mechanism of operation of the signal transduction components, and the availability of the mutants would facilitate the cloning of the regulatory genes.

Another area of interest is the understanding of the function of other copper-responsive genes (e.g., *Cpx1*) and biochemical activities (e.g., cupric reductase) in copper metabolism, the isolation of additional copper-responsive genes (e.g., encoding an assimilatory copper transporter), and the determination of whether all of these processes are regulated by a single copper-responsive signal transduction pathway. A problem that is open for investigation in the cyanobacterial and algal experimental systems is the metabolic fate of well-characterized and essential copper enzymes (e.g., cytochrome oxidase) under conditions of copper deficiency.

Acknowledgments

The work in my laboratory has been supported by research grants from the National Institutes of Health and the National Research Initiative Competitive Grants Program of the United States Department of Agriculture. I thank the members of my group, especially Kent Hill, Hong Hua Li, and Jeanette Quinn, for their contributions to the work described here and for their enthusiasm for the project. I also thank the National Institutes of Health for a Research Career Development Award.

References

Berghofer, Y., K. Agha-Amiri, and A. Klein. 1994. Selenium is involved in the negative regulation of the expression of selenium-free [NiFe] hydrogenases in *Methanococcus voltae*. *Mol. Gen. Genet.* 242:369–373.

Bishop, P. E., D. M. L. Jarlenski, and D. R. Hetherington. 1980. Evidence for an alternative nitrogen fixation system in *Azotobacter vinelandii*. *Proc. Natl. Acad. Sci. U.S.A.* 77:7342–7346.

Bohner, H., H. Bohme, and P. Boger. 1980a. Reciprocal formation of plastocyanin and cytochrome c-553 and the influence of cupric ions on photosynthetic electron transport. *Biochim. Biophys. Acta* 592:103–112.

Bohner, H., H. Merkle, P. Kroneck, and P. Boger. 1980b. High Variability of the Electron Carrier Plastocyanin in Microalgae. *Eur. J. Biochem.* 105:603–609.

Bovy, A. 1993. Metal-regulated expression of the ferredoxin, plastocyanin and cytochrome c553 genes in cyanobacteria. Ph.D. thesis, University of Utrecht, The Netherlands.

Bovy, A., J. de Kruif, G. de Vrieze, M. Borrias, and P. Weisbeek. 1993. Iron-dependent protection of the *Synechococcus* ferredoxin I transcript against nucleolytic degradation requires *cis*-regulatory sequences in the 5′ part of the messenger RNA. *Plant Mol. Biol.* 22:1047–1065.

Bovy, A., G. de Vrieze, M. Borrias, and P. Weisbeek. 1992. Transcriptional regulation of the plastocyanin and cytochrome c553 genes from the cyanobacterium *Anabaena* species PCC7937. *Mol. Microbiol.* 6:1507–1513.

Briggs, L. M., V. L. Pecoraro, and L. McIntosh. 1990. Copper-induced expression, cloning and regulatory studies of the plastocyanin gene from the cyanobacterium *Synechocystis* sp. PCC 6803. *Plant Mol. Biol.* 15:633–642.

Ghassemian, M., B. Wong, F. Ferreira, J. L. Markley, and N. A. Straus. 1994. Cloning, sequencing and transcriptional studies of the gene for cytochrome *c*-553 and plastocyanin from *Anabaena* sp. PCC 7120. *Microbiology* 140:1151–1159.

Gralla, E. B., D. J. Thiele, P. Silar, and J. S. Valentine. 1991. ACE1, a copper-dependent transcription factor, activates expression of the yeast copper, zinc superoxide dismutase. *Proc. Natl. Acad. Sci. U.S.A.* 88:8558–8562.

Hill, K., and S. Merchant. 1992. *In vivo* competition between plastocyanin and a Cu-dependent regulator of the cytochrome c6 gene. *Plant Physiol.* 100:319–326.

Hill, K., and S. Merchant. 1995. Coordinate expression of coproporphyrinogen oxidase and cytochrome c6 in the green alga *Chlamydomonas reinhardtii* in response to changes in copper availability. *EMBO J.* 14:857–865.

Hill, K., H. H. Li, J. Singer, and S. Merchant. 1991. Isolation and structural characterization of the *Chlamydomonas reinhardtii* gene for cyt c6: Analysis of the kinetics and metal specificity of its Cu-responsive expression. *J. Biol. Chem.* 266:15060–15067.

Hill, K. L., R. Hassett, D. Kosman, and S. Merchant. 1996. Regulated copper uptake in *Chlamydomonas reinhardtii* in response to copper availability. *Plant Physiol.* 112:697–704.

Ho, K. K., and D. W. Krogmann. 1984. Electron donors to P700 in cyanobacteria and algae. An instance of unusual genetic variability. *Biochim. Biophys. Acta* 766:310–316.

Howe, G., and S. Merchant. 1992. Heavy metal-activated synthesis of peptides in *Chlamydomonas reinhardtii*. *Plant Physiol.* 98:127–136.

Howe, G., and S. Merchant. 1993. Maturation of thylakoid lumen proteins proceeds posttranslationally through an intermediate *in vivo. Proc. Natl. Acad. Sci. U.S.A.* 90:1862–1866.

Howe, G., and S. Merchant. 1994. Role of heme in the biosynthesis of cytochrome c6. *J. Biol. Chem.* 269:5824–5832.

Howe, G., J. Quinn, K. Hill, and S. Merchant. 1992. Control of the biosynthesis of cytochrome c6 in *Chlamydomonas reinhardtii. Plant Physiol. Biochem.* 30:299–307.

Jungmann, J., H.-A. Reins, J. Lee, A. Romeo, R. Hassett, D. Kosman, and S. Jentsch. 1993. MAC1, a nuclear regulatory protein related to Cu-dependent transcription factors is involved in Cu/Fe utilization and stress resistance in yeast. *EMBO J.* 12:5051–5056.

Kim, H., and R. J. Maier. 1990. Transcriptional regulation of hydrogenase synthesis by nickel in *Bradyrhizobium japonicum. J. Biol. Chem.* 265:18729–18732.

Koide, S., H. J. Dyson, and P. E. Wright. 1993. Characterization of a folding intermediate of apoplastocyanin trapped by proline isomerization. *Biochem.* 32:12299–12310.

Laudenbach, D. E., M. E. Reith, and N. A. Straus. 1988. Isolation, sequence analysis, and transcriptional studies of the flavodoxin gene from *Anacystis nidulans* R2. *J. Bacteriol.* 170:258–265.

Laudenbach, D. E., S. K. Herbert, C. McDowell, D. C. Fork, A. Grossman, and N. A. Straus. 1990. Cytochrome c-553 is not required for photosynthetic activity in the cyanobacterium *Synechococcus. Plant Cell* 2:913–924.

Li, H. H., and S. Merchant. 1992. Two metal-dependent steps in the biosynthesis of *Scenedesmus obliquus* plastocyanin: Differential mRNA accumulation and holoprotein formation. *J. Biol. Chem.* 267:9368–9375.

Li, H. H., and S. Merchant. 1995. Degradation of plastocyanin in copper-deficient *Chlamydomonas reinhardtii. J. Biol. Chem.* 270:23504–23510.

Li, H.-M., S. M. Theg, C. M. Bauerle, and K. Keegstra. 1990. Metal-ion-center assembly of ferredoxin and plastocyanin in isolated chloroplasts. *Proc. Natl. Acad. Sci. U.S.A.* 87:6748–6752.

Lockau, W. 1981. Evidence for a dual role of cytochrome c-553 and plastocyanin in photosynthesis and respiration of the cyanobacterium, *Anabaena variabilis. Arch. Microbiol.* 128:336–340.

Merchant, S., and L. Bogorad. 1986a. Regulation by copper of the expression of plastocyanin and cytochrome c552 in *Chlamydomonas reinhardtii. Mol. Cell Biol.* 6:462–469.

Merchant, S., and L. Bogorad. 1986b. Rapid degradation of apoplastocyanin in Cu(II)-deficient cells of *Chlamydomonas reinhardtii. J. Biol. Chem.* 261:15850–15853.

Merchant, S., and L. Bogorad. 1987a. Metal ion regulated gene expression: Use of a plastocyanin-deficient mutant of *Chlamydomonas reinhardtii* to study the Cu(II)-regulated expression of cytochrome c-552. *EMBO J.* 6:2531–2535.

Merchant, S., and L. Bogorad. 1987b. The Cu(II)-repressible plastidic cytochrome c. *J. Biol. Chem.* 262:9062–9067.

Merchant, S., K. Hill, and G. Howe. 1991. Dynamic interplay between two Cu-titrating components in the transcriptional regulation of cytochrome c6. *EMBO J.* 10:1383–1389.

Moore, G. R., and G. W. Pettigrew. 1990. *The Cytochromes: Evolutionary, Structural and Physicochemical Aspects*, Springer-Verlag, Berlin.

Morand, L. Z., R. H. Cheng, and D. W. Krogmann. 1994. Soluble electron transfer catalysts of cyanobacteria. In *The Molecular Biology of Cyanobacteria*, ed. D. A. Bryant. pp. 243–269. Kluwer Academic Publishers, The Netherlands.

Nakamura, M., M. Yamagishi, F. Yoshizaki, and Y. Sugimura. 1992. The syntheses of plastocyanin and cytochrome c-553 and regulated by copper at the pre-translational level in a green alga, *Pediastrum boryanum*. *J. Biochem.* 111:219–224.

Nakamura, M., F. Yoshizaki, and Y. Sugimura. 1994. cDNA cloning of plastocyanin from a green alga, *Pediastrum boryanum*. *Plant Cell Physiol.* 35 (suppl.):s98.

Pau, R. N. 1989. Nitrogenases without molybdenum. *Trends in Biochem. Sci. U.S.A.* 14:183–186.

Privalle, C. T., and I. Fridovich. 1993. Iron specificity of the fur-dependent regulation of the biosynthesis of the manganese-containing superoxide dismutase in *Escherichia coli*. *J. Biol. Chem.* 268:5178–5181.

Quinn, J. M., and S. Merchant. 1995. Two copper-responsive elements associated with the *Chlamydomonas Cyc6* gene function as targets for transcriptional activators. *Plant Cell* 7:623–638.

Redinbo, M., T. O. Yeates, and S. Merchant. 1994. Plastocyanin: Structural and functional analysis. *J. Bioenerg. Biomemb.* 26:49–66.

Sandmann, G. 1986. Formation of plastocyanin and cytochrome c-553 in different species of blue-green algae. *Arch. Microbiol.* 145:76–79.

Sandmann, G., M. L. Peleato, M. F. Fillat, M. C. Lazaro, and C. Gomez-Moreno. 1990. Consequences of the iron-dependent formation of ferredoxin and flavodoxin on photosynthesis and nitrogen fixation on *Anabaena* strains. *Photosyn. Res.* 26:119–125.

Sandmann, G., H. Reck, E. Kessler, and P. Boger. 1983. Distribution of plastocyanin and soluble plastidic cytochrome c in various classes of algae. *Arch. Microbiol.* 134:23–27.

Tamilasaran, R., and D. R. McMillin. 1986. Absorption spectra of d10 metal ion derivatives of plastocyanin. *Inorg. Chem.* 25:2037–2040.

van der Plas, J., A. Bovy, F. Kruyt, G. de Vrieze, E. Dassen, B. Klein, and P. Weisbeek. 1989. The gene for the precursor of plastocyanin from the cyanobacterium *Anabaena* sp. PCC 7937. *Mol. Microbiol.* 3:275–284.

Wood, P. M. 1978. Interchangeable copper and iron proteins in algal photosynthesis: Studies on plastocyanin and cytochrome c-552 in *Chlamydomonas*. *Eur. J. Biochem.* 87:9–19.

Zhang, L., B. McSpadden, H. B. Pakrasi, and J. Whitmarsh. 1992. Copper-mediated regulation of cytochrome c-553 and plastocyanin in the cyanobacterium *Synechocystis* 6803. *J. Biol. Chem.* 267:19054–19059.

Zhang, L., H. B. Pakrasi, and J. Whitmarsh. 1994. Photoautotrophic growth of the cyanobacterium *Synechocystis* sp. PCC 6803 in the absence of cytochrome c553 and plastocyanin. *J. Biol. Chem.* 269:5036–5042.

Index